Psychiatry

Konstantinos N. Fountoulakis

Psychiatry

From Its Historical and Philosophical Roots to the Modern Face

Springer

Konstantinos N. Fountoulakis
3rd Department of Psychiatry, University Hospital AHEPA
School of Medicine, Aristotle University of Thessaloniki
Thessaloniki, Greece

ISBN 978-3-030-86540-5 ISBN 978-3-030-86541-2 (eBook)
https://doi.org/10.1007/978-3-030-86541-2

This Springer imprint is published by the registered company Springer Nature Switzerland AG
The registered company address is: Gewerbestrasse 11, 6330 Cham, Switzerland

Foreword

Almost from its beginning in the nineteenth century modern psychiatry has been kept apart from the mainstream of medicine and psychiatrists were usually separated from other specialist physicians as well as from those in general practice. The separation was bad for both: psychiatrists paid little attention to physical diseases and specialists in other disciplines shied away from dealing with mental health problems. General practitioners dealt with a large proportion of common mental disorders on their own without consulting psychiatrists who, in turn, did not pay much attention to what the general practitioners were doing.

The separation of psychiatry from general medicine was somewhat diminished with the advent of psychotropic medications: prescribing pills was closer to the practice of other doctors and the fact that the pills worked attracted psychiatrists to the investigations of biological bases of mental disorder. While engaged in research on the functions of the brain, they established links with other researchers to the benefit of both. The separation from general medicine in clinical work however remained in most services, strengthened by the stigma of mental illness attached to mental illness and to everything that touched it including health workers dealing with mental illness as well as the services which were used to provide care to people who had those disorders.

The more recent findings about the high mortality of people with mental illness who die 10–20 years sooner than those who do not have mental illnesses as well as epidemiological findings and clinical experience showing that mental and physical disorders are often comorbid helped to bridge the gap between psychiatry and medicine in general but psychiatry still remains separate from mainstream medical work and its potential is only partially used.

The book which Professor Fountoulakis wrote is, in addition to its intrinsic interest, also an important contribution to bridging the gap between general medicine and psychiatry. It draws attention to the common origins of psychiatry and other medical branches and describes the history of both; it makes an important contribution to the discussion of the concept of illness and the nature of impairment and disability; it describes the potential of modern psychiatry and the areas of its application; and it places psychiatry—and medicine in general—in the context of ethics, of social science, and of human rights issues in medicine.

It is therefore a pleasure to recommend this book to all those interested in the practical and philosophical issues related to psychiatry, to its relation to medicine in general, and to its service to people struck by mental illness. The book will not answer all questions nor is this its intention: its major appeal is that it stimulates thought and offers a panoramic view of psychiatry seen by a psychiatrist with an open mind and comprehensive knowledge of the field.

Geneva, Switzerland
November 27, 2020

Norman Sartorius

Prologue

The prologue is traditionally the part which leads the book. It is the part the reader should read first. But actually, it is often the part which is written last, after major and important parts of the book have been completed and the author has a better view of his work. This is the case with the current prologue. I have chosen to write it from scratch after completing the entire book. This was because I realized that authoring this book was a marvelous adventure which reshaped me in many ways, changed many of my views, and enhanced my ability to see and read between the lines.

As I was reviewing the text for one more time, I was thinking of my motives to start writing such a book and where eventually this has led me. And it seems that all started many many years ago. And all seem to becoming so personal. And they indeed are personal.

To begin with, I was born in spring 1966 in a rather small town called "Agios Nikolaos" (Saint Nicholas) at the eastern part of the Mediterranean island of Crete. Now a city of around forty thousand inhabitants, until the 1980s it had no more than 5–6000 with an economy almost exclusively based on tourism. It was and still is a beautiful place with a gorgeous lake in the middle of it and a marvelous promenade facing two small islands. Its history was unremarkable and it would have been known only as a fine place for vacations, if it wasn't for the "island."

Spinalonga (Fig. 1) became internationally famous because of the 2005 book *The Island* by Victoria Hislop. For me its history was known since my early years, and it was my first contact with the mirage of an asylum. Spinalonga is an artificially created island. During the Venetian occupation of Crete, around 1526, part of a small peninsula was cut and a fortress was built there, near the modern small town of Elounda, a few kilometers away from my home. It was used as a leper colony from 1903 to 1957 and was one of the last active leper colonies in Europe. So although the place had stopped being an asylum for more than a decade before I was born, and in spite of the fact that efficacious treatment for leprosy was in place since decades, the shadow and the memories of the asylum were still heavy in the region as I was coming of age. I was lucky enough and happy to organize in the area the 2015 International Congress on Neurobiology and Psychopharmacology and to organize a pilgrimage on the small island of Spinalonga.

My father was probably the first doctor with ENT specialty in Crete (and his sister the first female pediatrician), and since my early years I was lucky to enjoy an intellectual environment with access to his library which was rather big taking into consideration the place and the time. Among the books I read during my pre-adolescent period it was Freud's *Totem and Taboo*. My mother was chief nurse at the local general hospital and through them I came in contact with medicine and life in medicine. For me the local hospital and my father's private office were like home.

Two are the instances from that period of my life that keep coming to my mind. The first is a memory from when I was around 5 and my father was driving slowly as he was about to pass the gates of the hospital for his afternoon visit when a kid much older than me (I now think it was around 14 years old) in dirty clothes appeared from nowhere and started jumping and laughing around the car. The next few weeks I saw this kid a couple of times and then it disappeared. My father was patient enough to explain to me elementary things about mental retardation on the basis of the limited knowledge he had himself and eventually he told me that the boy had been taken to the mental hospital of Souda, a town at the opposite end of Crete and the only psychiatric facility the island had at that time. The asylum of Souda was shut down many years later in 2006 during the deinstitutionalization process and the psychiatric reformation by my dear friend Pavlos Theodorakis who now serves with the WHO. Today I know that a large number of people suffer from mental disorders and they manifest significant disability. I don't really remember extraordinary cases of fellow schoolmates with behavioral problems or overt psychopathology but there were clearly many kids with severe learning disabilities and borderline mental capacity and they were not receiving

Fig. 1 The Spinalonga island, photo by George Giannopoulos

the special care they needed. The more severe cases were of course hidden from public view. No such images come to my mind. Later, when I grew up and especially after obtaining the specialty of psychiatry and became professor, I was approached by several fellow citizens (I had by then moved to Thessaloniki at the opposite end of Greece) and realized that many people I thought I knew were suffering from mental disorders, but were managing to hide their problems, even their suicidal attempts from the local community.

The second was a public event, which only a few people will remember today and they probably want to forget. I do not remember the exact year but it should have been in the mid-1970s when the town experienced several days of excitement as in the central church a clearly psychotic woman was treated with exorcism. Various rumors were circulating concerning her possession by the devil and how the devil was reacting to the presentation of the cross and icons and that when he was getting in and out of her a blow of wind was sweeping the church and that in one instance a window glass was broken. To his honor, the president of the local medical association went to the church with the police having an order by the judge; the woman was taken and sent for psychiatric treatment. For me the whole story was absolutely beyond comprehension and as far as I remember my father soon abandoned any effort to explain it to me. I had never heard of anything similar before and never heard of any similar event afterwards in the area.

For several years I was embarrassed by this memory and felt ashamed for my town of origin. But I was not fair; exorcism is much more common than we think. Later I was shocked when I discovered a photo depicting the catholic priest Don Alessi holding the Penitent Ring of Saint Vicinius, round the neck of a boy named Leonardo Arti. The priest was praying to exorcise satan while the boy was tied to a chair. The ceremony took place in 1956 at Saisina, Romana, Italy. I was even more astonished to learn about the Anneliese Michel, a 23-year old woman suffering from epilepsy and psychosis, who died in 1976 due to malnourishment and dehydration after 67 (!!!) exorcism sessions that lasted more than a year. The procedure was officially approved by Bishop Josef Stangl who granted the priest Arnold Renz permission to exorcise according to the "Rituale Romanum" of 1614, and additionally he ordered total secrecy. It is not strange that the 1973 film "the Exorcist" received so many Oscar awards and currently there seems to be a worldwide boom in exorcisms which are performed even through Skype while a few years ago the U.S. Conference of Catholic Bishops translated the church's exorcism ritual from Latin into English for the first time.

But this is again unfair; through the years I often encountered a protective and caring attitude both from the established church and its servants, as well as from faithful people. I would easily say with certainty that I have never noticed neglect, rejection, or cruelty from this part of the establishment and from this social group. Unfortunately, this was not the case with many secular establishments either state, governmental, or private, including highly reputed schools, health care facilities, or individual persons. All too often, talks and teaching were merely a show-off and in sharp contrast to the real attitude and behavior.

But as we all know, memory focuses and chooses what to store and what to retrieve on the basis of inclination, preferences, and other kinds of cognitive biases. And this inclination of mine emerged clearly for the first time at the fourth year of my medical studies when in order to pass the lesson on history of medicine I could choose between exams and writing a small paper; without second thought I chose to write a text on the history of psychiatry. Some 30 years ago, this turned out to be a bitter disappointment. It was not an easy task in the pre-internet era. I received a disappointing "pass" (nobody ever failed the lesson so this was unimpressive) and decided to go on and take the exams where I excelled. Then, the same year, a fellow student of mine, now a successful gynecologist, lent me the book *Myth of Mental Illness* by Thomas Szatz. Having just done with the lesson of psychiatry, I read the book within a few days and understood nothing!!! At that time, and since this was my first contact with antipsychiatric texts, I was missing the key basic concept and was not able to put into the equation the possibility that mental patients might not be patients at all. In my eyes what I had seen in the psychiatric ward was crystal clear. I soon noticed that only a few of my fellow students were sharing my certainty. This was something I was facing for the next several years with fellow doctors; while I had no opinion on how they should operate or treat their "somatic" patients, on the contrary they had very specific and aggressive opinions on how I should be doing my work with mental patients. This was widespread during my early years as a specialist, but it quickly disappeared probably as a result of my growing authority but also as a result of the accumulated impact reality had over the years on their ideologically based opinions.

I spent my residency years in the 1990s, and the situation was not better. In 1989, the year of my graduation from medicine, a storm hit Greek Psychiatry. BBC broadcasted pictures of mental patients chained up and naked in the psychiatric asylum on the island of Leros. The world was shocked since Greece was a member of the European Union (then European Common Market) since 1983 but the pictures suggested a condition similar to the dark ages. Leros was named the "island of the damned" and it was regarded as the worst mental institution in the whole of Europe (Figs. 2 and 3). Leros is a small island in the southeast Aegean sea and in 1959 the Greek government decided to turn the installations of an abandoned Italian naval base left from the era of the Italian occupation of the Dodecanese into a mental asylum. At that time it was considered to be a huge improvement since it would decompensate the overcrowded central asylums, but with each passing year it was turned into a depot of forgotten souls in exile. This was a shameful situation for the whole country and action was taken toward psychiatric reform with the financial aid of Europe. A side effect was that this constituted a huge boost for psychosocial theories and for their followers in the country, and many were pressing toward changes similar to those imposed in Italy by Franco Basaglia some years before. Extreme antirational theories became dominant.

In this general atmosphere, many of my fellow residents were following exotic theories and psychotherapies, homeopathy, or some kind of vague admixture of contrasting beliefs and approaches in an impossible theory of everything. At that time I myself wanted to become a psychoanalyst as most psychiatrists classically

Fig. 2 The Leros Asylum

OBSERVER

Established 1791 | 10 SEPTEMBER 1989 | No. 10326

Revealed: Greece's mental patients held on 'concentration camp' island

Europe's guilty secret

Greek tragedy: Inmates wander around in building 16 of the institution on the island of Leros. Photograph by John Wildgoose.

Exclusive

John Merritt
Leros

1,300 lost souls left to rot

HUNDREDS of mentally ill and handicapped men, women and children have been left to rot in concentration camp conditions on a remote Greek island. The full horror of the plight of the lost souls on the Aegean island of Leros, 172 miles south-east from the mainland port of Piraeus, is revealed today after two *Observer* journalists gained access to the hospital.

The inmates, some of whom were brought to the island in chains, have been chosen from all over Greece, seemingly at random. Aged from childhood to over 80, most are designated 'schizophrenic, degenerate, cranial pathetics and epileptics.' Many are kept naked or tied to bare iron beds in a permanent state of filth and deprivation, waiting only for death. Sixty died in 1987, and 41 died last year, out of a total hospital population of about 1,300. The appalling conditions here have been known to the European Community and a large number of psychiatric specialists throughout Europe for five years. But the secret has been kept from the world at large.

There is just one psychiatrist, and the remaining 800 workers and guards have no medical qualifications. In a written report the psychiatrist stated last year: 'It is extremely difficult to describe how they [the inmates] live. They have no possessions, there is no sense of time change. … The patients reside in misery and contempt, mainly abandoned by psychiatry itself.'

In 1984 and in 1987, teams of experts, including British professors, acting for the Commission of the European Community, visited the island. They proposed a number of measures, including a ban on new admissions, foreign task forces of medical volunteers, and the transfer of all inmates with the objective of closing the institutions.

It now seems likely that only the deaths of all who are held there will bring about its end. Although the EC agreed on a priority aid package totalling almost £2 million from the European Social Fund, conditional on a matching sum from the Greek government, most of the money has remained untouched.

In eight days, international psychiatric experts will meet on Leros to decide what to do. Next month the 8th World Congress on Psychiatry will be held in Athens, when members will propose a symposium on 'Human rights in mental hospitals'.

As one European delegate to a previous meeting on the island said: 'In ancient Greece the entrance to Hades was in a cave on the Peloponnese peninsula. In modern Greece, it is on Leros.'

Naked and damned, page 17.

Fig. 3 The newspaper "Observer" presenting the Leros Psychiatric Hospital (Sept 10, 1989)

wanted at that time. But my problem was that I wanted to study systematically and in depth which was not the usual approach and the field seemed too unsystematic. Eventually I was recommended to read Fenichel's *Psychoanalytic Theory of Neurosis*. This again did not turn out well. Very early in the book, between pages 8 and 9, Fenichel writes: "Persons who 'do not believe in psychoanalysis' will not be convinced by this book. They can only inform themselves about what the teachings of psychoanalysis actually are." That was a real shock. Was I reading a religious or a scientific text after all? I read almost half of the book but with increasing disappointment. At that time I was feeling awfully alone. When asking something specific and was expecting a concrete answer with some kind of meaning in it, I was receiving criticism that my approach to things is too simplistic and naïve. Only decades later I learned that many others around the world had similar thoughts and feelings. Ironically, this was the “decade of the brain” as championed by Lewis Judd whom I met in 2007.

At that time, around me, intense discussions were taking place on the nature of psychiatry and mental illness, on the form, efficacy, and essence of therapy but they were also expanding to include wider philosophical questions concerning human beings and humanity. Politics were also part of the main menu, and at that time it was impossible to understand how clearly conflicting political, philosophical, social, and medical theories and views could be synthesized by the same person in a single concept of things. This book deals also with this phenomenon in psychiatry.

After I started teaching at the university I noticed that a significant minority of students were openly hostile to psychiatry and they tended to ask not concerning the elements of the lesson but the validity of the concept itself. Their questions seemed to stereotypically repeat in each group again and again year by year. “Do medications exert any real therapeutic effect on the cause or they simply suppress symptoms?” or “Is there any real cure in psychiatry?” The most devastating and difficult to explain answer is “isn’t it a choice to behave this way?” To be perfectly accurate, these questions seem to be reduced the last few years. I am not sure whether this is a good or bad sign, whether it reflects a maturity of our teaching or simply that mental health is no longer a priority in medical studies.

As for myself, I was slowly adapting and started developing a repertoire of answers. My favorite response to the question on the “treatment of the cause or suppression of symptoms” was to ask back whether antihypertensive medication constituted treatment of hypertension or insulin for diabetes mellitus. However my favorite analog was that of the use of diuretics in congestive heart disease. Responses to the questions on whether mental illness is a choice were less successful and this topic was proven to be more complex. Suggesting an analog to free-will issues by talking about efforts to quit smoking or to lose weight by dieting was only partially accepted by the students. It was clear that deeper work was urgently needed.

Another situation where I faced a similar atmosphere was and still is the various scientific meetings and congresses in my home city of Thessaloniki. In a city of approximately one million, it is expected that the audience is usually too heterogeneous and the vast majority of participants come from the social and psychological sciences. Most of them do not have any real background in the theory of science, and they find it difficult to follow when hard scientific arguments are in place. I remember once having much difficulty persuading people that simply designing a psychotherapeutic method in theory (no matter how perfect or careful such a design might be) does not mean it will be efficacious and needs clinical data for verification. It was obvious that empirical theory testing was out of their mind. I was frequently embarrassed when chairs of my speeches embraced my speech in a wider sociocultural context where everything goes. On the other hand, equally frustrating was the effort of many colleagues to utilize receptor profile in order to arrive at exotic psychopharmacological combinations in sharp contrast to the results of clinical trials. I was often referring to these efforts as "psychoanalytic pharmacotherapy."

No more to my surprise, patients seem to be increasingly educated by the internet and they ask deep and persistent questions on the differential diagnosis and the options for treatment. Many of these questions concern exotic or not-so-exotic alternative therapies as antiquated as Hippocratic dietary or as stupid as zeolite supplements.

So the next question we need to address is, does a clinical psychiatrist need to have special knowledge and training based on the history of psychiatry? Does he really need to have a background in the philosophical studies, to know social theories and be able to argue on the interface of all these with the scientific method?

In my opinion if the answer is "no" then he is unarmed against strong attacks aiming (some deliberately) to destroy his/her everyday clinical practice. But there is more in this. If psychiatry is to make a quantum leap into the future, it needs to face its past and elaborate on the content of its soul. And this is not a matter of a small group of scholars in ivory towers; it concerns all psychiatrists and even more, it concerns everybody who is professionally involved in mental health. Such education and training is currently not part of any curricula, neither pregraduate nor during residency. In fact, many curricula find it easier or more convenient to go smoothly along with a lay, old, and anti-scientific current, never to challenge it, and in this way their survival is secured without any hard times. However we live neither in the 1960s nor in the 1970s. Thanks to the efforts of pioneers and very brave people like Robert Spitzer, we find ourselves now in the position to be able to look ahead into the future. But to move on, we need to share a deep elaborating perspective on the soul of our discipline. Often several colleagues discussed with me my wanderings in the field of philosophy of science and the allied disciplines, and their standard view was that such intellectual adventures offer us the opportunity to enrich our discipline. My approach was exactly the opposite. My endeavor has nothing to do with the enrichment of our discipline and to put it raw, I am perfectly indifferent to such an enrichment. I am seeking to distill the circulating thoughts, ideas, conclusions, theories, or proposals with the use of the scientific way. If such a distillation leaves our discipline poorer, then so be it. Reliability and validity is what matters, not inflated eloquent verbalism.

I am not sure if this need for basic education and training in history and philosophy as well as in the allied sciences is good or bad for us. No other medical specialty needs to reflect on its history or basic concepts. No cardiosurgeon needs knowledge on the history of cardiosurgery or training on the philosophy of science in order to perform better when operating. On the other hand it is so beautiful that our discipline is so complex and appealing and this is because it focuses on our deepest questions and worries. It tries to answer questions like who and what we are, what we really want, are we "good" persons, and other existential questions. Only religions and religious-like philosophy cover a wider landscape, but on the other hand, we are scientists, not poets or eloquent literary authors. It is both a blessing and a curse.

At the center of our adventures is the question "What is mental illness?" It is a question which tantalizes psychiatrists, neuroscientists, psychologists, philosophers, patients and their families, and the sensitive and educated lay persons alike.

There are radically different theories and proposals about the basic nature of mental illness. The debate seems endless and through the history of psychiatry, the dominant approach swings endlessly between two extremes: a mindless neuroscience and a brainless psychology. There is nothing similar in any other medical specialty and no matter how charming this seems and is, it definitely contributed to a bad reputation for psychiatry among the medical sciences but also among scientists in general.

Psychiatry is similar to no other medical specialty; it is highly demanding in terms of interpersonal skills and resilience to interpersonal stress and needs a specific cognitive ability to understand delicate and complex concepts and interactions, and accept frustration which often emerges in all fields of medicine which deal with severe, capacitating, and horrible illnesses. The additional heavy burden of psychiatry is that frustration often reflects an experience with the depth of human existence. It would be very optimistic and would reflect a positivistic wishful thinking to say that most psychiatrists do possess these trait qualities. Experience says that many do not have these qualities at all, while many others are simply able to maintain a superficial management of survival and acceptance. What is clearly missing is any systematic work toward the fostering and development of these qualities and their embedment in the everyday clinical practice of the average clinician. With the gradual abandoning of psychoanalytical studies and training, and the adoption of more technocratic psychotherapies and pharmacotherapy, and also in combination with high work load, these qualities tend to be neglected.

Psychiatry is no longer the black sheep of medicine, and psychiatrists are no longer those medical graduates who wish to escape the true medical profession. There is no anxiety on how we will embed ourselves in general hospitals or how we are going to be accepted by other doctors. However there is still much to be done if psychiatry is to take its place among the other medical specialties in a way not only of equal status but a way which reflects the fact that mental disorders are the most devastating disorders the humankind suffers from. The sad fact is that if we improve the bulk of disability currently caused by mental disorders by say 10% it would be equal to the complete eradication of some very disastrous medical conditions, e.g., blindness. Thus the place of psychiatry should be determined by the fact that development in its area will have a tremendous and disproportional benefit on humanity.

This book was the end product of these life experiences, thoughts, and intellectual wanderings of mine. It includes a comprehensive history of psychiatry since antiquity and until today, with an emphasis not only on main events but also on the chain of events that led to a particular development. It mentions the role of the dye chemistry and industry in the development of modern pharmacotherapy, the disappointing fact that two out of three Nobel prizes directly awarded to psychiatry concerned treatment options which were proven huge disappointments (to

Wagner-Jauregg for pyretotherapy and to Egas-Moniz for psychosurgery while the third was to Arvid Carlsson for the discovery of dopamine). For the latter, it is astonishing that, in contrast to how it is considered today as a failed and oppressive method, at that time it was viewed as so advanced and efficacious medical treatment that prominent persons, including Rosemary, sister of the US President John F. Kennedy as well as Evita Peron, wife to Argentinian President Juan Domingo Peron, underwent it, unfortunately with poor results. Equally disappointing to me was that the names of a number of people with definite proven contribution in the improvement of the outcome of patients are unknown while the names of eloquent "zeros" are widely celebrated. Who knows the name of the men who introduced antidepressants? Has anybody heard of the brave Henry Anonymous? It seems our professional and scientific memory is determined almost exclusively by lay memory shaped by newspapers and lay magazines and books. Names of stars of psychosocial movements are given to hospitals and clinics but nobody honors those who really changed the face of psychiatry and the lives of millions of patients.

For me, another shocking realization was when I saw that very carefully and scientifically informed efforts and therapeutic proposals failed tragically. The most impressive such example is psychosurgery which was solidly based on knowledge of the time and experiments on primates. On the other hand, it is impressive that proposals with an entirely mistaken origin, misleading line of thinking, and wrong interpretation of initial results came to be proven as one of the most successful treatments in the armamentarium of psychiatry. This was the case of lithium. It is evident that the history of psychiatry was shaped both by visionaries and renegades, and also serendipity (finding by luck) played a major role. These dedicated people raised our discipline from the asylum alienist to a special kind of physician and relabeled mental illness from a stigma of outcasts to a major public health problem. The fact that today younger lay people might relate us more with psychopharmacological pills rather than with the psychoanalytic coach has both its positive and negative aspects.

This book also includes a comprehensive review and systematic elaboration on the definition and the concept of mental illness, the issue of free will as well as the state of the art of contemporary psychiatry. I thought it is of prime importance to discuss the issue of free will and the mind-brain problem extensively. This because many philosophers as well as lay persons consider mental illness to be the result of free choice, while the conceptualization of an immortal soul which is responsible for behavior plays a crucial role in the way we approach mental disorders and their treatments. Finally the book includes a description of the academic, social, and professional status of psychiatry and psychiatrists and a view of future needs and possible developments.

A last moment addition was the chapter on conspiracy theories. This was a consequence of the experience with the social media and the public response to the COVID-19 outbreak, which coincided with the final stage of the preparation of the book. Conspiracy theories are of course not the subject of psychiatry per se, but still they spring out of irrationality. I felt it would be important to delineate the relationship of them to mental health and to share my research efforts on this topic during the pandemic. In my opinion conspiracy theories and similar phenomena emerge out of deep human needs and fears, they seem to relate to depression or dysphoria and to some extent, since they can be an important tool in the hands of populists, they constitute a threat to democracy itself. Thus, their study is an excellent opportunity to dig deep into the relation among human psychology, mental health, society, and politics and to swim in intellectually dangerous waters.

I want to acknowledge and express my gratitude to my wife Katerina and sons Nikolas and Ioannis for their continuous support and understanding of my "workaholic" tendencies. My deceased now parents, Nikolaos and Olga, gave me valuable lessons of how life should be lived and what values should determine my thoughts. My parents in law Ioannis (deceased now) and Zoumboulia played a decisive role in the adventures of my scientific life.

I owe special gratitude to my teachers, those persons who took a young inexperienced psychiatrist living in a medium-sized city without any significant background and showed him the marvels of thinking. These were primarily Hagop S Akiskal, Hans Jurgen Moeller, Siegfried Kasper, and Norman Sartorius. I owe special thanks to the late Nobel Laureate Arvid Carlsson for his advice and support. One unexpected and extraordinary experience occurred more than 20 years ago when I submitted one of my first papers to the journal *Psychiatry Research* where Monte Buchsbaum was the editor-in-chief. To my surprise, he took my hand and led me through seven or eight revisions, guiding me step by step to re-analyze the data from scratch and re-write the whole manuscript. He did that for a colleague he did not know, from half the world away, without asking for anything in return. For the years to come, he served as a role model for me.

Finally, my thanks go to my students at the Faculty of Medicine, School of Health Sciences at the Aristotle University of Thessaloniki in Greece, especially to the most restless of them. Over the years with their merciless questions as well as their embrace of my activities, they made me look again at places I have never thought it was worth and meaningful to glance. I questioned all until then established beliefs; I tried to respond to all their questions and leave no dark spots in my teachings. This was a truly adventurous way to teach.

But above all, this book is dedicated to all our patients; its authoring was motivated by the desire to improve their lives and I am grateful to them for giving a wider perspective to my life.

Thessaloniki, Greece Konstantinos N. Fountoulakis

Contents

Chapter 1
What Is Psychiatry?

It is a question as old as the human race: Why do we succumb repeatedly to foolish temptations against our better judgment? It seems very strange for a creature which possesses reason and free will and has adequate cognitive abilities to assess the consequences of actions. Humans seem embarrassed when they face behaviors which do not seem to make sense, seem completely irrational. On the other hand, a significant part of the human population manifests a consistent pattern of rather irrational behaviors which inevitably lead to some kind of direct or indirect self-harm or harm to others while there are absolutely no signs of insanity present. Since the times of Plato and Aristotle, the approaches were balancing between morality and insanity, with many philosophers (including Plato himself) suggesting that what is moral is healthy and vice versa.

This was more or less the case until recently. Only after the renaissance and especially since the nineteenth century, insanity was recognized as a reason someone is not responsible for his actions. But this concerned people with severe psychotic disorders. Thus, historically, professionals and the public alike considered only markedly unusual and unexplained behaviors to be evidence of disturbed minds. This reflected a narrow view which suggested that milder behavioral issues were indicators of "moral inferiority," vanity, or dissipation rather than a mental disorder. Today, we know that this is not true, although the debate is often hot.

The rule was that instead of recognizing that mild behavioral disturbances may be early stages of potentially more severe disorders, or even they could be mental disorders by themselves, they treated them with ridicule and condemnation. These archaic and inhumane views have given way in recent years to more humane and psychologically sensitive attitudes. Mental illness as well as mental health is no longer a remote province of professional study tacked on as a postgraduate specialty for interested physicians. It is a necessary part of the training of psychiatrists, psychologists, social workers, teachers, nurses, and ministers, who serve the needs of the troubled, but it is also a significant part of the curricula of all colleges.

K. N. Fountoulakis, *Psychiatry*, https://doi.org/10.1007/978-3-030-86541-2_1

Regardless of the definition and whether it is narrow or wide, mental disorders were the focus of attention since the prehistoric times. So psychiatry has a long history. Accounts of persons with mental disorders as well as treatment efforts are dated back to ancient Mesopotamia, Egypt, and Greece. It was always part of medicine in spite of religious and spiritual beliefs of the time. In fact, it is one of the oldest fields of medicine, and it can be recognized as such in ancient myths and texts. As a separate specialty in the modern sense, it emerged together with other specialties of medicine during the last couple of centuries.

Its birth as a modern discipline can be traced to the late eighteenth and early nineteenth century, and the person who is considered to have defined it as a separate discipline within medicine is Philippe Pinel (1745–1826) in France, during and immediately after the French revolution. Maybe some of the most potent images in the history of psychiatry are the various paintings depicting Philippe Pinel while freeing mental patients from their chains (Fig. 1.1). While Pinel was neither the only

Fig. 1.1 Philippe Pinel (1745–1826). Ordering the removal of chains from female mental patients in 1795 at the Paris Asylum of Salpêtrière. Painting (1878). Up: Painting by Charles Louis Muller (1815–1892). Down: Painting by Tony Robert-Fleury (1837–1911)

nor the first to introduce this kind of reforms in the treatment and care of patients, with Vincenzo Chiarugi (1759–1820) and William Tuke (1732–1822) having similar ideas, it was Pinel's interventions those who survived time and politics while those of the others did not.

The term "psychiatry" was coined in Germany by Johann Christian Reil (1759–1813; Fig. 1.2) in 1808, and it was derived from the Greek words "*ψυχή*" (psyche meaning the soul) and "ιατρική" (iatrike meaning the discipline of medicine). Thus, a verbatim translation for psychiatry is "medicine of the soul."

The word "psyche" comes from ancient Greek, and it survives until today. While its contemporary meaning in modern Greek is exclusively that of the "soul," in the ancient language, it had a broader meaning. Etymologically, it probably comes from the verb "ψύχειν" (psykhein), which means "to blow, to cool," and in turn comes probably from the Proto-Indo-European root "bhes-." Subsequently, it includes the meaning "to breath" or "to be alive." The word also referred to butterfly, probably because of its beauty, of its flying ability, but also its metamorphosis from a nymph to a fully developed butterfly. Psyche is often depicted as a butterfly or with butterfly

Fig. 1.2 Johann Christian Reil (1759–1813)

wings, both in ancient as well as in modern art, while butterflies are included in the coat of arms of the British College of Psychiatrists (Fig. 1.3) [1].

The concept is probably similar but not identical in meaning with the Latin “animus” and the Hebrew “nephesh.” More specifically, the word “nephesh” in the Old Testament had been translated into “psyche” in the Septuagint (the translation of the Seventy from Hebrew into common Greek during the third and second century BC).

Psyche was a minor goddess in Greek and latter in Latin mythology. Her husband was *Έρως* (Eros or Cupid in Latin), and their daughter was *Ηδονή* (Hedonie or Volupta in Latin). There is a great myth concerning the conflict between Psyche with her family but also with the goddess Aphrodite (Venus) which was the mother of Eros and thus mother-in-law of Psyche. The story was beautifully narrated in the novel “Asinus aureus” (the Golden Donkey) by Lucius Apuleius (125–170 AD) which is the only ancient Roman novel in Latin to survive in its entirety. It is a novel loaded with Neoplatonic philosophical elements and reflects the ancient myths concerning the two principal forces in life, *ερως* and *θάνατος* (love and death), in a way

Fig. 1.3 Up: Sarcophagus from fourth century BC. Low right: Art by John Gibson (1790–1866) in 1853. Low left: The coat of arms, Royal College of Psychiatrists (former Royal Medico-Psychological Association)

similar to the basic forces and instincts described much later by Sigmund Freud (1856–1939). In many ways, it was the predecessor of the various stories concerning Faust and the famous novel (1808) by Johann Wolfgang von Goethe (1749–1832). The part of the story where Aphrodite orders her to be sent to a mountaintop and wed to a murderous beast, but saved by Eros, is also one of the origins of the later myth of St. George and the dragon and probably served as an inspiration for various stories and fairytales, including "The Beauty and the Beast." This narration constituted also the inspiration for a significant number of artists and their creations (Figs. 1.4 and 1.5).

In ancient and medieval times, there was no particular group of physicians specialized in the treatment of mental patients, although as early as the twelfth century, there were separate wards for mental patients, for example, in the hospital of Pantocrator (*Ξενών Παντοκράτορος*), founded by the Emperor Ioannis II Comnenus in Constantinople in 1136 AD (Fig. 1.6). Interestingly, in that hospital, mental patients were separated also from neurologic patients.

By the nineteenth century, beginning in France, the term "alienist" was used for a psychiatrist. The term comes from French word "*aliéné*" (meaning: insane) which in turn comes from the Latin "*alienus*" or "*alius*" meaning the foreigner, stranger, different (similar to Greek '*άλλος*'). The root comes from the Proto-Indo-European "al-" (meaning: beyond). Thus, an alienist of the time was a physician who acted as an intermediary between the society and mental patients and, by defining their relations, was the agent of the patients' alienation [2].

Fig. 1.4 The Wedding Banquet of Cupid and Psyche (1517) by Raphael and his workshop, from the Loggia di Psiche in Villa Farnesina, a renaissance suburban villa in the Via della Lungara, in the district of Trastevere, Rome, Italy

Fig. 1.5 Psyche Revived by Cupid's Kiss. Sculpture by Antonio Canova (1757–1822) in 1787

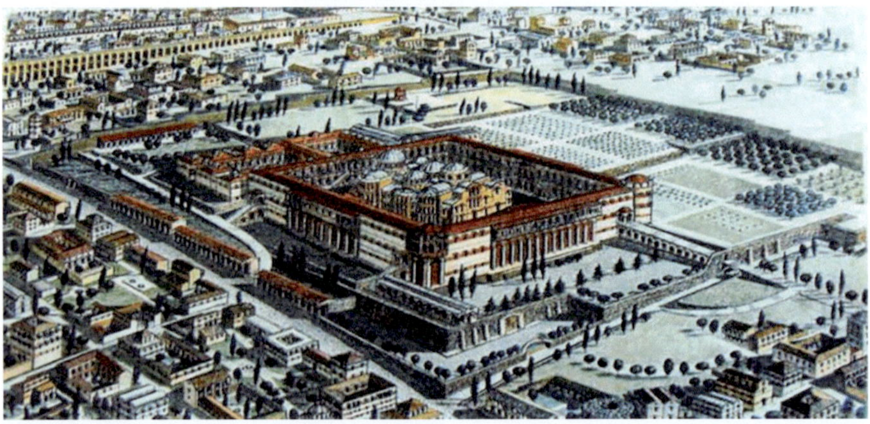

Fig. 1.6 Up: Xenon Pantokratoros, Constantinople. Art by George Anapniotis, architect (2011). Reproduced after permission. Low: The same building as it is today as Zeyrek Mosque

Johann Christian Reil changed the term into "psychiatrist" in 1808 and gave birth to psychiatry as a medical specialty. He also described the qualities needed: "perspicacity, a talent for observation, intelligence, good will, persistence, patience, experience, an imposing physique, and a countenance that commands respect" [3].

Psychiatry went on an independent pathway from the other medical specialties across Europe and America, except from Germany where it was closely linked to neurology following the tradition established by Wilhelm Griesinger (1817–1868).

Neurology separated from the rest of medical specialties later, in the mid-nineteenth century, having as its focus the diagnosis and treatment of those nervous disorders with gross anatomical abnormalities, in sharp contrast with the "psychological" disorders which were the focus of psychiatry.

This separation of two distinct medical specialties, with a common anatomical system to study and treat, reflects the nineteenth-century Cartesian dualistic approach of body and mind and would have a huge impact on the concept and treatment of mental disorders in the centuries to come.

In clinical practice, however, this was not the case. The distinction between psychiatrists and neurologists during much of the nineteenth century concerned almost exclusively the severity of mental patients each treated; psychiatrists were dealing with the most severe cases under long-term hospitalization in asylums, while the less severely ill outpatients were receiving diagnoses like hysteria or neurasthenia and were treated by neurologists as outpatients [4]. Subsequently, a clear somatic cause was considered to be behind the problems of those treated by neurologists, and this enjoyed a favorable reception by patients who could distinct themselves from the lunatics of the asylums. The term "nervous breakdown" became popular during the Edwardian age.

However, until now, no clear brain damage is identified, and this was already obvious a century ago, leading to the development of a precise psychiatric nosology by Emil Kraepelin (1856–1926) and of psychoanalysis by Sigmund Freud (1856–1939), on the basis of clinical wisdom alone. It is interesting that while for the general public psychiatry identifies with Freud, his theories are academically obsolete today while the legacy of the unknown-to-the-public Kreapelin is absolutely dominant today (Fig. 1.7).

During the first half of the twentieth century, psychiatry was shaped by the influence of psychoanalysis and the two world wars. Psychoanalysis pushed psychiatrists out of the asylums and shaped a profession not only widely accepted by the society but also financially profitable. Alienists/psychiatrists became "shrinks" according to the Hollywood jargon.

However, during the second half and until today, it was dominated by the development of precise tools and operationalized criteria for the diagnosis and the development of psychopharmacological agents for the treatment of mental disorders [5]. The contribution of Robert Spitzer (1932–2015) in this process was decisive. An additional important development was the introduction of a wave of pragmatic methods of psychotherapy as a substitute for psychoanalysis, pioneered mainly by Aaron Beck (1921–) and Albert Ellis (1913–2007). These approaches would eventually have a far greater impact on the discipline of psychology which currently is largely based as a profession on the implementation of these treatment options [6, 7].

At the end of the day, it seems that until today, the real heart of psychiatry lies in its relationship with the treatment of severe mental disorders; the modern psychiatrist is a technocratic and humanistic descendant, not of the psychoanalyst but of the nineteenth-century alienist [2].

Fig. 1.7 Left: Emil Kraepelin (1856–1926). Right: Sigmund Freud (1856–1939)

References

1. James FE. Psyche. Psychiatr Bull. 2018;15(7):429–31. https://doi.org/10.1192/pb.15.7.429.
2. Littlewood R, Lipsedge M. Aliens and alienists: ethnic minorities and psychiatry. Psychology Press; 1997.
3. Shorter E. A history of psychiatry: from the era of the asylum to the age of Prozac. New York: Wiley; 1997.
4. Wessely S, Lutz T. Neurasthenia. In: Berrios G, Porter R, editors. History of clinical psychiatry. London: Athlone Press; 1995.
5. Wilson M. DSM-III and the transformation of American psychiatry: a history. Am J Psychiatry. 1993;150(3):399–410. https://doi.org/10.1176/ajp.150.3.399.
6. Beck A. Cognitive therapy and the emotional disorders. New York: International Universities Press; 1976.
7. Ellis A. Rational psychotherapy. J Gen Psychol. 1958;59(1):35–49. https://doi.org/10.1080/00221309.1958.9710170.

Part I
The Historical Roots

Chapter 2
Introductory Remarks to the History of Psychiatry and Mental Illness

Those who know only their own generation always remain children.

Cicero (106–43 BC; Fig. 2.1)

Fig. 2.1 Marcus Tullius Cicero (106–43 BC) was a Roman statesman, orator, lawyer, and philosopher

K. N. Fountoulakis, *Psychiatry*, https://doi.org/10.1007/978-3-030-86541-2_2

History sheds light on the mistakes our predecessors have made in their scientific quests, but also it brings to our attention knowledge and wisdom which are now long forgotten or pushed aside by more favorable or operationalized pathways. Knowledge of these lost and long forgotten ideas not only protects from committing the same mistakes over and over again but also could provide with valuable insight which can not only radically change the way we perceive and interpret the present but also might give us tools to foresee the future.

In order to have a good historical account, one needs to assemble a complex mosaic which includes both facts and persons put in the wider background of tradition and culture and always with the correct historical environment, the dominant ethos of the time.

Both the author but also the reader should be cautious but also open to interpretations of past events. There are different standpoints reflecting different approaches:

- The "presentist" approach, that is, the interpretation of history from a present viewpoint.
- Another is the "historicist" approach, that is, the conceptualization exclusively within the contextual events of their time period and neglect of prior influences or contemporary implications.
- Finally, there is the "internalist" approach which considers events in the narrow frame of the author's discipline and neglects their broader historical context.

One particular trap is to consider that science, philosophy, and religion were always distinct from each other, were as clearly separated as they are today, and always had a meaning and context which were similar to the modern. This is maybe the most erroneous thing to believe. During antiquity, there was no clear delineation between them; in fact, more often, they reflected different faces of the same approach rather than distinct approaches. This is why, in ancient civilizations, although the etiopathogenesis is often metaphysically approached, treatment is often pragmatic and medically oriented.

To write specifically on the history of psychiatry and mental illness is a quite complex issue. Not only the definition of what constitutes disease and treatment is vague, but also it is obvious that diagnosis and treatment goes hand to hand with the development of theories on etiopathogenesis, while religious, sociocultural, and politico-economic forces exert a strong influence.

In psychiatry, theory rather than the evidence drives the therapeutic endeavors. But besides from "treatment" per se, also the attitudes of the society are also part of the history of psychiatric therapeutics. These attitudes reflect the care for the mentally ill and the investment in their well-being or, on the contrary, on their isolation and restrain.

During the course of history, often, many "stories" take place independently and in parallel to each other, while in other cases, developments in other fields and disciplines of science come to play a determining role in the course of events in mental health. The puzzle of history is never two-dimensional, and probably, it is not even 3D. It is comprised of many overlapping but often independent dimensions and trajectories, making the narration complex and difficult. To make things worse, "truth" is often hidden in anecdotal or unverified personal accounts, while "official records" can give only crude information.

Discovery by chance and in the absence of any theoretical background is not uncommon, and there is not a better example than the development of psychopharmacotherapy. This stroke of good luck is often called "serendipity," a term coined by Horace Walpole (1717–1797) in 1754 in a letter to Horace Mann (1706–1786) concerning the fortunate finding of the lost painting of "Bianca Cappello" created by Giorgio Vasari (1511–1574). He derived this term by referring to the story of *The Three Princes of Serendip* (Serendip is the ancient name of Sri Lanka/Ceylon) who according to that story made a number of lucky discoveries.

Many eminent scientist have acknowledged the role luck and chance played in their discoveries [1, 2], but it was the Nobel laureate Walter Cannon (1971–1945) who made it clear that chance is important only in the presence of a mind which is educated, trained, and open to recognize, accept, and appreciate the unexpected novelty [3].

In the chapters that follow, the author will attempt to give both the precise historical detail but also the essence of the amazing efforts of humanity to combat with mental illness since antiquity.

The author chose to present and discuss the events and discoveries in a chronological order rather than in the form of isolated "complete" stories of individual developments. This is of course a very difficult task because many events stretch through decades and are connected with events decades or even centuries before; however, this is the only way to uncover the effect of interaction among people, theories, and developments which become obvious as many pieces of the puzzle take their proper place, and they assemble a picture of the landscape concerning mental health in each time period and within the wider frame of the development and change of human society.

Knowing history is important. Knowing the chain of events that led to some of the most exciting developments in the history of human kind is of outmost importance and carries an important message: As already mentioned above, it is the rare exception rather than the rule that important discoveries happen by chance. This is a popular myth that spreads from penicillin to psychotropics. Not only in most cases it is far from truth, but often, it is deliberately produced so that a particular researcher will keep all the credits for himself. It is also a frequent effort to justify failures as "essentially unscientific," although most of them emerged and flourished within the laws of science before being rejected, as again the laws of science demand and predict.

Theoretically, there are two contrasting views of science:

- The first suggests that science evolves by progressively more adequate theorizing, and the theories are checked experimentally.
- The second view suggests that new developments in technology create a generalized boost also in other fields of science.

However, it is essential to realize that it was the painful chase for new knowledge and the agony of a great number of prominent people which led to today's miracles through the centuries. These developments are miraculous in spite of the fact that still they are not satisfactory enough. This is not an oxymoron if one considers where we started from and where we stand today.

At the end of the day, we should also consider that the development of treatment options followed the changing concept of madness through the ages. More particularly during the last couple of centuries, they followed the focus of mental health from the asylum insane to the drug abuser to the criminal psychopath and problematic personalities. It also reflects our changing opinion on human beings and humanity, on our society and the place of the individual within it, and on our wishes and predictions on our common future.

References

1. Dale H. Accident and opportunism in medical research. Br Med J. 1948;2(4574):451–5. https://doi.org/10.1136/bmj.2.4574.451.
2. Merbaum M, Lowe M. Serendipity in clinical research. In: Kendall P, Butcher N, editors. Handbook of research methods in clinical psychology. New York: Wiley; 1982. p. 95–123.
3. Cannon W. The way of an investigator. New York: W W. Norton; 1945.

Chapter 3
Antiquity

3.1 The Prescientific Period

The efforts to understand the nature of human kind and especially the nature of the human mind and behavior can be traced several thousand years ago. They are characterized by complex turns and unexpected influences and alliances, and of course, in the beginning, there were characterized by the lack of the scientific eye and method. These early attempts are sacred rather than scientific or philosophical and can be classified [1] into:

- Animistic
- Mythological
- Demonological (religiousness of late antiquity with strong ethical elements)

At the dawn of humanity, it was widely believed that supernatural phenomena were the cause of mental illness. They included possession by spirit or demons, witchcraft, or attack by evil angry spirits or deities. Treatments were mystical but also aggressive and brutal.

The first documented attempts to treat mental illness date back as early as 5000 BC. This is supported by the discovery of trephined skulls during excavations in areas which hosted ancient civilizations. The first report on trephination was made in 1863–1865 by the American anthropologist Ephraim George Squier (1821–1888) who observed a rectangular hole in a prehistoric skull during his visit to the Cuzco district of Peru (Fig. 3.1). In 1867, Paul Broca (1824–1880) confirmed that this hole was deliberately made. Of prime importance was the observation that there was a porous structure in the bone surrounding the opening, with signs of infection. This means that the subject had survived the operation, at least for a short

K. N. Fountoulakis, *Psychiatry*, https://doi.org/10.1007/978-3-030-86541-2_3

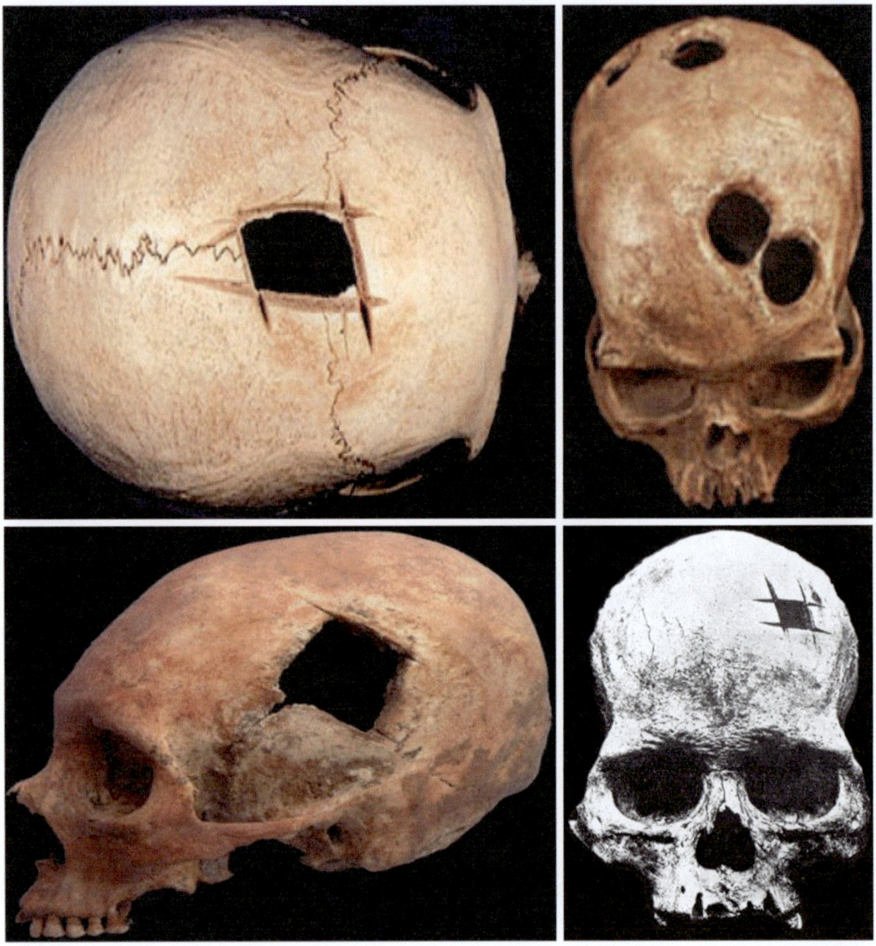

Fig. 3.1 Prehistoric skulls with marks of trephination, including the skull found by E.G Squier, E. G. (bottom row, right)

period of time [2]. The method of trephining includes the chipping of a hole (trephine) into the skull. The only available tools at that time were crude stone instruments while later more sophisticated tools made of metal appeared (Fig. 3.2). The idea was that evil spirits would leave the head of the patient through this hole. The method survived until the early twentieth century in some cultures, as a cure for a number of health problems including fractures of the skull, migraines, and mental illness (Fig. 3.3).

In most ancient societies, magical forces, malevolent deities, or demons were considered to be the source of mental illness to and "treatment" concerned religious or magical rituals performed mainly by priests. Christianity named these deities

Fig. 3.2 The bronze cutting instrument, known as Antze's tumi, which depicts the operation of trephining (from: Asenjo, 1963). (Courtesy of Charles C. Thomas, Publisher, Springfield, Illinois

"demons" (after the Greek word *δαίμων* which means lower rank deity) and attributed a definitely negative connotation to the word. However, even after the spread of Christianity, these beliefs were dominant in many societies, even among members of the higher socioeconomic class in Europe. The history of civilization suggests that credence in demons fluctuated in complex ways.

In ancient Egypt, history started with Imhotep who lived during the Third Dynasty of Egypt's Old Kingdom, under the rule of King Djoser (Zoser) (c. 3,000 BC), and he was a high priest. His name means "he who comes in peace." He was a man of many disciplines including astronomy, magic, and architecture and is believed to be the mastermind behind the "step pyramid." After his death, legends about him grew, and eventually, he rose as a demigod, at the center of a large cult with healing temples being erected in his honor at Memphis, Sakkara, Philae, and elsewhere. The oldest surviving medical document, the Edwin Smith Surgical Papyrus, bought in the Egyptian city of Luxor in 1862 by archeologist Edwin Smith (1822–1906), was written approximately 1000 years after Imhotep's death, but

Fig. 3.3 The Cure of Folly, by Hieronymus Bosch (1450–1516), depicts a trephination operation. Painting displayed in the Museo del Prado in Madrid, completed around 1494 or later

many historians believe that it contains his teachings [3]. The papyrus includes information pertaining to brain injury and its treatment. Smith's daughter donated the papyrus to the New-York Historical Society in 1906. The society asked the Egyptologist James Breasted (1865–1935) to translate it in 1920, which was made possible, thanks to the discovery of the Rosetta (Rashid) stone in 1799 by an officer of Napoleon under the name of Boussard. The stone served as a transcriptional lexicon between Egyptian and ancient Greek, and Jean-Francois Champollion (1790–1832) fully deciphered the stone in 1822. Utilizing these tools, it took breasted 10 years, and when the translation was ready, scholars saw that it contained among others, twenty-seven cases of head injury and a description of the brain itself.

Smith acquired a second papyrus which took its name from Georg Ebers (1837–1898), a German Egyptologist to whom Smith sold the papyrus in 1873 [4]. It is also known as the "Book of the heart" and includes descriptions of anger and sadness but in relationship with heart diseases (Fig. 3.4) [5].

It is also based on records from the Old Kingdom and even on more ancient ones (c. 3100–2890 BC). It contains almost 900 detailed and specific prescriptions for a number of medical conditions, including migraine and tremor. Interestingly, while

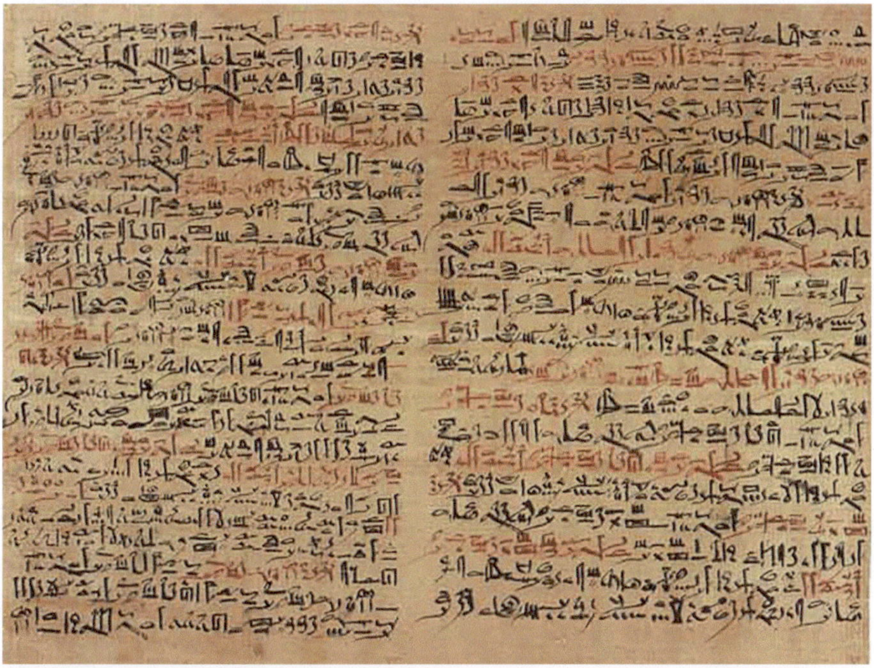

Fig. 3.4 Eber's papyrus (book of the heart); Egypt c. 3000 BC

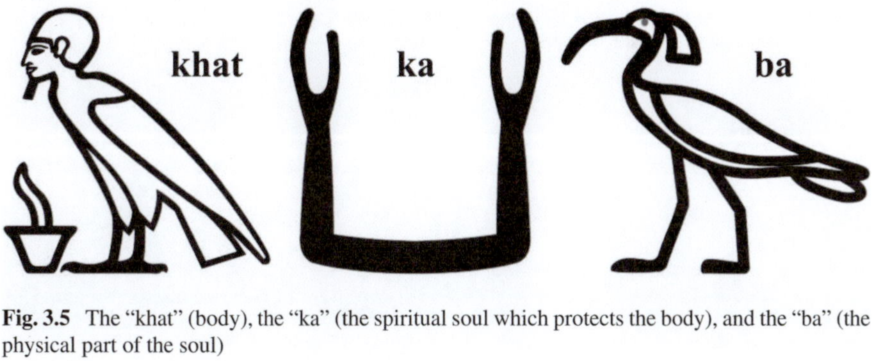

Fig. 3.5 The "khat" (body), the "ka" (the spiritual soul which protects the body), and the "ba" (the physical part of the soul)

some of the ingredients could be considered helpful even according to contemporary standards, one interesting characteristic is that 55 of these prescriptions contain urine or feces ("coprotherapy" from the Greek word "*κόπρος*" meaning the feces). The logic behind this was to make the body disgusting and intolerant for demons to possess. One prime guideline was that the treatment should be applied as soon as possible before the demons would inflict irreparable damage to the body.

The ancient Egyptians considered the heart, not the brain, to be the seat of mental life, and they conceptualized the body and the mind as an analogue of the river Nile valley. According to their beliefs, humans were comprised by the "khat" (body), the "ka" (which protects the body), and the "ba" (spirit and destiny) shown in Fig. 3.5. They thought that a system of channels ("*metu*") was connecting the heart with the rest of the body by carrying air, liquids, as well as solid matter. The conceptualization as an analogue to the Nile environment pointed to the importance of moisture but also to the balance of elements and mechanisms, leading to the first idea of the "equilibrium." While the valley of the Nile was suffering from too little (dryness) or too much (floods) water, as well as from extremes of temperature, the human body and mind were suffering in a similar way. Reasoning by analogy, as the channels of the river should be kept open, but not too open, so should the "metu." In this frame, bloodletting was practiced both with the use of leeches as well as with venesection. Cathartics were used for the bowels.

Egyptian mythology suggested that after death, the heart, which during life records all good and evil deeds, will be weighed against a feather to assess sin and guilt (Fig. 3.6). As a consequence, during mummification, the heart was treated with respect and was never removed from the body, in contrast to the brain which was extracted with the use of a hook through the nostrils and then discarded [6] while other internal organs were kept in sacred jars.

In ancient China, Guan Zhong (c. 720–645 BC) [7] in the Guanzi text recorded the existence of institutions for those somatically and mentally disabled in the frame

Fig. 3.6 At left, Ani (the deceased) and his wife Tutu enter the assemblage of gods. At the center, the jackal-headed god of the dead, Anubis, weights the heart of the deceased in the Duat using the feather of Maat as the measure in balance, observed by the goddesses Renenutet and Meshkenet, the god Shay, and Ani's own ba. At right, the monster Ammut, who will devour Ani's soul if he is unworthy, awaits the verdict, while the god Thoth prepares to record it. At top are the gods acting as judges: Hu and Sia, Hathor, Horus, Isis and Nephthys, Nut, Geb, Tefnut, Shu, Atum, and Ra-Horakhty. The baboon sitting atop the scales represents Thoth, the god of writing, and is overseeing the procedures. The winged figure to the left of the scales is Ani's ba or physical soul

of an early broad social welfare policy. As the system was strictly court-centered, inevitably, it was socially oriented as all Chinese philosophy and medicine. The Chinese philosophy of the time was preoccupied with the number "five." The analogues of the theories of Hippocrates and Empedocles include five basic elements: wood, metal, fire, earth, and water. There are five basic body senses operated by five basic organs, five tastes, five colors, and five emotions (sorrow, fear, anger, desire, and joy).

The rules of treatment are described in the Huangdi Neijing (Nei Ching) or Inner Canon of the Yellow Emperor (475 BC–200 AD) which also includes descriptions of neurological and psychiatric clinical symptoms and attributes them to excessive anger [8]. The treatment included herbs, acupuncture, or "emotional therapy" in the frame of what today we call "traditional Chinese medicine." It emphasized the relationship between body and psychological states, especially emotions. Overall, diseases were thought to comprise or were classified into five stages or elements and imbalance between Yin and Yang. In the fourth century BC, Shan Hai Jing (Classic of Mountains and Seas) included several medication treatments for mental disorders, for example, preparations from the plant Vitex cannabifolia for the treatment of mental unrest as well as preventive advice, for example, that eating too much meat from Mandarin duck (Aix galericulata; Fig. 3.7) causes madness [9]. Later, Hsun Tzu (or Xunzi 298–212 BC) proposed a view of the world and philosophical approaches with significant similarities to those of Aristotle.

In India, the Hindu system showed significant resemblance but also differences from the Chinese concepts for health and disease. In this system, mental health was part of the metaphysical and religion, although mental disorders are included in

Fig. 3.7 The Mandarin Duck (Aix galericulata), a perching duck species native to east Asia. It is medium-sized, at below 0.5 m longa with a 65–75 cm wingspan. In Chinese, Korean, and Japanese culture, they are considered to mate for life and thus are also called "lovebirds"

medical texts. There are descriptions of depression and anxiety in the scriptures known as Ramayana and Mahabharata. In general, mental illness was thought to be caused by metaphysical or supernatural entities or witchcraft. At the same time Polyvos developed the theory of bodily humors in Greece, there is a description of a similar concept in the text of the Charaka Samhita (900–200 BC) which is part of the Hindu Ayurveda (meaning "knowledge of life"). This text suggests that bad health is the result of the imbalance among three kinds of bodily fluids or forces called dosha. These were named the "Vata," the "Pitta," and the "Kapha" and are made from the combination of the five basic elements (earth, water, fire, air, and ether; Fig. 3.8). Their imbalance could be caused by inappropriate diet, excessive fear or joy, inappropriate physical activity, and disrespect toward the gods, teachers, or significant persons. Accordingly, the treatment included intervention at different levels: at the biological level with herbs and ointments, the metaphysical with charms and prayers but also at the interpersonal with moral or emotional persuasion, and shocking the person [10]. This is maybe the first time that shocking of the patient is recommended as it was believed this could produce a therapeutic effect. Shock therapy as a concept will be developed again in the west during the eighteenth century by Hermann Boerhaave and will be the dominant concept until the second half of the twentieth century.

The most important book in the Indian subcontinent was the "Susruta Samhita" (the Compendium of Susruta). Susruta (meaning "well heard") was one of the most important ancient Indian physicians; however, the name refers to many different persons, and the compendium is likely to be a collective work through several centuries. It was probably completed by the fourth century BC. Susruta is represented as a son of Brahmarshi Vishvamitra in the Mahabharata. The concept of mental disorder as this is reflected in the Susruta Samhita is based on demoniacal

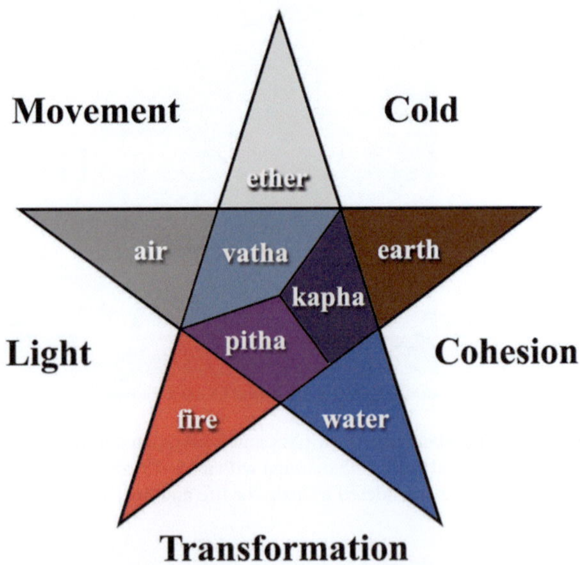

Fig. 3.8 The three doshas and the five great elements according to Ayurveda

possession; in this frame, it is interesting that both physical as well as psychological treatments are proposed [11]. During the reigns of King Ashoka, whose name means "without sorrow" (c. 268–232 BC; Fig. 3.9), several hospitals were established especially for the mentally ill [12]. King Ashoka himself was reported to suffer from frequent loss of consciousness of unknown cause although it is speculated they were either stress or epilepsy-related [13]. His view of the world included the adoption of Buddhism and respect for all live beings; he went on to create an early type of state health system for all people in this kingdom, and also, he established hospitals especially for animals.

It is interesting that skepticism flourished in India. The philosophical school of "Ajñana" was teaching radical skepticism and was a major rival of early Buddhism.

In ancient Mesopotamia, mental disorders were considered to be caused by divine spirits or the gods [14, 15], and in order to imply that the person is "seized," the word referring to "control" and also "hands" was used in order to denote a concept similar to modern "seizure." Thus, mental illnesses had names like "Hand of Ishtar" (Qāt Ištar) or "Hand of Shamash," but we do not know to which specific

Fig. 3.9 King Ashoka (c. 268–232 BC)

nosological entities they corresponded [14]. It is possible that therapists attributed some kind of interpretation to symptoms, especially hallucinations, more often with a spiritual or metaphysical meaning much similar to the folk interpretation of dreams [15]. Babylonians assigned the demon Idta for insanity as they did by assigning a demon for all diseases. They utilized exorcisms but also medications made from herbs and plants. They also utilized psychological treatments in the form of confessions and religious rituals.

In Hebrew tradition, both disease and cure come from God. Disease was considered as punishment for committing sin, while healers were priests who made appeals to God in order to cure sickness. In the Old Testament, King Saul was presumably suffering from manic depression and could find relief only to the music played by David and his harp. Eventually, he committed suicide after a military defeat. Also in the Old Testament, in the Book of Daniel, there is extensive description of the madness of King Nebuchadnezzar II of Babylon (605–562 BC; Fig. 3.10) which came from God as a punishment and lasted 7 years.

In ancient Persia, illness was attributed to demons, but interestingly, the prevention of disease included the improvement of hygiene and moral purity of the mind as well as of the body which could both be achieved through good deeds and thoughts.

Fig. 3.10 King Nebuchadnezzar II of Babylon (605–562 BC). Illustration from the Prize Bible (Wells Gardner, Darton & Co, c 1890)

3.2 The Dawn of the Scientific Mentality

However, the dawn of scientific era started in ancient Greece. Before this, there is a wealth of myths and stories on mental illness and its treatment. Melampus (name meaning "black feet") was a mythical figure who supposedly had cured the mania of Ifinoe and Ifianassa, the daughters of King Proetos of Argos. He also cured Ificlos, son of King Fylacos of Fylacae from his psychogenic impotency. In Iliad, Homer mentions the manic episode of Ajax, followed by deep depression and shame which results in the suicide of the hero. Interestingly, Ulysses had faked a mental disorder in order to avoid going to Troy. The hidden meaning of the myth of Jason and Medea and the argonauts has been the focus of work of many scholars (Fig. 3.11). It is interesting that their names suggest a relationship with medicine and treatment. Jason is the English version of the name "*Ιάσων*" (iason meaning the healer), while Medea corresponds to "*Μήδεια*" (pronounced Midia; comes from the Greek verb "*μέδομαι*" meaning take care, study, improvise, and predict). Medea relates also to the word Medusa (the gorgon) and to medicine (through Latin). He was of barbaric origin and considered a witch-priestess. One interesting aspect of the myth is that Jason left her to marry "*Γλαύκα*" (Glauca, meaning owl, the bird of wisdom). After this, Medea went to Thebes where she healed Hercules from the curse of Hera (bursts of anger similar to acute manic episodes during which he killed Iphitus, Hercules' best friend).

Fig. 3.11 Jason and Medea images on a Greek amphora from the National Archaeological Museum of Naples, c. fourth century BC. The depiction of the serpent is quite similar to the modern symbol of medicine, and the scene resembles the Old Testament scene of Adam, Eve, and the snake

The myth of Jason and Medea and the argonauts received a number of interpretations with most dominant that of the exploratory travelling of ancient Greeks in the black sea and the Caucasus area in search for gold. However, one cannot leave unnoticed the connotation of names: The "healer" goes east to bring eastern metaphysical tradition of how to take care and govern only to soon abandon the newly acquired habits for the sake of knowledge. The early depictions of Jason, Media, and the dragon-guard of the golden fleece seem too similar to images of Adam, Eve, and the serpent (Fig. 3.11).

In another line of myths, Apollo, son of Zeus, was the god of medicine. He was the teacher of centaur Chiron (half human, half horse) who further enriched healing practice with the discovery of medicinal herbs and hymns. According to Hesiod, in turn, Chiron taught Asclepius (Greek *Ασκληπειός*, Latin: Aesculapius) who was son of Apollo and the nymph Coronis [16] (Fig. 3.12). The name "coronis" (in Greek *κορωνίς*) is similar to Latin and French "coronis" and belongs to the same group of words with "crown" and "horn," meaning "something on the top of." This was an allegory, meaning that Asclepius was the ultimate of medical charisma. There seems to be a real person who was highly skilled in medicine, carried the name of Asclepius, and lived in Thessaly (near Pelion, the mountain of the centaurs) at the era of the Trojan war. According to myths, his wife was *Ηπιόνη* (Epione meaning the one who calms the pain). He had five daughters: *Υγεία* (Hygeia that is Health) the goddess of health, cleanliness, and sanitation; *Ιασώ* or *Ιησώ* (Iaso that is Healer; conveniently similar to *Ιησούς*-Jesus which however is of a Hebrew origin) the goddess of

Fig. 3.12 The great chryselephantine statue of Asklepios made by Thrasymedes of Paros as restored by M. Defrasse. The god was sitting on a throne, a large golden serpent rising up to his left hand; on his right lay a dog, and in front was an altar. From Caton, R: The temples and ritual of Asklepios at Epidauros and Athens. Two lectures delivered at the Royal Institution of Great Britain. London, C. J. Clay, 1900

recovery from illness; *Ακεσώ* (Aceso that is the Therapist) the goddess of the healing process; *Αίγλη* (Aegli that is Prestige) the goddess of beauty, splendor, glory, magnificence, and adornment; and *Πανάκεια* (Panacea) the goddess of universal remedy. Later and after the third century BC, *Τελέσφορος* (Telesphorus meaning the one who brings a successful result) was added as a son, probably with the adoption of a Celtic god. Possibly but not proven beyond doubt, Romans added a sixth daughter called Meditrina (possibly meaning "healer") which represented the magic spells and herbs. Asclepius was associated with the Roman and Etruscan deity Vediovis (probably meaning the God on earth). According to the myths, Asclepius managed to bring a man back from the dead; this provoked the anger of Zeus who killed him with a bolt of lightning.

Like Imhotep, also, Asclepius rose to become the demigod at the center of a healing cult in Greece. A number of temples, the "asclepieia" were dedicated to his worship and served as early hospitals (Figs. 3.13 and 3.14) [17]. In one of these temples in the Aegean island of Kos, it is believed Hippocrates may had begun his career. In these temples, treatment included the ritual purification followed by offerings or sacrifices to the god. Afterward, the patients would spend the night in the *άβατον* (avaton) or *άδυτον* (adyton), both names meaning the place where nobody can enter. It constituted the most sacred place of the temple, and there, the god would contact the patients through their dreams. The priests would interpret these dreams or visions and would accordingly prescribe the appropriate therapy. An interesting detail concerns the Aesculapian snakes, a particular type of nonvenomous snake, which were allowed to slither around freely on the floor in the rooms where the patients slept. In general, snakes were considered to be sacred because they emerge from the ground, and thus, they carry the healing powers of mother earth (gaea mother, the prototype representation of the female deity in Greek mythology). An additional reason snakes were considered sacred is that they shed their skin which is then rejuvenated; thus, they are a symbol of death and resurrection and eternal life [18]. Snakes were an important element of ancient Athenian

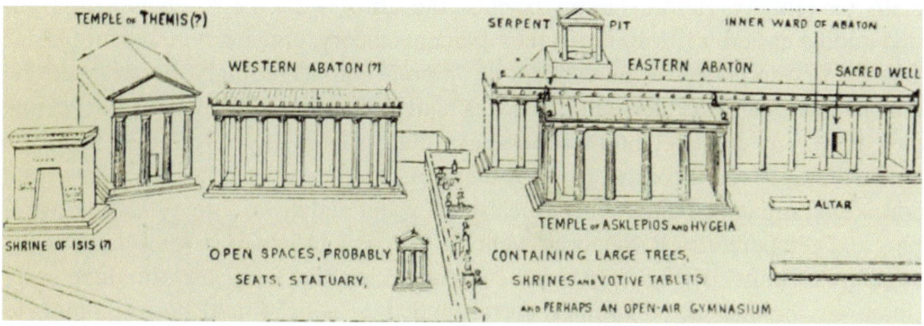

Fig. 3.13 Design of the Asclepeion of Athens based on ancient ruins, from Caton, R: The temples and ritual of Asklepios at Epidauros and Athens. Two lectures delivered at the Royal Institution of Great Britain. London, C. J. Clay, 1900

Fig. 3.14 The Tholos and the Sanctuary of Asclepius at Epidaurus, from the Wellcome Collection

worship; they represented their bonds with the land (crawl onto the ground), and they were the symbol of Athena, goddess of wisdom; a sacred snake was kept in the Parthenon before the Persian wars. It is interesting that a similar role (thought with a negative connotation) is reserved for snakes in the Genesis in the Old Testament. This is the origin of Asclepius' rod which was a snake-entwined staff, remaining until today as the symbol of medicine.

At the center of Greek philosophical thinking was the rejection of the belief that humans were the puppets of the gods [19] and in combination with an effort to comprehend the "*κόσμος*" (cosmos, that is the universe) by reducing it to its fundamental elements. The place of humankind within this new physical universe was also a question. Maybe the first human ever to specifically reject mysticism was Thales of Miletus (652–588 BC; Fig. 3.15) in Ionia, then Greek east coast of Asia Minor, today west Turkey, who argued that gods are not responsible for the weather or any events in nature. He could be considered to be the first scientist on the face of the earth. In this frame, he suggested that mental disorders are natural occurring events and should be studied and treated according to the scientific way.

Balancing between religion and science, it was Pythagoras (580–496 BC) who suggested for the first time that the brain is the organ where mental phenomena take place and proposed that the mathematical concepts of balance and ratio should be used to explain how and why physical characteristics of the body (e.g., moisture vs. dryness, cold vs. hot) lead to mental disorders. Pythagoras was mainly a mystic rather than a scientist, and since the basis of his philosophy was numbers and especially the tetractys (tetrad) simply because the sum 1 + 2 + 3 + 4 equals 10, he unified middle eastern theories into a four elements theory, organized in two bipols. He was the first to conceive the concept of "bipol," and accordingly, he arranged the characteristics and the elements. This was both radical and brilliant as a concept and paved the way for later philosophers, especially Empedocles (c. 494–c. 434 BC) who developed the theory of the four elements of the universe in fine balance and admixture, organized in two bipols: earth vs. water and air vs. fire, reflecting moisture and temperature. Pythagoras went further and proposed a four-components structure of humans (Fig. 3.16); one component was the body, and the three other (three was another sacred number) corresponded to "*νους*" (mind, corresponding to reason and intelligence), "*ψυχή*" (soul, corresponding to affect), and "*πνεύμα*" (spirit, corresponding to the immortal part). But the real structure of the tetractys reflects two bipols: mortality (body) vs. immortality (spirit) and reason (mind) vs.

Fig. 3.15 Thales of Miletus. Illustration from "Illustrerad verldshistoria utgifven av E. Wallis. volume I": Thales

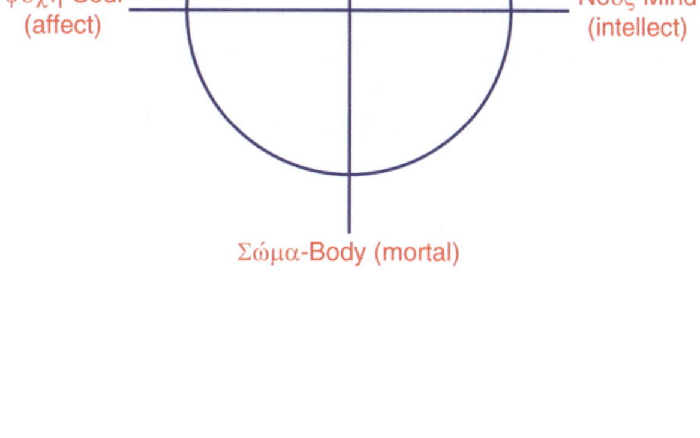

Fig. 3.16 The mind/body according to Pythagoras (tetractys)

passions (soul). Pythagorean metaphysical and religious beliefs also included immortality and the transmigration of souls, and these beliefs influenced his scientific approaches especially concerning mental phenomena and human behavior [20]. It is interesting that his proposal of the "spirit" in many aspects corresponds to what today is defined as metacognition.

The emergence of the scientific method and the retreat of metaphysics and religion were not without conflicts, bitterness, and blood. The ancient Greek philosophical spirit fought with tradition and its defenses in a ruthless war. Free and rational thinking about man, nature, and the gods was a challenge for the society of the time, in contrast to the romantic impression we have for the democracy of that time. In 432 BC in Athens, the famous law or decree of Diopeithes was passed concerning the public accusation "against persons who did not believe in gods or who taught doctrines regarding the heavens" (Plutarch, Pericles, 32.1). A series of trials against prominent philosophers followed, including that of Socrates himself in 399 BC on the basis of "*ασέβεια*" (asevia, meaning impiety). Before that and around 416 BC in Athens, the books of Protagoras were the first books in history to be publicly burned, an event not absolutely verified but mentioned by Diogenes Laertius and by Cicero in his work "de Natura Deorum."

3.3 The Period of Classic Antiquity in Greece

The first to attribute mental disorders to nervous system disorders rather than to spiritual or metaphysical forces was Alcmaeon of Croton who lived in the fifth century BC and was one of the most important natural philosophers of antiquity. During the same period and well within the frame of wider sociopolitical changes and especially with the dawn of democracy and the concept of the responsible citizen, the Hippocratic physicians recognized that on one hand, their patients had the right to choose any type of treatment and any therapist, and on the other hand, the physicians themselves rather than some kind of divinity should take responsibility for the treatment outcome.

The *Corpus Hippocraticum* (original name in Greek: *Ιπποκρατική Συλλογή)* was created by the scholars of the library of Alexandria around 300 BC, after Ptolemy Soter (367–283 BC), who was ruler of Egypt at that time and ordered them to collect books containing all of human knowledge, by compiling all classical medical documents they could find [20, 21]. These texts are probably the collective work of unknown authors, not of Hippocrates himself.

In them, the concept of the four essential body humors was introduced. These humors included the black bile (from the liver), the yellow bile (from the spleen), the phlegm (from the brain), and the blood (from the heart). It was believed that humors were secreted each one by a different organ and possessed different properties and characteristics. More importantly, it was believed that these qualities of humors varied with the seasons. Early versions of this theory might had existed in ancient Egypt or Mesopotamia, but the essence of this approach is ascribed to the school of Kos and to Polybus who was a pupil and son-in-law to Hippocrates (fourth century BC; Fig. 3.17) and author of the book "*Περί φύσεως ανθρώπου*" (Peri physeos anthropou meaning "On the Nature of Man"). This theory is in fact a microcosmic form of the macrocosmic theory of the four elements (earth, water, air, fire) and the four qualities arranged on two bipols (dry vs. wet and cold vs. hot) of the universe as first proposed by Empedocles (fifth century BC). Essentially, it was a model with two factors (or bipols), humidity and temperature [22], with wider metaphysical and philosophical implications as the deity of fire was Zeus and of air was Hera (male vs. female), while the deity for earth was Hades and of water (and sperm) was Nestis (death vs. life).

It is important to note that the brain was considered to be the seat of life. In these "Hippocratic texts," some of which are only remotely related to Hippocrates himself or his school, it is clearly hypothesized that behind all diseases exists a specific disequilibrium among the body humors. In terms of Pythagorean philosophy, and in

Fig. 3.17 Hippocrates (460–370 BC)

accord with previous Egyptian ideas, health was identified with "eucrasia" (in Greek *ευκρασία*; good admixture and balance) while in contrast, illness was identified with "dyscrasia" (in Greek *δυσκρασία*; bad admixture or imbalance). This concerns mental disorders also. According to this theory, yellow bile causes manic rage while black bile (which was under the influence of the planet Saturn and related to autumn) caused melancholia (in ancient Greek "*μέλας*" means black, "*χολή*" means bile), and this constitutes the very first biological model of a mental disorder [23]. It is also important that they believed that the black bile causes melancholia because it influences the brain. Long-lasting anxiety, fear, and moodiness were described as core characteristics of melancholia. Concerning mania, they believed that it is caused by excess of blood. Different patterns of body fluids were believed to underlie different "idiosyncrasies" (in Greek *ιδιοσυγκρασίες* meaning unique or characteristic admixtures of elements, temperaments). It is interesting to mention that it was thought that yellow bile was responsible for the choleric temperament (irritable, hostile, and given to rage) while phlegm for the phlegmatic (indolent, irresolute, and timid).

Basic to the Hippocratic approach was the notion that always there should be given time and priority to Mother Nature in order to heal the patient. This was an axion in spite of the availability of various treatment options. As a consequence of this principle and also because of the rejection of the metaphysical approach, later in the middle ages in Europe, the term "physician" emerged out of the Greek word "*φύσις*" meaning the Nature.

The Hippocratic authors described and classified mania and melancholia for the first time. These two concepts are the oldest ones in psychiatry. Their view of mania probably corresponds to what we consider "excited psychoses" today, while melancholia corresponds to anxious fearful depression rather than what we today consider "melancholic features."

The etymology of the word "mania" is interesting since it reflects the early concepts for this condition. It has the same root with the Greek words *μένος* (menos) and *μήνις* (minis) which both mean anger. It is also related to the verb *μένω* (meno=stay) and thus reflects a condition in which the state of the mind is stalled because of intrusion of anger and intense emotions. Interestingly, in ancient Greek, the word *μήνις*, which is related to the Latin word *mens* (=mind), also means crescent and thus relates the states of the mind with the phases of the moon. The Greek word for month (which corresponds to moon phases) is *μήνας* (minas), while the Latin word is *mensis*. This group includes also the words "menses" and "menstrual," and thus, they possess both a temporal/cyclical and an emotional meaning. The word *μαινάς* (=maenad) used for women who participated in the orgiastic rites of God Dionysus and *μηνίσκος* (meniscus of the knee) have a similar etymology. Peculiarly, in his book "On Acute Diseases," the Roman physician Caelius Aurelianus, who was a student of Soranus of Ephesus and member of the Methodist school, suggested several different etymologies. One of them relates mania to the Greek word *ανία* (ania=dysphoric boredom) while another to the word *λυμαίνειν* (lymainin=devastate, defile). The other options include variations like *μονούσθαι* (monousthae=seeks solitude), or *υπομονετικός* (ypomonetikos=having patience). It is obvious that the term "mania" belongs to a group of words coming from the Ancient Greek language and refer to mind and its states, intense emotions and especially anger, crescent and phases of the moon, loneliness, and patience [23].

On the other hand, the term "melancholia" refers to black bile (as mentioned above), the modern term "depression" was derived from the Latin verb "*deprimere*" (=to press down). Since the fourteenth century, the word "depression" was used to denote a negative psychological state, and this was officially done in written language for the first time in 1665 in Richard Baker's "Chronicle." The term is also used in other scientific disciplines like physiology and economics (e.g., "the great depression" of the 1930s). The French psychiatrist Louis Delasiauve (1804–1893) was the first to use the term in 1856. The term is equivalent to the German word "*niedergedruckt*," the English "*downcast*," the French "*abattement*," the Spanish "*abatid*," and the Italian *"essere giu*." The modern Greek translation of depression is "*κατάθλιψη*" (katathlipsi =overwhelming sorrow). The word "*κατά*" means among others "overwhelming," "down," and "against," while it is to be noted that in Greek, the verb *"θλίβω"* (thlivo) means not only that I cause sorrow but also that I break into pieces. In this sense, the word "κατάθλιψη" also means "total breaking" and "destruction" [23].

Ancient Greeks described melancholia as a state of aversion to food, despondency, sleeplessness, irritability, and restlessness, and they also recognized the close relationship between depression and anxiety. Also, they were the first to recommend "scientific," "non-metaphysical" therapeutic interventions, like personal hygiene, bathing, and dieting so as to keep the humors in balance. They also used purgatives, cathartics, and bleeding as a more aggressive attempt to reset the balance of humors in more difficult or emergency cases. At that period, psychological treatments were also available, including the interpretation of dreams as well as talking to encourage, console, and understanding of the illness (Plato's dialectics) [23].

In general, during the times of ancient Greece and the Roman Empire, religion, science, and philosophy were interplexed concerning the way mental illness was understood. The psychological vs. somatic approach has its roots in that time period, and essentially, it derives from the differences in the philosophical views of Plato vs. those of Aristotle (384–322 BC; Fig. 3.18). Plato (424–348 BC) was probably the first to talk about subconscious conflicts and cognitive biases. Both him and Socrates (c. 470–399 BC) believed that mental symptoms (corresponding what used to be Axis I in past versions of the DSM) are unrelated to personality (Axis II) but also suggested that one kind of mental illness is biological while another is closely related to morality. These ideas are clearly shown in Plato's "Phaedrus" and "Timaeus," part of the "Socratic Dialogues" [24–26].

Fig. 3.18 "The School of Athens" by Raphael (Raffaello Sanzio da Urbino; 1483–1520) c. 1511. At the center of the picture, Plato points toward heaven and Aristotle to the earth in a symbolic scene of the conflicting philosophical approaches. Around them gather other philosophers including the Greek Socrates, Zeno of Citium, Epicurus, Pythagoras, Antisthenes, Anaximander, Heraclitus, Diogenes of Sinope but also Averroes, Zoroaster, and possibly other more contemporary figures

Socrates was probably the first skeptic philosopher and disputed knowledge itself and the ways it can be obtained. The phrase "The only thing I know is that I know nothing (original in Greek *Έν οἶδα ὅτι οὐδὲν οἶδα* or in Latin translation "*ipse se nihil scire id unum sciat*"). Probably, he never said the exact phrase, and the phrase cannot be found in the works of Plato, the biographer of Socrates. However, it is a summary of Socratic teachings and approaches and quoted by Arcesilaus (c. 315–240 BC).

Plato was the first to consider logic and strict arguments as an absolute necessity for philosophical thinking. In this way, he was the first to conceive what in the early twentieth century was named "analytical philosophy" and was introduced by Bertrand Russell (1872–1970). More specifically, Plato considered mathematics and geometry as a prerequisite and the cornerstone of philosophy. He had been educated in geometry and mathematics by Theodoros from Kerenaea, who was a Pythagorean. Plato was so convinced that mathematics was absolutely essential for someone to become a philosopher that he put the inscription "*Μηδείς αγεωμέτρητος εισίτω μου την στέγην*" which means that nobody who is not familiar with mathematics should enter the academy.

While Plato was more a philosopher and argued on the supremacy of the brain, Aristotle (Fig. 3.19) was more an early scientist rather than a philosopher and stressed the need for experimental verification and the existence of observable empirical data in order to avoid subjective misinterpretations. According to Plato, the soul is immortal and contains three parts which he called "*γένη*" (genus which have nothing to do with genetics or biology but rather they reflect different "kinds"):

Fig. 3.19 Aristotle (384–322 BC). Statue in the campus of the Aristotle University of Thessaloniki Greece

the rational, the spirited, and the appetitive. They are completely independent, and "friendship" between these parts is necessary in order for the human soul to function well. However, Plato also believed in the supremacy of rational part over the others and especially over senses and observations. His theoretical concept was linking the soul to the realm of ideas; the soul possessed the knowledge but lost it when entering the mortal human body. To Plato, all experience was simply the stimulus for the soul to remember and recall the forgotten knowledge. For him, vice and mental health were one and the same thing, and for him, the only reason for someone to do something wrong or evil was essentially to suffer some kind of mental disorder [27, 28]. Many scholars trace the formal conceptualization of mental illness as back as to Plato and his work "*Πολιτεία*" (the Republic) [26, 29]. The summary of Plato's ideas suggests that there are inherent ideas and traits while experience is unreliable and serves only the awakening of the inner realm.

Aristotle accepted Plato's theory on the division of the soul into rational and nonrational part, but for him, the heart, not the brain, was the organ of mental life. Changes in temperature were the cause of mental illness; excessive temperature of the blood could lead to mental confusion, while cold could lead to melancholia. Aristotle was also an accomplished animal dissector, and it is reported he had experience of dissection with 49 animal species. This experience led to the observation that the heart was full of activity as well as blood, while the brain was still and bloodless [30]. According to Aristotle, the role of the brain was minor and supportive to that of the heart, mainly by assisting in the preservation of the homeostasis around the heart. In his book "*Περί Ψυχής*" (De Anima; On the Soul), he argues that experience rather than inherent ideas and traits shape the human mind. This belief was further elaborated by the Stoic school of philosophy (e.g., Zeno of Citium) as well as the Arabs (e.g., Avicenna) and later led to the concept of "tabula rasa" (blank slate) in the seventeenth century AD by John Lock and Thomas Hobbes.

In his work "Nicomachean Ethics," Aristotle presents a theory of happiness. According to his ideas, happiness is the ultimate end-purpose of human life, since it is self-sufficient and final, and "is always desirable in itself and never for the sake of something else" (Nicomachean Ethics, 1097a30–34). He talked about "eudaimonia" (in Greek "*ευδαιμονία*" meaning the possession by good spirit). For Aristotle, happiness encompasses the totality of one's life, and it is the ultimate value of life reflecting one's full potential as a human being. This is the first detailed elaboration of the issue in western culture, it differs radically from Buddhism, as it mainly concerns relation to others and social life while eastern philosophies focus more on internal feelings, impulses, and self-control. It is important because happiness will be adopted some 2000 years later, during the Enlightenment as the true purpose of life, and it will be utilized in several efforts to define mental disorder.

Both Plato and Aristotle consider mental health in a continuum with mental illness and tend to explain mental illness in terms of unresolved conflicts among the parts of the soul. In this sense, they were the first to develop an early psychodynamic theory. However, morality and ethics are dominant in their philosophical theories, and they both link mental health to virtue and mental disorder to vice [31–34].

In addition to Hippocratic somatic and medical treatments, the ancient Greeks developed three psychological treatments: (a) They were inducing sleep by using various methods and (b) interpreting dreams (by priests in temples, most often in the temples devoted to Aesculapius, the god of healing), and (c) they were using words to persuade, console, and gain knowledge of an illness (e.g., when using Plato's dialectics). Antiphon of Rhamnus is reported to have developed some kind of early cognitive-like therapy on the basis of the assumption that all mental symptoms could be alleviated though reasoning and discussion. He developed the method of "alypia" (in Greek *αλυπία* meaning "without distress") [35]. The same time, guardians were often assigned to prevent mental patients from harming themselves or others, while Plato held the family as the sole responsible for the patient. Fears of spreading of the contamination by the demons were widespread among laypeople.

A different story was that of Herophilos (335–280 BC), a Greek physician who lived in Alexandria of Egypt during the Hellenistic period. He was the first anatomist, and he is said to had performed more than 600 anatomic dissections in alive prisoners. He was much interested in the brain and considered it to be the site of intellect. He believed that neural transmissions occurred by means of pneuma (in Greek *πνεύμα* meaning the spirit) and that humoral disequilibrium prevented pneuma from reaching the brain and thus resulting in mental disorders [36, 37].

3.4 The Roman Period and Late Antiquity

In general, the Romans followed the psychiatric teachings of the Greeks. Roman medicine was influenced by Greek Stoic and Epicurean beliefs that passions and unsatisfied desires act on the soul to produce mental illnesses. Their cultured classes were practically oriented, and they supported a mechanistic interpretation of mental illness and opposed metaphysical beliefs. Cicero (103–43 BC) accepted Plato's proposal concerning the presence of cognitive biases. The most widely spread theory was that mental illness was the result of abnormal diameter of the pores inside the body, and any healing process should aim at restoring their proper size. This was very similar to the ancient Egyptian theories concerning the "metu." They also believed that arguments and reasoning could lead to the controlling of these illnesses, and the mental state of "*αταραξία*" would emerge (ataraxia meaning placidity, from proverb *α*- which is similar to un-, and the word "*ταραχή*" which means agitation). Some modern tranquilizers have been classified as "ataractics." Interestingly, they described "*laxum state*" resembling depression, "*strictum state*" for excitement and agitation, and "*mixtum*" for mixed states.

There was no social responsibility for the treating of the insane in Greece or Rome. Exceptions were a few patients admitted to Aesculapian sanctuaries and Roman soldiers admitted to military hospitals. But the big picture was that most of the seriously ill patients were kept under restraint at home by their families. The Romans were the first to have laws for appointing guardians to protect the holdings of the insane and for declaring them incompetent. It is interesting to note that according to Roman law, insanity was evaluated by judges, not by physicians. The public was afraid and hated the insane, with the exception of those thought to be divinely inspired. As a consequence, mental patients were beaten, imprisoned, and driven away. On the other hand, Romans were probably the first to officially recognize the need to take into consideration the mental state of those on trial. They utilized categories of mental competence: "non compos mentis," "fanaticus," "ideotus," and "furiosus" [38].

An interesting exception was the Greek physician Asclepiades of Bithynia (or Prusa) in Asia Minor (c. 124–40 BC). He practiced in Rome and rejected the humors theory. He identified two nosological entities: "phrenitis" and "catatonia." His approach led his pupil Themison of Laodicea to establish the Methodic school of medicine. He introduced humane treatments including diet and massages and was the first to free the insane from confinement. According to his teaching, digestion was a primary factor determining health. The size of the particles of food were seen as the main cause of indigestion. He believed that drugs were not the solution to cure an illness, and he was often prescribing food and wine followed by an enema [39]. In accord with these, it was the vitalistic approach of Aretaeus of Cappadocia (first century AD) and later the mechanical models of Aurelianus (255–320 AD), while in contrast to these biological/mechanistic approaches, Celsus (15 BC–30 AD) argued for the presence of an important cultural component; for the first time, he cited civilization as the cause of mental illness and

Fig. 3.20 Galen (131–201 AD)

believed that agricultural life close to nature was protective. This was the first time that civilization was considered the source of problems, an idea which Jean-Jacques Rousseau (1712–1778) would systematically pursuit seventeen centuries later. Soranus of Ephesus (98–135 AD), one of the main representatives of the Methodic school of medicine, criticized the harsh attitude of contemporary Romans toward the mentally ill.

The greatest physician during the Roman era was Galen of Pergamon, also known as Claudius Galenus (131–201 AD), who influenced medical thinking for almost 15 centuries (Fig. 3.20). Although his name derived from the Greek word "*γαλήνη*" meaning tranquility and peacefulness, Galen was quite the opposite as a character: elitist, competitive, and castigating and abusive to those with opposing views. He was teaching that disease is the result of external adverse influences, including bad diet and bad air. These external factors act on an existing vulnerability or predisposition, corresponding to an abnormality of a humor or pneuma. According to his teachings, psychological disorders cause physical disorders and vice versa. He argued that the correct treatment is the one which opposes what is unnatural in a disease (cooling the feverish, warming the chilled) and should be in accord with the patient's age, temperament, and way of life [40–42]. He rose medicine from informed craft to applied science by establishing the scientific method or research through dissection, observation, and clinical experimentation.

Galen rejected the Aristotelian hypothesis that the heart was the seat of the mind on the basis of his anatomical observation and mainly because he traced the nerves from the various sense organs entering the brain, instead of the heart. Additionally, he showed that the nerves responsible for the voice were connected with the brain and also that some kind of stupor could result from pressure on the brain while head injuries could cause changes in perception and cognition.

Concerning brain physiology, he combined the ideas of Plato with the pneumatism of Erasistratus of Alexandria. According to his beliefs, three kind of spirits (natural, vital, and animal) travel through hollow nerves to and from the brain and the ventricles [43]. According to this theory, natural spirits from the organs are turned in the heart into vital spirits, and some of them go to the brain where they are turned into animal spirits which constitute the spirits of the mind. This conversion supposedly takes place in the choroid plexuses or the "rete mirabile" (miraculous net; in Greek "*Θαυμάσιο δίκτυο*"), a network of fine arteries surrounding the pituitary gland. However, this was one of his cardinal mistakes since this network exists in many animals but not in humans and happened because Galen's knowledge of anatomy was based on animal, not human dissection. It would take centuries and the studies of Andreas Vesalius (1514–1564) to clarify such misbeliefs. Galen believed that after their production, these animal spirits were stored in the ventricles [44].

In his work "On Melancholia," Galen makes a description of fear and aggression inherent in the pathophysiology of depression. He proposed that mania could be either primary or secondary to other diseases, and his most important contribution in the field was the elaboration of the theory of humors.

Concerning pharmacology, Galen utilized the "mithidaticum," a preparation of 44 ingredients developed by the King Mithridates VI Eupator (135–63 BC), ruler of Pontus, a Hellenistic kingdom at the northern shores of Asia Minor. Mithridates developed it in order to protect himself from attempts of murder by poisoning. After the conquest of the kingdom by the Romans, doctors in Rome expanded the formula at will, and Galen himself had his own version which probably included over a hundred ingredients. He used this theriac in the treatment of three emperors, and one of them, Marcus Aurelius (121–180 AD), was taking it so often that he was probably addicted to the opium it contained [45]. Galen also used electricity from fish to treat headaches [46].

According to Aurelianus, his contemporary Soranus of Ephesus (first to second century AD) referred to the episodic nature of depression, linked depression, and aggression to suicide and described mixed episodes for the first time. Soranus probably rejected a relationship between mania and melancholia, and in his writings, he mentions that the followers of Themison were considering mania and melancholia to be different facets of the same disease. Themison of Laodicea (modern-day Syria) was a first-century BC physician, was the founder of the Methodic school of medicine, and was one of the most eminent physicians of his time; however, we know almost nothing concerning his life and works. Thus, it is widely considered that it was Aretaeus of Cappadocia (second century AD) the one who strongly connected melancholy with mania and made a description of manic episodes which was very close to the modern approach, including comorbid personality disorder, psychotic features, alcohol abuse, and seasonality. He suggested that mania was the end stage of melancholia, and he was also the first to observe that manic patients could be either euphoric or aggressive-irritable. Aretaeus lived in

Fig. 3.21 Saint Augustine of Hippo (354–430) in Gerard Seghers painting "The Four Doctors of the Western Church," created between 1600 and 1650

Alexandria (modern-day Egypt) and was a prominent representative of the school of Eclectics. In his writings, he considered mania to represent a worsening of the disease (melancholia) rather than a distinct type of disorder or episode [23].

Aurelius Augustine (354–430), better known as St. Augustine, was a key personality that shaped the transition from late antiquity to the middle ages (Fig. 3.21). This transition concerned mainly the rise of Christianity and the domination of religious and metaphysical beliefs in the way the reality is perceived and interpreted. St. Augustine was bishop of Hippo Regius in Roman, North Africa, in modern city of Annaba, in Algeria. He was bishop during the conquest of the area by the Vandals and died during the siege of his city. One of the difficult questions from the citizens he was obliged to answer was that Christianity prevailed by promising better times: How is it possible that now that all were turned into Christians the empire collapses? He was also obliged to cope with his inability to keep pace with the theological advancements of the time, which were taking place far away in a relatively safe Asia Minor and in the Greek language. It was very difficult or even impossible to translate from Greek to Latin because the vocabulary in Greek was much bigger and more sophisticated and provided the base for very complex verbal elaborations in dogma. As a result, he developed the concept that this world is evil, and not only it might be destroyed but, even more important, it is the will of the God to be destroyed. This is the first major rejection of rationality in history in an attempt to justify things and tailor reality to beliefs. Among other consequences, it led to a moralistic and hostile approach to mental disease. For him, the issue of insanity and diminished capacity was far more complex than it was for the ancient philosophers. Christian scholars were in the need to distinguish between legal offenses and sins. According to St. Augustine, one difference between the two was that "there cannot be a sin which is not voluntary" [38].

Nemesius (390–446 AD) was the Bishop of Emesa (in present day Syria) and author of the book "*Περί Φύσεως Ανθρώπου*" (De Natura Hominis or On Human Nature) in which he tried to combine the medical knowledge of his time with Christian philosophy and theology. He developed a hierarchical philosophical explanation of the world, reflecting mostly the ideas of Plato. He considered the brain ventricles to be the site of mental functioning.

The last important physician of antiquity who studied mania and depression was Alexander of Tralles in Lydia (sixth century AD) who wrote on the cyclical alternation of mania and melancholia, and in accord with Aretaeus of Cappadocia, he suggested that mania is a worse form of melancholia and that melancholia predisposes to suicidality but also to homicidality.

By then, in the Western Roman Empire, there was a gradual dominance of religious beliefs over the ancient secular legal tradition, while on the contrary, the Galen tradition was kept in the Eastern Roman Empire and was passed to the Arabs. The environment in the west became even worse with the fall of the Roman Empire and the rise of the Carolingian Europe. Much of the legal tradition was lost, and in 873, the witch hunt officially started [38].

References

1. Roccatagliata G. Storia de la psichiatria antica. Milano: Hoepli; 1973.
2. O'Callaghan M, Carroll D. Psychosurgery: a scientific analysis. Lancaster: MTP Press Limited; 1982.
3. Breasted J. The Edwin Smith surgical papyrus. Chicago: University of Chicago Press; 1930.
4. Ebbell B. The Papyrus Ebers: the greatest Egyptian medical document. Copenhagen: Levin and Munksgaard; 1937.
5. Okasha A, Okasha T. Notes on mental disorders in Pharaonic Egypt. Hist Psychiatry. 2016;11(44):413–24. https://doi.org/10.1177/0957154x0001104406.
6. Woollam D. Concepts of the brain and its functions in classical antiquity. In: Perrin M, editor. The brain and its functions. Oxford: Blackwell, Wellcome Foundation; 1958. p. 5–18.
7. Rickett W. Kuan tzu. In: Loewe M, editor. Early Chinese texts: a bibliographical guide. Berkeley: Society for the Study of Early China; Institute of East Asian Studies, University of California Berkeley; 1993. p. 244–51.
8. Veith I. The Yellow Emperor's classic of internal medicine. Revised paperback edition. Berkeley, LA: University of California Press; 1972.
9. Schiffeler JW. Chinese folk medicine: a study of the Shan-hai Ching. Asian Folkl Stud. 1980;39(2):41–83.
10. Bhugra D. Psychiatry in ancient Indian texts: a review. Hist Psychiatry. 1992;3(10):167–86. https://doi.org/10.1177/0957154X9200301002.
11. Abhyankar R. Psychiatric thoughts in ancient India. Mens Sana Monogr. 2015;13(1):59–69. https://doi.org/10.4103/0973-1229.153304.
12. Parkar SR, Dawani VS, Apte JS. History of psychiatry in India. J Postgrad Med. 2001;47(1):73–6.
13. Wig NN, Sharma S. Emperor Ashoka: did he suffer from von Recklinghausen's diseases? Indian J Psychiatry. 2015;57(1):95–7. https://doi.org/10.4103/0019-5545.148536.
14. Black J, Green A. Gods, demons and symbols of ancient mesopotamia: an illustrated dictionary. London: The British Museum Press; 1992.

15. Nemet-Nejat K. Daily life in ancient Mesopotamia. Santa Barbara, CA: Greenwood; 1998.
16. Edelstein E, Edelstein L, Ferngren G. Asclepius: collection and interpretation of the testimonies. Baltimore: Johns Hopkins University Press; 1998.
17. Majno G. The healing hand. Cambridge, MA: Harvard University Press; 1975.
18. Wickkiser B. Asklepios, medicine, and the politics of healing in fifth-century Greece: between craft and cult. Baltimore: Johns Hopkins Press; 2008.
19. Snell B. The discovery of the mind (Translation by Rosenmeyer TG). Oxford: Blackwell; 1953.
20. Sigerist H. A history of medicine, vol. 2. Early Greek, Hindu, and Persian medicine. New York: Oxford University Press; 1961.
21. Canfora L. The vanished library. Berkeley: University of California Press; 1990.
22. Fountoulakis K. The role of temperament. In: Fountoulakis K, editor. Bipolar disorder: an evidence-based guide to manic depression. Berlin: Springer; 2015b. p. 163–79. https://doi.org/10.1007/978-3-642-37216-2.
23. Fountoulakis K. Historical perspective. In: Fountoulakis K, editor. Bipolar disorder: an evidence-based guide to manic depression. Berlin: Springer; 2015a. p. 1–25. https://doi.org/10.1007/978-3-642-37216-2.
24. Ahonen M. Plato on madness and mental disorders. In: Mental disorders in ancient philosophy. Cham: Springer; 2014. p. 35–67. https://doi.org/10.1007/978-3-319-03431-7_3.
25. Ahonen M. Ancient philosophers on mental illness. Hist Psychiatry. 2019;30(1):3–18. https://doi.org/10.1177/0957154X18803508.
26. Seeskin K. Plato and the origin of mental health. Int J Law Psychiatry. 2008;31(6):487–94. https://doi.org/10.1016/j.ijlp.2008.09.004.
27. MacIntyre A. A short history of ethics. South Bend, IN: University of Notre Dame Press; 1998.
28. Sidgwick H. Outlines of the history of ethics for English readers. London: Macmillan and Co. Ltd.; 1888.
29. Kenny AJP. Mental health in Plato's Republic. Proc Br Acad. 1969;55:229–53.
30. Clarke E. Aristotelian concepts of the form and function of the brain. Bull Hist Med. 1963;37:1–14.
31. Charland L. Character: moral treatment and the personality disorders. In: Radden J, editor. The philosophy of psychiatry: a companion. International perspectives in philosophy and psychiatry. New York: Oxford University Press; 2004. p. 64–77.
32. Harcourt E. Attachment theory, character and naturalism. In: Peters J, editor. Aristotelian ethics in contemporary perspective. London: Routledge; 2012. p. 114–29.
33. Megone C. Aristotle's function argument and the concept of mental illness. Philos Psychiatry Psychol. 1998;5(3):187–201.
34. Megone C. Mental illness, human function, and values. Philos Psychiatry Psychol. 2000;7(1):45–65.
35. Guthrie WKC. A history of Greek philosophy, vol. 3. The fifth century enlightenment, part 2. Socrates. Cambridge: Cambridge University Press; 1971.
36. Debru A, Von Staden H. [Herophilus, or the art of medicine in ancient Alexandria]. Rev Hist Sci Paris. 1991;44(3–4):435–45.
37. Dobson JF. Herophilus of Alexandria. Proc R Soc Med. 1925;18(Sect Hist Med):19–32.
38. Fulford KWM, Davies M, Gipps RGT, Graham G, Sadler J, Stanghellini G, Thornton T. The next hundred years: watching our Ps and Q. In: Fulford KWM, Davies M, Gipps RGT, et al., editors. The Oxford handbook of philosophy and psychiatry. Oxford: Oxford University Press; 2013. https://doi.org/10.1093/oxfordhb/9780199579563.001.0001.
39. Rawson E. The life and death of Asclepiades of Bithynia. Class Q. 1982;32(2):358–70. https://doi.org/10.1017/s0009838800026549.
40. Haas LF. Claudius Galen 131-201 AD. J Neurol Neurosurg Psychiatry. 1991;54(4):287. https://doi.org/10.1136/jnnp.54.4.287.
41. Jackson SW. Galen—on mental disorders. J Hist Behav Sci. 1969;5(4):365–84. https://doi.org/10.1002/1520-6696(196910)5:4<365::aid-jhbs2300050408>3.0.co;2-9.

42. Maroth M. [Galen as mental healer]. Sudhoffs Archiv; Zeitschrift fur Wissenschaftsgeschichte Beihefte. 1993;(32):145–55.
43. Temkin O. On Galen's pneumatology. Gesnerus. 1951;8(1–2):180–9.
44. Rocca J. Galen and the ventricular system. J Hist Neurosci. 1997;6(3):227–39. https://doi.org/10.1080/09647049709525710.
45. Trancas B, Borja Santos N, Patricio LD. [The use of opium in Roman society and the dependence of Princeps Marcus Aurelius]. Acta Med Port. 2008;21(6):581–90.
46. Kellaway P. The part played by electric fish in the early history of bioelectricity and electrotherapy. Bull Hist Med. 1946;20(2):112–37.

Chapter 4
The Middle Ages

The centuries that followed the fall of the Greco-Roman world were characterized by the domination of religious-metaphysical approaches and the regression of reasoning and scientific thinking. Vice was transformed into a spiritual failing of human nature, a metaphysical sin [1, 2].

Among the sciences, it was particularly medicine who retreated. Priests and the church in Europe took over much of the health field and especially mental health. As a consequence, the treatment was administered either by physicians or, more frequently, by priests or peculiar persons claiming to possess special powers. The therapeutic procedure included a variety of different practices, from exorcism of demons by magical techniques to confession to a priest, in association with church activities of penance and medicinal treatments recommended by Hippocrates and Galen. Of course, there were also time-honored folk remedies and a variety of esoteric medicines, for example, bark of a tree from Paradise. As centuries passed, many priests and monks were formally trained as physicians, and monasteries became quasi-science centers. Thus, all the more often, religious approaches were combined with scientific treatments by the same therapist on the same patient [3].

The tenth and eleventh century AD are characterized by the domination of Arab scholars in medicine. Abu Bakr Muhammad ibn Zakariyya al-Razi, also known by his Latinized name Rhazes (854–925), promoted the scientific approach, accepted the four elements theory of the Hippocratic school, and opposed to metaphysical explanations. Najab ud-din Unhammad (870–925) proposed 30 mental disorders grouped into nine categories. Ibn Sina or Abu Ali Sina, also known as Avicenna (980–1037) and considered to be the "Galen of Islam," wrote the "Canon of Medicine." He further developed the theory of the temperaments and observed that the emergence of anger, restlessness, and violence marked the transition from melancholia to mania. He localized perception in the frontal part of the frontal ventricles and the representation of experience to the latter part of the same ventricles. Furthermore, he argued that "rational imagination" takes place in the third ventricle and retention in the fourth ventricle.

K. N. Fountoulakis, *Psychiatry*, https://doi.org/10.1007/978-3-030-86541-2_4

Ishaq Ibn Imran developed an early biopsychosocial model by suggesting that melancholia is caused by "damaged father's sperm" in interaction with temperament and circadian rhythms. Rabbi Moses ben Maimon or Maimonides (1135–1204) tried to combine and reconcile faith and reason. The prevailing attitude in the Islamic world of that time was that society is responsible for the kindly care of the insane. Accordingly, it were the Arabs who first built hospitals with psychiatric divisions in Baghdad (750 AD) and Cairo (873 AD); they also built special insane asylums in Damascus (800 AD), Aleppo (1270 AD), and the Muslim-ruled Spanish city of Grenada (1365 AD).

At that time, the Christian attitudes toward the madness fluctuated between rejection and toleration. In the Byzantine Empire (East Roman Empire), a number of hospitals were caring for the insane. It is interesting that in Byzantium, the prevailing attitude toward medicine was primarily secular, probably as a direct heritage of the Greco-Roman world. This was in spite of the acceptance of a metaphysical and demonological explanation for mental illness in the writings of Michael Psellos (1017–1078; meaning "stutterer") which was the most prominent scholar of his time. Sources suggest the presence of hospitals and charity foundations called *ξενώνες* (xenon's, which means shelter for strangers) for the insane as early as the seventh century AD, that is, the same time they appeared in the Arab world. However, the exact structure and function of them is largely unknown. The most prestigious of institutions was the hospital of the monastery of the Pantocrator (*Ξενών της Μονής Παντοκράτορος,* meaning the Hostel of the Almighty Christ Monastery), founded by the Emperor Ioannis II Comnenus in Constantinople in 1136 AD (Fig. 4.1). It included five wards, one of them dedicated to mental patients and interestingly a separate one for neurologic patients. The patients were offered both

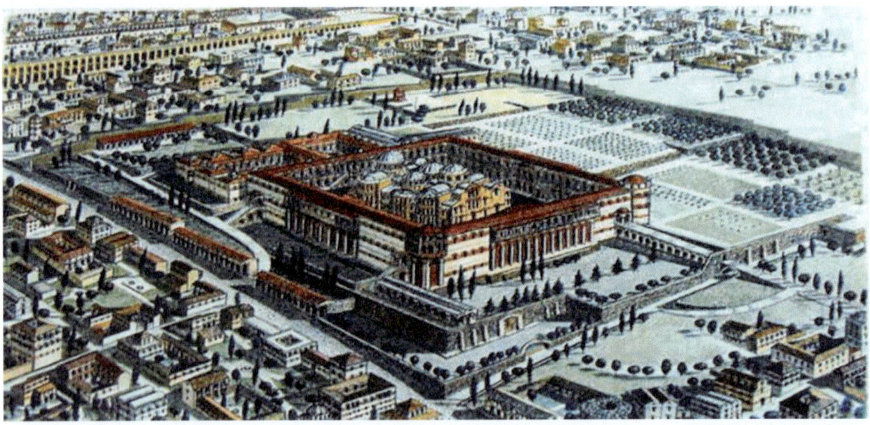

Fig. 4.1 Xenon Pantokratoros, Constantinople. Image by George Anapniotis, architect (2011)

religious/spiritual as well as formal medical treatment. It is interesting that the hospital included two chapels (one for men and another for women) with a full-time priest and lector each. Also, a priest was always to be available to hear confession. The medical treatment included frequent baths, ointments, vegetarian diet, a large list of pharmaceutical substances, music therapy, occupational and physical therapy, psychotherapy, and spiritual support [4, 5]. What survives today of this complex was turned into the Zeyrek Mosque.

Life during the medieval period was difficult with the collapse of the safety and security mechanisms of the Roman Empire. This was not only because to frequent and violent wars and unrest but also because of pandemics. Fear was the predominant feeling, and this led to the flourishing of mysticism. Interestingly, there are reports on the occurring of quasi-epidemic mass cases of mania and agitated behavior since the tenth century, named St. Elmo's fire and was due to ergotism from the ingestion of the alkaloids produced by the Claviceps purpurea fungus that infects rye and other cereals. It is also called "St. Anthony's fire" (Fig. 4.2). There is the possibility that several witchcraft cases, including the famous Salem cases, to had had their origin in such cases. Catastrophic events led people to consider mental disorder as the consequence of affiliation with the devil, and thus one of the particular sins that attracted the wrath of God. Treatment was by exorcism while supposed witches were burned alive until 1782. Mental patients were also incarcerated in madmen's towers and expelled from towns.

Fig. 4.2 Matthias Grünewald (1470–1528): "Isenheim Altarpiece" c. 1516, Colmar, Musée d'Unterlinden. Left: The Meeting of St. Anthony Abbot and St. Paul in the Wilderness. Right: The Temptation of St Anthony. In the detail, a patient suffering from advanced ergotism is shown

Fig. 4.3 Upper image: Saint Dymphna church, Geel. Lower image: Saint Dymphna's well

An interesting case was the Flemish village at Geel during the fourteenth century. There, at the healing shrine of Saint Dymphna (Fig. 4.3), all mental patients were accepted with compassion. According to hagiography, Saint Dymphna, was an Irish princess, and her legend is a typical one of a medieval virgin martyr. She sought refuge in Flanders, being pursuit by her pagan father whose sexual approaches she refused. After her martyrdom in the town of Geel, the church where she was buried was named in her honor, and she was canonized as protector saint of the insane and the victims of sexual abuse. Mental patients started gathering in Geel to seek cure, and in 1270, that Hendrik III Berthout founded a hospital, or gasthuis, under the spiritual patronage of the Virgin Mary and Saint Dymphna called the Sint-Dimpnagasthuis (Fig. 4.4). Similar places existed in Spain and Portugal, but eventually, they constituted rare exceptions.

However, bypassing the centuries, an explanation concerning the etiopathogenesis of mental illness emerged and suggested that insane persons lacked a mind because of God's favor and thus they needed to be cared for. This was at least partially the consequence of progress and change in scientific thinking, with Albert Magnus (1193–1280; Fig. 4.5), and Roger Bacon (1214–1292; Fig. 4.6), emphasizing the supremacy of the empirical method and pointing to an orientation outside of the church and faith.

Fig. 4.4 The Mental Hospital of Geel, early twentieth century

Fig. 4.5 Saint Albertus Magnus, a fresco by Tommaso da Modena (1352), Church of San Nicolò, Treviso, Italy

Fig. 4.6 Roger Bacon observing the stars at Oxford University

Thomas Aquinas (1225–1274; Fig. 4.7) on the other hand tried to reconciliate faith and science and established Scholasticism. According to his views, the essence of human existence was the immortal soul which consisted of intellect and will. This was the early mentioning of the importance of "free will" and as such is somewhat a precursor to modern existential theories.

Fig. 4.7 Saint Thomas Aquinas (1225–1274)

References

1. MacIntyre A. A short history of ethics. South Bend, IN: University of Notre Dame Press; 1998.
2. Sidgwick H. Outlines of the history of ethics for English readers. London: Macmillan and Co. Ltd.; 1888.
3. Jackson S. Care of the psyche: a history of psychological healing. New Haven, CT: Yale University Press; 1999.
4. Miller T. The birth of the hospital in the Byzantine Empire. Baltimore: Johns Hopkins University Press; 1997.
5. Thomas J, Constantinides A, Constable G. Byzantine Monastic Foundation documents: a complete translation of the surviving founders' typika and testaments. Washington, DC: Dumbarton Oaks Research Library and Collection; 2000.

Chapter 5
Renaissance

For didactic reasons, the year 1401 was chosen as the year marking the transcendence from the middle ages to the renaissance. This year is proposed by several authors because at that time, the rival geniuses Lorenzo Ghiberti (1378–1455) and Filippo Brunelleschi (1377–1446) competed for the contract to build the bronze doors for the Baptistery of the Florence Cathedral (Fig. 5.1) [1].

It is interesting that around that period of change between epochs, and not before the fourteenth century, that hospitals in Western Europe started caring for the mad. Until that time, they were doing this solely for the physically ill. Usually, it is said that the first mental hospital in the European continent appeared in Valencia, Spain, in 1409, and was a consequence of Islamic influence. This is not true because there were the Byzantine hospitals before, but also in Western Europe, there are records of the presence of an asylum in Hamburg, Germany, in 1375. Around 1400, the Bethlem Hospital in London started treating mental patients, and two centuries later, it was turned into an exclusively mental asylum (Fig. 5.2). In popular culture, it was known as "Bedlam" (a place of uproar) and constituted the prototype for the asylum approach in the treatment and care of the insane. It became a special and unique institution, pioneering in the asylum management and care of the insane [2]. Soon after the establishment of "Bedlam," other countries began to follow and founded their own mental health facilities. San Hipolito was built in Mexico 1566 and claims the title of the first asylum in the American continent. La Maison de Charenton was the first mental facility in France, founded in 1641 in a suburb of Paris. Constructed in 1784, the Narrenturm or Lunatics' Tower in Vienna became a showplace (Fig. 5.3). The same period, the first mental asylum in the Indian subcontinent was the Mandu Hospital built by Mahmud Khilji (1436–1469) at Dhar in the Madhya Pradesh state [3]. The word asylum comes from the Greek word "*άσυλον*" (asylon) which means "what cannot or should not be violated." It is constituted from the prefix *α*- (corresponds to English un-) and the verb συλώ (sylo) which means "plunder."

K. N. Fountoulakis, *Psychiatry*, https://doi.org/10.1007/978-3-030-86541-2_5

Fig. 5.1 The Gates of Paradise, Baptistery of the Florence Cathedral, created by Lorenzo Ghiberti in 1422 after 21 years of work

Fig. 5.2 The Bethlem Hospital in London (Bedlam) by William Henry Toms: illustration from William Maitland's history of London, published in 1739

Fig. 5.3 The Narrenturm or Lunatics' Tower in Vienna

Treatments were following the beliefs of that time period. They included, among other methods, purging and bloodletting as well as showers with either hot or ice-cold water to shock the patients' minds back into a normal state. One particular belief of the time was that patients were able to choose rationality over insanity. This led to the adoption of techniques aiming to intimidate and to shock, including blistering, physical restraints, and threats. Furthermore, most of the attendants were males, and the criterion to get the job was their physical strength and probably their defiant attitude. Persons of problematic personality and criminal or otherwise disturbed background were overrepresented in the staff.

This horrible situation was portrayed with sensational details in the works of the painter William Hogarth (1697–1764). Many asylums were open to the curious public for their amusement, resembling a human zoo or freak show. This situation was utilized in Hogarth's cartoons work, especially in the "Rake's Progress" in order to create an allegory for the society and to question which of the persons are truly mad. For Hogarth, the entire world was essentially a madhouse (Fig. 5.4). There is at least some dispute on whether these paintings reflect reality or in fact reflect an exaggeration as it is the case with modern depictions of mental health and hospitals in films and novels [4, 5].

While during the antiquity there was a distinction between "black" and "white" magic, with the rise of Christianity, all witchcraft was considered to be black and subsequently harmful, and it was taken as proof of a covenant with Satan. The same period, there was a profound shift in the accusatorial methods with the emergence of the inquisitorial method and the fusion of maleficium (means wrongdoing) with that of heresy. In this dark environment, in 1486, two monks, Henry Kramer (1430–1505) and James Sprenger (1436–1495), published the "Malleus Maleficarum" (malleus meaning hammer and maleficarum is plural to maleficus meaning the evil, wicked, or witchcraft) as a handbook on how to recognize and identify witchcraft. This resulted in the persecution of mental patients as being witches [6].

Fig. 5.4 William Hogarth "A Rake's Progress, Plate 8, In The Madhouse." The engrave includes the insane, with the central person being chained by two members of staff and two elegantly dressed women who came to have fun by watching the lunatics. On the wall, there is the emblem of Britannia. Soane Museum, London

However, as times were changing and the empirical method was gaining ground, experimentation and observations became more appropriate and more accurate. Jacopo Berengario da Carpi (1450–1530) studied human bodies and reported that he could not find the rete mirabile or other anatomical elements described by Galen and Arab scholars. While social attitudes were changing with Juan Luis Vives (1492–1540) rejecting the philanthropic approach in order to argue for the responsibility of the society for the medically disabled, in natural sciences, a real revolution was under way. In 1543, Nicolaus Copernicus (1473–1543) published "De revolutionibus orbium coelestium" (Concerning the Revolutions of Heavenly Bodies) and challenged the theory that the earth is the center of the solar system.

The same year, Andreas Vesalius (1514–1564; Fig. 5.5) published "De humani corporis fabrica" (On the Workings of the Human Body) with Book VII dealing exclusively with the brain [7]. Vesalius realized that a problem with Galen's anatomy was that it was based entirely on animal dissection, and thus his assumptions were often not in accord with actual human anatomy. One case of such a false belief was the existence of the "rete mirabile" in humans. On the other hand, Vesalius pointed out that human ventricles are almost identical in shape to those of other primates although there is an immense difference in cognitive abilities. Vesalius was a dedicated Christian and died after a pilgrimage in Jerusalem on the Greek island of Zante and buried in the cemetery of the Church of Santa Maria delle Grazie, which was completely destroyed by the earthquake of 1953.

Fig. 5.5 Portrait of Vesalius created by Jan van Calcar; from Page xii of "De humani corporis fabrica" (1534 edition)

Theophrastus Bombastus Von Hohenheim, better known as Paracelsus (c. 1493–1541), rejected the theory of body humors, but his medical ideas were often confused, contradictory, and influenced by philosophy, astrology, magic, and myths. Consequently, his treatment approach suggested that the physician must study the instinct which is dominant in a patient and the star that corresponds to it. He was teaching that emotions can cause mental illness, and this is why he is often considered to be the first psychosomatist, which is of course not true. He also proposed psychotherapy, venesection, and even trephination (perforation of the skull; Fig. 5.6). One of his main proposals concerned the use of sulfur which possesses sedative properties, and also, it could heal through "sympathy" which he considered to be one of the main forces of the universe. He was the first physician to develop a systematic classification of mental disorders into "lunatici" (from Latin "Luna" meaning the moon; referred to reactions to the phases of the moon), "insani" (from Latin "in-sanus" meaning not healthy; referred to inherited mental conditions), "vesani" (from Latin "ve-sanus" meaning decline in health; referred to as a result of poisoning), "melancholici" (from the Greek "melancholia"; referred to temperamental origin), and "obsessi" (from Latin "ob-sedeo" meaning possessed; referred to possessed by the devil). The progress this classification brought was that it

Fig. 5.6 Cutting the Stone, also called The Extraction of the Stone of Madness or The Cure of Folly, by Hieronymus Bosch (c. 1494)

recognized additional mental disorders to those accepted at that time, which were exclusively of metaphysical origin until Paracelsus' proposal.

Marko Marulić Splićanin, or better known by his Latin name Marcus Marulus Spalatensis (1450–1524), was a Croatian poet and Renaissance humanist who coined the term "psychology" for the first time from the Greek words "ψυχή" (psyche meaning the soul) and "λόγος" (logos meaning the study and the knowledge) in his book "Psichiologia de ratione animae humanae" which was published c. 1520 and only its title is saved [8, 9].

Then 77 years after its first publication, Johann Weyer (1515–1588), who is considered by many to be the father of modern psychiatry, attacked the Malleus Maleficarum with his 1563 book De Praestigiis Daemonum (Delusion of Witches). The book had six editions between 1563 and 1583. Weyer did not deny the reality of Satan or its role on the development of mental illness. However, he rejected the accusatory attitude against mental patients. He considered them to be victims instead of collaborators of Satan. He also argued that mental illness was a natural phenomenon without any spiritual significance, and to support this, he described the pharmacological properties of various plants and their extracts including belladonna and opium. As a result, his clinical books were included in the prohibited Index of the Roman Catholic Church for several centuries [6].

Around the same period, Juan Huarte (1530–1588) was the first to focus attention on individual differences which would later lead to the development of psychological testing during the nineteenth century. The general environment in philosophy continued to change. Francis Bacon (1561–1626; Fig. 5.7) strongly supported empiricism. He further elaborated on the ideas originally conceived by Plato on false baseline assumptions which hold in bondage the mental capacity to comprehend and named them "idols," thus describing in more detail what later in twentieth century would had been defined as "cognitive biases." In 1602, Felix Platter (1536–1614) published his classification of mental diseases, but he failed to connect mania with depression. He was perhaps the first to propose the term "alienation" which gave the label "alienists" to psychiatrists for the next four centuries.

Fig. 5.7 Francis Bacon, portrait from Sylva Sylvarum: or a Naturall Historie in Ten Centuries. London; printed by J. H. for William Lee, 1631

In 1621, Robert Burton (1577–1640) wrote the first English-speaking text on the field of mood disorders, the “Anatomy of Melancholy” (Fig. 5.8), which was a scholarly review of medical and philosophical wisdom accumulated in the past. Among other things, Burton created a list of causes of melancholia (including among others the influence of planet Saturn, autumn, heredity, diet, alcohol, biological rhythms, and intense love) but also mentioned the existence of “causeless” melancholias. Interestingly, like others before him, he considered men to be at a higher risk. He was chronically depressed himself. There is some speculation that he died by hanging himself [10].

During the late middle ages, science and medicine gradually gain ground, and eventually in the early seventeenth century, the scientific approach, free of metaphysical and religious influences, started being dominant again in Europe. This did not happen overnight; it was rather a gradual progress with compromises and setbacks, and centuries have passed until the metaphysical was completely expelled from the understanding of mental phenomena.

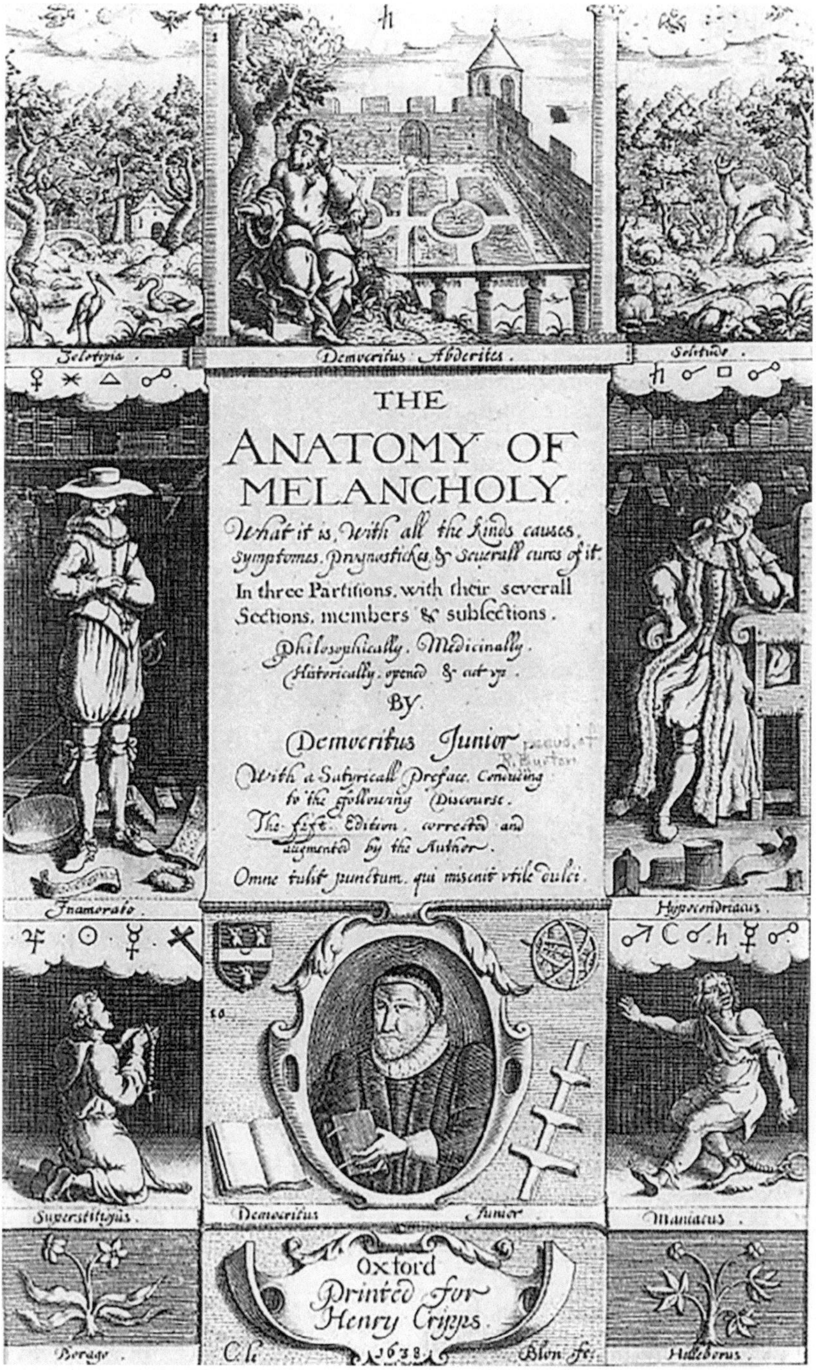

Fig. 5.8 The front cover of the 1638 edition of Robert Burton's The Anatomy of Melancholy

References

1. Walker P. The feud that sparked the renaissance: how Brunelleschi and Ghiberti changed the art world. New York: Perennial-Harper Collins; 2003.
2. Andrews J, Briggs A, Porter R, Tucker P, Waddington K. The history of Bethlem. London: Routledge; 1997.
3. Parkar SR, Dawani VS, Apte JS. History of psychiatry in India. J Postgrad Med. 2001;47(1):73–6.
4. Parry-Jones W. The trade in lunacy. London: Routledge and Kegan Paul; 1972.
5. Porter R. Madness: a brief history. Oxford University Press; 2002.
6. Fulford KWM, Davies M, Gipps RGT, Graham G, Sadler J, Stanghellini G, Thornton T. The next hundred years: watching our Ps and Q. In: Fulford KWM, Davies M, Gipps RGT, et al., editors. The Oxford handbook of philosophy and psychiatry. Oxford University Press; 2013. https://doi.org/10.1093/oxfordhb/9780199579563.001.0001.
7. O'Malley C. Andreas Vesalius of Brussels, 1514–1564. Berkeley: University of California Press; 1964.
8. Fernando V. The sciences of the soul: the early modern origins of psychology. Chicago: University of Chicago Press; 2011.
9. Lapointe FH. The origin and evolution of the term "psychology". Rivista Critica di Storia della Filosofia. 1973;28(2):138–60.
10. Aubrey J. In: Bennett K, editor. Brief lives with an apparatus for the lives of our English mathematical writers. Oxford: Oxford University Press; 2015.

Chapter 6
The Early Modern Era

While there is not specific year which distinguishes between the renaissance and the early modern period, it is generally considered that René Descartes (1596–1650; Fig. 6.1) was born during the last decades of the renaissance and died during the first years of the early modern era. In a pivotal moment in the history of modern psychiatry and neurosciences, in his books "The Description of the Human Body" (1647) and "Passions of the Soul" (1649), he proposed the mind-body dichotomy (Cartesian dualism) [1]. This was essentially a rejection of the Aristotelian approach which was dominant at that time, especially in religious circles, and in accord with the ideas of Plato [2]. His assumption suggested that mind was distinct from matter but could influence matter (through an unknown way). He also thought that "animal spirits" travel through the nerves. Throughout his life, Descartes experienced a threatening atmosphere because of his radical ideas, and this was the reason he chose to delay the publication of his ideas with his cardinal work being published only months before his death. Because of these fears, he changed 24 different addresses in 13 different towns, and in general, he kept his address secret.

It is believed that Descartes was impressed and inspired by the spectacular animated fountains in the Royal Gardens of Paris, with the moving figures of Neptune, Diana, and other mythological entities. These mechanical wonders in a way inspired him on how to consider the human body. Overall, Descartes was greatly impressed by the "automatons" of the time which were manufactured mainly as toys, but also, efforts were made to use them for practical as well as military purposes. Such devices were on display throughout Paris, and it was a fashion that lasted for centuries (Fig. 6.2). He modeled the human body after these devices and suggested that the brain/body and the mind are completely separate [3]. At that time, there was no knowledge of how the senses (vision, hearing, smell, etc.) function, and thus, he was mainly influenced by theology and physics [4, 5].

According to his theory, the two substances, body and mind, are essentially distinct, and each of them can exist apart from the other. However, he gave priority to the mind by suggesting that it can exist without the body, but the body cannot exist

K. N. Fountoulakis, *Psychiatry*, https://doi.org/10.1007/978-3-030-86541-2_6

Fig. 6.1 Portrait of Rene Descartes painted by Frans Hals the Elder

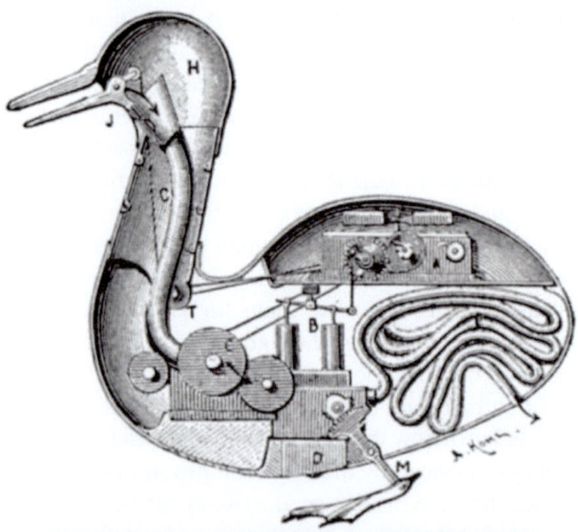

INTERIOR OF VAUCANSPN'S AUTOMATIC DUCK.
A, clockwork; B, pump; C, mill for grinaing grain; F, intestinal tube; J, bill; H, head; M, feet.

Fig. 6.2 An example of an "automaton" was the Canard Digérateur, or Digesting Duck, was an automaton in the form of a duck, built by Jacques de Vaucanson in 1739 The picture is an imagery design of how this duck would look like, from the January 1899 edition of Scientific American without the mind [6]. This was essentially a compromise between the Catholic Church and the scientific community, allowing medicine to progress in issues pertaining to the physical body, but the same time kept the realm of human behavior (and mental health) still in the "jurisdiction" of religion. This is known as "Cartesian dualism" (or Mind-Body Dualism) and exerted a profound influence on western philosophy and culture [7–10].

However, Descartes could not answer the riddle of the mind-body interaction which was evident through basic functions (e.g., hunger, sexual desire, etc.) as well as the effect of substances on behavior (e.g., alcohol) [11]. Thus, he proposed that the pineal gland was serving as connector between them [12–15] and that essentially, the pineal gland is the "siege de I'ame" (seat of the soul) for a number of reasons including the "rete mirabile" and the fact it is a singular structure [16]. He argued that the sensory signals were carried from the sensory organs to the pineal gland by the animal spirits, and although the function of the gland was under God's will, the animal spirits were often distorting the process, and humans were obliged to control passions though their will [14]. In this way, Descartes defended free will. The suggestion that the pineal gland was under the control of God was in accord with the widely held belief that God was everywhere and was filling the space among humans, and thus any connection and communication among humans were essentially mediated by God. At that time, the way sensory organs function was not well understood, and the knowledge on how the brain processes perceptual information and creates internal representations of the environment was even more limited. God was filling all the gaps in knowledge, and literally divine substance was filling all the space among humans, connected them to each other, and mediated the perception and the experience of the surroundings. This was an interesting early metaphysical concept of what later would be identified as electromagnetic radiation and fields. Again, this concept would later serve as the basis for Georg Wilhelm Friedrich Hegel (1770–1831) to suggest the existence of a "social mind" ("Volksgeist" or spirit) in the sense of a complete and independent entity, in the form of a collective mind with individual persons being merely its agents. This world-mind he believed was controlling the function and the evolution of humanity as well as human society (see also Chap. 8).

The prevailing concept of the time was that there is a divine determinism guiding the universe, and according to Aristotle, this constituted the "final cause" [17]. Descartes and his contemporaries Johannes Kepler (1571–1630) and Galileo Galilei (1564–1642) expelled the "final cause" from the physical universe (res extensa), but as a compromise, Descartes kept the mind (or res cogitans) under the jurisdiction of religion and the church and supported the beliefs concerning the immortality of the soul, as well as the validity of divine determinism concerning behavior and mental life [6, 18–20]. Descartes remained a faithful Catholic through his entire life, and this had a profound influence on his ideas and work [21]. It should be noted, however, that Cartesian "dualism" is probably not in accord with the Christian teaching and dogma and reflects more oriental religions which consider the body to constitute the shell or even the prison of the soul.

Another important element in Descartes' philosophy was the concept of the presence of innate ideas and knowledge, and this was also attributed to God [14]. He also distinguished six basic passions: wonder, love, hatred, desire, joy, and sadness [22, 23].

It is interesting what within this largely theological frame, Descartes believed that mathematics and mechanics could explain the function of the mind [24]. In spite of all the efforts of Descartes, nothing could stop the Catholic Church from prohibiting his books in 1663 [2].

In accord with his dualistic approach to the body-mind problem, Descartes denied that animals had "soul," "reason," and "intelligence" and considered them to be similar to the automatic machines of the time ("automata") [24]. This was widely accepted and one of the consequences it was that it justified the cruel treatment of animals until the emergence of the theories of Charles Darwin in the mid-nineteenth century.

In the long term, dualism was problematic, caused, and still causes damage, merely because it was so successful and was eventually adopted by the church and the lay culture. It is still a dominant approach for many philosophers and mental health professionals even in a vague and unsystematic way (see also Chap. 16) [25–28].

It is interesting how Descartes' remains were treated after his death, and this reflects the need of human psychology and nature for divinity, heroes, martyrs, and the metaphysical, in an often inhuman and offending way. In an admixture of admiration for his contribution to science, philosophy, and the society and probably because of a metaphysical need for relics, his fingers were taken for posterity and bones for jewelry rings, and his skull was stolen, and today rests in the Musée de l'Homme, with a skull of a prehistoric man to its right and a 3D reconstruction of the skull of the athlete Lilian Thuram to its left and among the remains of saints and various ethno-anthropological pieces [29] (Fig. 6.3). On the skull surface, the names of several former owners are engraved, as if, in this essentially "sacrilegious" way, they had managed to grasp some of Descartes' fame or brilliance.

The mechanistic concept concerning animals that they do not have a soul but instead they are a kind of organic "automata" enjoyed wide acceptance, and this is evident from a story implicating Father Nicolas Malebranche (1638–1715) who, in public, abused a pregnant dog specifically in order to show that it is a soulless automaton, a mere machine, not deserving sentiments or treating it as a living being. In sharp contrast, Julien Offray de La Mettrie (1709–1751) in his book "L'Homme-machine" (The Man Machine) argued that humans are superior to animals only because of their better-developed brains and not because animals lack metaphysical/supernatural souls.

Thomas Willis (1621–1675; Fig. 6.4) is considered by many the father of neurology as he was the one to have coined the term "neurology" itself and gave to the term "psychology" its current meaning. In contrast to Descartes, he adopted a monistic approach to mind and brain and was the first to propose a hierarchical organization of the brain into levels of function from back and low to frontal and higher. This theory of hierarchical organization was the basis for

Fig. 6.3 Up: Descartes skull in the Musée de l'Homme. The names of several former owners together with lyrics are engraved on the skull surface (picture by J. C. Domenech, MNHN, Paris). Down: Descartes skull with a skull of a prehistoric man to its right and a 3D reconstruction of the skull of the athlete Lilian Thuram to its left (picture by D. Ponsard, MNHN, Paris, after permission)

Fig. 6.4 Thomas Willis (1621–1675) by unknown artist from his book "Opera omnia, nitidius quàm vnquam hactenus edita, plurimùm emendata, indice rerum copiosissimo, ac distinctione characterum exornata," c.1694

the idea on the evolution and dissolution of the nervous system proposed by Hughlings Jackson (1835–1911; Fig. 9.29) who rejected localization of symptoms in brain areas and argued that mental and nervous disorders were a matter of nervous system de-evolution or dissolution in a reversed Darwinian way (see also Chap. 9). A group of remarkably talented and radically thinking men who called themselves "the virtuosi" had a profound impact on Willis' way of thinking and career [30]. The team included Richard Lower (1631–1691) who performed the first blood transfusion, Robert Boyle (1627–1691) who described the laws of the gases and many consider him to be the founder of modern chemistry, and Thomas Millington (1628–1714) and Sir Christopher Wren (1632–1723) who draw the illustrations for Willis' books and later became famous as the architect of St. Paul's Cathedral in London. They all assisted Willis in his major work, the book "Cerebri Anatome" published in 1664 and played an important role in founding the Royal Society of London, a famous club of philosophers. Willis, after his first book, went on to write a book on his ideas concerning the mind-body problem and published the "De anima brutorum" in 1672 [31].

Willis was a faithful Christian and friend to the Archbishop of Canterbury with whom he frequently discussed his ideas and theories. Subsequently, he was very careful with issues pertaining to mind and soul. Interestingly, his work "De anima brutorum" distinguishes the immortal human soul from the "soul of brutes" (animals) which he also calls "corporeal," "body," or "material" soul. He suggested that humans possess both kinds of souls but animals only the inferior and mortal one. This corporeal soul was held responsible for what today we call higher cognitive functions. Thus, Willis proposed a significantly different mind-body model in comparison to that of Descartes but on the same dualistic principle and maybe very close to the Pythagorean concepts. One interpretation of his theory would be that he rearranged the Pythagorean tetractys into a linear hierarchy. He also followed the Aristotelian ideas when he suggested, before John Lock (1632–1704) who was his student, that nothing is in the brain that is not first in the senses [30]. He was one of the fathers to what today is called "biological psychiatry" since he argued that all mental phenomena and disorders stem from nerve transmission rather than blood vessels. He confirmed the frequent coexistence of mania and depression in the same patient.

The important implication of Willis' approach was that his ideas permitted animals to think and act somewhat rationally; thus, they were not simple automata without a soul. By this, Willis succeeded in a remarkable achievement, that is, to secure several issues simultaneously; he was free to research brain and his functions and free to study aspects of human behavior. Animals were not mere soulless things (automata), and the same time, the Christian Church was fully satisfied with all these.

In this frame, he felt comfortable to suggest that mental disorders are essentially diseases of the corporal soul and need medical attention and care. In his text in the "De anima brutorum," Willis describes a relationship between mania and melancholia and their alternation in the form of bipolar disorder. Furthermore, he rejected the theory of the wandering womb as the cause of hysteria by arguing that hysteria could be observed also in men and also by reporting no womb abnormalities in autopsies of hysterical females. Instead, he pointed to the back of the brain as the location of the abnormal function responsible of the symptoms of hysteria [30, 32]. The word "hysteria" comes from the Greek word "υστέρα" meaning the womb, because it was originally believed that it was caused exclusively in females as a result of abnormal movements and contractions of the womb.

In the 1656, an edict of the King Louis XIV (1638–1715) of France created a new hospital administrative organization for Paris. It consisted of asylums for the insane among which was the Hospital Pitié-Salpêtrière (Fig. 6.5). The name "Salpetriere" comes from the French word for saltpeter, which is a key ingredient of gunpowder. Originally, it was built to be an arsenal, but its location near the center of Paris and the frequent accidental explosions made the residents of the neighborhood to send a petition to King Louis XIII to store munitions to a safer location, which was done in 1634 [33]. However, in these asylums, other people were also detained including indigents, orphans, prostitutes, homosexuals, aged persons, and the chronically ill. Similar institutions and similar practices also occurred in Germany but not in other European countries. Especially in England, the state paid little attention to the matter, and most mental patients were either kept at home or left to wander in the streets without any care. In this environment, the Salpetriere was converted into a hospice for female patients while the Bicetre, a second large Paris hospital, served males. During the nineteenth century, they came to house over 8000 patients each, and this was pretty much the same picture for all asylums throughout Europe.

During this period, the first systematic clinical work and observations appeared. Thomas Sydenham (1624–1689) worked on the definition of health and disease and proposed that it is very rare a single cause to lead to the development of a syndrome. He believed in the presence of multiple interacting etiologies and supported the

Fig. 6.5 The hospital Pitié-Salpêtrière in Paris during the rule of Luis XIV from page 1201 of "Paris à travers les siècles. Histoire nationale de Paris et des Parisiens depuis la fondation de Lutèce jusqu'à nos jours, etc.," published in 1879

syndromal approach vs. the disease approach. Theophile Bonet (1620–1689) introduced the term "folie maniaco-melancolique" in 1684 [34], while during the same period, Francois Sylvius (1614–1672) described the brain fissures and gyri.

In parallel, the philosophical discussions on the nature of human mind and its function continued. Benedict (Baruch) Spinoza (1632–1677; Fig. 6.6) was a leading philosopher of the time. Because of his radical ideas, the Jewish religious authorities issued a "herem" (similar to Christian excommunication) while his books were put on the Catholic Church's Index of Forbidden Books. He rejected Cartesian dualism and suggested that body and mind are one and viewed them to be two aspects of the same fundamental phenomenon (neutral monism). Further, he suggested that psychological function is a natural phenomenon, and subsequently, the human mind is subject to the laws of nature. In an ethical extension of these views, the concepts of "good" and "evil" are merely the products of the emotions of pleasure and pain. John Locke (1632–1704; Fig. 6.7) rejected the presence of "innate ideas" according to Descartes and suggested that only experience shapes the human mind. Similar concepts were expressed by Thomas Hobbes (1588–1679; Fig. 6.8), but both Locke and Hobbes seemed embarrassed when discussing the possible total lack of inherent ideas. Hobbes believed that "atoms" were streaming through the nerves. Both philosophers utilized the "tabula rasa" theory to promote the idea of universal human rights. The reason why there was such a tendency to deny the presence of inherent innate ideas and predetermined elements in human existence was because at that time, they were connected with the presence of God and they mediated God's will

Fig. 6.6 Benedict (Baruch) Spinoza (1632–1677) portrait by unknown artist, c. 1665

Fig. 6.7 John Locke (1632–1704) portrait by **Sir Godfrey Kneller c. 1697**

Fig. 6.8 Thomas Hobbes (1588–1679)

and the way God was supporting human existence but also controlling human beings. Rejecting God and promoting human freedom and humanism went inevitably through the idea of the "tabula rasa."

Gottfried Wilhelm Leibniz (1646–1716; Fig. 6.9) was a prominent polymath and philosopher. His philosophical thinking was a rather fragmented admixture of rationalism but also of Neoplatonic ideas together with Christian religiosity, with influences from Chinese philosophy and religion, leading to a kind of optimism which characterized his ideas. His work includes many conflicting ideas, but his metaphysical tendencies produced specific solutions to the problems and filled the gaps in the theories of Descartes, Spinoza, and others. He rejected Locke's and Hobbes' ideas and argued that the human mind is comprised of inherent "monads" that is functional units or substances which are pre-synchronized and whose coordinated function produces mental phenomena (monadology). His approach was essentially monistic but more resembling the idealistic concept (mind superior to matter) with divinity (God) being present everywhere and acting as the substance that connects monads but also individual human beings. Additionally, Leibniz believed in the existence of perceptions below the threshold of awareness, thus arguing for the existence of the subconscious. He named them "petites perceptions" in contrast to the conscious which he named "apperceptions." He considered the two classes to exist on a continuum. He proposed a number of principles he believed they govern mental life: the presence of a continuum from the unnoticed petite (subliminal)

Fig. 6.9 Gottfried Wilhelm Leibniz (1646–1716) portrait by Christoph Bernhard Francke c. 1695

perceptions to the clearly perceived, apperception, logical inference, the principle of causality, and the principle of purpose (teleology). In spite of the fact that much of his work reflects an idealistic monism, he was also an advocate of psychophysical parallelism and argued that mind and brain do not interact but rather function harmonically and independently from each other [35–38].

References

1. Descartes R, Cottingham J, Stoothoff R, Murdochtrans D. The philosophical writings of René Descartes (1641). Cambridge: Cambridge University Press; 1984.
2. Mendler R. False truths: the error of relying on authority. Lanham: Hamilton Books; 2014.
3. Cottingham J. 'A brute to the brutes?': Descartes' treatment of animals. Philosophy. 1978;53(206):551–9.
4. Cunning D. The Cambridge companion to Descartes' meditations. Cambridge companions to philosophy. Cambridge: Cambridge University Press; 2014a. https://doi.org/10.1017/cco9781139088220.
5. Watson RA. What moves the mind—an excursion in Cartesian dualism. Am Philos Q. 1982;19(1):73–81.
6. Dicker G. Descartes: an analytic and historical introduction. Oxford: Oxford University Press; 2013.
7. Haas LF. Rene Descartes 1596-1650. J Neurol Neurosurg Psychiatry. 1992;55(3):176. https://doi.org/10.1136/jnnp.55.3.176.
8. Shampo MA, Kyle RA. Rene Descartes (1596-1650). JAMA. 1975;234(12):1247. https://doi.org/10.1001/jama.234.12.1247.
9. Urban E. On matters of mind and body: regarding Descartes. J Anal Psychol. 2018;63(2):228–40. https://doi.org/10.1111/1468-5922.12395.
10. Van Gijn J. Rene Descartes (1596-1650). J Neurol. 2005;252(2):241–2. https://doi.org/10.1007/s00415-005-0775-2.
11. Cunning D. The Cambridge companion to Descartes' meditations. Cambridge: Cambridge University Press; 2014b.
12. Berhouma M. Beyond the pineal gland assumption: a neuroanatomical appraisal of dualism in Descartes' philosophy. Clin Neurol Neurosurg. 2013;115(9):1661–70. https://doi.org/10.1016/j.clineuro.2013.02.023.
13. Lokhorst GJ, Kaitaro TT. The originality of Descartes' theory about the pineal gland. J Hist Neurosci. 2001;10(1):6–18. https://doi.org/10.1076/jhin.10.1.6.5629.
14. Shiraev E. A history of psychology: a global perspective. Thousand Oaks: Sage Publications Inc; 2010.
15. Smith CU. Descartes' pineal neuropsychology. Brain Cogn. 1998;36(1):57–72. https://doi.org/10.1006/brcg.1997.0954.
16. Finger S. Descartes and the pineal gland in animals: a frequent misinterpretation. J Hist Neurosci. 1995;4(3–4):166–82. https://doi.org/10.1080/09647049509525637.
17. Sangiacomo A. The ontology of determination: from Descartes to Spinoza. Sci Context. 2015;28(4):515–43. https://doi.org/10.1017/S0269889715000319.
18. Duncan G. Mind-body dualism and the biopsychosocial model of pain: what did Descartes really say? J Med Philos. 2000;25(4):485–513. https://doi.org/10.1076/0360-5310(200008)25:4;1-A;FT485.
19. Janiak A. Space and motion in nature and scripture: Galileo, Descartes, Newton. Stud Hist Phil Sci. 2015;51:89–99. https://doi.org/10.1016/j.shpsa.2015.02.004.
20. Shapiro L. Princess Elizabeth and Descartes: the union of soul and body and the practice of philosophy. Br J Hist Philos. 1999;7(3):503–20. https://doi.org/10.1080/09608789908571042.

21. Rodis-Lewis G. Descartes his life and thought. New York: Cornell University Press; 1999.
22. Kirkeboen G. Descartes on emotions, reason, and the adaptive unconscious: the pioneer behind the caricature. Hist Psychol. 2019;22(1):17–39. https://doi.org/10.1037/hop0000109.
23. Stone MH. Modern concepts of emotion as prefigured in Descartes' "passions of the soul". J Am Acad Psychoanal. 1980;8(4):473–95. https://doi.org/10.1521/jaap.1.1980.8.4.473.
24. Kogge W, Richter M. Synthetic biology and its alternatives. Descartes, Kant and the idea of engineering biological machines. Stud Hist Phil Biol Biomed Sci. 2013;44(2):181–9. https://doi.org/10.1016/j.shpsc.2013.03.013.
25. George B, Carpentier A, Evans J, Apuzzo ML. Rene Descartes: mind, reason, and challenging assumptions. Neurosurgery. 2000;47(1):244–9. https://doi.org/10.1097/00006123-200007000-00054.
26. Nierenberg AA. Failure to integrate brain and mind (or blaming Descartes to understand a very modern dilemma about explanatory models, psychiatric treatment, and underfunding of clinical care and research). CNS Spectr. 2009;14(11):599–600. https://doi.org/10.1017/s109285290002383x.
27. Sessa B. Let's not give in to Descartes: we need pharmacology and psychotherapy. J Psychopharmacol. 2009;23(2):227–8. https://doi.org/10.1177/0269881108093884.
28. Ventriglio A, Bhugra D. Descartes' dogma and damage to western psychiatry. Epidemiol Psychiatr Sci. 2015;24(5):368–70. https://doi.org/10.1017/S2045796015000608.
29. Shorto R. Descartes' bones: a skeletal history of the conflict between faith and reason. New York: Doubleday Publishing Group, a division of Random House, Inc.; 2008.
30. Isler H. Thomas Willis (1621–1675): doctor and scientist. New York: Hafner; 1968.
31. Finger S. Origins of neuroscience: a history of explorations into brain function. Oxford: Oxford University Press; 2001.
32. Spillane J. The doctrine of the nerves. Oxford: Oxford University Press; 1981.
33. Micale MS. The salpetriere in the age of Charcot: an institutional perspective on medical history in the late nineteenth century. J Contemp Hist. 1985;20(4):703–31. https://doi.org/10.1177/002200948502000411.
34. Theophile Bonet. Physician of Geneva (1969). JAMA. 1620–1689;210(5):899. https://doi.org/10.1001/jama.1969.03160310087021.
35. Fancher R, Schmidt H. Gottfried Wilhelm Leibniz: underappreciated pioneer of psychology. In: Kimble G, Wertheimer M, editors. Portraits of pioneers in psychology, vol. V. Washington, DC: American Psychological Association; 2003. p. 1–17.
36. Kyle RA, Shampo MA. Gottfried Wilhelm Leibniz. JAMA. 1981;245(11):1105. https://doi.org/10.1001/jama.245.11.1105.
37. Russell B. History of western philosophy: collectors edition. London: Routledge; 2013.
38. Verhave T. Contributions to the history of psychology. 3. G. W. Leibniz (1646-1716) on the association of ideas and learning. Psychol Rep. 1967;20(1):111–6. https://doi.org/10.2466/pr0.1967.20.1.111.

Chapter 7
The Age of Enlightenment

The Age of Enlightenment (or Age of Reason) was an intellectual and philosophical movement that dominated the eighteenth-century Europe, a period also called the "Century of Philosophy." Traditionally, it is considered that it lasts from 1715 (beginning of the reign of Louis XV) to the French revolution in 1789. It was a vivid period with ideas spreading and circulating with meetings at scientific academies, Masonic lodges, literary salons, and coffeehouses while the circulation of books, journals, and pamphlets increased exponentially. The basic ideas challenged the absolute authority of the monarchy and the church on the basis of reason as the primary source of knowledge and with an emphasis on the scientific method.

During that period, a number of eminent philosophers developed theories, ideas, and interpretation of the human nature, its relationship with the universe and divinity, as well as of the way humans can perceive and understand the world. While these theories and ideas are now obsolete and appear almost strange today, at that time, the prevailing belief was that humans exist, perceive, and understand the world and communicate to each other only because and through God. This was not new of course, since the concept that God was present everywhere and was the essence of the universe and of human existence was dominant since centuries (see previous chapters).

An interesting person which seems to mark the transition for the previous period to the age of reason was Emanuel Swedenborg (1688–1772). He was an authority in mathematics, astronomy, and mining before he turned to the life sciences in 1736 in the frame of his metaphysical interests and particularly concerning the body-mind issue. Between the years 1740–1745, he was the first to extend the ideas of Thomas Willis and suggested that different functions are located in different areas of the brain. Thus, he conceived the localization principle of brain function many years before Gall and more a century before Broca (see also Chaps. 8 and 9). His main arguments were that senses are not mixed and different head injuries seem to relate to distinct effects on cognitive functions. He even linked the frontal lobes with higher intellectual function. However, his work is not widely known, and he is not

K. N. Fountoulakis, *Psychiatry*, https://doi.org/10.1007/978-3-030-86541-2_7

recognized widely, mainly because all his ideas were developed and embedded not within a scientific but rather in a religious doctrine. His books were mostly theological rather than scientific or philosophical. In 1744, he began having mystical experiences, and after his death, the Swedenborgian (New Jerusalem) Church was founded in England in 1784 based on his teachings [1–4].

The debate of passions vs. reason started with David Hume (1711–1776; Fig. 7.1) which in his 1738 book "A Treatise of Human Nature" pushed toward a total naturalistic science of man and proposed that passion rather than reason governs human behavior. This is in accord with modern neuroscientific knowledge; today, we know that the vast majority of human decisions and especially of common everyday decisions are not based on analytical reason but rather are the result of sub- or unconscious procedures involving impulses, habits, or established routines. However, Hume rejected the existence of innate ideas, thus agreeing with Locke and Hobbes. So for Hume, the struggle was pure passion vs. pure reason without any restrains, limitations, or modifiers for either of the two.

Around the same time period, Julien de La Mettrie (1709–1751), a French philosopher, published the book "L' Homme Machine" (The Human Machine) in 1748 where he adopted a completely mechanical explanation concerning how the human body and brain work. In the same line of thinking, François Marie Arouet, better known as Voltaire (1694–1778; Fig. 7.2), rejected any metaphysical element in life and especially any belief in God after the great Lisbon earthquake of 1755 and the

Fig. 7.1 David Hume (1711–1776), portrait by Allan Ramsay, c. 1766

Fig. 7.2 Voltaire (1694–1778), portrait by Nicolas de Largillière, c. 1724

death of more than 30,000 people (Fig. 7.3). He was one of the most important minds of all time. His ideas had early elements of humanistic and existential theories, including the concept of self-actualization. The interplay and tight embrace between scientific understanding of the human beings and sociopolitical ideas were extremely strong. High in the agenda of the period was freedom from the two major sources of oppression: the absolute monarchies and the church. The first was rather easy to criticize; the second was more difficult and demanded a radically different understanding of what humans are. The dominant idea was that of the free human being, who makes decisions on the basis of pure reason. No "innate ideas" or other cognitive restrains were acceptable since at that time, they were implying a divine intervention. The elaboration on the concept of "blank slate" and the conception of society as oppressive by definition led Jean-Jacques Rousseau (1712–1778; Fig. 7.4) to argue that people are inherently "noble savages," while civilization and society have deleterious effects because they both lead to the neglecting of emotional life. Historically, the idea that the current civilization is evil appeared in the writings of St. Augustine (354–430 AD; see Chap. 3) while the deleterious effect of civilization was later adopted by Sigmund Freud.

Immanuel Kant (1724–1804; Fig. 7.5), who was essentially a metaphysical philosopher, in his 1781 book "Critique of Pure Reason" [5], rejected the Cartesian dualism and agreed on this issue with Spinoza. He essentially also rejected

Fig. 7.3 The 1755 Earthquake, by João Glama Strobërle. Angels holding a fiery swords fly above the city, personifying divine judgment. This event and the surrounding religious interpretations made Voltaire to reject any belief in God (National Museum of Ancient Art, Lisbon, Portugal)

Fig. 7.4 Jean-Jacques Rousseau (1712–1778) portrait by Maurice Quentin de La Tour, c. 1753 (Musée d'art et d'histoire de Genève)

Fig. 7.5 Immanuel Kant (1724–1804)

empiricism in favor of logical reasoning and rational philosophy, although some scholars suggest that he was only delineating the limits of empiricism, not completely rejecting it [6]. Such a compromise is obvious in his approach to the nature of the mind and its relationship with brain. According to him, on one hand, the mind can know only what the senses allow ("phenomenon" from the Greek "φαινόμενον" meaning what one can sense) but not the meaning and the qualities of the thing ("noumenon" from the Greek "νοούμενον" meaning what one can think of, understand, and interpret, what the object stands for). On the other hand, in his work "Anthropology from a Pragmatic Point of View," which is a compendium of his lectures during 1772–1773, he rejects the structural and topographical approach to mind, suggesting that the brain as a whole acts as the substrate for the mind to develop and interact with history, culture, and the environment, a concept similar in many ways to Locke's "tabula rasa" [7]. This concept considers the human mind "in context," in what we today would call a "network of interactions" with infinite feedback loops between the mind itself, other minds, the environment (synchronically), and history (diachronically). Thus, with "anthropology," Kant suggested the presence of a contingent element in knowledge which is considered to be both transcendental and contingent. However, although this is a pragmatic approach, Kant never really managed to bridge theoretically these two facets of knowledge. Reading Kant with the eyes of the modern scientist, this incompleteness of these theories is not a negative element. Kant recognized the limitations, accepted them, and probably refused to theorize eternally and without ground for solid reason.

Andrés Piquer-Arrufat (1711–1772) described the alternation of manic and depressive episodes in King Ferdinand VI of Spain (1713–1759; Fig. 7.6) after his death in 1759. It was reported that also his father Philip V (1683–1746) was suffering from bipolar disorder. Ferdinand suffered from melancholic episodes throughout his life, and after his wife's death, he developed psychotic nihilistic depression with suicidal tendencies and died in a state of "fits." He also had frequent episodes of hypersexuality. Some historians suggest peculiar and uncontrolled hypersexuality was a characteristic of his family, the house of Bourbons [8]. Piquer-Arrufat considered both mania and melancholia to constitute a single diagnostic entity. He also described mixed states, seasonality, and rapid cycling and included his approach in a medical textbook he published in 1764 [9].

The lack of real advances in the development of treatment options for mental disorders led the alienist of the time to improvise. It is not known when and where the use of alcohol as a sedative agent began, but documents suggest its widespread use during the nineteenth century. Other sedatives included opium grains, unguents, and laudanum, treatments which were already standardized by Thomas Sydenham during the seventeenth century.

Fig. 7.6 King Ferdinand VI of Spain (1713–1759) (Museo del Prado)

One of the peculiar devices designed during the eighteenth century and very popular during the early nineteenth century for the treatment of insanity was the spinning chair [10–13]. In spite of the fact that the Dutch psychiatrist Herman Boerhaave (1668–1738) was the first to use the "gyrating chair," the concept of "spinning therapy" is credited to Joseph Mason Cox (1763–1818). It is interesting that Boerhaave was the first to suggest that mental illness is caused by abnormally fixed electrical arrangements in the brain. Several others contributed to the theory that spinning was therapeutic, as well as to the design and improvement of the device (Fig. 7.7). The gyrating chair became popular in Europe and North America. The rationale behind it was to shake the body and especially the blood and in this way to restore equilibrium. Among its enthusiastic supporters were Erasmus Darwin (1731–1802; which was the grandfather of Charles Darwin) and Benjamin Rush (1745–1813; Fig. 7.8). Erasmus Darwin was a physician, philosopher, and scientist

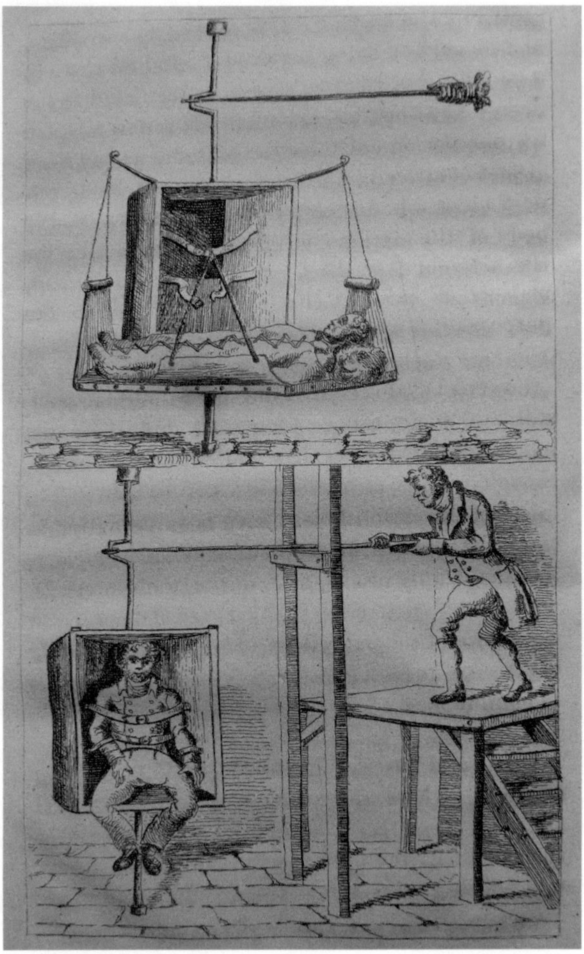

Fig. 7.7 Hallaran's gyrating chair, from William Saunders Hallaran: Practical Observations on the Causes and Cure of Insanity. Cork, 1818

Fig. 7.8 Benjamin Rush (1745–1813) portrait by Charles Willson Peale c. 1818

and developed the spinning-couch device. His theory was that sleep could cure disease and that spinning around could easily induce sleep. Only Benjamin Rush (1745–1813) showed some interest in this idea but for different reasons; he thought that spinning could be therapeutic because it might reduce brain congestion. The idea was based on the belief of the time that "madness" was an arterial disease, an inflammation of the brain. Overall, Benjamin Rush believed that insanity was caused by irritation of brain blood vessels [14] and was also using the tranquilizing chair (Fig. 7.9). Another supporter of this method was William Saunders Hallaran (1765–1825), who was physician superintendent at the county and city of Cork Lunatic Asylum. He was the first to distinguish between mental insanity and organic delirium. For the treatment, he used emetics and purgatives, digitalis and opium, shower bath, exercise, and occupational therapy. He reported remission of acute mania with spinning treatment. His observations that some of his patients enjoyed the rotatory experience led him to propose the swing as a mode of amusement [15, 16]. Ernst Horn (1774–1848) reported in 1818 cures of hysterical patients with therapy utilizing the spinning chair at the psychiatric wards of the Charite Hospital in Berlin [17].

Some of these treatment methods were not only peculiar but often horrifying and dangerous. One such example is deliberate near-drowning (waterboarding) which is often used today as an interrogation technique. That was also its main use at that time. This technique was used as a treatment option for mental illness for more than

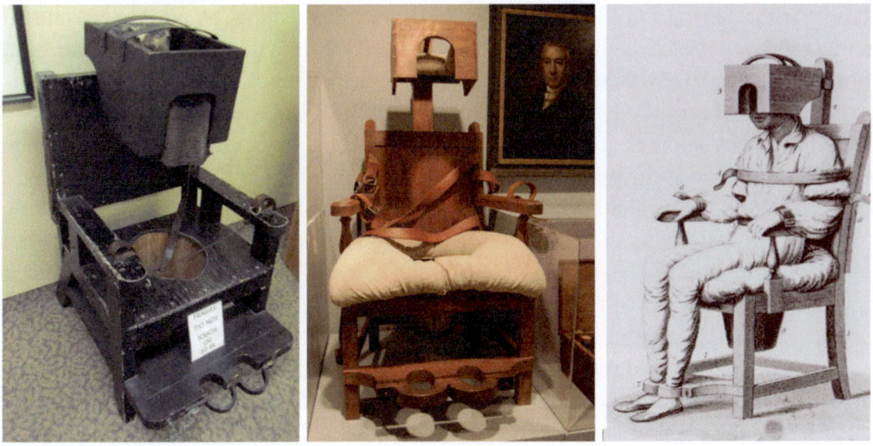

Fig. 7.9 The tranquilizing chair (left: Glore Psychiatric Museum; middle: reproduction of the chair at Public Hospital Museum at Williamsburg, Virginia, on the basis of the early nineteenth-century drawing at the right)

two centuries. It was introduced by Jan Baptista van Helmont (1580–1644), who observed remission of mental symptoms in a patient after he was revived following an accidental near-drowning. Hermann Boerhaave recommended it only for the most desperate cases. Hieronymus David Gaubius (often mentioned as Jerome Gaub; 1705–1780) who was a student of Hermann Boerhaave, attributed its efficacy to bodily changes induced by anxiety and by the shock because of the realization of imminent death [18]. In this way, the theoretical frame of "shock treatment" emerged again in the western world after its first appearance in India during the fifth century BC (see also Chap. 3). The core concept was that shock caused a number of bodily events which in turn affected the "common sensorium" which was believed to be the place where mind and body meet. In turn, this was triggering ideas in the mind which could not be produced without the experience of extreme affective states like horror. Jean-Joseph Menuret de Chambaud (1739–1815) wrote in the Encyclopédie by Denis Diderot (1713–1784) and Jean-Baptiste le Rond d'Alembert (1717–1783) that "in mania therapy is directed to the body, in which it aims to produce a shock and a deep disturbance." Even Philippe Pinel (1745–1826) believed that eliciting powerful emotions like fear had a role in the frame of moral therapy. It is interesting that the therapeutic methods used by Francis Willis (1718–1807) to treat the mad King George the Third of England included a partially psychological approach which was essentially a mixture of bullying, encouragement, and "fixing with the eye" [19]. Ironically, the king might had suffered from what today we would diagnose as organic psychotic disorder as a result of acute intermittent porphyria [20].

Around that time, a method which is still popular appeared. An otherwise-not-described-in-detail form of hypnosis was introduced by the Austrian physician

Franz Anton Mesmer (1734–1815; Fig. 7.10). He claimed that by putting patients in some kind of a trance, he had successfully treated hysteria and many other illnesses. For this purpose, he was also using the healing properties of an invisible and impalpable fluid that he called "animal magnetism" [21–25].

While many sources suggest that the straitjacket (Fig. 7.11) appeared in France in 1790 and it was invented by an upholsterer named Guilleret, for the Bicêtre Hospital, this is probably not true. In 1772, the physician David MacBride (1726–1778) gave a detailed description of it, and his writings imply that it was well known in England by that time [26]. William Cullen (1710–1790; Fig. 7.12) praised it not only as a means of restraint, but he considered it to have a therapeutic effect, and also he even suggested a physiological rationale for its use and supposed effect [27]. Straitjacket is a garment shaped like a jacket with overlong sleeves. Once the arms are inserted into the straitjacket's sleeves, they are then crossed across the chest while the ends of the sleeves are tied to the back of the wearer. In this way, it is ensured that the arms are kept close to the chest with as little movement as possible. The straitjacket was typically used to restrain agitated or aggressive persons.

Fig. 7.10 Franz Anton Mesmer (1734–1815) by Pujos/Legrand - Musée de la Révolution française

Fig. 7.11 Canvas straitjacket for restraining adult patients, made in London, England, c.1930–1960, and used at the Frenchay Hospital, Bristol, taken after permission from http://www.sciencemuseum.org.uk

Although in English the word "straitjacket" is the most common spelling, the term "straitjacket" is also frequently used, while in Scotland, the term "strait-waistcoat" is used, but it is generally deemed archaic. Straitjackets are also known as camisoles. The French name is "camisole de force," but Philippe Pinel in his translation of the works of Cullen in 1785 uses the term "chemisette serrèe."

However, in spite of intellectual and theoretical advances in the efforts to understand mental life and mental illness, the practical outcome was poor, and the everyday life of patients was still miserable. Disappointment from the lack of effectiveness of treatment options was the prevailing attitude during the eighteenth century. The insane inmates in public mental asylums were regarded as having incurable diseases and were subjected to physical restraints, beatings, and constant fear (although in some private asylums, treatment was more lenient). While the most impressive depictions concerning mental health and asylums were horrible, some scholars argue that, for example, the paintings by William Hogarth (1697–1764; Fig. 5.4) could in fact constitute an exaggeration rather than reflecting reality in a similar way modern depictions of mental health and hospitals in films and novels are negative [19, 28]. At any case, there is no doubt that this was the darkest hour. The worst incident happened during the French revolution, during the so-called "September

Fig. 7.12 William Cullen (1710–1790)

Massacres" in 1792. At that time, there was a fear that foreign and royalist armies would attack Paris and that the inmates of the city's prisons would be freed and join them. The action was undertaken by 150–300 Sansculottes, Guardsmen, Marseillais, Gendarmes, and Fédérés with the support of the Cordeliers, the insurrectional Commune, and the Revolutionary sections of Paris. Half the prison population of Paris, that is, more than 1200 prisoners were murdered. The Salpêtrière asylum was stormed during the night of September 3, 1792, by a mob from the impoverished working-class district of the Faubourg Saint-Marcel. As a consequence, 25–35 out of a total of 186 female mental patients and other inmates were murdered after being dragged in the streets of Paris while still chained (Fig. 7.13). Also in the Bicêtre, 160–170 male patients were killed (Fig. 7.14) [29].

But then, toward the end of the century, harsh treatment and restrains were abolished, and the asylums started being respectful to the patients' human rights. These changes happened almost simultaneously in various European countries including Italy in 1789 with the physician Vincenzo Chiarugi (1759–1820; Fig. 7.15) in the hospitals of Santa Maria Nuova and Bonifazio in Florence and in England in 1796 with the layman and Quaker William Tuke (1732–1822; Fig. 7.16), who founded the York Retreat in York. In France, Jean-Baptiste Pussin (1746–1809) was the first to remove the chains from the inpatients of the Bicètre Hospital in Paris. Although this is usually credited to Phillippe Pinel (1745–1826), the fact is that Pussin was the first to free the insane in 1797, after Pinel had left for the

Fig. 7.13 Massacre at Salpêtrière in September 3, 1792. At the center, a commissioner reads the register of the insane while the sansculottes armed with axes, masses, and clubs execute them one by one. In the foreground, two men search the corpses to seize any valuables

Salpêtrière. Pinel did remove the chains from patients at the Salpêtrière 3 years later, but this happened only after Pussin joined him there. It is believed that the Pinel myth was fabricated by Pinel's son, Scipion Pinel (1795–1859), together with Pinel's foremost pupil, Jean-Étienne Dominique Esquirol (1772–1840). In order to fabricate the myth, Scipion Pinel even ordered relevant paintings to be created (Figs. 7.17 and 7.18). There was also a political flavor in this story, since the glorification of Pinel happened during the years of Bourbon monarchy and included also a fictional story of confrontation between Pinel and a paraplegic but powerful and dedicated Jacobin named Georges Couthon during the era of the French revolution [30].

At the end of the eighteenth century, the discussion on the concept and usefulness of the term "syndrome" appeared. During that period, probably the most important and enduring influence on psychiatry was from the work of the Swedish Carl Linnaeus (1707–1778), a botanist, physician, and zoologist who is considered to be the father of modern taxonomy (Fig. 7.19). At that time, the need for gathering and organizing the various independent observations and reports into a comprehensive classification system was evident for medicine, and Linnaeus' book "Species Plantarum" published in 1753, which is considered to be the starting point of modern botanical nomenclature, created a scientific earthquake and pushed physicians toward a similar direction. François Boissier de Sauvages (1706–1767; Fig. 7.20) had followed the path of Linnaeus in his three-volume 1771 work "Nosologie Methodique" which was published several years after his death. The model he proposed dominated psychiatric nosology for centuries. A few years later, it was followed by the 1789 "A Treatise of the Materia Medica" by William Cullen (1710–1790; Fig. 7.12). Cullen repeated the suggestions of Aretaeus of Cappadocia, and in his book "First Lines of the Practice of Physic" (repeated editions through the years 1777–1784), he writes that the same state of brain

Fig. 7.14 Right: The executions at the Bicêtre Hospital in September 3, 1792. At a screened window appears a head planted at the end of a spike. Left: The executions at the Chatelet prison. Catalogue of the National Library of France

dysfunction when moderate causes melancholia, and when extreme, it causes mania. Cullen conceptualized mental illness in the frame of "dynamic neurophysiology" and was not following the taxonomic approach inspired by the work of Linnaeus [19]. In 1781, he coined the term "neurosis" to denote "all diseases of the nervous system not accompanied by fever" [31, 32]. Until that time, "neurotic" was any substance having an effect on the nervous system (stimulants, "pepperuppers," tonics, etc.).

However, for others, the impressive success of classification of plants and animals served as an example for the rest of science. Taxonomy and classification were considered to be a valuable tool which could reveal underlying common mechanisms and processes, and even it could determine the choice of treatment options. In his 1798 book "Nosographie Philosophique" (Philosophical nosography), Pinel developed a new classification of mental illnesses into mania, melancholia, idiocy, and dementia. Following the same path, in 1801, in his book "*Traite Medico-philosophique de l'Alienation Mentale*" (A medical-philosophical treatment of insanity), he proposed treatment methods [33].

Pinel's classification served as the basis for an asylum-based nonviolent and non-medical treatment, tailored to the needs of the individual patient. This method was

Fig. 7.15 Vincenzo Chiarugi (1759–1820). Stipple engraving by C. Lasinio, c. 1804. Credit: Wellcome Collection

in fact developed initially by Pussin, but it was described for the first time by Pinel in this book and was essentially a "milieu therapy" which also included education, reasoning, and persuasion. Thus, Pinel rejected the use of bloodletting, shock treatment. and corporal punishment and put the emphasis on compassionate individualized care. He proposed a method called "le traitement moral" which in French means "mental therapy" and has nothing to do with moral or morality. It was translated in English as "moral treatment of insanity," and under this name, it became known worldwide.

One major change in popular culture during the eighteenth century was the acceptance of happiness as a major element of life [34]. This was in accord with the ideas of Aristotle who talked about "eudaimonia" (in Greek *"ευδαιμονία"* meaning the possession by good spirit) in his work "Nicomachean Ethics" and considered it as the ultimate end-purpose of human life (see also Chap. 3). Both the Aristotelian concept as well as the philosophical approaches during the Enlightenment differentiated between "happiness" and "hedonism" (from the Greek word *"ηδονή"* meaning the lust or pleasure). In fact, happiness was considered to be the aim of life, mainly in order to downgrade the role of religious-based meanings in life. Happiness was a more pragmatic as well as humanistic proposal, but the meaning of happiness remained open to debate, leaving space for religious as well as secular

Fig. 7.16 Portrait of William Tuke (1732–1822), Wellcome Library, London

Fig. 7.17 Philippe Pinel (1745–1826) ordering the removal of chains from patients at the Salpêtrière asylum in 1800. Painting by Tony Robert-Fleury, c. 1876

Fig. 7.18 Philippe Pinel (1745–1826) ordering the removal of chains from patients at the Bicêtre in 1792. In fact, it was Jean-Baptiste Pussin (1746–1809) not Pinel who removed the chains at the Bicêtre. Painting by Charles Louis Müller c. 1840–50

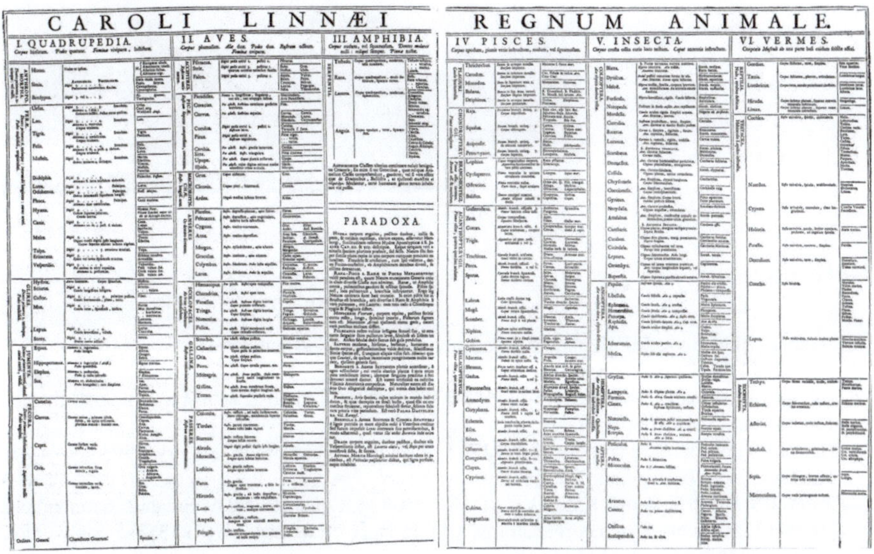

Fig. 7.19 Carolus Linnaeus's first edition (1735) of Systema Naturae

interpretations. For many, happiness constituted a basic human right, and society started changing toward a structure that could allow the pursue of happiness. Change was radical; not only the seeking of happiness became perfectly legitimate, but on the other hand, not being happy was now considered to be a problem in need for a solution. Further on this path, the concept of "happiness" would be later utilized in several efforts to define mental disorder (see also Chap. 13).

Fig. 7.20 Françoise Boissier de Sauvages (1706–1767)

References

1. Akert K, Hammond MP. Emanuel Swedenborg (1688-1772) and his contributions to neurology. Med Hist. 1962;6:255–6.
2. anonymous. Emanuel Swedenborg (1688-1772) natural scientist, neurophysiologist, theologian. JAMA. 1968;206(4):887–8.
3. Haas LF. Emanuel Swedenborg (1688-1772). J Neurol Neurosurg Psychiatry. 1993;56(4):343. https://doi.org/10.1136/jnnp.56.4.343.
4. Tubbs RS, Riech S, Verma K, Loukas M, Mortazavi M, Cohen-Gadol A. Emanuel Swedenborg (1688-1772): pioneer of neuroanatomy. Childs Nerv Syst. 2011;27(8):1353–5. https://doi.org/10.1007/s00381-011-1422-0.
5. Kant I (1998) Critique of pure reason (P. Guyer and A. W. Wood, Eds. and Trans.). Cambridge University Press, Cambridge.
6. Kant I (2000) Critique of the power of judgment (P. Guyer and E. Matthers, Eds. and Trans.). Cambridge University Press., Cambridge.
7. Kant I. Anthropology from a pragmatic point of view. In: Cambridge texts in the history of philosophy. Cambridge: Cambridge University Press; 2006.
8. Green V. The madness of kings (personal trauma and the fate of nations). Stroud: Alan Sutton Publishing Ltd; 1994.
9. Perez J, Baldessarini RJ, Cruz N, Salvatore P, Vieta E. Andres Piquer-Arrufat (1711-1772): contributions of an eighteenth-century spanish physician to the concept of manic-depressive illness. Harv Rev Psychiatry. 2011;19(2):68–77. https://doi.org/10.3109/10673229.2011.565251
10. Cox J. Practical observations on insanity. London: Baldwin and Murray; 1804.

11. Wade NJ. The original spin doctors--the meeting of perception and insanity. Perception. 2005;34(3):253–60. https://doi.org/10.1068/p3403ed.
12. Wade NJ, Norrsell U, Presly A. Cox's chair: 'a moral and a medical mean in the treatment of maniacs'. Hist Psychiatry. 2005;16(61 Pt 1):73–88. https://doi.org/10.1177/0957154X05046093.
13. Winter L, Wollmer MA, Laurens J, Straumann D, Kruger TH. Cox's chair revisited: can spinning alter mood states? Front Psych. 2013;4:132. https://doi.org/10.3389/fpsyt.2013.00132.
14. Noll R. The Encyclopedia of schizophrenia and other psychotic disorders. 3rd ed. New York: Facts on File Inc.; 2007.
15. Breathnach CS. Hallaran's circulating swing. Hist Psychiatry. 2010;21(81 Pt 1):79–84. https://doi.org/10.1177/0957154X09342760.
16. Kelly BD. Dr William Saunders Hallaran and psychiatric practice in nineteenth-century Ireland. Ir J Med Sci. 2008;177(1):79–84. https://doi.org/10.1007/s11845-007-0046-6.
17. Harsch V. Centrifuge "therapy" for psychiatric patients in Germany in the early 1800s. Aviat Space Environ Med. 2006;77(2):157–60.
18. Rather LJ. Mind and body in eighteenth century medicine: a study based on Jerome Gaub's De Regimine mentis. California: University of California Press; 1965.
19. Porter R. Madness: a brief history. Oxford: Oxford University Press; 2002.
20. Hunter R, MacAlpine I. Three hundred years of psychiatry 1535–1860. London: Oxford University Press; 1963.
21. Ellis H. Franz Mesmer: pioneer in the treatment of functional disease or charlatan? Br J Hosp Med. 2015;76(3):170. https://doi.org/10.12968/hmed.2015.76.3.170.
22. Forrest D. Mesmer. Int J Clin Exp Hypn. 2002;50(4):295–308. https://doi.org/10.1080/00207140208410106.
23. Makari GJ. Franz Anton Mesmer and the case of the blind pianist. Hosp Community Psychiatry. 1994;45(2):106–10. https://doi.org/10.1176/ps.45.2.106.
24. Parish D. Mesmer and his critics. New Jersey medicine: the journal of the Medical Society of New Jersey. 1990;87(2):108–10.
25. Spiegel D. Mesmer minus magic: hypnosis and modern medicine. Int J Clin Exp Hypn. 2002;50(4):397–406. https://doi.org/10.1080/00207140208410113.
26. Macbride D. A methodical introduction to the theory and practice of physic. Edinburg: W. Strahan; 1772.
27. Cullen W. First lines of the practice of physic. Worcester, Massachusetts: Isaiah Thomas; 1790.
28. Parry-Jones W. The trade in lunacy. London: Routledge and Kegan Paul; 1972.
29. Bosworth M. Anatomy of a massacre. Violence against Wom. 2016;7(10):1101–21. https://doi.org/10.1177/10778010122183766.
30. Weiner D. 'Le geste de Pinel': the history of a psychiatric myth. In: Micale M, Porter R, editors. Discovering the history of psychiatry. New York: Oxford University Press; 1994. p. 232–47.
31. Tuke DH. A dictionary of psychological medicine giving the definition, etymology and synonyms of the terms used in medical psychology with the symptoms, treatment, and pathology of insanity and the law of lunacy in Great Britain and Ireland, vol. 2. Philadelphia: P. Blakiston, sons, & co., Phil; 1892.
32. Zilboorg G, Henry GW. A history of medical psychology. New York: Norton; 1941.
33. Pinel P. In: Todd W, editor. A treatise on insanity. Oxford: Oxford University; 1806.
34. McMahon DM. The pursuit of happiness in history. In: Boniwell I, Davis S, editors. Oxford handbook of happiness. Oxford: Oxford University Press; 2012.

Chapter 8
Nineteenth Century: The Romantic First Half

By the nineteenth century, the term "alienist" was used for a psychiatrist, beginning in France. The term comes from French word *"aliéné"* (meaning: insane) which in turn comes from the Latin *"alienus"* or *"alius"* meaning the foreigner, stranger, and different (similar to Greek *"άλλος"* meaning the other). The root comes from the Proto-Indo-European "al-" (meaning: beyond). Thus, an alienist of the time was a physician who acted as an intermediary between the society and mental patients. By defining their relations was the agent of the patients' alienation [1]. Johann Christian Reil (1759–1813; Fig. 8.1) changed the term into "psychiatrist" in 1808 and gave birth to "psychiatry" as a medical specialty [2]. He coined the term from the Greek words *"ψυχή"* (pronounced psihi meaning soul) and *"ιατρική"* (pronounced iatriki meaning medicine and healing). He also described the qualities needed: "perspicacity, a talent for observation, intelligence, good will, persistence, patience, experience, an imposing physique, and a countenance that commands respect" [3]. Reil, who was also the founder of rational psychotherapy, recognized the therapeutic value of institutional surroundings, music, psychodrama, and occupational therapy. He also found the first psychiatric journal.

The nineteenth century begun also with the wider adoption that the rights of patients should be respected and that chains and inhuman conditions of living were unacceptable. Joseph Guislain (1797–1860; Fig. 8.2) in Belgium arranged for patients with completely different mental disorders to live together so that, according to his opinion, they would balance one another's deficits, and hopefully, this could produce beneficial effects. In the meantime, it was the work of Dorothea Lynde Dix (1802–1887; Fig. 8.3), a Massachusetts schoolteacher, who in 1841 investigated the widespread neglect and brutality in mental institutions and, through a vigorous program of lobbying state legislatures and the United States Congress, created the first mental asylums in the United States. Benjamin Rush (1745–1813) became gradually the father of American psychiatry and described "tristimania" (from Latin *tristis* = sorrow) as a form of melancholia with sadness as the predominant feature.

K. N. Fountoulakis, *Psychiatry*, https://doi.org/10.1007/978-3-030-86541-2_8

Fig. 8.1 Johann Christian Reil (1759–1813), by Alphonse-Léon Noel (drawer) and de Frey (lithographer) c. 1811

Back in France, Jean-Philippe Esquirol (1772–1840; Fig. 8.4) had succeeded Pinel and was the first to clearly point that melancholia was a disorder of the mood (of "passions") with "partial insanity" (monomania). He used the word "lypemania" (from Greek λύπη = sorrow) to describe a kind of depressive insanity. He also grouped mental disorders into five general syndromes: lypemania, monomania, mania, dementia, and imbecility/idiocy. Both Rush and Esquirol used the term "mania" to denote insanity and not elevated mood. Esquirol soon became the prototype of the psychiatric specialist, and the school he founded determined the basic characteristics of the new medical specialty. He was the first to adopt a clinical-descriptive approach and remained extremely cautious of any psychological interpretation, to the extent he was criticized of creating a "psychiatry without psychology" [4].

In the late eighteenth and early nineteenth century, one of the most important advances in science and technology concerned electricity. Back in 1745, the "Leyden Jar" (Fig. 8.5) was invented independently by the German cleric Ewald Georg von Kleist (1700–1748) and by the Dutch scientist Pieter van Musschenbroek (1692–1761) of Leiden (Leyden). This device was the first successful attempt to store a high-voltage electric charge, and thus, it is the original form of a capacitor (condenser) and made easy the application of electrotherapy which was utilized already since the era of Galen. Now, no electric ray or fishes were necessary since the electric spark could be produced artificially. The next step was the development

Fig. 8.2 Joseph Guislain (1797–1860), bust made by Antoine Van Eenaeme, c. 1860, Belgian Royal Academy of Medicine

Fig. 8.3 Dorothea Lynde Dix (1802–1887), photo kept in the US Library of Congress

Fig. 8.4 Jean-Étienne Dominique Esquirol (1772–1840), portrait by Ch. Bazin, lithography by de Delpech, c.1838

Fig. 8.5 Discovery of the Leyden Jar by Pieter van Musschenbroek (1692–1761). from Augustin Privat Deschanel (1876) "Elementary Treatise on Natural Philosophy, Part 3: Electricity and Magnetism," D. Appleton and Co., New York, translated and edited by J. D. Everett, p. 570, Fig. 382 on Google Books

of "friction machines" (electrostatic generators) in the eighteenth century. One who quickly caught the utility of these developments was Abbe Jean-Antoine Nollet (1700–1770; Fig. 8.6) who used the devices for spectacular demonstrations but also for therapeutic reasons, especially on paralyzed body parts. Luigi Galvani

Fig. 8.6 Abbe Jean Antoine Nollet (1700–1770)

(1737–1798) developed his theory on animal electricity, and in 1803, his nephew and assistant Giovanni Aldini (1762–1834) managed to produce movements of the eyelids, lips, and eyes by experimenting on the exposed brains of oxen as well as of recently executed criminals [5]. Following these findings, in 1817, Karl August Weinhold (1782–1829) suggested that the brain was essentially an organic battery with attached wires. He went as far as to insist that he had managed to revive a dead kitten by removing its brain and filling the empty cavity with different metals (bimetallic electricity).

This whole idea triggered much enthusiasm and astonishment in the general public but also among scholars and intellects; it even inspired Mary Shelley (who was 19 years old at that time; 1797–1851) to write her masterpiece gothic horror novel "Frankenstein, the modern Prometheus" in 1816. In that novel, Dr. Frankenstein creates a human body through patchwork of remains from different human corpses and revives it with electric current gathered from a thunder [6]. After all, this was the modern version of the ancient story of the Titan Prometheus (in Greek "*Προμηθέας*" meaning "who thinks in advance, predicts"). That story while in ancient Greece concerned the stealing of fire from the gods by Prometheus, who gave it to humans (and was brutally punished by Zeus for this), its later Roman version suggested that Prometheus stole fire from the sun and used it to give life to human corpses. This later version of the myth was what Mary Shelley had in mind and historically was a Neoplatonic tale used to counterbalance Christian theology. As a consequence, Prometheus, Lucifer (who brings the light), and devil became one and the same figure in Christian tradition.

Fig. 8.7 Johann Christian August Heinroth (1773–1843), lithograph by Carl Lutherer c. 1835

Johann Christian August Heinroth (1773–1843; Fig. 8.7) was the first to occupy a professorship in psychiarty in 1811 at the University of Leipzig. He developed a theory of mind which included three structures corresponding to a lower (animalistic instincts), an intermediate (intelligence and self-awareness), and a superior (conscience) level of functioning. He suggested that conflict could exist between the priorities and the functioning of these layers, and this could constitute a very early psychodynamic conceptualization, bringing together the ideas of Plato concerning the "genera" ("*γένη*") and their often conflicting interactions with the topographic concept of the brain by Thomas Willis (see also Chaps. 3 and 6). In 1818, he retrieved and revived the term "paranoia" from ancient Greek texts including the works of Euripides, Aeschylus, Aristophanes, Aristotle, and Plato (from "*παρά*" meaning "beside, by" and "*νους*" meaning the mind) as well as from the works of de Sauvages [7, 8]. He labeled as "paranoia" the disorders of the intellect and "paranoia ecstasia" the disorders of feeling, and he introduced the concepts of "Wahnsinn" (insanity) and "Verrucktheit" (madness). He classified mental disorders into "hyperthymias," "asthenias," and "hypo-asthenias." In "hyperthymias," he included "melancholia erotica," "melancholia metamorphosis," and "melancholia saltans." Clearly mixed states dominate his works, and this is also the case in the writings of Joseph Guislain (1797–1860). Jean-Pierre Falret (1794–1870; Fig. 8.8) and Jules Gabriel Francois Baillarger (1809–1890; Fig. 8.9) established the connection between depression and mania. In 1851, Falret defined it as a single illness and gave it a name ("folie circulaire" or "folie à double forme"). In 1858, Frank Richarz introduced the term "agitated depression." Following them, Henry Maudsley (1835–1918; Fig. 8.10) coined the term "affective disorder" to describe milder forms of melancholia without psychotic features.

Another development at the dawn of the nineteenth century in Europe was that a new field of science emerged as the product of the combination of philosophy and life sciences. It was sociology and focused on "humanity as a group of individuals" rather than on each person separately and later led to the development of concepts like collective mind and mass psychology.

Jean Itard (1775–1843; Fig. 8.11) reported on Victor of Aveyron (c. 1788–1828; Fig. 8.12), a 12-year-old boy found living as a savage in the forests around Aveyron [9] possibly suffering from some type of autism [10], while later in 1825, he described the first case of Tourette syndrome in Marquise de Dampierre [11]. Later,

Fig. 8.8 Jean Pierre Falret (1794–1870)

Fig. 8.9 Jules Gabriel Francois Baillarger (1809–1890), Collection Bibliothèque de l'Académie nationale de médecine

in 1838, together with Esquirol, they introduced radical ideas in legislation and are considered to be the fathers of the new legal requirements in the Napoleonic Code in 1838 that not only provided an asylum in every region of France but also included detailed provisions for patient care. This law was in force until the late twentieth century. Even before that, the Napoleon criminal code of 1810 stipulated that "no crime or delict exists if committed in a state of dementia," with the word "dementia" standing for "mental illness."

During this period, unusual treatment options were frequent. During the eighteenth century, Mesmer had attributed mental illness to an imbalance of universal magnetic fluids and used devices in an effort to restore the equilibrium. He conducted a number of fascinating experiments and gained popularity (see Chap. 7). One of his followers, Amand-Marie-Jacques de Chastenet, the Marquis de Puysegur (1751–1825), performed public sessions of mesmerism to cure patients in group sessions, often in the squares of villages or other public places. He noticed that the real healing power was not magnetism but the healer's will to impose suggestions on the patients. During the nineteenth century, another mesmerist, the Scottish James Braid (1795–1860), believed that the true source of healing was the induction

Fig. 8.10 Henry Maudsley (1835–1918), the Wellcome Library c. 1881

of artificial sleep, and subsequently, he developed techniques to induce it. Thus, he discovered "hypnotism" a method of inducing sleep which took its name from the Greek word *"ύπνος"* ("hypnos" meaning sleep). In 1843, he published the book "Neurohypnology: or the Rationale of Nervous Sleep, Considered in Relation with Animal Magnetism." Another treatment which was usual at that time was the use of leeches for the treatment of nervous diseases in order to relieve cerebral congestion because of excess of blood. Often, they were applied to the temples to treat epilepsy. In 1813, Thomas Mapleson, which was serving as the official cupper to the English king, wrote of cupping particularly efficacious in the treatment of brain disorders.

Starting in the late eighteenth century, Franz Josef Gall (1758–1828; Fig. 8.13) accepted the theory of Johann Lavater (1741–1801) that the physical shape of the body and face could directly reflect specific psychological characteristics and eventually founded "phrenology." Phrenology was controversial at that time and today is considered to be an absolute "pseudoscience." However, it should be noted that Gall's work overall promoted psychology as a separate discipline, and it would be unfair to remember him exclusively because of phrenology.

Initially, Gall developed the theory of "Organology" (mind is composed of independent entities hosted by the brain) and the method of "Cranioscopy" (assessing personality and psychological characteristics by studying the external shape of

Fig. 8.11 Jean Itard (1775–1843), portrait by unknown artist, Institut National de Jeunes Sourds de Paris

Fig. 8.12 Victor, the salvage of Aveyron (1788–1828), Gallica Digital Library

Fig. 8.13 Franz Josef Gall (1758–1828)

the skull; Fig. 8.14). Therefore, he is considered the first to theorize on the localization of brain function (although in reality it was Emanuel Swedenborg; see also Chap. 7). Because of the opposition by the Catholic Church, he was prohibited to give lectures in his home country Austria, so he was obliged to move to Paris; however, also there, he was treated as a threat to the French culture [5]. Jean Pierre Flourens (1794–1867; Fig. 8.15), considered by many to be the father of experimental brain science, was a great opponent of Gall and argued against the localization of brain function. He supported his ideas with a number of experiments. The Gall-Flourens debate is a unique example of two men of science, one with the right ideas but ill methodology and reasoning while the other had the wrong idea but followed a strict and correct methodological approach. The reason why Flourens failed to detect localization in his experiments was probably because of the problematic study samples and possibly also because of the influence of his own opinion biases.

After his death, Gall's body was buried without a religious ceremony, and his skull together with his own collection of skulls is now at the Musée de l'Homme, as he had requested, while the major part of his collection is with the Rollett museum in Baden near Vienna [12].

The term "phrenology" was given to Gall's theories by his research partner Johann Gaspar Spurzheim (1776–1832; Fig. 8.16) [12]. Spurzheim went on to propose a significantly different system of cranioscopy and phrenology, which was

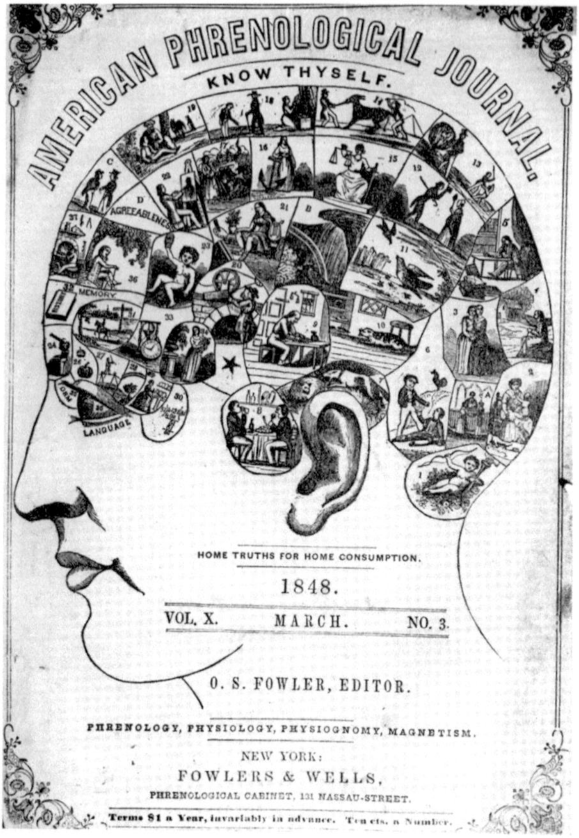

Fig. 8.14 Picture of the 1848 edition of American Phrenological Journal published by Fowlers and Wells, New York City. It shows the localization of the various psychological domains on the skull

more appealing to the general public and with more profound sociopolitical implications. Also, his skull is part of the Harvard Medical School collection.

By the mid-nineteenth century, phrenology was no longer an appealing theory, and its supporters were waning dramatically. The problem with Gall's ideas was that they were based on the study of the skull and never of the brain itself. He often neglected opposing data and was cherry-picking concerning observations. However, through his work, he significantly contributed to the methods of anatomical study of the brain which previously involved unsystematic, crude, and random slicing of the tissue [13–15].

As a consequence of the previously mentioned developments, at that time, the concept of "individual differences" emerged. While some type of mental and psychological testing existed already in ancient China for the selection of the imperial civil servants already two millennia BC, and this need for testing was also advocated by Plato concerning the military, no real progress in the field had occurred. But then, in 1816, the astronomer Friedrich Bessel (1784–1846; Fig. 8.17) read the book Zeitschrift für Astronomie [16], where he noticed the story of David Kinnebrook (1772–1802) who was an assistant astronomer at the Greenwich

Fig. 8.15 Jean Pierre Flourens (1794–1867), the father of experimental brain science

observatory in England and was fired on grounds of incompetence in 1796. The specific reason was that his longer reaction time by 800 msec when judging stellar transits in comparison to that of his superior; this difference led to erroneous astronomic calculations [17]. Bessel went on to study the issue and concluded that there are consistent differences in reaction times among individuals and suggested that each astronomer has a "personal equation" factor when recording stellar transit times [18]. This incident probably gave birth to experimental psychology, surprisingly from a man who was neither a physician nor a philosopher or a psychologist [19].

In 1822, Antoine Laurent Jessé Bayle (1799–1858) provided a comprehensive description of general paresis, often referred to as "maladie de Bayle." During that period, many patients living in asylums were suffering from this kind of organic mental disorder of syphilitic origin. This discovery that in reality they were suffering from a brain inflammation exerted significant influence on the conceptualization of mental disorders for the decades to come [20]. For the first time, the idea that an exogenous agent was causing inflammation of the brain gained solid scientific support.

The beginning of the nineteenth century was also very productive for philosophy. Georg Wilhelm Friedrich Hegel (1770–1831; Fig. 8.18) developed "absolute

Fig. 8.16 Johann Gaspar Spurzheim (1776–1832), portrait c. 1832

Fig. 8.17 Friedrich Bessel (1784–1846), portrait by Christian Albrecht Jensen c. 1839

Fig. 8.18 Georg Hegel (1770–1831)

idealism" which is an extreme form of "idealism." Idealism refers to the concept that the experience of reality by humans is a subjective mental construct. Absolute idealism breaks the boundaries between opposing concepts with "sublation" (in German: Aufhebung, integration without elimination or reduction), and thus, Hegel is considered as the originator of the thesis, antithesis, and synthesis triad. Essentially, idealism argues that the mind is superior and controls the body/brain. According to this view, matter (and the environment) is a product of the mind [21]. By following this line of thinking, from Plato to Leibniz and Kant, and by accepting a higher order of existence for the human soul in comparison to the material world, Hegel subsequently put absolute emphasis in human self-determination and free will. However, what is unique with Hegel is that instead of rejecting the Cartesian and Kantian dualism, he introduced a synthesizing approach of mind and brain (monism). He started with the definition of "being" and "nothing," and while Kant argued that humans can have knowledge only of a "finite world of appearances" (subjective), Hegel developed the idea that the "absolute" (monistic approach to the mind-brain problem) excludes multiplicity (dualism); therefore, subjective and objective are one but in the sense that the mind determines matter. In simple words, this means that the mind creates reality, and subsequently, reality is not perceived by the brain, but it is an absolute and complete product of the mind. Thus, "reality" was defined as "what determines itself." On the contrary, "finite" things (according to

Descartes) define each other by their borders and are essentially independent from the mind of the observer. Behind, this is idea that the "finite" (object) has to become "infinite" (subjective creation) in order to reach reality (subjectively created) [22]. In this frame, the brain and the mind are one with properties which are not under the laws of nature (a synthesis leading to monism). Hegel synthesized mind and body dualism in "true infinity," thus suggesting that it is possible to conceive it rather than accepting it without understanding. This idea was first introduced by Plato which thought that "ethics" transcend body restrictions. According to Hegel, the mind and the brain are neither incompatible nor compatible, but mind self-transcends the brain as a progressively developing and self-perfecting whole [22, 23]. See also Chap. 16 for details on idealism.

Hegel also put the foundations of social psychology with his work during the early years of the nineteenth century. His radical proposal was for the existence of a "social mind" ("Volksgeist" or spirit) in the sense of a complete and independent entity, in the form of a collective mind with individual persons being merely its agents. This world-mind, he believed, was controlling the function and the evolution of humanity as well as human society. In this frame, individual freedom was considered to be a schism from this collective, and thus instead of being "freedom of the individual," it was "submission of the group." Absolute priority was given to the needs of the group rather than the individual [24–26]. Hegel's philosophy is essentially an attempt to understand and conceptualize the entire universe in a unified theory grounded in Christian faith. His theories, especially the "social mind," were used by various political theorists ranging from the far right to the far left [27–29] and also inspired scientific research [23, 30–34]. The resemblance of Hegel's philosophy to religious beliefs with reference to a personalized God is obvious.

During the same period, the exactly opposite idea was developed by Søren Aabye Kierkegaard (1813–1855; Fig. 8.19) who argued that life is characterized by the struggle against two opposing forces: the wish for immortality and the reality of death. Connection with others is essential, and lack of it creates alienation which is a disturbing human condition. He believed that individuals become what they make of themselves through a series of choices. He divided choices into "inauthentic" (when individuals allow themselves to be shaped by the demands and expectations of the masses in what he called the "tyranny of the plebs") and "authentic" (when the individual assumes full responsibility of fate and existence). "Existential anxiety" develops when the person struggles to make authentic choices and avoid inauthentic ones.

Thus, in the early nineteenth century, a bipol of philosophical theories concerning human existence and consciousness and free will as well as the relationship of the individual with others and with the environment were developed. Hegel's theory suggested that individual humans are part of a collective (social) whole with metaphysical or even divine characteristics, with freedom constituting a violation of the natural order of things, while on the other hand, Kierkegaard argued that submission to the choices of the others results in "inauthentic" mental life, and the struggle between the wish for freedom and authenticity vs. keeping the connection with others creates existential anxiety.

Fig. 8.19 Søren Aabye Kierkegaard (1813–1855) portrait c. 1840

Karl Marx (1818–1883; Fig. 8.20) built his psychological theory on Hegel's work by radically modifying it. He maintained the monistic approach but substituted idealism with dialectic materialism; thus, matter and not any metaphysical mind was considered to have priority. Marx suggested that material labor is what connects the subjective human mind with the objects in the environment, instead of Hegel's collective social soul. By arguing on the basis of dialectic materialism, and by also incorporating some elements from the theories of Kierkegaard, Marxism proposes that the human nature exists as a function of labor and only through "meaningful labor" the person obtains the object and becomes fully human. Otherwise, humans are alienated from the environment with dire consequences [35, 36].

In this complex environment, with many fascinating socio-philosophical theories appearing and triumphing, many psychiatrists in France and Germany expressed caution concerning the biological etiopathogenesis of mental disorders. In Germany, the "Psychiker" school appeared, as a result of the influence of various religious, philosophical, and social theories, and adopted a radical dualistic position that there is an absolute distinction between mind and body, and even more, there is a primacy

Fig. 8.20 Karl Marx (1818–1883) c. 1875

of the soul over body. This was a post-Kantian German long-standing philosophical tradition, which argues for a sharp distinction between the “sciences of mind” (Geisteswissenschaften) and the “natural sciences” (Naturwissenschaften) [37]. This approach was championed among others by Johann Christian August Heinroth (1773–1843; Fig. 8.7), Karl Wilhelm Ideler (1795–1860), and Johann Gottfried Langermann (1768–1832). In this frame, mental disease had nothing to do with the body or biological phenomena, and therefore, the term “disease” has only a metaphoric value for psychiatry.

By following these sociopsychological avenues, psychiatry went on an independent pathway from the other medical specialties across Europe and America. However, the “Psychiker” school triggered a strong opposition by other psychiatrists in Germany, with Wilhelm Griesinger (1817–1868; Fig. 8.21) as the leading figure. He was a 28-year-old young and inexperienced German psychiatrist in 1845 when he wrote his book “Mental Pathology and Therapeutics” (*Die Pathologie und Therapie der psychischen Krankheiten)* which is considered to be “a scientific metaphysics of the mind combined with a practical, empirically tested medical therapeutics” [38–40]. This book and his conceptualization of mental disorders as neurobiological brain disorders would shape German and continental European psychiatry until today.

Fig. 8.21 Wilhelm Griesinger (1817–1868). Lithograph by von Georg Engelbach, c. 1865

Overall, Griesinger's work proposed an early biopsychosocial model and paved the way to modern-day psychiatric clinical practice. It is true that both his contemporaries as well as earlier histories viewed him as a synthesizer rather than an innovator. This was mainly because at that period, psychology was still part of philosophy [41], but Griesinger was not following the paradigm of experimental psychology which was dominant at that time [42–44].

The basis of his approach had existential elements and viewed the mind in a dynamic relationship with the brain, the environment, as well as the person's own cultural and personal past experience [38, 44, 45]. However, he went beyond classical idealism and included elements of pragmatic medical practice. Thus, he accommodated both "speculative philosophy" and "scientific empiricism" and argued for the need of a "healthy environment" as a prerequisite for mental health and wellness [42]. His approach essentially abandoned any absolute standard for identifying mental illness or its etiology. The major contribution of Griesinger was the reform of treatment of mental illness and of the asylum system and his promotion of the need to reintegrate patients back into the society. He was adamant in placing psychiatry within medicine and the sciences rather than moral therapeutics [45].

As a consequence, paradoxically, Germany did not follow the "Psychiker" school path, and psychiatry remained closely linked to neurology, following the tradition established by Griesinger. Neurology separated from the rest of medical specialties later, in the mid-nineteenth century, having as its focus the diagnosis and treatment of those nervous disorders with gross anatomical abnormalities, in sharp contrast with the "psychological" disorders which were the focus of psychiatry. This separation of two distinct medical specialties, with a common anatomical system to study and treat, reflects the nineteenth-century Cartesian dualistic approach of body and mind and would have a huge impact on the concept and treatment of mental disorders in the centuries to come.

In the area of public mental health, a number of changes characterize the first half of the nineteenth century. Most asylums were not under medical supervision before 1800, and in the UK, this was made mandatory by law only in 1828. On the other hand, being under medical supervision did not necessarily mean good care since many of the best places were run by laypeople such as the Quakers who run the Retreat at York. John Conolly (1794–1866; Fig. 8.22) started in 1839 a movement to abolish mechanical restraints but later also chemical restraints (bromide and chloral hydrate) at the New Middlesex County Lunatic Asylum at Hanwell in West London. His actions led in 1841 the Association of Medical Officers of Hospitals for the Insane (which later evolved into what today is known as the Royal College of Psychiatrists) to accept the nonrestraint practice as the official and acceptable care for the insane. Together with Robert Gardiner-Hill (1811–1878; Fig. 8.23), they attempted to replace mechanical and chemical restrains with surveillance and occupational therapy [46] and tried to transform asylums into self-sufficient communities with their own farm laundries and workshops.

In 1843, in the UK, the case of Daniel M'Naghten (1813–1865; Fig. 8.24) who, while suffering from persecutory delusions, committed murder, influenced English and American legal criteria for acquittal on the basis of insanity. He was a Scottish woodturner who assassinated English civil servant Edward Drummond in the mistaken belief he was Robert Peel, the Tory Prime Minister. He was ruled not guilty by reason of insanity and was committed to an insane asylum for the rest of his life [47–52]. His defense attorney was Sir Alexander Cockburn (1802–1880) (later Lord Chief Justice; Fig. 8.25) who utilized the hearing to denounce earlier similar trials. Sir Nicholas Conyngham Tindal (1776–1846; Fig. 8.25) as Chief Justice of Common Pleas was responsible for the inception of the verdict. This was followed by the 1845 Lunacy Act in England and Wales which recognized the status of "patients" for all persons with mental disorders (lunatics).

The same year 1843, James Braid (1795–1860) in England proved that the results of Mesmer's method were caused not by animal magnetism but by a subjective impression. He used the term "neurohypnotism" which was later shortened to "hypnosis" from the Greek word *"ύπνος"* (hypnos) which means sleep.

During the mid-nineteenth century, another unusual device was designed. This was the Utica Crib (Fig. 8.26). It took its name and was popularized in the United States in 1846 through its use at the New York State Lunatic Asylum at Utica. It was similar to a crib but with way less space. It had caged lid, and it was designed for

Fig. 8.22 John Conolly (1794–1866)

adults. Patients were confined in this narrow space for hours in order to calm down and sedate. When put inside, they could barely turn over. Reports suggested that overall, it was efficacious in calming down agitated patients, but it fell out of fashion after an interview by William Alexander Hammond (1828–1900) to the New York Sunday Herald in which he described the crib as a "barbarous and unscientific instrument." Later, Hammond would be the first to prescribe lithium. There were reports of deaths of patients after confinement in this crib.

In 1847, the Austrian Baron Ernst Von Feuchtersleben (1806–1849) coined the term "psychosis." While the ancient Greek word "ψύχωσις" originally meant "animation" as it was composed by the word "psyche" (soul) and suffix "-osis" (meaning the filling or the energizing, like apotheosis), German alienists of the time used the term to denote abnormal mental conditions [53]. By that time, the Greek suffix "-osis" was linked with disease and came to denote the presence of a pathological condition, for example, sclerosis, cirrhosis, or asbestosis [54]. It seems that the term "psychosis" was chosen as a euphemism in order to alleviate the stigma related to insanity. As Henry A. Davidson writes, educators "made valiant efforts to persuade medical students and practitioners to say 'psychotic' instead of 'insane,' promising that families, and the general public, would accept this with equanimity, whereas they would recoil at the harshness of the word 'insane'" [55].

Fig. 8.23 Robert Gardiner-Hill (1811–1878)

Around the same time, in 1847, Alfred Baring Garrod (1819–1907; Fig. 8.27) announced that he had discovered uric acid in the blood of gouty patients. Since, at least in the laboratory, lithium could dissolve uric acid crystals isolated from the kidneys, using lithium to dissolve urate in the body seemed a logical step. Garrod discovered that gouty uric acid deposits in finger joints are soluble in vitro in a lithium solution, and Alexander Ure (early nineteenth century–1866; Fig. 8.28) in 1843 introduced lithium into medicine by showing that in vitro, a uric acid bladder stone lost weight in a lithium carbonate solution. It is interesting that according to the works of Armand Trousseau (1801–1867) in France and Alexander Haig in the UK, mania and depression were thought to be related to the uric acid nosology (brain gout). Although Garrod made lithium treatment of gout widely known in 1859, unfortunately, it was proven that the levels of lithium needed to dissolve urate in the body were toxic [56–59].

In 1848, an interesting event occurred. Probably, it was not the first of its kind, but certainly, it was the first to be described in detail, to be registered in scientific journals, and to enjoy wide publicity. Chances are the descriptions were not accurate, and they were tailored to fit contrasting theories and approaches; nevertheless,

Fig. 8.24 Daniel M'Naghten (1813–1865) photo by Henry Hering c. 1856; Bethlem hospital museum archives

Fig. 8.25 Left: Sir Alexander James Edmund Cockburn (1802–1880) defended successfully Daniel M'Naghten (1813–1865). Right: Sir Nicholas Conyngham Tindal (1776–1846) who, as Chief Justice of Common Pleas, was responsible for the inception of the special verdict "Not guilty by reason of insanity" at the trial of Daniel M'Naghten

Fig. 8.26 The Utica Crib: Left as it has survived in the National Museum of Psychology, University of Akron, Ohio, US. Top right: In a late nineteenth century drawing. Bottom right: The Utica asylum

Fig. 8.27 Alfred Baring Garrod FRS (1819–1907), c. 1870

Fig. 8.28 Alexander Ure (early nineteenth century–1866)

it rose as a pivotal medical case report in the wider field of neurosciences. Phineas Gage (1823–1860; Fig. 8.29) was working in the construction of the railway for the Rutland and Burlington Railroad in New England south of the village of Cavendish, Vermont. During the process of preparing the blasting of a rock, an accidental explosion led an iron bar, which was approximately 3.2 cm thick and more than 1 meter long, to enter the left side of his face from below and through the upper jaw, and after passing behind the left eye and through the left frontal lobe, it fractured the top of the skull and went out completely. After surviving consequences of his injury, Gage tried to return to work after a few months, but it seemed that his intellect and emotional stability had deteriorated, and it was impossible for him to coordinate his team like before. Friends and relatives commented that his "personality" had completely changed. After repeated failures to regain control over his life, Gage died probably during a status epilepticus at the age of 37 [60–63].

Fig. 8.29 Left: Phineas P. Gage (1823–1860) shown holding the tamping iron which injured him (digitally enhanced and manually retouched; photograph by Jack and Beverly Wilgus of daguerreotype originally from their collection and now in the Warren Anatomical Museum, Center for the History of Medicine, Francis A. Countway Library of Medicine, Harvard Medical School. Enlarged and retouched by Joe Haythornthwaite). Right: The skull of Gage and the iron bar at the Warren Anatomical Museum; photographed c. 1868

References

1. Littlewood R, Lipsedge M. Aliens and alienists: ethnic minorities and psychiatry. London: Psychology Press; 1997.
2. Marneros A. Psychiatry's 200th birthday. Br J Psychiatry. 2008;193(1):1–3. https://doi.org/10.1192/bjp.bp.108.051367.
3. Shorter E. A history of psychiatry: from the era of the asylum to the age of Prozac. New York: Wiley; 1997.
4. Zilboorg G. A history of medical psychology. New York: Norton; 1941.
5. Finger S. Origins of neuroscience: a history of explorations into brain function. Oxford: Oxford University Press; 2001.
6. Finger S, Law MB. Karl august Weinhold and his "science" in the era of Mary Shelley's Frankenstein: experiments on electricity and the restoration of life. J Hist Med Allied Sci. 1998;53(2):161–80. https://doi.org/10.1093/jhmas/53.2.161.
7. Lewis A. Paranoia and paranoid: a historical perspective. Psychol Med. 1970;1(1):2–12. https://doi.org/10.1017/s0033291700039969.
8. Schifferdecker M, Peters UH. The origin of the concept of paranoia. Psychiatr Clin North Am. 1995;18(2):231–49.
9. Lane H. The wild boy of Aveyron. Cambridge: Harvard University Press; 1976.
10. Bettelheim B. Feral children and autistic children. Am J Sociol. 1959;64(5):455–67.
11. Kushner HI. Medical fictions: the case of the cursing marquise and the (re)construction of Gilles de la Tourette's syndrome. Bull Hist Med. 1995;69(2):224–54.

12. Finger S, Eling P. Franz Joseph Gall. New York: Oxford University Press; 2019. https://doi.org/10.1093/oso/9780190464622.001.0001.
13. Livianos-Aldana L, Rojo-Moreno L, Sierra-Sanmiguel P. F.J. gall and the phrenological movement. Am J Psychiatry. 2007;164(3):414. https://doi.org/10.1176/ajp.2007.164.3.414.
14. Temkin O. Gall and the phrenological movement. Bull Hist Med. 1947;21(3):275–321.
15. Van Wyhe J. The authority of human nature: the Schadellehre of Franz Joseph gall. Br J Hist Sci. 2002;35(124 Pt 1):17–42. https://doi.org/10.1017/s0007087401004599.
16. Lindenau B, Bohnenberger J (1816) Zeitschrift für Astronomie und verwandte Wissenschaften Tübingen.
17. Mollon JD, Perkins AJ. Errors of judgement at Greenwich in 1796. Nature. 1996;380(6570):101–2. https://doi.org/10.1038/380101a0.
18. Hoffmann C. Constant differences: Friedrich Wilhelm Bessel, the concept of the observer in early nineteenth-century practical astronomy and the history of the personal equation. Br J Hist Sci. 2007;40(3):333–65. https://doi.org/10.1017/s0007087407009478.
19. Boring E. A history of experimental psychology. New York: Appleton-Century-Crofts; 1950.
20. Brown EM. French psychiatry's initial reception of Bayle's discovery of general paresis of the insane. Bull Hist Med. 1994;68(2):235–53.
21. Solomon R. In the Spirit of Hegel. Oxford: Oxford University Press; 1985.
22. Miller A. Hegels science of logic. Amherst: Humanities Press International Inc.; 1993.
23. Mills J. Hegel on projective identification: implications for Klein, bion, and beyond. Psychoanal Rev. 2000;87(6):841–74.
24. Fuller BW. Back to Hegel? On Gillian Rose's critique of sociological reason. Br J Sociol. 2018;69(2):265–85. https://doi.org/10.1111/1468-4446.12281.
25. Gasparyan D. Mirror for the other: problem of the self in continental philosophy (from Hegel to Lacan). Integr Psychol Behav Sci. 2014;48(1):1–17. https://doi.org/10.1007/s12124-013-9247-x.
26. Yamamoto K. An alternative analysis of the discourse by Descartes, Kant and Hegel in terms of the ethical structure of the Kanun. Coll Antropol. 2015;39(2):317–34.
27. Michelini F. Hegel's notion of natural purpose. Stud Hist Phil Biol Biomed Sci. 2012;43(1):133–9. https://doi.org/10.1016/j.shpsc.2011.05.012.
28. Ter Meulen R. Solidarity, justice, and recognition of the other. Theor Med Bioeth. 2016;37(6):517–29. https://doi.org/10.1007/s11017-016-9387-3.
29. v d Luft E. Hegel, the Nazis, and child euthanasia. Arch Pediatr Adolesc Med. 1998;152(6):606–7.
30. Chessick RD. The relevance of Hegel's phenomenology of Spirit to the process of contemporary psychoanalysis. J Am Acad Psychoanal Dyn Psychiatry. 2008;36(4):677–706. https://doi.org/10.1521/jaap.2008.36.4.677.
31. Marchetti I, Koster EH. Brain and intersubjectivity: a Hegelian hypothesis on the self-other neurodynamics. Front Hum Neurosci. 2014;8:11. https://doi.org/10.3389/fnhum.2014.00011.
32. Michelini F, Wunsch M, Stederoth D. Philosophy of nature and organism's autonomy: on Hegel, Plessner and Jonas' theories of living beings. Hist Philos Life Sci. 2018;40(3):56. https://doi.org/10.1007/s40656-018-0212-3.
33. Saks V, Monge C, Guzun R. Philosophical basis and some historical aspects of systems biology: from Hegel to Noble - applications for bioenergetic research. Int J Mol Sci. 2009;10(3):1161–92. https://doi.org/10.3390/ijms10031161.
34. Zwart H. The oblique perspective: philosophical diagnostics of contemporary life sciences research. Life Sci Soc Policy. 2017;13(1):4. https://doi.org/10.1186/s40504-017-0047-9.
35. Lefever D, Lefever J. Marxian alienation and economic organization: an alternate view. Am Econ. 1977;21(2):40–8.
36. Marx K. Critique of Hegel's dialectic and philosophy in general. In: Easton L, Guddat K, Marx K, editors. Writings of the young Marx on philosophy and society. Indianapolis: Hackett Publishing Company, Inc.; 1997. p. 314–47.

37. Matthews E. Mental disorder: can Merleau-Pondy take us beyond the ‘mind-brain’ problem? In: Fulford KWM, Davies M, Richard GTG, et al., editors. The Oxford handbook of philosophy and psychiatry. Oxford: Oxford University Press; 2013. https://doi.org/10.1093/oxfordhb/9780199579563.001.0001.
38. Arens K. Wilhelm Griesinger: psychiatry between philosophy and praxis. Philos Psychiat Psychol. 1996;3(3):147–63.
39. Arens K. Wilhelm Griesinger: philosophy as the origin of a new psychiatry. In: Fulford KWM, Davies M, Richard GTG, et al., editors. The Oxford handbook of philosophy and psychiatry. Oxford: Oxford University Press; 2013. https://doi.org/10.1093/oxfordhb/9780199579563.001.0001.
40. Marx OM. Wilhelm Griesinger and the history of psychiatry: a reassessment. Bull Hist Med. 1972;46(6):519–44.
41. Arens K. Structures of knowing: psychologies of the nineteenth century. Boston studies in the philosophy of science, vol. 113. Dordrecht: Reidel; 1989.
42. Doerner K. Madmen and the bourgeoisie: a social history of insanity and psychiatry. Oxford: Basil Blackwell; 1981.
43. Ellenberger HF. The discovery of the unconscious: Th e history and evolution of dynamic psychiatry. New York: Basic Books; 1970.
44. Wahrig-Schmid B. Der junge Wilhelm Griesinger im Spannungsfeld zwischen Philosophie und Physiologie. Tübingen: Gunter Narr; 1986.
45. Verwey G. Wilhelm Griesinger: Psychiatrie als ärztlicher Humanismus. Nijmegen: Arts & Boeve; 2004.
46. Porter R. Madness: a brief history. Oxford: Oxford University Press; 2002.
47. Coleman AH, Davidson AT. M’Naghten rule: the right or wrong of criminal law. J Natl Med Assoc. 1978;70(8):599.
48. Diamond BL. Isaac ray and the trial of Daniel M’Naghten. Am J Psychiatry. 1956;112(8):651–6. https://doi.org/10.1176/ajp.112.8.651.
49. Quen JM. An historical view of the M’Naghten trial. Bull Hist Med. 1968;42(1):43–51.
50. Rosenberg CE. From M’Naghten to Hinckley: historians and the possibly insane offender. Essay review. Bull Hist Med. 1983;57(4):621–4.
51. Sauer RH, Mullens PM. The insanity defense: M’Naghten vs. ALI. 1976;4(1):73–5.
52. Slater E. The M’Naghten rules and modern concepts of responsibility. Br Med J. 1954;2(4890):713–8. https://doi.org/10.1136/bmj.2.4890.713.
53. Tuke DH. A dictionary of psychological medicine giving the definition, etymology and synonyms of the terms used in medical psychology with the symptoms, treatment, and pathology of insanity and the law of lunacy in Great Britain and Ireland, vol. 2. Philadelphia: P. Blakiston, sons, & co., Phil; 1892.
54. Zilboorg G, Henry GW. A history of medical psychology. New York: Norton; 1941.
55. Davidson HA. Psychiatry and euphemistic delusion. Am J Psychiatry. 1953;110(4):310–2. https://doi.org/10.1176/ajp.110.4.310.
56. Johnson FN, Amdisen A. The first era of lithium in medicine. An historical note. Pharmacopsychiatria. 1983;16(2):61–3. https://doi.org/10.1055/s-2007-1017450.
57. Marmol F. Lithium: bipolar disorder and neurodegenerative diseases possible cellular mechanisms of the therapeutic effects of lithium. Prog Neuro-Psychopharmacol Biol Psychiatry. 2008;32(8):1761–71. https://doi.org/10.1016/j.pnpbp.2008.08.012.
58. Shorter E. The history of lithium therapy. Bipolar Disord. 2009;11(Suppl 2):4–9. https://doi.org/10.1111/j.1399-5618.2009.00706.x.
59. Strobusch AD, Jefferson JW. The checkered history of lithium in medicine. Pharm Hist. 1980;22(2):72–6.
60. Damasio H, Grabowski T, Frank R, Galaburda AM, Damasio AR. The return of Phineas gage: clues about the brain from the skull of a famous patient. Science. 1994;264(5162):1102–5. https://doi.org/10.1126/science.8178168.

61. Harlow J. Passage of an iron rod through the head. Boston Medical and Surgical Journal. 1848;39:389–93.
62. Macmillan MB. A wonderful journey through skull and brains: the travels of Mr. Gage's tamping iron. Brain Cogn. 1986;5(1):67–107. https://doi.org/10.1016/0278-2626(86)90062-x.
63. Van Horn JD, Irimia A, Torgerson CM, Chambers MC, Kikinis R, Toga AW. Mapping connectivity damage in the case of Phineas gage. PLoS One. 2012;7(5):e37454. https://doi.org/10.1371/journal.pone.0037454.

Chapter 9
Nineteenth Century: The Realistic Second Half

Since the middle of the nineteenth century, there is an observable acceleration of developments in all facets and aspects of science and human life in general. New breaking-through technical equipment, new scientific theories, and even new disciplines appeared. Developments followed each other at an increasing speed, and different disciplines were influencing and triggering each other in a marvelous network with increasing complexity. As a consequence, the narration of the history of the period becomes very difficult, and often, it will seem almost fragmented; only later the significance of previous events and their interaction with other contemporary developments can be appreciated in full.

At that time, the debate concerning the presence or not of a "unitary psychosis theory" (einheitspsychose) appeared in the works of Joseph Guislain (1797–1860), Ernst Albrecht von Zeller (1804–1877), Wilhelm Griesinger (1817–1868; Fig. 8.21), Heinrich Neumann (1814–1888), Karl Ludwig Kahlbaum (1828–1899; Fig. 9.1), Ewald Hecker (1843–1909), Emanuel Mendel (1839–1907), and George M. Robertson [1–4]. The debate was eventually concluded with the description of manic depression and dementia praecox by Emil Kraepelin (1856–1926) in the early twentieth century [5].

In the next few decades, Kahlbaum would introduce the terms "hebephrenia" from Greek "*ήβη*" (ivi, meaning adolescence) and "*φρην*" or "*φρεν*" ("phrin" or "phren," meaning mind), catatonia from Greek "*κατά*" ("kata" meaning excessively) and "*τόνος*" ("tonos" meaning tone, flexibility, and reactivity of bodily and psychological function), and cyclothymia from Greek "*κύκλος*" ("kiklos" meaning cycle) and "*θυμικόν*" ("thymikon" meaning affective and emotional function) but with a different meaning in comparison to the modern. He attempted to classify mental disorders according to their longitudinal course and outcome. He gave his name to "Kahlbaum's syndrome" which refers to catatonia with prominent mood symptoms and stereotypical rhythmic repetition of words and sentences [6].

After 1850, a number of natural remedies extracted from plants including Atropa belladonna and Mandragora officinarum were increasingly used in asylums mainly

K. N. Fountoulakis, *Psychiatry*, https://doi.org/10.1007/978-3-030-86541-2_9

Fig. 9.1 Karl Ludwig Kahlbaum (1828–1899)

to calm down agitated and "difficult" patients. Starting in 1857, salts of bromine were used to treat epilepsy but also in mental patients, while in 1859, chloral hydrate, which is a sedative agent, was discovered. In 1882, paraldehyde was introduced as a major sedative for asylum patients. A number of novel discoveries were crucial in this increasingly pharmacological approach. In 1844, the Irish physician Francis Rynd (1801–1861) had invented the hollow needle, and in 1853, Charles Pravaz (1791–1853) and Alexander Wood (1817–1884) developed the first syringe and used it to make the first recorded subcutaneous injections to treat neuralgia.

This invention increased the popularity of opium which could now be injected. Richard Freiherr von Krafft-Ebing (1840–1902; Fig. 9.2) which was at the time professor of psychiatry in Graz, Austria, recommended opium in injections and as suppositories, especially in cases of melancholia and acute anxiety. Von Krafft-Ebing published several books, but the two most important were the "Lehrbuch der Psychiatrie" in 1879 and the "Psychopathia Sexualis" in 1882. He was the first to discuss issues of sexuality in scientific terms, and although he had the tendency to identify sexual perversion everywhere, his contribution to the opening of the field to scientific research was of outmost importance. He introduced the term "masochism" after the author Leopold Ritter von Sacher-Masoch (1836–1895), and he

Fig. 9.2 Richard Freiherr von Krafft-Ebing (1840–1902)

argued for the presence of pervasive psychopathology in all aspects of non-acceptable sexuality of Victorian times. Interestingly, while concerning sexuality he was in full accord with contemporary cultural preferences, he challenged the authority of the Catholic Church by suggesting that martyrdom (the desire for sanctity) is psychologically associated with hysterical features and masochism [7].

Already during the first decade of the second half of the nineteenth century, a number of new concepts, ideas, and approaches appeared. Claude Bernard (1838–1878; Fig. 9.3) was one of the greatest biologists of its time, and he observed that often the adaptive reactions could be more harmful than the insult. This observation led to the theory that many symptoms, especially mental, in essence reflect compensatory adaptations. It was the first time that a dynamic mechanism of action and reaction was proposed in a scientific way to explain the formulation of symptoms. He was also the first to propose the need for blind experiments in order to test hypotheses. In 1853 in the UK, Walter Cooper Dendy (1794–1871) was the first to introduce the term "psychotherapeia" (psychotherapy) from the Greek words "*ψυχή*" (soul) and "*θεραπεία*" (therapy, treatment). He defined it as a method to treat and prevent with means of psychological influence. Auguste Comte

Fig. 9.3 Claude Bernard (1838–1878)

(1798–1857) is considered to be the father of the doctrine of positivism and the first to coin the terms "sociology" and "altruism" in the 1850s. According to his theories, there is a coexistence of both altruistic and egocentric instincts, and social evolution is dominant and overall called for a science-based new social approach. He exerted major influence on many social theorists and philosophers including Karl Marx ([8], reprint). In 1854, in the United States, Thomas Kirkbride (1809–1883) published his work on what would be known as the Kirkbride Plan and proposed a standardization for buildings that housed mental patients [9].

In 1857, Benedict-Augustin Morel (1809–1873; Fig. 9.4) proposed degeneration as the main cause of mental disorders and considered it to be the consequence of both organic and social factors. He considered this degeneration to accumulate over the generations. He was the first to coin the term "dementia praecox." Unfortunately, his theories triggered a deep therapeutic nihilism in mental health professionals of the time and led to the further stigmatization of both patients, their families, and mental health professionals [10].

The first organized collective action and advocacy appeared in the UK in 1845 with the founding of the 'Alleged Lunatics' Friend Society by patients with

Fig. 9.4 Benedict-Augustin Morel (1809–1873)

adverse experiences from the treatments of that time [11]. Since 1851 and until 1860, patients committed to a New York Lunatic Asylum published an annual literary magazine, "The Opal," which painted mostly a rather benign picture of asylum life but also allowed the expression of dissatisfaction and even radical liberatory rhetoric sometimes [12, 13]. This was the first ever attempt to produce culture with mental illness as the essential source and to develop a kind of community identity, the predecessor of what later would develop into "Mad Pride" and "Mad Identity" (see Chap. 24). Following this, in 1859, George Robinson (1821–1875) published his book "On the Prevention and Treatment of Mental Disorders" in which he was arguing on preventive strategies outside the asylums and into the community. This was probably the first conceptualization of the model of social and community psychiatry, almost a century before its emergence [14].

Gustav Theodor Fechner (1801–1887; Fig. 9.5) is considered to be the father of psychophysics with his groundbreaking 1860 book "Elemente der Psychophysik" [15, 16]. He was preceded by Ernst Heinrich Weber (1795–1878; Fig. 9.6) who is

Fig. 9.5 Gustav Theodor Fechner (1801–1887)

considered to be the father of experimental psychology as was the first to quantitatively measure mental processes. One of Weber's major contributions was the concept of the "Just-Noticeable Difference" which suggests that humans do not perceive absolute differences between objects but the ratio of this difference to the magnitude of things compared (relative not absolute difference). He went on to describe in detail the threshold of this phenomenon for the different senses and developed what was later labeled as Weber's Law [17]. This was modified by Fechner and became the Weber-Fechner Law which determines that the intensity of a sensation increases arithmetically in relation with a geometrical increase of the stimulus. In 1838, the Fechner color effect was described, and in 1878, he developed the notion of the median. In 1876, he published the book "Vorschule der Aesthetik" where he used mathematical analysis to uncover the psychological mechanisms behind aesthetics. In this frame, he also described the concept of the "golden section" (or golden ratio) which corresponds to 0.62 (21 divided by 34) [18]. This ratio can be found in many masterpieces of art and contributes to their harmony and aesthetic value (Fig. 9.7). Fechner even published a satirical work on the comparative anatomy of angels [19].

Fig. 9.6 Ernst Heinrich Weber (1795–1878)

Paul Briquet (1796–1881; Fig. 9.8) published his book "Traite Clinique et Therapeutique a l'Hysterie" in 1859 and eventually gave his name (Briquet's syndrome or hysteria Briquet) to what today is labeled somatization disorder. He argued that neither married women were at higher risk nor men were at a lower risk to develop hysteria.

Paul Broca (1824–1880; Fig. 9.9) elaborated on the issue of the localization of brain function and the specialized role for the hemispheres. He argued that the frontal lobes were responsible for the highest of all cognitive functions (e.g., judgment, reflection, and abstraction), and subsequently, their volume could be indicative for higher intelligence. This came to solve the riddle of Cro-Magnon skulls with larger size in comparison to modern humans but with smaller frontal areas. It also served some racist concepts that the intelligence of Europeans was higher in comparison to other human races although some of them might have larger skull volumes [20, 21]. Such a debate emerged when in 1861, the anatomist and anthropologist Louis Pierre Gratiolet (1815–1865) presented to the Societe d'Anthropologie in Paris a skull of a Mexican Indian which was larger in comparison to European skulls [22].

In 1863, Gustave Dax sent a manuscript to the Academie de Medecine in Paris [23, 24] arguing that his long deceased father, Marc Dax (1770–1837), was the first

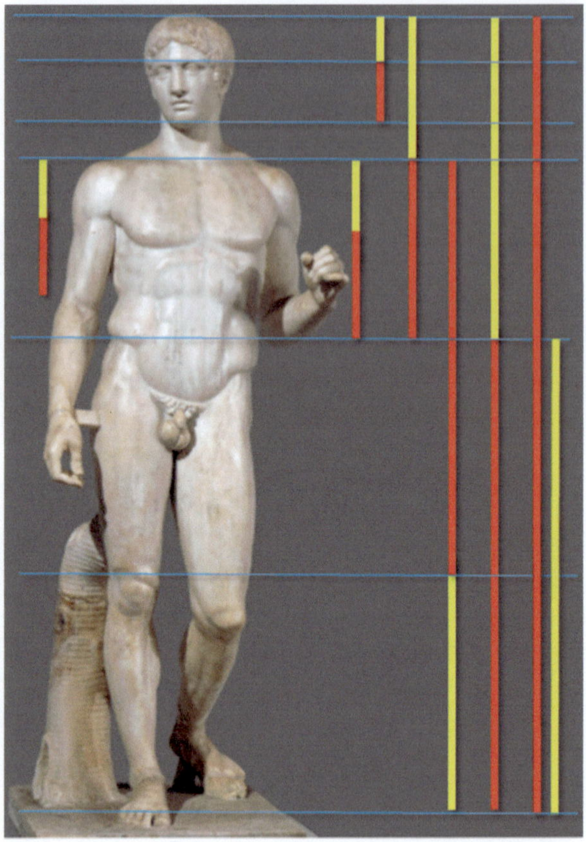

Fig. 9.7 The statue "doryphoros" (spear bearer) by Polykleitos c.440 BC. The harmony of the statue is based entirely on the utilization of the "golden ratio." In the figure, the golden ratio is obtained by dividing the length of the yellow lines to that of the red lines

to had recognized the special role of the left hemisphere for the production of speech and had presented his idea in a lecture during a congress in 1836. As already said, neither Dax nor Broca were the first to conceive that brain function is localized; however, over the years, only the name of Broca was credited with the advances in this particular field. Later in 1874, Carl Wernicke (1848–1905; Fig. 9.10) identified the location of a brain damage which lefts speech fluent but not meaningful. At that particular time period, when the distinct role of the two hemispheres was discussed, Robert Louis Stevenson (1850–1894) began to work on the famous "Strange Case of Dr. Jekyll and Mr. Hyde," published in 1886. Also in 1895, the Scottish psychiatrist Lewis Bruce published a clinical case study of a male with two distinct consciousnesses, each seemingly related to one of the hemispheres and completely unaware of each other [25, 26].

And then came the era of the great revolutions. The first was done by Charles Robert Darwin (1809–1882; Fig. 9.11) who published his theory of evolution in his 1860 book "On the Origin of Species" [27], and he continued working during the 1860s and the 1870s. He proposed a common ancestry for all species of life, and while at that time this as an exceptionally radical idea, today it is considered to be a

Fig. 9.8 Paul Briquet (1796–1881), Bibliothèque de l'Académie nationale de médecine of France

fundamental concept and a triumph of science, a unifying theory of the life sciences [28]. Central to his theory was the concept of "natural selection," a process through which evolution leads to more fit species. Overall, his work made the study of animals a model for the study of elements of human behavior [29]. His book "The Expression of the Emotions in Man and Animals" [30] triggered an interest in emotions from an evolutionary perspective. In that book, Darwin stressed the universal nature of emotions and the connection of mental states to the neurological organization of movement. Central to his understanding was a shared human and animal ancestry in sharp contrast to the contemporary claims that there were divinely created human muscles to express uniquely human feelings. Darwin's original suggestion was that emotions evolved via natural selection and therefore have cross-culturally universal counterparts, a proposal confirmed almost a century later. Furthermore, he argued that animals undergo emotions comparable to those of humans.

However, philosophy of the time continued to follow theoretical pathways and approaches, without any contact with hard scientific data. Charles Sanders Peirce (1839–1914) built on the ideas of Arthur Schopenhauer (1788–1860) and Friedrich

Fig. 9.9 Paul Broca (1824–1880)

Wilhelm Nietzsche (1844–1900) and, as a mathematician and philosopher of logic, conceived the mind as determined by its underlying function. The concept was rather too abstract and premature, taking into consideration the level of neurobiological knowledge of the time. However, to some extent, it constituted the basis of the works of William James and John Dewey.

William James (1842–1910; Fig. 9.12) is considered to be the father of American psychology and was a strong advocate of "functional psychology" which is based on the idea that process rather than structure is important in order to understand mental phenomena. This was a cardinal contribution and maybe the next important idea after Descartes, that is, "process" rather than "substance" is what constitutes the mind [31]. The theory was of course based on dualism and substituted the metaphysical "soul" with a mechanistic function, in an astonishing analogy to what today we would call hardware vs. software in computers.

James together with Peirce, they established the philosophical school of "pragmatism." Among his most influential ideas was the conceptualization of the "material," "social," "spiritual" self and the "pure ego" which provides the continuity of the self between past, present, and future and forms an identity. He considered this "pure ego" to be free of substance, and as such, it could not be the focus of scientific research [32]. Such a position was in accord with a version of psychophysical

Fig. 9.10 Carl Wernicke (1848–1905)

parallelism (a concept developed around the same period by Wundt), but James refused to further elaborate on the possible mode of interaction and the way mind and brain are synchronized. He argued that such an endeavor was premature and impossible given the current status of knowledge. He developed a two-stage model of free will with the two factors as "chance" and "choice" [33]. Later, in a series of lectures in 1884, he delineated and defined the terms and the ground on which future discussions on the issue of "free will" would take place, by talking on "hard" and "soft" determinism (later called compatibilism) [33, 34].

In the meantime and during the American civil war, Acting Assistant Surgeon Jacob Mendez Da Costa (1833–1900), while serving at Turner's Lane Hospital in Philadelphia, reported he had encountered with over 400 soldiers which manifested unusual heart problems, most usually racing heartbeat, without having been physically wounded in any way. Da Costa thought this was a heart problem with the "heart damaged by ill use" and named the syndrome "soldier's heart" [35, 36]. This was the first time that psychological trauma due to war experience was reported and was documented during the wars of the twentieth century [37–39], and today, we know that it is often accompanied by organic heart problems [40].

Fig. 9.11 Charles Robert Darwin (1809–1882)

During the same decade, Pliny Earle (1809–1892) was the first to become professor on the topic of mental health in the United States. In 1866–1867, he was the first person to address an audience exclusively of mental patients in any other than a religious discourse. He also pointed to the need for accurate hospital records and exposed the "curability delusion" of its time, that is, the false belief held by psychiatrists concerning the cure and favorable course of their patients. In 1871, he visited 46 institutions for the insane in Europe and wrote a memoir [41].

The second revolution occurred in Philadelphia, in 1870, and paved the way for the great psychopharmacological developments almost a century later. It was the first psychiatric indication for lithium which came from the neurologist Silas Weir Mitchell (1829–1914). Mitchell recommended lithium as anticonvulsant and hypnotic [42] and later for "general nervousness" [43]. Then in 1871, William Alexander Hammond (1828–1900; Fig. 9.13) was maybe the first to prescribe a modern and effective psychotropic agent, and this was lithium [44]. At that time, he was professor of Diseases of the Mind and Nervous System at the Bellevue Hospital Medical College in New York. A few years later, the Danish psychiatrist Carl Lange (1834–1900; Fig. 9.14) used lithium in the treatment of recurrent brief depression in 1886 while his brother Frederik Lange (1842–1907) used lithium in the treatment of 35 melancholic depressive patients (including some milder forms of bipolar disorder) in 1894 [45].

Fig. 9.12 William James (1842–1910) c. 1903

However, in spite of encouraging results, by the turn of the twentieth century, the "brain gout" theory of mood disorders disappeared as a medical entity, and the use of lithium in psychiatry was abandoned. This was also at least partially the result of the weakness of Danish psychiatrists to establish an innovative method among the colliding rocks of the French, German, and British psychiatry.

In 1869, Sir Francis Galton (1822–1911; Fig. 9.15) coined the phrase "nature vs. nurture" [46], and later, he was the first to suggest that all observed personality features are already incorporated in human language; thus, the study of language could give valuable information concerning personality function; this idea was later called "psycholexical theory" [47].

Jean-Martin Charcot (1825–1893; Fig. 9.16), who is considered to be the father of clinical neurology, became the head at the deteriorated Salpetriere women's hospital in 1852 (the same hospital where Pinel had released patients from their chains almost a century earlier). At that time, the state of the hospital was problematic with more than 5000 inmates and many of its buildings in ruins. The inmates were thrown together in an almost random way, and the system was in a desperate need for radical intervention. Together with Edmé Félix Alfred Vulpian (1826–1887; Fig. 9.17), they studied the clinical pictures of hospitalized women, and eventually in the 1870s and 1880s, Charcot published his studies on hysteria which were to

Fig. 9.13 William Alexander Hammond (1828–1900)

define the concept of hysteria for the next several decades. Charcot believed that only hysterics could be hypnotized and that hysteria could be equally found in males as in females. An important element was that at that time, Charcot also became interested in medical photography and used this new tool to document his observations (Fig. 9.18).

Charcot tried to identify different types of hysteria and to separate hysteria from epileptic fits, but this was proven difficult especially because many "hysteric" patients had seen epilepsy and they were mimicking it. Eventually, he came to believe that a hybrid of the two disorders, "hystero-epilepsy," exists. However, Joseph Delboeuf (1831–1896), after visiting Salpetriere, was convinced that all the observed remarkable pictures and elaborations were in fact performance by the patients guided by the staff and motivated by rewards and overestimations and misinterpretations by Charcot. This seemed to him especially true concerning the famous patient Marie "Blanche" Wittmann (1859–1913; the "Queen of Hysterics"; Fig. 9.19) [48–53].

During the 1880s, Charcot used hypnosis to study female patients and concluded that hypnosis is a state that occurs only in hysterics, but this was rejected by Ambroise-Auguste Liébeault (1823–1904) and Hippolyte Bernheim (1840–1919), who showed that most people could be hypnotized. They also showed that hypnosis could constitute a psychotherapeutic tool in neurotic disorders.

Fig. 9.14 Carl Georg Lange (1834–1900)

Although his ideas were heavily criticized later, at his time, Charcot was greatly appreciated and attracted a number of individuals as pupils, which later shaped neurology and psychiatry, including Sigmund Freud, Pierre Janet, Joseph Babinski, and Georges Gilles de la Tourette [54]. A beautiful painting of Charcot and his eminent pupils together with Wittmann was the "Clinical Lesson at the Salpêtrière," created by Pierre Aristide André Brouille in 1887 (Fig. 9.20). Ironically (?), a picture with a patient with tetanus in a similar body posture with that of Wittmann is included hanging on the wall in the left side of the painting.

Although seemingly unrelated, the third significant revolution was happening in the fields of physiology and basic pharmacology during the second half of the nineteenth century. While Wilhelm von Waldeyer (1836–1921; Fig. 9.21) coined the term "neuron" and promoted the "neuron theory" or the "neuron doctrine," the mechanism through which neurons could communicate was unknown. Until that time and for many decades to come, electrophysiology was considered to be the dominant function of nerve cells and the brain. In 1877, Emil Du Bois-Reymond

Fig. 9.15 Sir Francis Galton (1822–1911) c. 1850–1860

(1818–1896; Fig. 9.22) argued that nerves can excite muscles not only by means of electric current but also biochemically [50]. Notions of a receptive substance were first outlined by John Newport Langley (1852–1925) in 1878 [55]. These ideas were picked up by Alexander Crum Brown (1838–1922) and Sir Thomas Richard Fraser (1841–1920) in Edinburgh, who drew attention to the differences between L and D-hyoscine, the isomers of atropine. However, an even more important development came from an unexpected field: The Nobel Laureate Paul Ehrlich (1854–1915; Fig. 9.23) used a variety of dyes to stain different bacteria and tissues. Out of this work came the idea that agents related to the dyes might chemically bind selectively and specifically, and thus Ehrlich conceived the notion of the "magic bullet." This was to become the dominant therapeutic metaphor for the second half of the twentieth century, and it gave early currency to the idea of a receptor. In the same line of thinking, another Nobel Laureate, Hermann Emil Louis Fischer (1852–1919; Fig. 9.24) created the metaphor of the "lock and the key" concerning the reaction between messenger substances and receptors, in 1890 [56]. In 1893, George Oliver (1841–1915) was trying to invent an instrument to measure the diameter of arteries and utilized the extract of various animal glands in an effort to manipulate artery diameter. To the surprise of the scientific community, the experimental instrument

Fig. 9.16 Jean-Martin Charcot (1825–1893)

documented that the injection of adrenal gland extract caused a dramatic narrowing of the arteries diameter and a concomitant rise in blood pressure. An interesting detail was that Oliver performed these experiments on his younger son [50]. Approximately the same period, Theodor Meynert (1833–1892), a student of Wilhelm Griesinger, argued in favor of a neurobiological origin of mental disorders and entered in bitter conflicts with psychologically oriented authors whose approaches he considered to be unscientific.

The work of Wilhelm Dilthey (1833–1911), during the last quarter of the nineteenth century, concerned questions of scientific methodology. In an era of idealism being the dominant philosophical movement in Germany, Dilthey was mainly an empiricist [57]. He strongly rejected using also in human sciences a model formed exclusively from the natural sciences (Naturwissenschaften) and argued that the "scientific explanation of nature" (erklären) should be accompanied by a theory of how the world is perceived by humans. He then proposed the development of a separate model for the human sciences (Geisteswissenschaften). He based this thoughts on the ideas of the school of "Romantic hermeneutics" which were suggesting the use of the processes of "understanding" and "interpretation." These two processes were supposedly bridging the individual-psychological and the social-historical components of the world. Furthermore, Dilthey put a particular emphasis

Fig. 9.17 . Edmé Félix Alfred Vulpian (1826–1887)

Fig. 9.18 Albert Londe's photograph of a male Salpêtrière patient

Fig. 9.19 Marie "Blanche" Wittmann (1859–1913), the "Queen of Hysterics"

on "historicality" [58, 59], which he perceived as "a series of world views" as well as of a shared world [60]. Central to his arguments was the idea that the natural sciences utilize a "cause and effect" conceptualization, while in contrast, in human sciences, the aim is to understand the relationships between the part and the whole, that is, to understand human expressions of life [60]. For him, understanding was the key for the human sciences.

It is interesting that in psychology, he advocated on a double approach, that of the "explanatory psychology" based on a system of causality and that of "descriptive psychology," aiming to uncover how different mental processes converge in the "structural nexus of consciousness." He proposed the presence of an "objective mind" (*objektiver Geist*) in accord with Hegel's ideas for the presence of a collective mind, but while for Hegel, this was an entity though in a metaphysical concept, for Dilthey was the result of the existence of commonsense meanings and a core communality connecting the individuals into a cultural unity, but not a real entity by itself. By putting the emphasis on "commonsense meanings" as the basis for communication among individual persons, Dilthey's theory predicted that the more distance there is between simple and complex concepts, the more

Fig. 9.20 The painting "A Clinical Lesson at the Salpêtrière" by Pierre Aristide André Brouillet, c. 1887. The painting shows Jean-Martin Charcot demonstrating hypnosis on Marie "Blanche" Wittmann, who is supported by Dr. Joseph Babiński. Note the similarity to the illustration of opisthotonus (tetanus) on the back wall

insecurity of interpretation and understanding emerges, as a common process among many individual persons. His ideas had a profound influence on Karl Jaspers.

In 1879, Wilhelm Maximilian Wundt (1832–1920; Fig. 9.25), who was the first person ever to call himself a psychologist, founded the first laboratory for psychological research at the University of Leipzig, Germany. He also established the first academic journal for psychological research, the "Philosophische Studien" in 1881. In philosophical terms, he was opposing the theories of John Locke and English empiricism (sensualism) and was somewhat more in line with Immanuel Kant and Johann Friedrich Herbart (1776–1841), by arguing that many elements of intellect seem to be inherent, exist independently of, and are not included or caused by sensory input. He was far more influenced by the work of Gottfried Wilhelm Leibniz, and this influence is reflected specifically in almost all of his work. Wundt's contribution includes the conceptualization of mental phenomena as changing processes of consciousness and of humans as "motivated and thinking subjects." He introduced the notion of "Psychophysical parallelism" to argue that mental and physical phenomena are neither identical nor convertible into one another and simply coexist in a parallel way. He went further on to describe principles of mental causality (creative synthesis, relational analysis, mental contrasts, and heterogeny of purpose) in contrast to the natural causality of neurosciences. In principle, his ideas were reflecting a kind of dualism, as he believed that psychology cannot be reduced to

Fig. 9.21 Wilhelm von Waldeyer (1836–1921)

physiology, and this constituted a direct influence from Leibniz and was in accord with his contemporary William James [61–71].

In 1881, came the first attack on experimental neuroscience from the side of animal rights activists. While animal experiment led to significant advances in all fields of medicine, including neuroscience, the possible abusive nature of this kind of experimentation raised ethical questions already since the late eighteenth century. The opposition became stronger especially after the public demonstrations of experimenting on animals for the amusement of laypeople, and especially with the replication of the experiments of the "virtuosi" Robert Boyle (1627–1691) which manipulated the pressure of the air and induced asphyxia in animals [72–74]. A number of philosophers including Voltaire (1694–1778), Jean-Jacques Rousseau (1712–1778), and Arthur Schopenhauer (1788–1860) criticized the Cartesian concept of the "automaton" and argued for the rights of animals, thus not asking for human compassion only. Ethical questions concerned especially vivisection, that is, the cut of animals while alive. In general, animal experiments were considered justifiable after taking into consideration the benefit humans could gain concerning medical knowledge. During the nineteenth century, the explosion in medical

Fig. 9.22 Emil Du Bois-Reymond (1818–1896)

knowledge with the abandonment of the vitalistic theories and the replacement of the soul as the basis of life by biochemical mechanisms was based on animal experimentation, but as a consequence, it also demanded an ever increasing number of such experiments [75].

In Britain, the openly anti-vivisectionist positions of influential jurists, politicians, literary figures, clergymen, distinguished members of the medical community, and even of Queen Victoria contributed to an unfriendly environment for animal-based medical research and made Britain unable to catch up with the speed of medical developments in France and central Europe [76, 77]. On the contrary, Francois Magendie (1783–1855) and Claude Bernard (1813–1878; Fig. 9.3) in France went to extremes with vivisection, often causing unnecessary (and according to some, intentional) suffering to animals. Magendie became the iconic archvillain of the anti-vivisection movement [78]. As a consequence, the first law to protect animals passed in the UK in 1822 as the "Cruel Treatment of Cattle Act" while in 1876, a more specific "Cruelty to Animals Act" passed. This was the first law to regulate animal experiments, endured for 110 years, and followed the adoption of animal studies also by British physiologists during the second half of the nineteenth century. This change in scientific practice led to further intensification of the

Fig. 9.23 Paul Ehrlich (1854–1915), Nobel laureate 1908

anti-vivisectionist movement and resulted in the founding of the Victoria Street Society for the Protection of Animals Liable to Vivisection (later known as the National Anti-Vivisection Society) in 1875, led by the Irish feminist, suffragist, and animal advocate Frances Power Cobbe (1822–1904) [50, 79].

During the International Medical Congress of 1881 in London, and while a declaration on the need for animal experimentation was issued, the neuroscientist David Ferrier (1843–1928; Fig. 9.26) came under attack by the Victoria Street Society after an animal experiment he conducted during the congress and in front of an audience of foreign scientists. The society took Ferrier to court and the trial gained much publicity. It was the first time a neuroscientist was accused in such an aggressive way by animal rights activists. Although Ferrier was found not guilty, the psychological impact of this attack on the ranks of experimental neuroscientists was enormous [80].

At that time, during the late nineteenth century, a dark landscape in mental health was dominant. Mental asylums changed from places of progress and therapeutic intervention to places of custodial care, similar to prisons, but for people who have never committed any crime. The quality of the staff and the provided care were desperate. A dark picture is painted in the writings of the eminent psychoanalyst

Fig. 9.24 Hermann Emil Louis Fischer (1852–1919), Nobel laureate 1902

Ernest Jones (1879–1958) who describes alienists of the late nineteenth century as "glorified jailers" whose medical and ethical skills were not important for their recruitment [81, 82].

Partially, this happened because they were overcrowded and underfunded but also partially because psychiatrists and the other health professionals had developed a pessimistic view concerning mental illness and its treatment, after a century of failures with moral treatment and other traditional interventions. Already since 1857, Benedict-Augustin Morel (1809–1873; Fig. 9.4) argued for the degenerative nature of mental disorders causing therapeutic nihilism and spreading pessimism [10]. The result was similar across the European continent and the Americas. There were large institutions which hosted great numbers of insane people in unsanitary conditions, with multiple somatic diseases, and where despair and abuse was the rule rather than the exception. Therapeutic efforts had disappeared, and now the exclusive aim of the asylums was seclusion of patients.

This situation also defined much of the distinction between psychiatrists and neurologists during much of the nineteenth century. It concerned almost exclusively the severity of mental patients each treated; psychiatrists were dealing with the most

Fig. 9.25 Wilhelm Maximilian Wundt (1832–1920)

severe cases under long-term hospitalization in asylums, while the less severely ill outpatients were receiving diagnoses like hysteria or neurasthenia and were treated by neurologists as outpatients [83]. Subsequently, a clear somatic cause was considered to be behind the problems of those treated by neurologists, and this enjoyed a favorable reception by patients who could distinct themselves from the lunatics of the asylums. The term "nervous breakdown" became popular during the Edwardian age, and George Miller Beard (1839–1883; Fig. 9.27) coined the term "neurasthenia" in 1869. Eventually, no brain damage was identified, and progress seemed to face a dead end.

But at this dark moment, the wheels of history were set in motion once again. A number of seemingly unrelated advancements in distant fields occurred, and later, their consequences would converge to give birth to unimaginable progress several decades later. In one of them, in 1883, August Bernthsen (1855–1931; Fig. 9.28) in his laboratory in Heidelberg synthesized the phenothiazine nucleus. This nucleus gave the impression it consisted by three rings, but in fact, it consisted by two phenyl rings held together with a nitrogen and a sulfur atom. Bernthsen worked in the industry of dyes and studied especially dyes of the acridine and azine groups. One of them was methylene blue, which is both a dye and an antimalarial drug, is a

Fig. 9.26 David Ferrier (1843–1928)

phenothiazine derivative, and was synthesized in 1876 by Heinrich Caro [84]. Two years later, in 1885, in Japan, Nagayoshi Nagai (1844–1929) isolated ephedrine and phenylethylamine from Ephedra vulgaris and synthesized methamphetamine from ephedrine in 1893. The Romanian Lazar Edeleanu (1861–1941) synthesized amphetamine in Berlin in 1887. Ephedrine was rediscovered by Ku Kuei Chen (1898–1988) who developed his own method to isolate it in 1923. Ku Kuei Chen introduced ephedrine to the west and later served as head of pharmacological research at Eli Lilly and Company for 34 years. The same time, a then unknown researcher in Austria named Sigmund Freud (1856–1939), with a series of articles between the years 1885–1887, suggested that cocaine should be used as an antidepressant.

In the meantime, in March 1884, John Hughlings Jackson (1835–1911; Fig. 9.29) delivered the Croonian lectures at the Royal College of Physicians. The topic was on the evolution and dissolution of the nervous system, and in these lectures, Jackson rejected localization of symptoms in brain areas and argued that mental and

Fig. 9.27 George Miller Beard (1839–1883) c. 1870

nervous disorders were a matter of nervous system de-evolution or dissolution. He borrowed the term "dissolution" from the evolutionist Herbert Spencer (1820–1903). This assumption was based on the theoretical proposal that the nervous system is structured in a hierarchical way determined through Darwinian evolution. Essentially, this was an elaboration on the ideas of Thomas Willis (1621–1675; see also Chap. 6). This hierarchical structure included the more basic components of the primitive brain and the spinal cord, while phylogenetically newer components are built on top of more ancient ones. Eventually, he considered the frontal lobes to be the most advanced areas of the brain. According to Jackson, when higher-level centers are damaged, lower-level assume control, and as a result, symptoms emerge. These symptoms include those resulting from the loss of inhibition by higher-level centers ("positive symptoms") while another group of symptoms concerned loss of function because of the dysfunction of higher centers ("negative symptoms") [85–88]. This positive-negative distinction of symptoms was first conceptualized by Sir John Russell Reynolds (1828–1896) [89], and eventually, it was used to develop theories of schizophrenia in the late twentieth century.

Clinical psychopathology continued to enjoy major advancements during the last couple of decades of the nineteenth century. Paul Julius Moebius (1853–1907)

Fig. 9.28 August Bernthsen (1855–1931) c. 1883

was a German neurologist who had studied philosophy and theology. He wrote essays on the psychological profiles of Goethe, Rousseau, Schopenhauer, and Nietzsche. His major contribution to psychiatric classification efforts was his proposal on the distinction between exogenous and endogenous mental disorders. Emanuel Mendel (1839–1907) was the first to officially describe hypomania in 1881, as a milder form of mania, and this was in sharp contrast to the previous traditions which was focusing on chronic patients with poor prognosis. George M. Robertson in 1890 described the "hilarious" and the "furious" types of mania [4]. In 1889, Sergei Korsakoff (1853–1900; Fig. 9.30) described a syndrome of dysmnesia and confabulations in alcoholics, which would later hold his name as "Korsakoff's syndrome." This syndrome is characterized by inability in new learning and sustain attention as well as a tendency to falsify recollections (confabulations) as a means of compensating for memory gaps [90]. Karl Ludwig Kahlbaum (1828–1899; Fig. 9.31) and especially his pupil Ewald Hecker (1843–1909; 2.121) developed the concepts of hebephrenia and cyclothymia and proposed the existence of more than one discrete psychiatric disorder, in sharp contrast to the concept of "unitary psychosis," thus preparing the way for the separation between schizophrenia and manic depression. An interesting figure of

Fig. 9.29 John Hughlings Jackson (1835–1911) c. 1895

that time was Cesare Lombroso (1836–1909; Fig. 9.32) which was a supporter of the theory that criminals are hereditarily degenerate (criminal atavism). Additionally, he suggested that body measures and anthropological signs including shape of hands, eyes, or the jaw could serve as signs to unravel this underlying genetic deficit. His theory involved also the autonomic nervous system, and this led to the building of the first lie detector. He also believed that genius is closely related to madness, and in his 1888 book "The Man of Genius," he suggested that artistic genius was a form of hereditary insanity [91–96]. In 1892, Arnold Pick (1851–1924; Fig. 9.33) in Prague made the initial description of what would be named as Pick's disease and today is more often labeled as frontotemporal dementia.

In 1890, the Lunacy Act in England and Wales demanded that two medical certificates were now required for the detention of all patients. This is the first official recognition that there is both a need, but also it is possible to judge who should and should not be held responsible for their own actions. This development was controversial and caused much debate already since it emerged and the discussion continues until today [97].

Following the advances in clinical psychopathology, therapeutics enjoyed much attention during these last couple of decades of the nineteenth century. Some of them were preludes to the future while others were simply peculiar. Nevertheless,

Fig. 9.30 Sergei Korsakoff (1853–1900)

"efforts to find a cure" were ever increasing, heralding that the pessimism, which led asylums to decay, had not hampered research efforts in the field of mental health and therapeutics.

In 1889, Thomas Claye Shaw (1841–1927), a psychiatrist in a London asylum and lecturer in psychological medicine at St. Bartholomew's Hospital, asked surgeon Harrison Cripps to perform trephination in a patient suffering from neurosyphilis in order to relieve intracranial pressure. The reported success leads to a wave of similar operations on ground that relieving intracranial pressure was therapeutic, which lasted for a few years [98, 99]. The first modern account of psychosurgery was by Gottlieb Burckhardt (1936–1907) in 1891, who was the director of an institution for the insane at Prefargier, Switzerland. Burckhardt built on the idea of the Dutch psychiatrist Herman Boerhaave (1668–1738; see Chap. 7) who suggested that abnormally fixed electrical arrangements in the brain are responsible for mental illness. He reported that he conducted operations to excise regions of the cerebral cortex of the frontal, temporal, and parietal lobes. He also reported that the results were not overly encouraging; one patient experienced epileptic convulsions and died a few days after the operation, one improved temporarily but later committed suicide, two patients did not show any change, while another two patients became "quieter" [98, 100, 101].

Fig. 9.31 Left: Karl Ludwig Kahlbaum (1828–1899), c. 1890. Right: Ewald Hecker (1843–1909)

Fig. 9.32 Cesare Lombroso (1836–1909)

Fig. 9.33 Arnold Pick (1851–1924) c. 1920

One surprising development came just before the turn of the century. In 1896, Neil Macleod, a British physician in Shanghai, used sodium bromide to cure a female patient addicted to morphine and reported full remission of addiction after a long bromide-induced sleep. Subsequently, he reported success in eight out of nine patients he treated with bromide [102–104]. His work was replicated in the United States, and this method was about to be widely known as "deep sleep therapy." Earlier terms included "bromidism" and "bromide sleep." The enthusiasm concerning the early and believed-to-be miraculous treatment made the medical world to oversee the death of one of these nine patients which was attributed to pneumonia and considered death to be unrelated to bromide treatment [105]. The first death

Fig. 9.34 Richard Maurice Bucke (1837–1902)

attributed to the treatment itself was reported by Archibald Church who suggested that bromide should be given only at daytime and in facilities with hospital equipment, trained nurses, and a competent resident physician [106].

Together with true progress, a number of unfortunate and ethically problematic treatment approaches were introduced during the late nineteenth century. Richard Maurice Bucke (1837–1902; Fig. 9.34) while superintendent at the London Asylum for the Insane in Ontario, Canada, he advocated gynecological surgery on female patients. The basis of his theory was that there was some kind of "damage" or "displacement" of female reproductive organs that was producing insanity. Along with the local gynecologist Alfred Thomas Hobbs, they found that more than 85% of female patients had some kind of gynecological problem, probably due to the poor general health mental patients had at that time and to the harsh lives they were living. Subsequently, they performed 228 surgeries on female patients of which 16 hysterectomies, 12 removals of ovaries and tubes, 22 operations involving the replacing and retaining of the uterus in the normal position, 30 operations on the cervix, 21 minor uterine diseases, and 8 operations for vaginal lesions. Surprisingly, they reported impressive success since 71 out of 106 patients which were operated between 1895 and 1898 "either recovered their mental health or this was improved." However, the medical community was skeptical, and after Bucke's death, his theory and methods were quickly forgotten [107]. Later in his life, Bucke wrote a mystical book under the title "Cosmic Consciousness: A Study in the Evolution of the Human Mind" which was published in 1901.

Another interesting story concerns the manufacture of the electric vibrator which appeared in the late nineteenth century. Originally, it was a medical instrument for the alleviation of pain and the treatment of various disorders including hysteria, amenorrhea, arthritis, neuralgia, neurasthenia, irritability, indigestion, and constipation (Fig. 9.35). Joseph Mortimer Granville (1833–1900) is considered to be the inventor [108], but the first to use it was Marie Gabriel Romain Vigouroux (1831–1911) in 1878 at the Salpêtrière hospital in Paris [109]. Whether sexual massage was a treatment method used for centuries for the treatment of hysteria and eventually led to the manufacture of the vibrator is a matter of debate [109, 110]. Vigouroux established also a unit of electrotherapy at the Salpêtrière in the 1870s (Fig. 9.36).

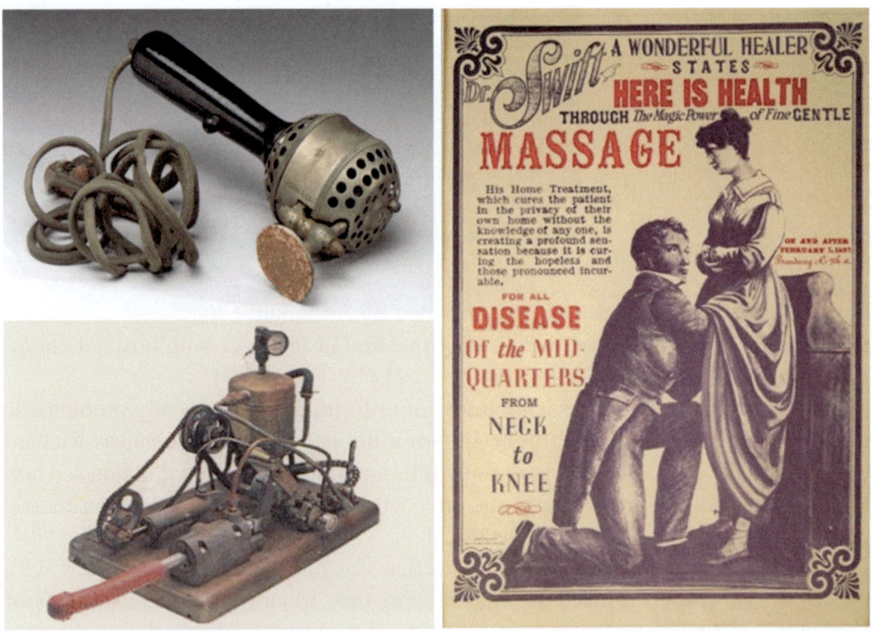

Fig. 9.35 Left up: Early vibrator devices designed for medical use (England c. 1909), the Science Museum Group Collection. Left down: Vibrator prototype named the "Manipulator" from the V.S.R. Institute's Museum of Unnatural Wonders. It was rejected by the US Patent Office in 1891. Right: Advertisement of vibrators

Fig. 9.36 Marie Gabriel Romain Vigouroux, (1831–1911) performing electrotherapy at the Salpêtrière; gravure by Daniel Vierge for 'Le Monde illustré' in 1887

Fig. 9.37 David Émile Durkheim (1858–1917)

Amidst sociopolitical and economic turmoil of the second half to the nineteenth century, social science flourished exponentially. One of the fathers of modern social science is considered to be David Émile Durkheim (1858–1917; Fig. 9.37) [111]. He developed an evolutionary theory of human society from "mechanical solidarity" to "organic solidarity." At first, he proposed that the total of beliefs and sentiments in a society create what he called "collective or common consciousness" and suggested that it has a life of its own. This was in accord with Hegel's ideas and had a strong metaphysical character. However, since Durkheim was not religious, the nature of this "consciousness" and "life" in his theories remains undetermined, but such controversies were the rule rather than the exception in philosophical thinking. Further on, his theory suggested that with increasing size, human societies are characterized by more complex interpersonal interactions and are more psychologically demanding. This leads the society to evolve from the phase of mechanical to organic solidarity (during which people are more self-sufficient, the society is less organized and fewer life options are available. Where people are much more interdependent, society is more organized and specialization of labor and cooperation is the rule rather than the exception). During this later phase, Durkheim suggested that the use of force and repression is essential in order to keep society together. A characteristic difference between these two phases according to Durkheim is the function of law, which in mechanical society is more punitive and aims to set an example while in organic society focuses on the person and on the repairing of the damage done [111, 112].

Concerning human psychological function, Durkheim argued that humans are inherently egoistic, but the collective consciousness (i.e., the shared values and the common belief system as well as human emotions) leads to social integration. As already said, Durkheim believed that this collective consciousness is literally alive [112]. It is through this collective consciousness that humans become aware of one another as social beings and differ from wild animals [112]. This seems to come directly from the teachings of Socrates as they were formulated in the writings of Plato, and it is in accord with Descartes, Leibniz, and Georg Hegel and proposes an alternative to the function of labor as proposed by Karl Marx.

When collective consciousness breaks down, then the state of "anomie" emerges, while when people are doing work, they are unsuited for the state of "forced division of labor" occurs. In both instances, the society is destabilized and people are unhappy [112]. In this frame, he suggested that crime could serve to release social tensions and exerts a cleansing or purging effect in society [113]. In his book "Suicide" in 1897, he proposed that the state of "anomie" is related to suicidality [114] and suggested that differences in observed suicide rates between Catholics and Protestants were due to the overall stronger social control among Catholics. He believed that suicide is a socially determined event rather than the result of stressful events on the person's life [111, 112]. He went further to create a normative theory of suicide and proposed the presence of four different types of suicide: egoistic (low level of social integration), altruistic (too much social integration), anomic (insufficient amount of social regulation), and fatalistic (too much social regulation).

A number of authors have criticized Durkheim's theories on suicidality in terms of ecological fallacy [115–118]; however, several others supported the sociological explanation of suicidality he introduced [115, 119–121].

During the last years of the century, Gerardus Heymans (1857–1930) followed Fechner's idea of psychic monism and identified three fundamental qualities of temperament: activity level, emotionality, and susceptibility to external vs. internal stimulation. Subsequently, he described eight character types (amorphous, apathetic, nervous, sentimental, sanguine, phlegmatic, choleric, and impassioned) which create what came to be known as "Heymans' cube" [122].

Finally, two major developments occurred just before the end of the nineteenth century. A young Viennese physician named Sigmund Freud (1856–1939; Fig. 9.38) began treating neurotic patients with hypnosis. Later, he developed the free thought association and the interpretation of dreams. In 1896, he coined the term "psychoanalysis" to define this new method of treatment. The term comes from the Greek word "ψυχή" (psichi, meaning soul) and "ανα" (ana, means again) and the verb "λύω" (lio, meaning open, free, solve, disassemble). Gradually, he started developing a theory on the structure and functioning of human mental life with special emphasis on the "unconscious" which was a concept known to philosophers since the seventeenth century, but it was introduced to psychiatry for the first time by Freud. However, the so-called "Dynamic Psychiatry" was originally conceived by Pierre Janet (1859–1947) at the Salpêtrière in Paris, and like Freud, he was also a student of Jean-Marie Charcot [123].

One year later, in 1897, Charles Scott Sherrington (1857–1952; Fig. 9.39), who later in 1932 would receive the Nobel Prize in Medicine and Physiology, coined the

Fig. 9.38 Sigmund Freud (1856–1939)

Fig. 9.39 Charles Scott Sherrington (1857–1952)

term “synapsis” (after the suggestion of English classical scholar Arthur Woollgar Verrall (1851–1912), for the junction between two neurons (later simplified to “synapse”). The term comes from the Greek words “συν” (“syn,” meaning “together”) and “άπτω” (“apto” meaning “connect”) from the Proto-Indo-European root “ap-” (meaning “to touch”).

References

1. Angst J. Historical aspects of the dichotomy between manic-depressive disorders and schizophrenia. Schizophr Res. 2002;57(1):5–13. https://doi.org/10.1016/s0920-9964(02)00328-6.
2. Berrios GE, Beer D. The notion of a unitary psychosis: a conceptual history. Hist Psychiatry. 1994;5(17 Pt 1):13–36. https://doi.org/10.1177/0957154X9400501702.
3. Moller HJ. Systematic of psychiatric disorders between categorical and dimensional approaches: Kraepelin’s dichotomy and beyond. Eur Arch Psychiatry Clin Neurosci. 2008;258(Suppl 2):48–73. https://doi.org/10.1007/s00406-008-2004-3.
4. Robertson G. Does mania include two distinct varieties of insanity, and should it be subdivided? J Nerv Ment Dis. 1890;36:338–47.
5. Kraepelin E. Manic-depressive insanity and paranoia. Edinburgh: Livingstone; 1921.

6. Peralta V, Cuesta MJ, Serrano JF, Mata I. The Kahlbaum syndrome: a study of its clinical validity, nosological status, and relationship with schizophrenia and mood disorder. Compr Psychiatry. 1997;38(1):61–7. https://doi.org/10.1016/s0010-440x(97)90055-9.
7. von Krafft-Ebing R. Psychopathia Sexualis: the classic study of deviant sex. Reprint ed. New York: Arcade; 2011.
8. Comte A. The positive philosophy of Auguste Comte freely translated and condensed by Harriet Martineau (original work published in 1855). New York: AMS Press; 1974. Reprint.
9. Kirkbride T. On the construction, organization, and general arrangements of hospitals for the insane. Philadelphia: Lindsay & Blakiston; 1854.
10. Morel B. Traite des degenerescences physiques, intellectuelles et morales de l'espece humaine: text and atlas. New York: Arno Press; 1857.
11. Hervey N. Advocacy or folly: the alleged Lunatics' friend society, 1845-63. Med Hist. 1986;30(3):245–75. https://doi.org/10.1017/s0025727300045701.
12. Reiss B. Letters from Asylumia: the opal and the cultural work of the lunatic asylum, 1851-1860. Am Lit Hist. 2004;16(1):1–28. https://doi.org/10.1093/alh/ajh001.
13. Tenney L. Who fancies to have a revolution here? The opal revisited (1851–1860). Radical Psychol. 2006;5.
14. Robinson G. On the prevention and treatment of mental disorders. London: Longman Brown Green Longman's and Robert's; 1859.
15. Fechner G. Elemente der Psychophysik 1860 (English transl by Helmut E. Adler). New York: Holt, Rinehart & Winston; 1966.
16. Robinson DK. Gustav Fechner: 150 years of Elemente der Psychophysik. Hist Psychol. 2010;13(4):409–10. https://doi.org/10.1037/a0021640.
17. Anonymous. Ernst Heinrich Weber (1795-1878) leipzig physiologist. JAMA. 1967;199(4):272–3.
18. Fechner GT. Vorschule der Aesthetik, Cambridge library collection - art and architecture, vol. 1. Cambridge: Cambridge University Press; 2013. https://doi.org/10.1017/cbo9781139854580.
19. Marshall ME. Gustav Fechner, Dr. Mises, and the comparative anatomy of angels. J Hist Behav Sci. 1969;5(1):39–58. https://doi.org/10.1002/1520-6696(196901)5:1<39::aid-jhbs2300050105>3.0.co;2-c.
20. Finger S. Origins of neuroscience. New York: Oxford University Press; 1994.
21. Finger S. Origins of neuroscience: a history of explorations into brain function. Oxford: Oxford University Press; 2001.
22. Schiller F. Paul Broca: founder of French anthropology, explorer of the brain. New York: Oxford University Press; 1979.
23. Finger S, Roe D. Gustave Dax and the early history of cerebral dominance. Arch Neurol. 1996;53(8):806–13. https://doi.org/10.1001/archneur.1996.00550080132021.
24. Roe D, Finger S. Gustave Dax and his fight for recognition: an overlooked chapter in the early history of cerebral dominance. J Hist Neurosci. 1996;5(3):228–40. https://doi.org/10.1080/09647049609525672.
25. Bruce L. Notes of a case of dual brain action. Brain Behav Immun. 1895;18:54–65.
26. Harrington A. Medicine, mind, and the double brain: a study in nineteenth-century thought. Princeton: Princeton University Press; 1987.
27. Darwin C. On the origin of species by means of natural selection: or the preservation of the favoured races in the struggle for life. On the origin of species by means of natural selection: or the preservation of the favoured races in the struggle for life. London: John Murray; 1860. https://doi.org/10.1037/14088-000.
28. Coyne J. Why evolution is true. Oxford: Oxford University Press; 2009.
29. Kandel E. The brain and behavior. In: Kandel E, Schwarts J, Jessell T, editors. Principles of neural science. New York: McGraw-Hill; 2000. p. 5–18.
30. Darwin C. The expression of the emotions in man and animals. The expression of the emotions in man and animals. London: John Murray; 1872. https://doi.org/10.1037/10001-000.

31. James W. The principles of psychology. New York: Henry Holt; 1890.
32. Cooper WE. William JAMES'S theory of the self. Monist. 1992;75(4):504–20.
33. Doyle B. Free will: the scandal in philosophy. Cambridge: Information Philosopher; 2011.
34. McDermott J. The writings of William James: a comprehensive edition. revised ed. Chicago: University of Chicago Press; 1977.
35. Tick E. PTSD: the sacred wound. Health Prog. 2013;94(3):14–22.
36. Wooley CF. Jacob Mendez DaCosta: medical teacher, clinician, and clinical investigator. Am J Cardiol. 1982;50(5):1145–8. https://doi.org/10.1016/0002-9149(82)90434-9.
37. Crocq MA, Crocq L. From shell shock and war neurosis to posttraumatic stress disorder: a history of psychotraumatology. Dialogues Clin Neurosci. 2000;2(1):47–55.
38. Dyde S. The chief seat of mischief: soldier's heart in the first world war. J Hist Med Allied Sci. 2011;66(2):216–48. https://doi.org/10.1093/jhmas/jrq026.
39. Howell JD. "Soldier's heart": the redefinition of heart disease and specialty formation in early twentieth-century Great Britain. Med Hist Suppl. 1985;5:34–52. https://doi.org/10.1017/s0025727300070502.
40. Pollard HB, Shivakumar C, Starr J, Eidelman O, Jacobowitz DM, Dalgard CL, Srivastava M, Wilkerson MD, Stein MB, Ursano RJ. "Soldier's heart": a genetic basis for elevated cardiovascular disease risk associated with post-traumatic stress disorder. Front Mol Neurosci. 2016;9:87. https://doi.org/10.3389/fnmol.2016.00087.
41. Frosch WA, Richardson D. Pliny Earle, M.D., 1809–1892. Am J Psychiatry. 2004;161(11):1989. https://doi.org/10.1176/appi.ajp.161.11.1989.
42. Mitchell S. On the use of bromide of lithium. Am J Med Sci. 1870;60:443–5.
43. Mitchell S. Clinical lecture on nervousness in the male. Med News. 1877;35:177–84.
44. Mitchell PB, Hadzi-Pavlovic D. Lithium treatment for bipolar disorder. Bull World Health Organ. 2000;78(4):515–7.
45. Lenox RH, Watson DG. Lithium and the brain: a psychopharmacological strategy to a molecular basis for manic depressive illness. Clin Chem. 1994;40(2):309–14.
46. Galton F. Hereditary genius. London: Macmillan; 1869.
47. Galton F. Measurement of character. Fortnightly Rev. 1884;36:179–85.
48. Chertok L. On the centenary of Charcot: hysteria, suggestibility and hypnosis. Br J Med Psychol. 1984;57(Pt 2):111–20. https://doi.org/10.1111/j.2044-8341.1984.tb01589.x.
49. Faber DP. Jean-Martin Charcot and the epilepsy/hysteria relationship. J Hist Neurosci. 1997;6(3):275–90. https://doi.org/10.1080/09647049709525714.
50. Finger S. Minds behind the brain: a history of the pioneers and their discoveries. New York: Oxford University Press; 2000.
51. Goetz CG. Charcot, hysteria, and simulated disorders. Handb Clin Neurol. 2016;139:11–23. https://doi.org/10.1016/B978-0-12-801772-2.00002-3.
52. Koehler PJ. Charcot, la salpetriere, and hysteria as represented in European literature. Prog Brain Res. 2013;206:93–122. https://doi.org/10.1016/B978-0-444-63364-4.00023-5.
53. Liberini P, Faglia L. Hysteria--100 years after Charcot. Can J Psychiatr. 1993;38(3):237–8. https://doi.org/10.1177/070674379303800318.
54. Boller F, Dalla Barba G. The evolution of psychiatry and neurology. In: Jeste D, Friedman J, editors. Psychiatry for neurologists. Totowa: Humana Press; 2006. p. 11–5. https://doi.org/10.1007/978-1-59259-960-8_2.
55. Maehle AH. "Receptive substances": John Newport Langley (1852-1925) and his path to a receptor theory of drug action. Med Hist. 2004;48(2):153–74. https://doi.org/10.1017/s0025727300000090.
56. Lemieux RU, Spohr U. How Emil Fischer was led to the lock and key concept for enzyme specificity. Adv Carbohydr Chem Biochem. 1994;50:1–20.
57. Rickman H. Wilhelm Dilthey, pioneer of the human studies. Berkeley: University of California Press; 1979.
58. Dilthey W. The formation of the historical world in the human sciences. Princeton: Princeton University Press; 2002.

59. Koslowski P, editor. The discovery of historicity in German idealism and Historism. New York: Springer; 2006.
60. Palmer R. Hermeneutics. Evanston: Northwestern University Press; 1969.
61. Araujo Sde F. Why did Wundt abandon his early theory of the unconscious? Towards a new interpretation of Wundt's psychological project. Hist Psychol. 2012;15(1):33–49. https://doi.org/10.1037/a0024478.
62. Bringmann WG, Balance WD, Evans RB. Wilhelm Wundt 1832-1920: a brief biographical sketch. J Hist Behav Sci. 1975;11(3):287–97. https://doi.org/10.1002/1520-6696(197507)11:3<287::aid-jhbs2300110309>3.0.co;2-l.
63. Greenwood JD. Wundt, Volkerpsychologie, and experimental social psychology. Hist Psychol. 2003;6(1):70–88.
64. Leary DE. Wundt and after: psychology's shifting relations with the natural sciences, social sciences, and philosophy. J Hist Behav Sci. 1979;15(3):231–41. https://doi.org/10.1002/1520-6696(197907)15:3<231::aid-jhbs2300150304>3.0.co;2-v.
65. Mulberger A. Wundt contested: the first crisis declaration in psychology. Stud Hist Phil Biol Biomed Sci. 2012;43(2):434–44. https://doi.org/10.1016/j.shpsc.2011.11.002.
66. Popple AV, Levi DM. Wundt versus Galton--two approaches to gathering psychophysical measurements. Perception. 2000;29(4):379–81. https://doi.org/10.1068/p2904ed.
67. Titchener EB. Wilhelm Wundt, I832-I920. Science. 1920;52(1352):500–2. https://doi.org/10.1126/science.52.1352.500.
68. Wassmann C. Physiological optics, cognition and emotion: a novel look at the early work of Wilhelm Wundt. J Hist Med Allied Sci. 2009;64(2):213–49. https://doi.org/10.1093/jhmas/jrn058.
69. Wong WC. Retracing the footsteps of Wilhelm Wundt: explorations in the disciplinary frontiers of psychology and in Volkerpsychologie. Hist Psychol. 2009;12(4):229–65. https://doi.org/10.1037/a0017711.
70. Wundt W. Principles of physiological psychology (Reprint of original 1874 publication). New York: Kraus Reprint; 1969.
71. Ziche P. Neuroscience in its context. Neuroscience and psychology in the work of Wilhelm Wundt. Physis Riv Int Stor Sci. 1999;36(2):407–29.
72. Festing S. Animal experiments: the long debate. New Sci. 1989;121(1649):51–4.
73. Maehle A, Tröhler U. Animal experimentation from antiquity to the end of the eighteenth century: attitudes and arguments. In: Rupke NA, editor. Vivisection in historical perspective. London: Croom Helm; 1987.
74. Maehle AH. Literary responses to animal experimentation in seventeenth- and eighteenth-century Britain. Med Hist. 1990;34(1):27–51. https://doi.org/10.1017/s0025727300050250.
75. Elliot P. Vivisection and the emergence of experimental physiology in nineteenth-century France. In: Rupke NA, editor. Vivisection in historical perspective. London: Croom Helm; 1987.
76. Berkowitz C. Disputed discovery: vivisection and experiment in the 19th century. Endeavour. 2006;30(3):98–102. https://doi.org/10.1016/j.endeavour.2006.07.001.
77. Preece R. The history of animal ethics in western culture. In: Blazina C, Boyra G, Shen-Miller D, editors. The psychology of the human-animal bond: a resource for clinicians and researchers. New York: Springer; 2011.
78. Guerrini A. Animal experiments and antivivisection debates in the 1820s. In: King CK, Goodall JR, editors. Frankenstein's science: experimentation and discovery in romantic culture. Burlington: Ashgate; 2008. p. 1780–830.
79. Franco NH. Animal experiments in biomedical research: a historical perspective. Animals (Basel). 2013;3(1):238–73. https://doi.org/10.3390/ani3010238.
80. French R. Antivivisection and medical science in Victorian society. Princeton: Princeton University Press; 1975.
81. Shepherd B. A war of nerves: soldiers and psychiatrists 1914–1994. London: Pimlico; 2002.

82. Turner TH. The Early 1900s and before…. In: Freeman H, editor. A century of psychiatry. London: Mosby; 1999.
83. Wessely S, Lutz T. Neurasthenia. In: Berrios G, Porter R, editors. History of clinical psychiatry. London: Athlone Press; 1995.
84. Bernthsen A. Zur Kenntnis des Methylenblau und verwandter Farbstoffe. Chemische Berichte (Berichte der Deutschen Chemischen Gesellschaft). 1883;16:2896–904.
85. Berrios GE. Positive and negative symptoms and Jackson. A conceptual history. Arch Gen Psychiatry. 1985;42(1):95–7. https://doi.org/10.1001/archpsyc.1985.01790240097011.
86. Berrios GE. The factors of insanities: J. Hughlings Jackson. Classic text no. 47. Hist Psychiatry. 2001;12(47 Pt 3):353–73. https://doi.org/10.1177/0957154x0101204705.
87. Berrios GE. Historical epistemology of the body-mind interaction in psychiatry. Dialogues Clin Neurosci. 2018;20(1):5–13.
88. An Introduction to the Life and Work of John Hughlings Jackson: Introduction. Med Hist Suppl. 2007;(26):3–34.
89. Pearce JM. Positive and negative cerebral symptoms: the roles of Russell Reynolds and Hughlings Jackson. J Neurol Neurosurg Psychiatry. 2004;75(8):1148. https://doi.org/10.1136/jnnp.2004.038422.
90. Korsakov SS. Korsakov's psychosis (1970). JAMA. 1853–1900;212(10):1700.
91. Carra G, Barale F. Cesare Lombroso, M.D., 1835-1909. Am J Psychiatry. 2004;161(4):624. https://doi.org/10.1176/ajp.161.4.624.
92. Gatti U, Verde A. Cesare Lombroso: methodological ambiguities and brilliant intuitions. Int J Law Psychiatry. 2012;35(1):19–26. https://doi.org/10.1016/j.ijlp.2011.11.004.
93. Gibson M. Forensic psychiatry and the birth of the criminal insane asylum in modern Italy. Int J Law Psychiatry. 2014;37(1):117–26. https://doi.org/10.1016/j.ijlp.2013.09.011.
94. Mazzarello P. Cesare Lombroso: an anthropologist between evolution and degeneration. Funct Neurol. 2011;26(2):97–101.
95. Palmer K. In memoriam: Cesare T. Lombroso. Neurodiagn J. 2014;54(2):105–9.
96. Rust RS. Cesare Lombroso MD, PhD (1917-2013). Pediatr Neurol. 2014;51(1):5–9. https://doi.org/10.1016/j.pediatrneurol.2014.04.020.
97. Hervey N. A slavish bowing down: The Lunacy Commission and the psychiatric profession 1845-4860. In: Bynum W, Porter B, Shepherd M, editors. Anatomy of madness, Vol II: institutions and society. London: Rutledge; 1985.
98. Berrios GE. The origins of psychosurgery: Shaw, Burckhardt and Moniz. Hist Psychiatry. 1997;8(29 Pt 1):61–81. https://doi.org/10.1177/0957154X9700802905.
99. Shorter E. A historical dictionary of psychiatry. Oxford: Oxford University Press; 2005.
100. O'Callaghan M, Carroll D. Psychosurgery: a scientific analysis. Lancaster: MTP Press Limited, Falcon House; 1982.
101. Stone JL. Dr. Gottlieb Burckhardt--the pioneer of psychosurgery. J Hist Neurosci. 2001;10(1):79–92. https://doi.org/10.1076/jhin.10.1.79.5634.
102. Macleod N. Morphine habit of long standing cured by bromide poisoning. Br Med J. 1897;2(1906):76–7. https://doi.org/10.1136/bmj.2.1906.76.
103. Macleod N. Cure of morphine, chloral, and cocaine habits by sodium bromide. Br Med J. 1899;1(1998):896–8. https://doi.org/10.1136/bmj.1.1998.896.
104. Macleod N. The bromide sleep: a new departure in the treatment of acute mania. Br Med J. 1900;1(2038):134–6. https://doi.org/10.1136/bmj.1.2038.134.
105. Bromide sleep. Merck's Arch. 1900;2(3):110–1.
106. Church A. The treatment of the opium habit by the bromide. Illinois Med J. 1900;50(7):291–7.
107. Cook R. Dictionary of Canadian biography, vol. 13. Toronto: University of Toronto Press; 1994.
108. Mortimer J. Nerve-vibration and excitation as agents in the treatment of functional disorder and organic disease. London: Churchill; 1883.
109. Maines R. The technology of orgasm. Baltimore: Johns Hopkins University Press; 1999.

110. King H. Galen and the widow: towards a history of therapeutic masturbation in ancient gynaecology. EuGeStA J Gender Stud Antiquity. 2011;1:205–35.
111. Calhoun C. Classical sociological theory. Hoboken: Wiley-Blackwell; 2002.
112. Allan K. Explorations in classical sociological theory: seeing the social world. Thousand Oaks: Pine Forge Press; 2005.
113. Durkheim E. The rules of sociological method (1895). In: Appelrouth S, Desfor L, editors. Classical and contemporary sociological theory: text and readings. Thousand Oaks: Pine Forge Press; 2007. p. 95–102.
114. Durkheim E. The division of labour in society (translation by Simpson G). New York: The Free Press; 1993.
115. Berk BB. Macro-micro relationships in Durkheim's analysis of egoistic suicide. Sociol Theory. 2016;24(1):58–80. https://doi.org/10.1111/j.0735-2751.2006.00264.x.
116. Freedman D. The ecological fallacy; 2002. https://www.stat.berkeley.edu/~census/ecofall.txt.
117. Selvin H. Durkheim's suicide: further thoughts on a methodological classic. In: Nisbet R, editor. Émile Durkheim. Englewood Cliffs: Prentice Hall; 1965. p. 113–36.
118. van Poppel F, Day LH. A test of Durkheim's theory of suicide--without committing the "ecological fallacy". Am Sociol Rev. 1996;61(3):500–7. https://doi.org/10.2307/2096361.
119. Gibbs JP, Martin WT. A theory of status integration and its relationship to suicide. Am Sociol Rev. 1958;23(2):140–7. https://doi.org/10.2307/2088997.
120. Inkeles A. Personality and social structure. In: Merton R, Broom L, Cottrell L, editors. Sociological today. New York: Basic Books; 1959. p. 249–76.
121. Johnson BD. Durkheim's one cause of suicide. Am Sociol Rev. 1965;30(6):875–86. https://doi.org/10.2307/2090966.
122. Millon T. Masters of the mind: exploring the story of mental illness from ancient times to the new millennium. Hoboken: Wiley; 2004.
123. Brown P. Pierre Janet (1959–1947). In: Freeman H, editor. A century of psychiatry. London: Mosby; 1999.

Chapter 10
Twentieth Century, the First Half: The Pre-psychopharmacologic Era

During the first half of the twentieth century, advances were accelerated and the field expanded dramatically, making very difficult the narration of events and developments in a simultaneous temporal and cause-and-effect relationship. The main aim of the author was to keep the narration of events in a chronological order and a secondary was to cluster them by topic, when such a clustering does not profoundly disturb the chronological order in the narration.

The twentieth century began with the publication of "The Interpretation of Dreams" by Sigmund Freud, in 1900. It triggered a revolution which spread beyond psychiatry to philosophy, sociology, and lay culture, often with malignant effects [1, 2]. It should be recognized, however, that the major contribution of psychoanalysis was that it spread a rejection of the metaphysical in the wider human culture concerning the interpretation of behavior. However, on the other hand, it was not scientific in nature, not even in terms of the early nineteenth century. That first book of Freud was followed by the publication of the "Three Essays on the Theory of Sexuality" in 1905. The psychoanalytical theory was now in place, and in the years that followed, several prominent psychoanalysts elaborated on this theory, including Pierre Janet (1859–1947), Alfred Adler (1870–1937), Sandor Ferenczi (1873–1933), Carl Gustav Jung (1875–1961), Karl Abraham (1877–1925), Otto Rank (1884–1939), Karen Horney (1885–1952), Sandor Rado (1890–1972), and the controversial and often irrational Wilhelm Reich (1897–1957) (Figs. 10.1 and 10.2).

In sharp contrast to psychoanalytical approaches, during the same period, Edmund Gustav Albrecht Husserl (1859–1938; Fig. 10.3) founded the philosophical school of phenomenology [3, 4]. His basic proposal was that humans experience representations of objects, not objects themselves. This was an elaboration on the ideas proposed by Kant. Such an approach suggests that there is no objective perception of nature but a subjective psychological representation through the senses and after process by the brain (transcendental subjectivity). Starting in 1900–1901, he developed the "epoche" method (from the Greek word "*αποχή*" meaning "to abstain" from everyday beliefs and common biases) so that the internal

K. N. Fountoulakis, *Psychiatry*, https://doi.org/10.1007/978-3-030-86541-2_10

Fig. 10.1 Group photo 1909 in front of Clark University, Worcester, Massachusetts. Front row: Sigmund Freud, G. Stanley Hall, and Carl Jung. Back row: Abraham A. Brill, Ernest Jones, and Sándor Ferenczi

Fig. 10.2 The inner circle of devoted followers "Committee") of the Vienna Psychoanalytic Society. Left to right seated: Freud, Sàndor Ferenczi, and Hanns Sachs. Standing: Otto Rank, Karl Abraham, Max Eitingon, and Ernest Jones. Berlin, 1922 Becker Maas, photographic studio. Prints and& Photographs Division. Library of Congress

representation of objects would be as less subjective as possible. This constitutes a major contribution to the world of scientific inquiry, since it dictates that the observer should be genuinely unbiased. Building on Leibniz "monads," he rejected the concept of solipsism (the self is all that can be known to exist) and proposed that there is a "sphere of ownness," but beyond it, the transcendental ego finds itself always paired with another ego, which refers to another "monad." This "a priori" connection is called "transcendental intersubjectivity" and scientifically is of problematic origin and nature, but it implies that the "self" as we conceive it cannot exist in absolute solitude. In Descartes, it is God that bridges different beings, but according

Fig. 10.3 Edmund Gustav Albrecht Husserl (1859–1938)

to Husserl, community between persons is an inherent property and ability of the ego itself, mainly by modeling of the "other" with high degrees of speculation and conjectural reciprocity. Additionally, he defined mental life and mental functions in terms of intentionality, a concept he borrowed from Franz Brentano (1838–1917). According to this approach, in contrast to physical phenomena, mental phenomena are characterized by having intentionality toward an object [4–10].

Edward Bradford Titchener (1867–1927; Fig. 10.4) followed the ideas of Wundt and especially built on his concepts of association and apperception. He suggested that both ideas and perceptions are products of sensations and described four properties for sensations: intensity, quality, duration, and extent. He also believed that distinct basic components of the mind determine mental functions. This was later called "structuralism" and was in sharp contrast with both the other schools of psychology, that is, "functionalism" and "behaviorism" [11–15].

In 1903, Pierre Janet (1859–1947; Fig. 10.5) introduced the term "psychasthenia," and the same year, barbital was discovered by Emil Fischer (1877–1947) and Joseph von Mering (1849–1908) at Bayer laboratories (Fig. 10.6). While barbituric acid was first synthesized in 1864, by Adolf von Baeyer (1835–1917), barbital was

Fig. 10.4 Bradford Titchener (1867–1927)

Fig. 10.5 Pierre Janet (1859–1947)

the first compound of this group with medical usefulness since experiments showed it induced sleep to dogs.

Fig. 10.6 Left: Emil Fischer (1877–1947). Right: Joseph von Mering (1849–1908). In 1903, they discovered barbital, the first compound of its group with medical usefulness

In 1904, the French Ministry of Education asked Alfred Binet (1857–1911; Fig. 10.7) to develop a method to identify students in need for additional supervision and help with their classroom work. He went on to devise the first operationalized IQ test, the Binet-Simon test. Revisions of the test were published in 1908 and 1911 [16–18].

A few years after the experiment of George Oliver and the discovery of adrenaline, in 1907, Walter Dixon (1871–1931) electrically stimulated the vagus nerves in a frog and reported that this resulted in the slowing of the heart rate. This "electrical" effect of the vagus nerve on the heart could be transmitted to the heart of a second frog through an extract from the first heart, and this effect could be blocked by atropine [19].

From 1903 to 1936, in Russia and USSR, Ivan Petrovich Pavlov (1849–1936; Fig. 10.8) showed how conditioned reflexes influence normal and pathological thought [20–22]. He received the Nobel Prize in 1904. Pavlov proposed that there are two fundamental processes in the nervous system, that is, inhibition and excitation. In 1932, he coined the term "first-signal system" to denote the perception and process of the sensorium in the brain hemispheres; this system is common in humans and animals. In humans however, there is a second-signal system which concerns verbal processes. He believed that under stress, behavior regresses from the processes of the second-signal system to those of the first-signal system, and this regression could be reversed through some kind of verbal therapy [23–26].

Fig. 10.7 Alfred Binet (1857–1911)

During the same period, Ernst Kretschmer (1888–1964) and later William H. Sheldon (1899–1977) introduced the constitutional or structural, as opposed to functional, models of psychopathology. He proposed that there is an association between specific body types and personality traits and mental disorders and described four body types: the pyknic, the athletic, the asthenic, and the dysplastic. They proposed that body types correlate with personality, temperament, and mental health and suggested that thin, asthenic individuals are prone to introversion, schizoid tendencies, and schizophrenia, while short and round individuals are be extrovert and with cyclothymic personalities and manic depression (Fig. 10.9). Sheldon added some experimental sophistication and measurement data to Kretschmer's hypothesis and proposed the existence of three basic morphological dimensions (endomorphy, mesomorphy, and ectomorphy) and three corresponding temperament clusters (viscerotonia, somatotonia, and cerebrotonia) Their theories are not of academic value anymore but continue to influence laypeople [27, 28].

William McDougall (1871–1938) put the emphasis on instincts and argued against behaviorism. He proposed the presence of three functional dispositions of instincts (intensity, persistency, and affectivity), and in line with Heymans, he

Fig. 10.8 Left: Ivan Petrovich Pavlov (1849–1936). Right up: One of the many dogs Pavlov used in his experiments (possibly Baika), Pavlov Museum Ryazan, Russia. Note the saliva catch container and tube surgically implanted in the dog's muzzle. Right down: Pavlov with assistants and students at the Imperial Medical Military Academy of St. Petersburg

subsequently argued for the existence of eight "tempers" (steadfast, fickle, unstable despondent, anxious, hopeful, placid, and sluggish). He also promoted research in parapsychology [29, 30].

In 1908, a patient named Clifford Beers (1876–1943; Fig. 10.10) published the book "A Mind That Found Itself" which constituted his memoirs of the bad treatment the author had received as a mental patient in three asylums [31, 32]. The next year, he founded the National Committee for Mental Hygiene, which essentially triggered "the mental hygiene movement" launched by Adolf Meyer (1866–1950; Fig. 10.11). However, although appealing and popular to the lay public, the idea that healthy living could help in the prevention of mental disorder and that prevention itself was possible, had no scientific support, and was more or less based on wishful thinking. The only areas such approaches could have an effect was the prevention of syphilis, since many mental patients of the time were in fact suffering from organic brain disorder secondary to this infection. The reduction of alcohol consumption was also a target since alcoholism was increasing out of control in the United States during that period. However, the war on alcoholism was no easy endeavor as shown by the failure of the alcohol prohibition in the United States during the years 1920–1933. The war on drugs has roots in this hygiene movement also.

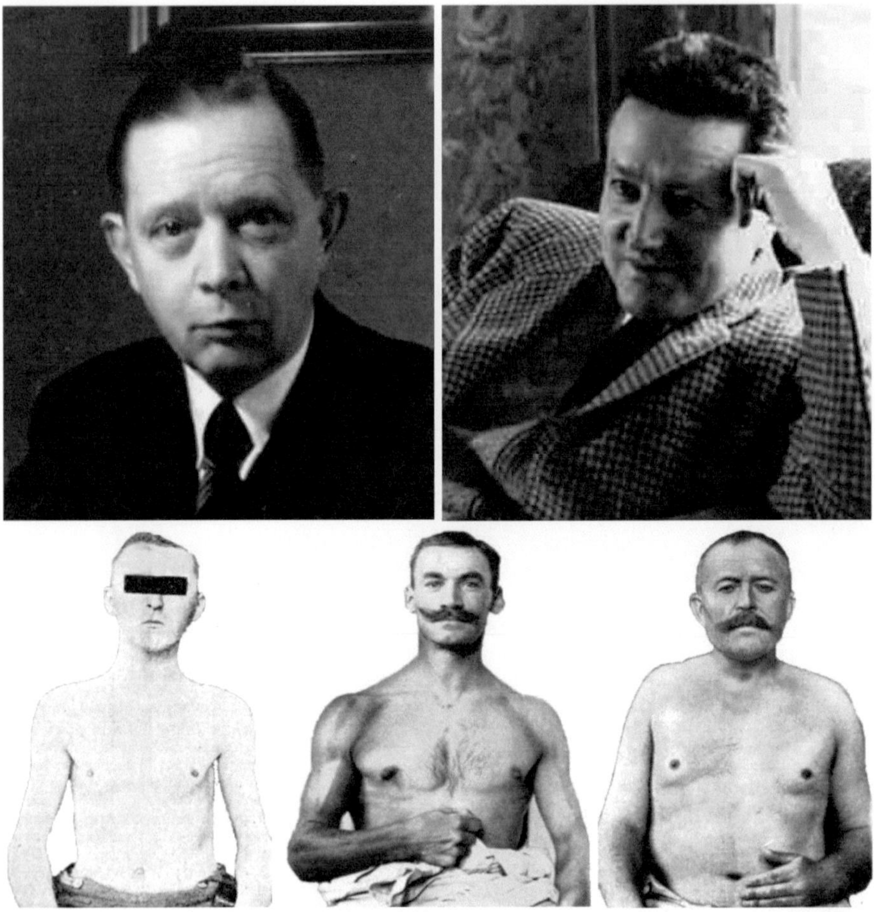

Fig. 10.9 Up left: Ernst Kretschmer (1888–1964). Up right: William H. Sheldon (1899–1977). Down, left: Asthenic type, schizoid psychopath (schizoid personality disorder). Down, middle: Athletic type with schizophrenia. Down, right: Pyknic type. Circular. (Pictures from Kretschmer, E: Körperbau und Charakter. Untersuchungen zum Konstitutionsproblem und zur Lehre von den Temperamenten. (1921)

The work of Meyer also gave rise to another field, that of "psychobiology," which appeared approximately the same time with psychoanalysis [33]. The term was introduced in its modern meaning in 1914 by Knight Dunlap (1875–1949) [34]. Adolf Meyer was a contemporary of Kraepelin and Freud who took his separate way in conceptualizing mental illness. He did not attempt to solve the mind-body conflict but considered both of them in the frame of a single and undivided apparatus with behavior being the response of this apparatus to external events. The treatment proposed by Meyer included commonsense counseling, social service, psychotherapy administered in the community, child-guidance clinics, and

Fig. 10.10 Clifford Beers (1876–1943)

continuous care in the community for discharged asylum and hospital patients. This theory greatly influenced Aubrey Lewis (1900–1975) and subsequently British psychiatry [35].

In 1908, the first International Psychoanalytical Congress took place in Salzburg, Austria (Fig. 10.12), while at the "opposite" pole of the scientific spectrum, in 1909, Korbinian Brodmann (1868–1918) published his work on cortical cytoarchitecture, and later, he described what is today known as Broadmann's brain areas (Fig. 10.13). In his studies, he described the structure and cellular composition of 52 specifically defined areas of the cortex, which are included into 11 histological groups.

One of the most important developments which sealed the future of clinical psychiatry was the resolution on the debate concerning the classification of psychotic illness. This was eventually concluded with the description of manic depression and dementia praecox by Emil Kraepelin (1856–1926; Fig. 10.14) [36]. The method used by Kraepelin was ingenious even for modern standards. He used a sophisticated card-filing system to collect data concerning the long-term course of his patients, and in this way, he arrived at the conclusion that there are two distinct major mental illnesses defined mainly by the episodic vs. chronic course, the family history, and the resulting severity of disability. As said, Kraepelin named the more

Fig. 10.11 Adolf Meyer (1866–1950)

Fig. 10.12 Third International Psychoanalytical Congress, 1911, in Weimar Germany. Sigmund Freud, Carl Jung, Karl Abraham, Otto Rank, and Sándor Ferenczi in the middle. What is interesting in this photo is the pyramidal shape of the gathering with Freud in the middle, whose figure emerges as taller and dominant in comparison to others

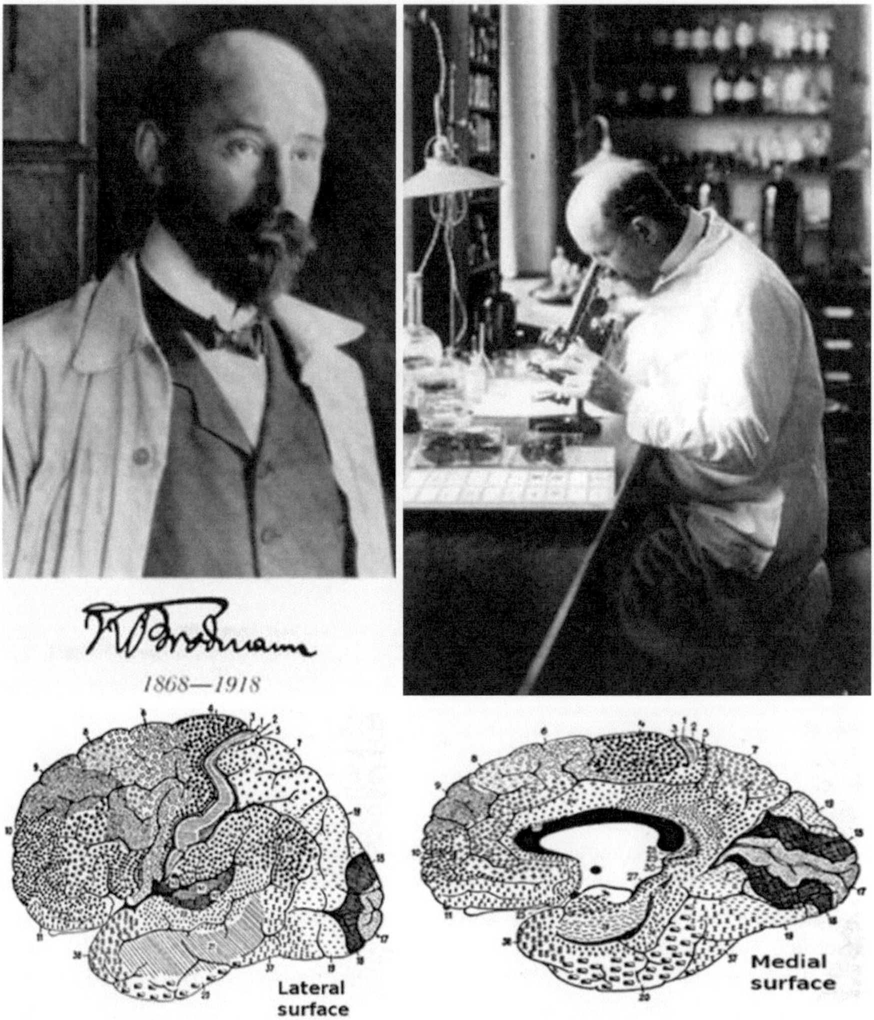

Fig. 10.13 Up: Korbinian Brodmann (1868–1918). Down: Broadmann's areas in the brain

severe and chronic form of mental illness with the name "dementia praecox," and it was the Swiss psychiatrist Paul Eugen Bleuler (1857–1939; Fig. 10.15) who gave it the name "schizophrenia" during a speech in 1908 after the Greek words "*σχίζω*" ("shizo" meaning split, tear apart) and "*φρην*" or "*φρεν*" ("phrin" or "phren," meaning mind) [37, 38]. He also coined the term "affective illness," and in this way, he broadened the concept of manic depression. He also pointed to the relationship between schizophrenia and manic depression and suggested they lie on a continuum.

Fig. 10.14 Emil Kraepelin (1856–1926)

In clinical terms, Kraepelin suggested that depression is characterized by lowered mood and physical and psychomotor retardation, while on the contrary, mania is characterized by elevation and acceleration of these same processes. He also described "involutional melancholia" as a form of agitated mixed state. His pupil Wilhelm Weygandt (1870–1939) published the first book on mixed clinical states [39]. Weygandt considered stupor to be the most important element of the mixed states.

Kraepelin conceptualized three basic dimensions (fundamental symptoms) of manic depression. These concern mood (euphoria vs. depression), cognition (flight of ideas vs. thought slowing), and physical activity (hyperactivity vs. inhibition). He proposed that these three dimensions fluctuate independently, and the combination of their phases could thus produce six different types of affective episodes: (a) depressive or anxious mania (depressed affect with flight of ideas and hyperactivity), (b) excited depression (depressed affect with thought poverty and hyperactivity), (c) mania with thought poverty (manic affect with thought poverty and hyperactivity), (d) manic stupor (manic affect with thought poverty and inhibition), (e) depression with flight of ideas (depressed affect with flight of ideas and inhibition), and (f) inhibited mania (manic affect with flight

Fig. 10.15 Eugen Bleuler (1857–1939)

of ideas and inhibition). The way of interaction of the basic dimensions is shown in Fig. 10.16.

The same period, a significant contribution to the basic modern understanding of the various clinical pictures was done by Karl Jaspers (1883–1969; Fig. 10.17) [40, 41]. Jaspers was both a psychiatrist and philosopher, and his intellectual multiplicity is reflected in his work on psychopathology which brings together a rigorous approach to classification combined with an emphasis on empathic understanding and a descriptive phenomenology [42]. He was influenced by Wilhelm Dilthey, Max Weber, and Edmund Husserl, and thus, he accepted a distinction between the natural and phenomenological attitudes, and subsequently between explanation and understanding [43–46]. However, Jaspers followed a pragmatic approach and adopted philosophic thoughts and proposals in order to clarify basic assumptions of the human nature and their implications for understanding psychopathology [47]. He sharply rejected Freudian theories and called them pseudoscience. Jaspers started from the viewpoint that the clinical assessment should be as free of hypothetical and

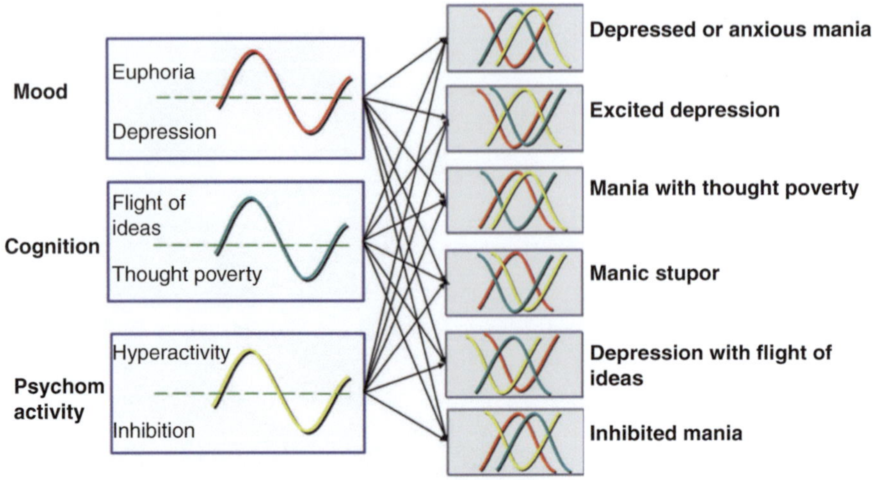

Fig. 10.16 Graphic representation of the three separate components of mood episodes (mood, cognition, and psychomotor activity) and their independent fluctuation which gives rise to six different types of mood episodes. (Created by author)

Fig. 10.17 Karl Jaspers (1883–1969) c. 1910 at the time he wrote his seminal psychopathology book

biased assumptions as possible, and thus he adopted the phenomenological approach. His preoccupation with the reliability and validity of the psychopathological assessment brought him closer to the theories of Nietzsche and accepted that all knowledge of the world is simply an interpretation, and it is never a safe asset but a limited term [48]. Thus, the important problem he was obliged to resolve was how the examiner/observer could overcome the subjective nature of perception and interpretation of the world, as this was a fundamental in existential philosophy. Simply describing the external behavior is not enough for clinical practice; a clinical interpretation is necessary. As a solution, he rejected the radical existential ideas and suggested that objective assessment of the patient could be achieved by empathetic understanding through mental simulation of the patient status by the examiner. In other words, the examiner should try to be in the patient's shoes by mentally playing various scenarios. Of course, no matter how skillful or clever the examiner is, this process always has an inherent and inevitable subjective component. Inevitably, this subjective nature includes also a strong moral/ethical aspect [49].

Returning to Kraepelin, another contribution of his was the introduction of Alzheimer's disease in 1910. He gave the disease the name of his late colleague Alois Alzheimer (1864–1915; Fig. 10.18). A few years before, in 1904, Alois Alzheimer, while working in Munich under the professorship of Kraepelin, reported on the presence of "cerebral plaques" in the autopsied brain of a female patient named "Augusta D" (Augusta Deter, c. 1850–1906; Fig. 10.18), who was admitted in the psychiatric department in Munich in November 25, 1901. Probably, Augusta D was not suffering from what today we label as Alzheimer's disease since she was suffering from reduced comprehension, memory loss, disorientation, aphasia, unpredictable behavior, paranoid ideation, intense jealousy toward the husband, and auditory hallucinations [50]. Some authors argue that Kraepelin manipulated the original report by Alzheimer so as to fit into a nosological entity which would be directly and specifically antagonistic to the one proposed by Pick a few years ago.

The same period, Franz Boas (1858–1942; Fig. 10.19) [51] rejected the idea of race as a biological concept and human behavior as the result of biological characteristics which were popular at that time under the theories of scientific racism [52]. He instead developed the theory of "cultural relativism" by arguing that all human populations around the world have a complete and fully developed culture which fits their needs. As a consequence, he suggested that the observed differences in human behavior are the result of cultural differences acquired through social learning rather than of biology. Thus, humans experience the environment through the eyes of their own culture, and their behavior reflects culturally acquired norms.

Margaret Mead (1901–1978; Fig. 10.19) was one of the students of Franz Boas. In her famous book 'Coming of Age in Samoa: A Psychological Study of Primitive Youth for Western Civilization' [53] she argued that young girls experienced smoother transitions to adulthood as a result of the society's sexual permissiveness during childhood. This was considered to be a living example

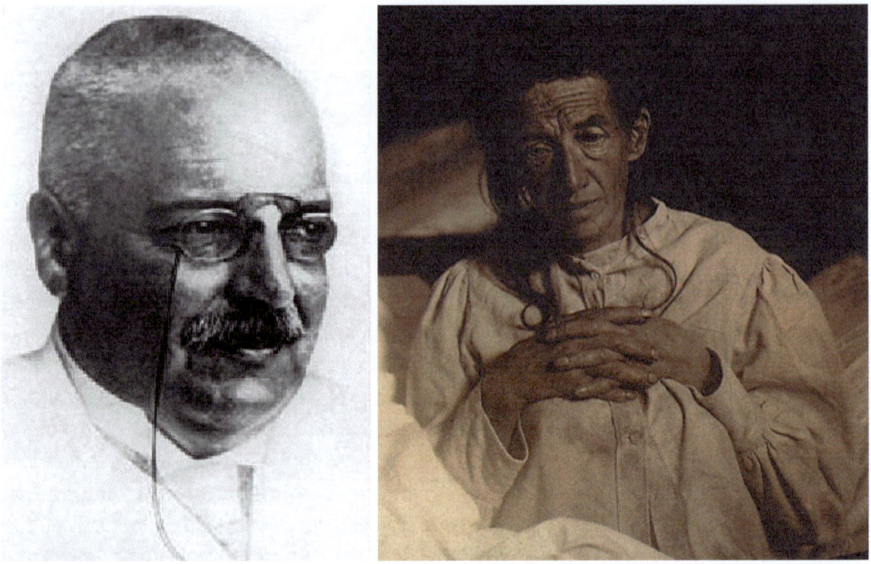

Fig. 10.18 Left: Alois Alzheimer (1864–1915). Right: Augusta D (c1850–1906), the first Alzheimer's disease patient

Fig. 10.19 Left: Franz Boas (1858–1942). Right: Margaret Mead (1901–1978)

of cultural relativism and influenced the 1960s sexual revolution. However the data on which her theories were based as well as their interpretation have been the focus of intense critique and debate and probably her critics were right [54–57].

In terms of treatment, a popular method of treatment for mental illness at the beginning of the twentieth century was hydrotherapy (Fig. 10.20). Because water can be heated or cooled to different temperatures, it was considered to be an effective tool for the treatment of insanity. When applied to the skin, mainly through changing its temperature, it could elicit various reactions in the body very quickly. Hydrotherapy took the form of baths, packs, or sprays. Warm continuous baths, lasting from hours to days, in quiet rooms and under dim light, were used to calm down insomnic, suicidal, or aggressive patients. For the baths, the water temperature typically varied between 33 and 36 °C (92–97 °F). The packs consisted of sheets dipped in varying temperatures of water. They were wrapped around the patient for several hours. Sprays were like showers and used either warm or cold water. Excited and aggressive patients were treated with cold water, that is, with temperature of 8–20 °C (48–70 °F). This was considered to slow down the brain blood flow and subsequently to decrease both mental and physical activity.

In 1910, a second attempt to introduce psychosurgery happened. Ludvig Puusepp (1875–1942) at the Psycho-Neurological Institute in St. Petersburg, Russia, reported that he had operated on three manic-depressive patients. The procedure was developed in collaboration with the prominent neurologist Vladimir Bekhterev (1857–1927) [58] and included the cutting of the connections between the parietal and frontal lobes. Overall, they considered the whole attempt a failure [59–61].

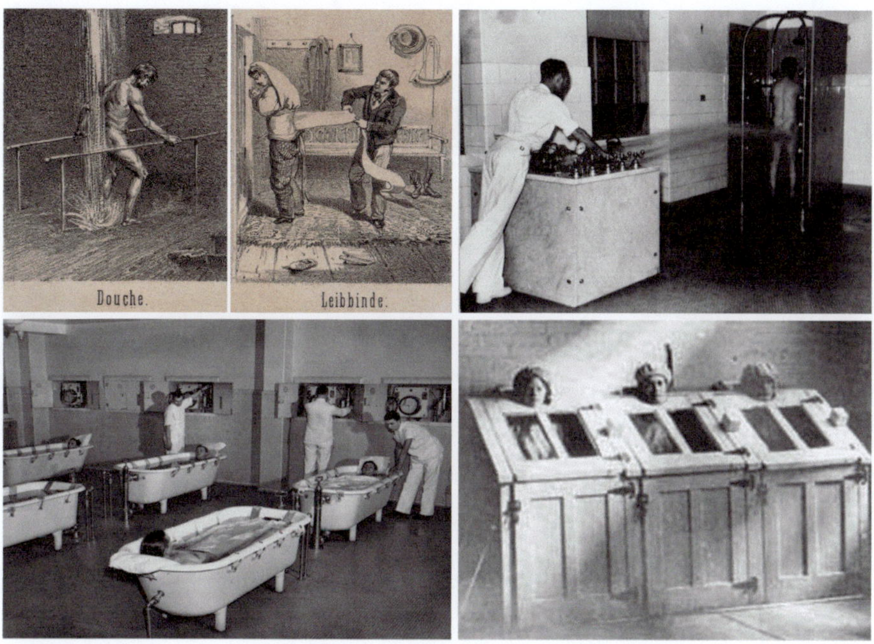

Fig. 10.20 Hydrotherapy

The same year, 1910, the term "receptor" appeared for the first time and was used by Henry Dale (1875–1968) and George Barger (1878–1939), ironically in an article attacking the concept [62]. In 1911, Otto Loewi (1873–1961), according to his autobiography, abruptly woke up during the night with a new idea for an experiment. He run to his laboratory, and 2 h later, at five in the morning, he had completed this experiment. It concerned the way the vagus nerve exerted its effect on the heart. Loewi prepared the heart of a frog and stimulated the vagus nerve still attached on it, and then after the vagus effect was in place, he rinsed the heart with Ringer's solution and applied it on the heart of a second frog whose vagus innervation had been cut. The second heart immediately became slower, suggesting the presence of a chemical agent in the solution. Still, it was not proven that this agent was originated from the vagus nerve of the first frog, or it was a by-product within the heart. The interesting comment by Loewi himself was that the design of the experiment was somewhat impulsive, and he had never performed it if it had given time for thought and concluded (verbatim): "Sometimes, in order to become a discoverer one has to be a naive ignoramus!" [63].

In 1936, Henry Dale and Otto Loewi (Fig. 10.21) shared the Nobel Prize. Sadly, 2 years later in 1938, the Nazis annexed Austria and Loewi was arrested and stripped of his property and spend months in jail before managing to escape to the United States. Later, Henry Dale wrote that luck ("serendipity") played an important role in his successful career [64].

The work of Carl Gustav Jung (1875–1961; Figs. 10.1 and 10.22) influenced a wide range of disciplines from psychiatry to philosophy and anthropology and to

Fig. 10.21 Left: Henry Dale (1875–1968). Right: Otto Loewi (1873–1961). Both received the Nobel Prize in 1936 for their work on neurotransmission

Fig. 10.22 Carl Gustav Jung (1875–1961)

religious studies. In 1910, he was the first head of Freud's newly founded International Psychoanalytic Association. However, soon, Jung developed his own and very distinct theory, that of analytical psychology, which focused on the development of self as the main task of human psychological development [65]. His theory included a number of novel concepts, including persona, the archetypal phenomena, the collective unconscious, the psychological complex, as well as extraversion and introversion. The concept of "collective unconscious" probably came directly from Hegel's notion of Volksgeist and was of prime importance since it was considered to include memories, ideas, and other psychological components, all inherited. By focusing on the collective unconscious, Jung de-emphasized the importance of libido, and this caused a bitter conflict with Freud in 1912 when his book on the psychology of the unconscious was published. The next year, it seems that he experienced some kind of a psychotic episode with visual and auditory hallucinations, and from then on, his thoughts included a more spiritual essence with a mystical and occult perspective and with an interest in paranormal phenomena and alchemy [66, 67]. What Jung seemed to have eventually embraced, a form of refined

Fig. 10.23 George Herbert Mead (1863–1931)

metaphysical approach was exactly what Freud was trying to eliminate from psychiatry and psychology.

George Herbert Mead (1863–1931; Fig. 10.23) proposed that the individual can only be understood in the context of the group's behavior [68, 69]. His major contribution was the idea that the mind ("I") and self ("me") emerge from the social interaction and communication (social behaviorism and the symbolic interactionist school of sociology), and they are not reducible to neurophysiology. In this frame, humans are first objects to other humans ("I" responses to others), and afterward, they become objects to themselves by incorporating the perspective of others ("me" emerges out of the assuming of the attitude of others) [70, 71]. According to Mead, the dialogue between the "I" and the "me" creates the thought process, and existence in community comes before the emergence of individual consciousness.

And then came the Great War which heralded a new era for medicine mainly because of the use of newer and deadly technologies for the first time from both parties. This had an enormous impact both on the somatic as well as the mental health of veterans. Not only head injury was much more common, but also psychological shock (shell shock) tormented the troops. There was now an urgent and massive need for a reliable and valid assessment and diagnosis and subsequently a demand for treatment and restoration of previous functioning in cases with grave disability. Several authors argue that it was the Great War the real cause for the transformation of psychiatry from an asylum-based and asylum-oriented discipline to a psychotherapeutic science who would treat the average person suffering from a common mental disorder as a consequence of adverse influences from the environment [72].

During WWI, Robert Woodworth (1869–1962) constructed the first self-report personality test, the Woodworth Personal Data Sheet. The instrument was mainly a test of emotional stability and resilience (i.e., neuroticism) and was meant to be useful for the military, but it was later also used in industry [73].

In practical terms, however, the first important breakthrough happened in 1917 in Austria, when Julius von Wagner-Jauregg (1857–1940; Fig. 10.24) introduced malaria-induced fever for the treatment of syphilitic general paresis. During the last couple of centuries, asylums were full with organic psychotic disorders due to neurosyphilis which is prominent during the later stages of the disease. It affects almost half of syphilitic patients after 10–20 years if left untreated, and at that time, there was no efficacious treatments available.

Jauregg observed that one of his patients, under the name of Hilda, suffering from general paresis and psychosis, recovered from her psychotic symptoms after a fever which was the result of a respiratory infection. Jauregg tried various methods in order to induce fever, and eventually, he came to conclude that the best way would be through infecting them with malaria. This method was named "pyrotherapy" from Greek "πυρ" (meaning fire). It is to be noted that the mortality rate of this method was approximately 15% [1]. Julius von Wagner-Jauregg became the first psychiatrist to receive the Nobel Prize in 1927.

However, the general picture during the first half of the twentieth century was gloomy. The statistics showed that in Great Britain, the asylum population grew from 16,000 to 98,000 between 1860 and 1910, three times faster than population growth. In the United States, it grew from 188,000 in 1910 to 850,000 by 1945. It was in this frame that psychoanalysis began to dominate the everyday clinical practice and established itself as both a medical treatment as well as a major cultural current. In the minds of the laypeople, psychoanalysis became the main feature characterizing psychiatry and started replacing asylums as the characteristic image.

At the same time (1912–1937), in the southern United States, the use of niacin or nicotinic acid in foods cured pellagra which used to be a major cause of dementia in this region.

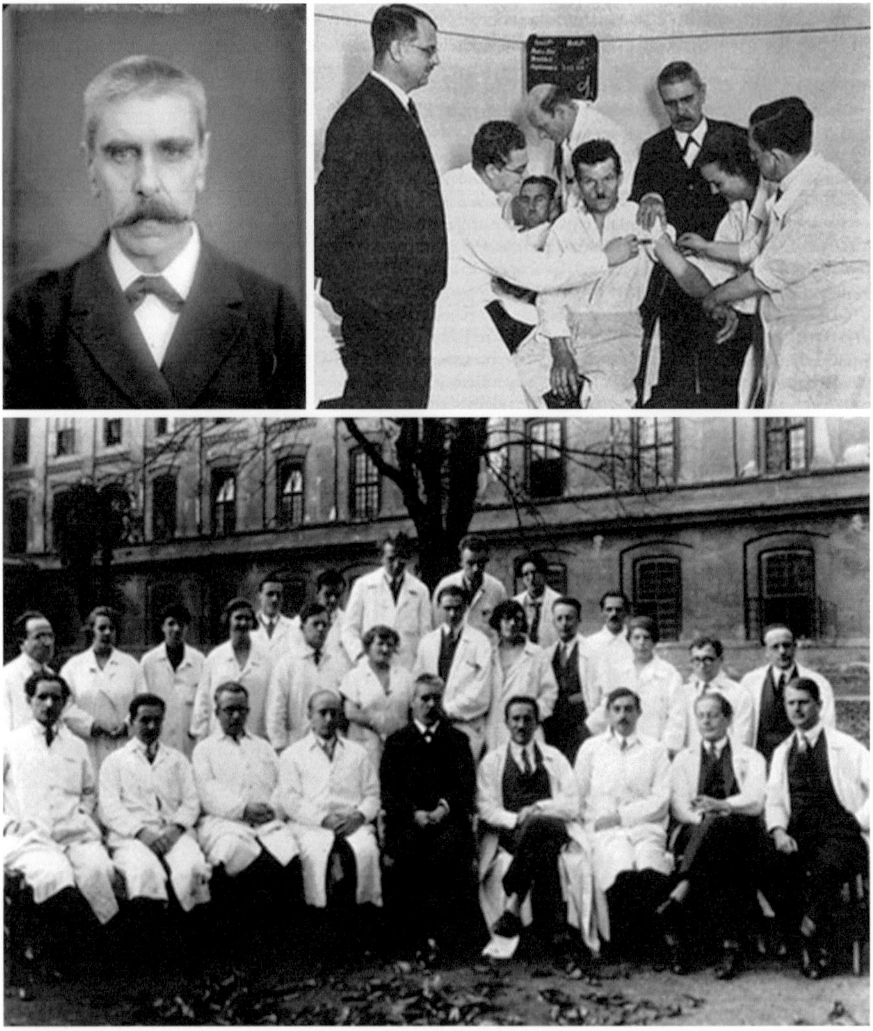

Fig. 10.24 Up left: Julius von Wagner-Jauregg (1857–1940). Up right: Wagner-Jauregg (center right in black jacket) watching a transfusion from a malaria patient (rear of the group) to a neurosyphilis victim (center) in 1934 (Institut für Geschichte der Medizin, Vienna, Austria). Down: Julius Wagner-Jauregg's staff in 1927. Front row (left to right): Max Weissmann, Bernhard Dattner, Heinrich Kogerer, Heinrich Herschmann, Julius Wagner-Jauregg, Josef Gerstmann, Paul Schilder, Heinz Hartmann, Otto Kauders. Second row (left to right): Erwin Stengel, Edith Vincze, Lydia Sicher, Annie Reich, Ludwig Horn, Clara Strassky, Robert Stern, Fannie Halpern, Otto Isakower, Alexandra Adler, Hans Hoff, Edward Bibring. Back row (left to right): Gottfried Engerth, Friedrich Stumpfl, Stefan Betlheim, Ludwig Eidelberg, Edith Klemperer, and Ernst Haase

Previously, before WWI, in 1913, in Turin, Italy, Giuseppe Epifanio used phenobarbital instead of bromide to induce sleep. "Bromide sleep" was already losing ground because it was considered dangerous and toxic. After the war, in 1922, in Zurich, the psychoanalyst Jakob Klaesi-Blumer (1883–1980) used aprobarbital (aprobarbitone) during sessions of psychoanalysis in patients with schizophrenia in a method called "narcoanalysis" [74]. However, sleep therapy manifested high mortality rate (close to 5%) mainly due to pulmonary embolism and bronchopneumonia. In spite of that, its reputation as effective treatment was so strong that when chlorpromazine was first marketed, its trade name was "Hibernal," a reference to some kind of "hibernation therapy," while narcoanalysis was used both in combat stress casualties during WWII and by the CIA for the rehabilitation of brain-washed prisoners of the Korean war. In the same line of reasoning, electric current was used to induce sleep, and the method was called "electronarcosis" [75–79].

Edward Lee Thorndike (1874–1949; Fig. 10.25) became very influential in a wide range of areas and disciplines, including psychology and education but also social science. He had major contributions to reinforcement theory and behavior

Fig. 10.25 Edward Lee Thorndike (1874–1949)

analysis and advocated for empirical research and the unbiased interpretation of findings. Thus, he paved the way toward behaviorism. He rejected sudden insight in favor of gradual learning. One of his major concepts was the "law of effect" which suggests that behaviors which are followed by positive consequences are more likely to be repeated in the future. This implied that reward has greater effect than punishment. Additionally, he developed a number of concepts including the laws of effect, use, disuse, readiness and recency, response by analogy, associative shifting, identifiability, and others [80, 81]. He believed that the observed differences between male and female parental behavior were due to biology rather than culture, but his arguments and conclusions often had a misogynist attitude.

Maybe the most important development in the field of philosophy occurred during the first two decades and throughout the first half of the twentieth century. Bertrand Russell (1872–1970), George Edward Moore (1873–1958), and Ludwig Wittgenstein (1989–1951) came to see philosophy not as the product of a foggy idealism but rather of a painstaking analytical thinking leading to the description of the logical structure of reality (Fig. 10.26) [82–84]. This was later called "analytical philosophy," has logical rigor in writing as its cornerstone, and is not widely embraced by laypeople and the general public; it is rather dry and technical, without any lyrical profundity. Analytical philosophy is in contrast with the so-called continental philosophy which includes all philosophical movements developed in continental Europe that do not utilize strict logical rules in the arguing and the use of language [85]. Again, this was not entirely new. It was Plato who was the first to consider geometry as a prerequisite and the cornerstone of philosophy. Plato had been educated in geometry and mathematics by Theodoros from Kerenaea who was a Pythagorean. Plato was so convinced that mathematics was absolutely essential for someone to become a philosopher that he put the inscription "*Μηδείς*

Fig. 10.26 Left: Bertrand Russell (1872–1970). Center: George Edward Moore (1873–1958). Right: Ludwig Wittgenstein (1989–1951). The founding fathers of analytical philosophy

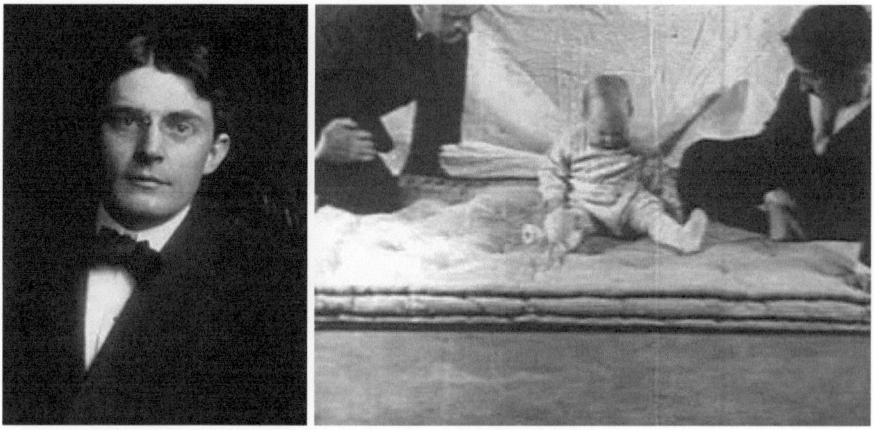

Fig. 10.27 Left: John Broadus Watson (1878–1958). Right: Little Albert

αγεωμέτρητος εισίτω μου την στέγην" which means that nobody who is not familiar with mathematics should enter the academy.

In 1920, John Broadus Watson (1878–1958; Fig. 10.27), an American psychologist who established the psychological school of behaviorism, published the "Little Albert" experiment. In that experiment, Watson exposed a 9-month-old boy to a white rat and other furry objects and conditioned exposure to a loud sound behind the boy's head. In this way Watson claimed he caused "experimental neurosis" to the boy because he reported that eventually, the animal and the fairy objects acted as triggers of fear [86]. However, the truth about the outcome of the "Little Albert" experiment has been questioned as well as his integrity and ethical standards [87]. Watson had an early life full of turmoil before succeeding in science, suggesting an unusual personality. Additionally, some authors suggest that the analyses of Watson's film footage of Albert suggested substantial behavioral and neurological deficits probably of congenital origin; the boy might had died in 1925 at the age of six with hydrocephalus [88]. However, following this theory, Mary Cover Jones (1897–1987; Fig. 10.28) did the "Little Peter" experiment in 1924, during which she managed to eliminate fear of rabbits in a 3-year-old boy after conditioning the rabbit with a candy [89]. She is today considered the mother of behavioral therapy.

During the same period, Charles Horton Cooley (1864–1929) developed a theory which was later called "introspectionist sociology." His major contribution was the development of the concept of the "looking glass self" which suggests that the "self" grows out of interpersonal interactions within the society and is based on the internalization and utilization of the perceptions others have concerning the person [90]. This was in accord with the ideas of his contemporary William McDougall (1871–1938) who argued that a human mind devoid of its social context is a theoretical and abstract concept that does not exist [29, 91]. This kind of sociological

Fig. 10.28 Mary Cover Jones (1897–1987) (photo by Paul Bishop). (License obtained from Paul Bishop)

approach was in conflict with the dominant behavioral theories of the time and had some influence also on parapsychology [29].

In 1922, Hermann Rorschach (1884–1922; Fig. 10.29) developed the first projective psychological test which was a set of inkblots referred to as the Rorschach test, in order to assess the unconscious part of personality [92]. Jean Piaget (1896–1980; Fig. 10.30) conceptualized four stages of cognitive development from infancy to adulthood: sensorimotor, preoperational, concrete operations, and formal operations [93–95]. Kurt Schneider (1887–1967; Fig. 10.31) considered personality as independent of mental disorders, and he was the first to conceptualize the distinction which later led to the axis I vs. axis II in the DSM. He published his ideas for the first time in the 1920s [96] and continued to revise them until the late 1950s [97]. He described ten personalities: the hyperthymic, depressive, insecure (sensitives and anankasts), fanatic, attention-seeking, labile, explosive, affectionless, weak willed, and the asthenic personality.

During the early twentieth century, there was a change in the initial idea behind somatic treatments. While during the antiquity it was the restoration of the equilibrium of somatic fluids, after the renaissance, it was more an effort to shock the brain or the person (or both). In 1922, the first controlled trial in psychiatry occurred [98]. Nicholas Kopeloff (1890–1959), a bacteriologist at the New York State Psychiatric Institute, then on Ward's Island, and the psychiatrist Clarence O. Cheney (1887–1947) tested the popular theory that teeth and tonsils infection

Fig. 10.29 Hermann Rorschach (1884–1922) c. 1910

was the cause of mental illness [99]. They performed operations in 27 patients and used 33 more as controls. The patients were not randomized. They reported no difference between the two groups [100]. However, in spite of these negative findings, the theory that toxins as a result of infection could be the cause of mental illness continued to be widespread.

In 1926, the Parisian psychiatrist Constance Pascal introduced the word "shock" to psychiatry. As in charge of the departmental asylum in Châlons-sur-Marne, she argued that mental illness originated from "mental anaphylactic reactions." This "autointoxication" was considered to be the result of toxins or infections, including infected teeth or toxins leaking from the gut. There was a vogue for teeth extraction but also for colectomy. As a treatment to this, it was believed that you needed to shock the brain and the autonomic nervous system back into equilibrium. She

Fig. 10.30 Jean Piaget (1896–1980)

proposed the use of colloidal gold, milk, or vaccines or fever therapy as means to shock the brain [101, 102]. The concept of "shock" was not new and was first conceived in ancient India (see also Chap. 3), and shock was achieved through a variety of methods including spinning chairs, seclusion, or intimidation, but in the twentieth century, the term became more "medicalized" starting from "insulin shock therapy." The zenith of the "shock" concept was with ECT which in the popular culture is always referred to as "electroshock."

In this frame, in 1926, at the University of Vienna Department of Psychiatry under the direction of Wagner-Jauregg, Edith Klemperer (1898–1987; Fig. 10.24) gave insulin to patients with delirium tremens. Insulin had been isolated only years before and took its name from the Latin "insula" meaning the "island" because it was an extract from the islets of Langerhans. The next year, in 1927, at the University

Fig. 10.31 Kurt Schneider (1887–1967) (photo: Max Plank Institute fur Psychiatrie)

of Budapest, Dezsö Miskolczy (1894–1978) under Karl Schaffer (1864–1934) gave high doses of insulin, and in 1928, Julius Schuster proposed treating psychosis with both shock and insulin. In 1929, at the mental hospital in Lausanne-Céry, Switzerland, Hans Steck (1891–1980) began the "hypoglycemic" treatment of psychosis. The difference was that he was avoiding inducing deep comas and convulsions. In 1931–1932 in Basel, John Staehelin (1891–1969) carried out insulin metabolism experiments in patients with schizophrenia. He reported significant behavioral changes in the fasting patients after receiving insulin.

In parallel, important development were occurring in the field of philosophy of mind. In 1927, Martin Heidegger (1889–1976; Fig. 10.32) published his book "Being and Time" [103]. The core of his theory concerned the notion of "dasein," or "being there" or "being in the world." Heidegger considered this as the central issue of human existence [104]. Those who cannot accomplish "being there" because of having dreadful feelings, they end up in a conventional and inauthentic life full of worthlessness and guilt. Existential anxiety is considered to be the result of fear of loss of "dasein," while depression comes from nonacceptance of the inevitable human fate of death.

Fig. 10.32 Martin Heidegger (1889–1976) in 1960

In 1930, the Mental Treatment Act in Great Britain declared no distinction between mental and physical treatments and subsequently between mental and physical illness.

In the 1930s, James Papez (1893–1958; Fig. 10.33) described the "Papez circuit" in the brain and its possible functions on emotional production and regulation. He proposed a neural structure which includes three parts: the hippocampus, the hypothalamus, and the cingulate gyrus. This system would be named "limbic system" 20 years later by Paul MacLean (1913–1990; Fig. 10.33).

The same period, the first family studies were conducted by the German psychiatrist Ernst Rudin (1874–1952) with little success and with much complications because of his work with the Nazi eugenics program. His students Johann Lange and Hans Luxenburger (1894–1976) devised the method of comparing monozygotic and dizygotic twins. They both had a difficult balance of loyalty since they worked with high status in Nazi research institutions and supported the eugenics tasks and program, but simultaneously, they adopted a criticizing attitude. Aaron Rosanoff

Fig. 10.33 Left: James Papez (1893–1958). Right: Paul MacLean (1913–1990)

(1879–1956) and Franz Kallmann (1897–1965) worked on the same pathway in the United States at that period.

Also, in the 1930s, a fascinating story was that of Princess Alice of Battenberg, later Princess of Greece and Denmark (1885–1969; Fig. 10.34). She was the mother of Prince Philip, Duke of Edinburgh, mother-in-law of Queen Elizabeth II, and a great-granddaughter of Queen Victoria. She was congenitally deaf. She married Prince Andrew of Greece and Denmark in 1903, and in 1930, she was diagnosed with schizophrenia. She was admitted to the sanatorium in Kreuzlingen, Switzerland, under the care of Ludwig Binswanger (1881–1966). In that same asylum, Vaslav Nijinsky was treated, and probably, they both coexisted there. The interesting element is that Binswanger (who later developed "*daseinsanalyse*") consulted Sigmund Freud who suggested that the psychotic symptoms were the result of sexual frustration, and therefore, he recommended X-raying her ovaries in order to destroy them together with her libido. It is unknown whether this unusual treatment was ever carried out [105]. In 1994, Princess Alice of Greece was announced "Righteous Among the Nations" for having hidden a Jew family in her house in Athens during WWII and, in 2010, was named a Hero of the Holocaust by the British Government.

In 1931, Eric Lindemann (1900–1974) introduced "narcoanalysis" and more specifically the "Amytal interview." According to this method, intravenous amytal was given to patients during sessions of psychoanalysis so that it was made easier

Fig. 10.34 Princess Alice of Greece (1885–1969) c. 1906

for them to drop their defenses and talk freely, thus retrieving more easily the buried material from the "unconscious" [106, 107].

Starting in 1913–1914 in Vienna and working through the 30s and the 40s in New York, Jacob Levy Moreno (1889–1974; Fig. 10.35) developed one of the most fascinating and creative models of group treatment, the psychodrama. This is a psychotherapeutic method which utilized techniques borrowed by theatrical acting in order to stimulate the expression of unconscious material through improvised dramatizations, role-playing, and other spontaneous playacting [108]. Moreno also worked on sociometry, group psychotherapy, sociodrama, and sociatry [109], and his 1934 book "Who Shall Survive?" includes maybe the earliest graphical depictions of social networks [110, 111]. Interestingly, he asked that his epitaph should read "the man who brought laughter to psychiatry."

Slowly but steadily, the environment changed in favor of patients. These changes were obvious also in the changes in vocabulary and terminology. "Psychiatry" was adopted universally as the term to define the field. Subsequently, there were changes in terms like "asile d'aliénés" to "hôpital psychiatrique" or "psychiatre" for

Fig. 10.35 Jacob Levy Moreno (1889–1974)

"aliéniste" or from "asylum," "attendant," and "lunatic" to "hospital," "nurse," and "patient." All these changes took part in the 1920s and 1930s.

Following the observations and suggestions concerning lithium and gut, a number of beverages and remedies included lithium as their component in the late nineteenth and early twentieth century (Fig. 10.36; see also Chap. 9). Charles Leiper Grigg (1868–1940) introduced a lemon-lime soft drink in 1929 under the label "Bib-Label Lithiated Lemon-Lime Soda" which soon changed to 7UP. However, by 1948, lithium has been removed from all beverages because of cases of toxicity, and its free marketing was prohibited [112]. In parallel, a number of remedies with lithium were marketed in the early twentieth century and were mostly indicated for the control of renal calculi and "uric acid diathesis" [113].

The turning point in the field of psychopharmacology came surprisingly from India, and it is largely forgotten (Fig. 10.37). The first report on the psychotropic properties of reserpine was published in a local Indian journal in 1931 by Kaviraj Gananath Sen and Kartick Chandra Bose [115] and went unnoticed [114]. Sen came from the tradition of Ayurvedic medicine, while Bose was a graduate of the Calcutta Medical College. Building on their work, the cardiologist Rustom Jal Vakil (1911–1974) [116] received the Lasker Award together with Nathan Klein in 1957.

In 1934, Manfred Sakel (born Menachem Sokol; 1900–1957; Fig. 10.38) injected 45 units of insulin to a patient with schizophrenia. This first administration

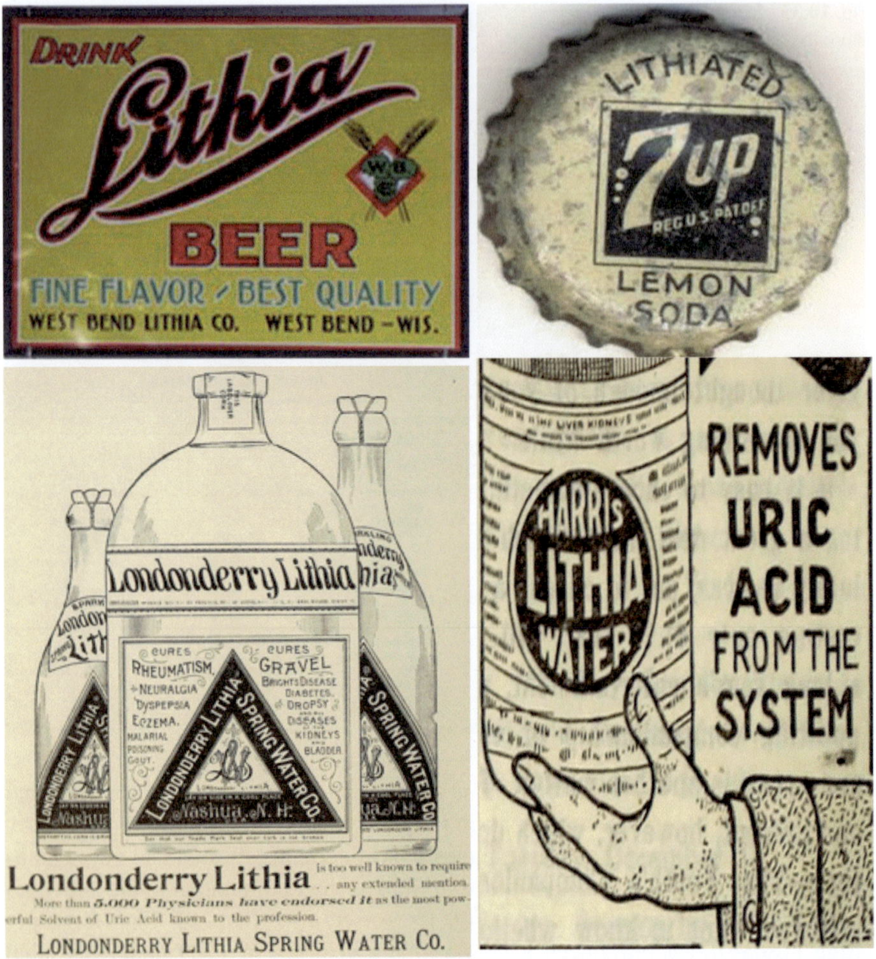

Fig. 10.36 Example of beverages which included lithium as their component in the late nineteenth and early twentieth century. Left upper corner: lithium beer. Right upper corner: The original 7Up which included lithium. Bottom: Example of remedies with lithium which were marketed in the late nineteenth and early twentieth century and were mostly indicated for the control of renal calculi and "uric acid diathesis"

produced no therapeutic effect. The administration of three injections of 40 units each followed the next morning, and the patient became less agitated. A fourth injection of 50 units in the afternoon induced a hypoglycemic shock and coma. To the surprise of the staff, when the patient spontaneously recovered from the coma, the psychotic symptoms had vanished. However, this was only temporary, and psychosis appeared again within hours. Insulin treatment continued for the following

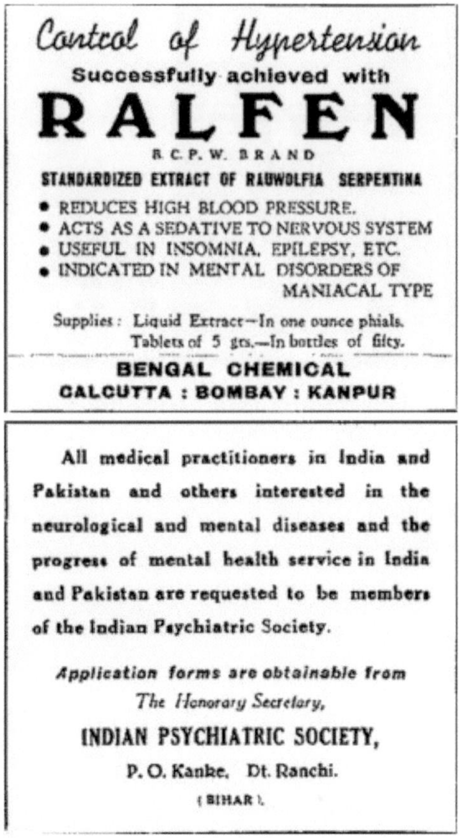

Fig. 10.37 Advertisement on the therapeutic properties of reserpine in India (from [114])

few days without any remarkable results; however, after 3 weeks, the patient experienced an epileptic grand mal, and after this, he manifested a full remission. One of the patients who were treated by Sakel himself with insulin coma was the famous Russian ballet star Vaslav Nijinsky (1889–1950) who was suffering from schizophrenia. It is reported that Nijinsky responded slowly but satisfactory to this kind of treatment.

Sakel reported that from the first 58 recent-onset patients he treated, 86% remitted (full or good remission or fit for work). From his 46 chronic patients, nearly half experienced remission. However, these impressive results were not replicated by other researchers, and the statistics and the true results were disputed. For example, Joseph Wortis (1907–1995), who introduced insulin coma in the United States, reported that in Vienna, he witnessed one death and 18 remissions (72%) in 25 patients [117].

The big story of the development of convulsive therapy began in Budapest, when Ladislas Meduna (1896–1964; Fig. 10.39) made a chance observation under the microscope [118]. It is interesting to note that his work was entirely unrelated to

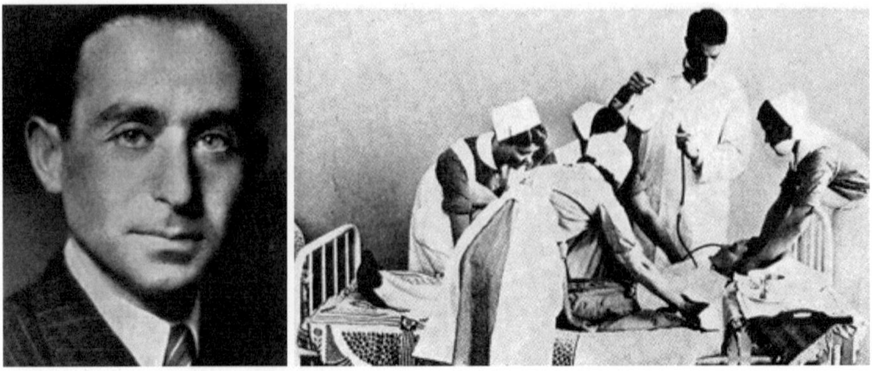

Fig. 10.38 Left: Manfred Sakel (1900–1957). Right: insulin shock treatment in Lapinlahti Hospital, Helsinki, in 1950s

Fig. 10.39 Ladislas Meduna (1896–1964)

Sakel's, and while Sakel's work was "empirical," Meduna's approach had a rigorous scientific base [119]. On the basis of the pathological study of the brains of patients with schizophrenia and epilepsy, Meduna concluded that the glia system played an important role in epilepsy and schizophrenia, with conflicting mechanisms and functions within it. In other words, his idea was that the way of glia functioning in epilepsy precluded the way of functioning in schizophrenia and vice versa. By extending the argument, Meduna suggested that by inducing epilepsy, one could cure schizophrenia [120]. At the clinical level, at that time, it was widely believed that epilepsy was rare in patients with schizophrenia, and even more, it had been reported by Julius Nyirö and Albin Jablonszky that in epileptic patients who developed psychosis, the convulsions disappeared [121].

Initially, Meduna used simple camphor, and in 1934, he injected with it the first patient who was for the last 4 years in catatonic stupor. Two days and five injections later, catatonia disappeared. The next few patients were injected with camphor or with pentylenetetrazol (Cardiazol or metrazol). High success rates were reported [122]. More specifically, in [123] monograph, he reported that from 110 patients (including 62 with poor prognosis), 80% manifested remission [123].

Cardiazol (metrazol) was more preferable to camphor because it was more soluble and faster in action, but unfortunately, both agents were problematic. Not only they were inducing a dreadful anxious feeling before the convulsions, but also they were triggering seizures at a random rate for the next few days. The reports were suggesting that Cardiazol therapy was superior to insulin therapy mainly because it was safer. Insulin therapy had a high death rate ranging from 2 to 10% depending on the setting. But as mentioned, the biggest problem with Cardiazol was the significant apprehension it induced to patients, with anecdotal reports suggesting that Cardiazol treatment was popular in Nazi Germany exactly because of the terror it caused and the subsequent discipline it imposed. In 1939, Phillip Polatin (1905–1980) and coworkers at the New York State Psychiatric Institute reported that vertebral fractures occurred in 43% of the patients after Cardiazol treatment [124].

Whether insulin treatment exerted its therapeutic effect through coma or through seizures is still unclear. Sakel used the term "wet shock" to refer to comas (because hypoglycemic comas include sweating) and "dry shock" to refer to convulsions and suggested that both were therapeutic. He was rather in favor of comas and, in 1950, during the Paris World Psychiatry Congress, said that convulsions are toxic, but this position might had been the result of his conflict with Meduna. There was a bitter confrontation between Sakel and Meduna during that congress concerning the comparisons of the two treatment methods, the origin of ideas, and the underlying mechanisms with particular focus on the need or not of coma and convulsions [118]. Sakel might suffered himself from a severe mental illness, probably bipolar disorder, and this affected his attitude toward his work and peers, although this is not solidly documented [125]. The scientific story of these developments was also obscured by the political environment of the time. Both Sakel and Meduna were Jews living in Austria during the rise of the Nazi Germany, while Julius Wagner-Jauregg, who was against their treatment methods, was a known anti-Semitist. It is worth mentioning that also Edith Klemperer (1898–1987; Fig. 10.24), a member of the Vienna group under Jauregg and the first ever psychiatrist to use insulin for the treatment of patients, was also a Jew and was forced to leave Austria for the United States when the Nazis came. The fascinating story of Manfred Sakel, his personality, scientific contribution, and controversial practice, has been the focus of extensive historical research [118, 125–127].

Overall, at that time, both Cardiazol and insulin treatments seemed to be efficacious but rather too complex, demanding, and risky for the average psychiatric hospital. Eventually, the insulin coma therapy was abandoned completely in the 1960s after a number of studies which failed to prove its efficacy [128–131].

In 1935, Myron Prinzmetal (1908–1987), a Los Angeles internist (later, he became famous as a pioneer cardiologist), and Wilfred Bloomberg (1905–1987), a Boston neuropsychiatrist, compared Benzedrine with ephedrine in nine narcolepsy patients. It was a crossover study in which all patients were started on a sodium chloride placebo and then received all Benzedrine, followed by ephedrine and followed again finally by Benzedrine. The patients were not randomized and served as their own controls [132].

The same year, Bill Wilson (1895–1971) and Bob Smith (1879–1950), both alcoholics, established the Alcoholics Anonymous (AA) and developed the 12-step group psychotherapy for the treatment of alcoholism based on spiritual and character development. This model has since evolved into a model for organizations covering a wide range of addictive or similar behaviors.

Again in 1935, the Portuguese neurologist António Egas Moniz ([133]; Fig. 10.40) introduced prefrontal leukotomy. Moniz was professor of neurology and started his career in neuropsychiatry with a scandalous book on sexology in 1901. He was also a politician supporting a republican form of government, which led to several brief periods in jail. In 1900, he was elected to parliament and was reelected several times over the next 15 years. He served as Ambassador to Spain and Minister of External Affairs and represented Portugal at the 1918 Versailles Peace Conference. He retired from politics in 1926 when Salazaar became dictator. At the age of 51, he started his research career in neuropsychiatry. Leukotomies were performed under general anesthesia by the neurosurgeon Pedro Almeida Lima (1903–1985). The word "lobotomy" and "leukotomy" come from the Greek *λοβός* (lobos meaning lobe), *λευκός* (lefkos meaning white; corresponds to white matter), and *τομή* (tomi meaning cut, transection). The leukotomy was initially performed with injections of ethanol, but later, the method of choice was section of the fibers. For this, they introduced a surgical instrument called a "leukotome." The diagnoses of the 20 patients, aged 27–62 years, which were initially operated, included schizophrenia, bipolar disorder, panic, and agitation. Moniz reported that one-third of patients improved significantly and another third somewhat improved. There were no deaths and Moniz considered the adverse events of leukotomy to be transitory while he reported that no patient deteriorated afterward [133–135].

Moniz made contradictory statements concerning how he went on to develop such a dangerous and controversial treatment method, and it is important to note that this happened at a time all previous reports on psychosurgery were negative. Also,

Fig. 10.40 António Egas Moniz (1874–1955)

much of the scientific community of the time considered frontal lobe to be a "silent" lobe, although some isolated cases of remission of mental disorders after neurosurgical removal of brain tumors or damaged brain tissue had already been reported [59]. Interestingly, following the example of Julius Wagner-Jauregg with malaria fever, in France, Maurice Ducosté (1875–1956) injected 5 ml of malarial blood directly into the frontal lobes of 100 patients with syphilitic paresis through holes drilled into the skull [136], while in Switzerland, François Ody (1896–1957) removed the entire right frontal lobe of a patient with catatonia [134, 136]. The determining factor was that at that time, the results of brain research on primates had begun to appear. Moniz probably came in contact with John Farquhar Fulton (1899–1960) and Carlyle F. Jacobsen (1902–1974) (Fig. 10.41) during the Second International Congress of Neurology which was held in London in 1935 and discussed with them the results of their experiments concerning frontal lobectomies and subsequent changes in behavior and intellectual function in two chimpanzees [137]. It is interesting to note that Moniz started practicing leukotomy in mental patients 3 months after this congress, and he utilized a similar technique [134, 138–142].

Fig. 10.41 Left: John Farquhar Fulton (1899–1960). Right: Carlyle F. Jacobsen (1902–1974). Their work on primates inspired Moniz to develop prefrontal lobotomy

Reception of the method was reserved. José de Matos Sobral Cid (1877–1941), who had sent the first patients for leukotomy from his own hospital in Lisbon, denounced frontal lobe surgery. He declared that the patients were not cured but instead were "diminished" and had suffered a "degradation of personality" after the operation.

The first prefrontal leukotomy in the United States was performed at the George Washington University Hospital on September 14, 1936, by the neuropsychiatrist Walter Freeman (1895–1972) and the neurosurgeon James W. Watts (1904–1994) [140, 141]. In 1937, Freeman and Watts (Fig. 10.42) adapted Lima and Moniz's surgical procedure and created the Freeman-Watts technique, also known as the Freeman-Watts standard prefrontal lobotomy, which they styled the "precision method" [143–147]. However, this method still required drilling holes in the scalp, so surgery had to be performed in an operating room by trained neurosurgeons. The difficult and problematic methodology kept leukotomy from being widely accepted and applied. In 1937, Amarro Fiamberti (1894–1970) devised the transorbital procedure whereby the frontal lobes were accessed through the eye sockets [148]. This was a much easier procedure which included puncturing the thin layer of the orbital bone at the top of the eye socket and then injecting alcohol or formalin or inserting a hypodermic needle into the white matter of the frontal lobes through this aperture [149].

Following the ideas of Sir Francis Galton, Gordon Allport (1897–1967; Fig. 10.43) and Henry S. Odbert published their pivotal psycholexical study in 1936 by using the Webster's New International Dictionary as source, and they isolated

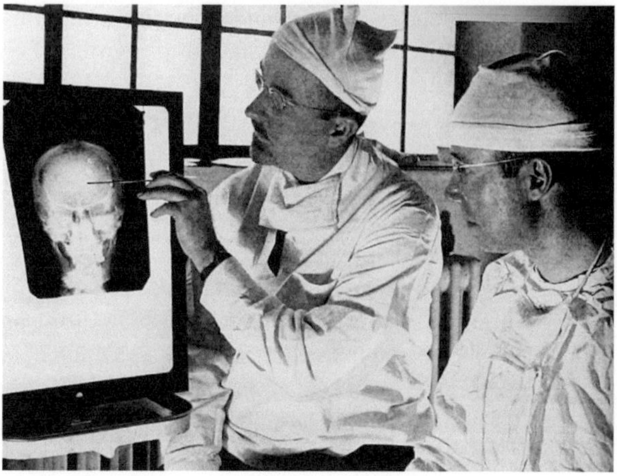

Fig. 10.42 Walter Freeman (left) and James W. Watts (right) study an X-ray before a psychosurgical operation, c. 1941

Fig. 10.43 Gordon W. Allport (1897–1967)

17,953 individual words related to descriptions of personality among 400,000 words. Then they separated these words in four groups (columns) corresponding to traits, states, social evaluations, and "miscellaneous." A significant limitation of the whole approach was that the average agreement between different reviewers concerning the classification of words was 47% [150, 151].

In 1937, Charles Bradley (1902–1979) reported that treatment with the Benzedrine had improved the behavior of a group of children who had been hospitalized because of "behavior disorders" [152, 153]. Occasional reports of the use of stimulants in the treatment of behavioral problems in children and adults followed. The same year, the first trials concerning the treatment with amphetamines of a group of depressive-like patients appeared [154] but also of clinically depressed patients [155]. In 1938, Sir Aubrey Lewis (1900–1975) proposed a continuum model from anxiety to psychosis [156].

The most important development in the area of therapeutic convulsions occurred in Italy, in 1938, when Ugo Cerletti (1877–1963) and Lucio Bini (1908–1964) developed electroconvulsive therapy (ECT) which concerned the induction of epileptic seizures through the application of electroshock (Fig. 10.44). Initially, Cerletti was interested in the hippocampus and its role in epilepsy. He started his research in 1931 by inducing epileptic seizures in dogs and afterward doing pathologic examination of targeted brain tissue. One of the problems he encountered concerned the method to induce seizures, and by 1934, he had concluded that applying electric current was the safest method. He was utilizing 50 Hz and 125 V at a duration of 1.5 s. Following this development, it was Lucio Bini who built the device that could specifically and conveniently apply electric current to cause seizures in dogs. As in 1935 Meduna announced the therapeutic effect of seizures in schizophrenia, Cerletti considered the possibility to use electric-induced seizures for this purpose. An additional advantage, as Cerletti already knew from his dog experiments, was that the electric method was also inducing an instantaneous loss of consciousness; therefore, the patients would not experience all the negative experience with insulin and Cardiazol. The main issue was that already from the beginning, the whole idea of applying electrical current for therapeutic reasons and especially to cause seizures was culturally problematic. Not only the fascist political environment of Italy at that time was an adverse connection, but also a bad idea was the visit of the Rome pig slaughterhouse by Cerletti and Bini in order to

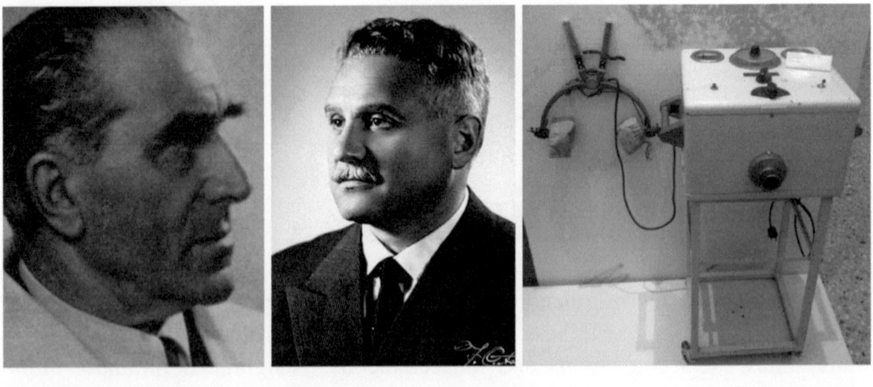

Fig. 10.44 Left: Ugo Cerletti (1877–1963). Center: Lucio Bini (1908–1964). Right: Cerletti's ECT machine preserved at Museo di Storia della Medicina in Rome

establish the safety parameters. However, eventually, history would record the first successful ECT treatment in a 39-year-old psychotic patient named Enrico X in Rome in April 1938 [125, 157].

David Wechsler (1896–1981; Fig. 10.45) developed the Wechsler Adult Intelligence Scale (WAIS) in 1939 [158], the Wechsler Intelligence Scale for Children (WISC) in 1949 [159], and the Wechsler Preschool and Primary Scale of Intelligence (WPPSI) in 1967. He divided the concept of intelligence into verbal and performance (nonverbal) domains.

In October 1939, soon after the outbreak of WWII, a Führer decree (Fig. 10.46) ordered doctors to kill patients with incurable illnesses. In this frame, approximately 270,000 mental patients were killed by physicians and medical staff. But this was not a sudden or surprise development. Under the Nazi regime, eugenics and philosophies of "social Darwinism" became dominant and were established as official state position and ideology [160, 161].

As Nazism was spreading around Germany, and as psychoanalysis had a strong Jewish component, many psychoanalytical books were banned or burned and psychoanalysts fled Europe. Freud fled for London in 1938 after paying the Nazis about 200,000 USD for an exit visa. As the cousin of Hermann Göring, Dr. Matthias Heinrich Göring (1879–1945) took over the German Society for Psychotherapy, and he turned it into the "Reich Institute for Psychological Research and Psychotherapy" and purged all Jewish and psychoanalytic elements. These migrant psychoanalysts transferred the capital of psychoanalysis from Vienna and continental Europe to the United States, where gradually it became the dominant force of American psychiatry for the decades to come [1, 2, 162].

Fig. 10.45 David Wechsler (1896–1981) seated at a table, conducting some sort of test on a patient, c. 1955

Fig. 10.46 Left: Hitler's order to grant euthanasia to incurably sick patients. In English: "Reichsleiter Bouhler and Dr. Brandt are assigned to be professionally accountable in charge to extend the legal authority of specific named Medical Doctors, so that they may grant euthanasia to as far as is humanly possible to tell incurably sick patients after most critical adjudication of their medical condition—A. Hitler". Center: Nazi propaganda showing "how much it cost" the German state to take care of the disabled. This discourse even appeared in children's math textbooks with arithmetic problems about the subject c. 1936. Right: Five disabled Jewish prisoners, photographed for propaganda purposes, who arrived in Buchenwald after "Juni-Aktion"

Mental patients and especially handicapped children were among those exterminated in an organized and systematic way (Operation T4), and undoubtedly, several psychiatrists were involved. One such example is the Austrian Heinrich Gross (1915–2005), the head of the Spiegelgrund clinic in Vienna, who, in 1997, came under investigation for his actions resulting in the murder of approximately 800 children during WWII and was stripped from the honors awarded to him so far but never convicted [163]. Another example is Dr. Ernst Baumhard (1911–1943) who was director of the Hadamar Psychiatric Institute near Wiesbaden, Germany (code-named Facility-E), where at least 15,000 patients were killed [164, 165]. In Würzburg, the director was Dr. Werner Heyde (1902–1964; Fig. 10.47), the psychiatric expert for the Gestapo who committed suicide after his arrest in 1964 [166]. Extermination was done either directly or through inclusion in fatal research projects [167]. The masterminds behind the Operation T4 were the physician Karl Brandt (1904–1948) and the SS officers Philipp Bouhler (1899–1945) and Viktor Hermann Brack (1904–1948) (Fig. 10.48). Two of them were convicted for crimes against humanity and were executed. The role of the prominent psychiatrist and geneticist Ernst Rüdin (1874–1952) remains controversial. It is interesting that the Austrian psychiatrist Dr. Irmfried Eberl (1910–1948; Fig. 10.47) was the first commandant of the Treblinka death camp. He committed suicide after his arrest [168]. While Germany led the field of psychiatric research for centuries, this stopped with the rise of the Nazis and the Third Reich in 1933.

In the meantime, the first double-blind placebo-controlled trial took place in 1939, when Leonard Dub and Louis Lurie at the Longview State Hospital in Cincinnati gave amphetamine or a placebo to a study sample which comprised of 48 female inpatients suffering from depression or schizophrenia. They utilized a cross-over design, and for the first time in psychiatry, the trial was double-blind. It

Fig. 10.47 Left: Werner Heyde (1902–1964). Right: Irmfried Eberl (1910–1948). Both were psychiatrists actively involved in the Nazi extermination projects

Fig. 10.48 Left: Karl Brandt (1904–1948) in 1946 during the Nuremberg trial. Center: Philipp Bouhler (1899–1945). Right: Viktor Hermann Brack (1904–1948) in 1946 during the Nuremberg trial. They were the masterminds behind the T4 extermination program

demonstrated the benefits of amphetamines in treating depression but not schizophrenia [98, 169]. Also, Andrew Salter (1914–1996) introduced a Pavlovian model of hypnotherapy and self-hypnosis training and later the Conditioned Reflex Therapy [170].

The MMPI was developed in the early 1940s by Starke Rosecrans Hathaway (1903–1980) and John Charnley McKinley (1891–1950), specifically with an atheoretical empirical criterion approach to operationalize clinical phenomena, and by the late 1950s, it was the most widely used measure of personality and psychopathology [171].

As the United States entered WWII, the military noticed that the rates of rejecting recruits for psychiatric reasons varied unacceptably among centers. As a result, a committee was established in 1941 headed by William Menninger (1899–1966), who was brother of Karl Menninger and cofounder of the Menninger Clinic, in order to provide with guidance on the issue of mental health evaluation. The committee eventually produced the War Department Technical Bulletin "M*edical 203*" *in 1943 which constitutes a* landmark in clinical psychiatry.

During both world wars, the psychological trauma of horrific experiences on soldiers was observed and described, and its presence was well established. The first reliable description was published by Charles Myers (1873–1946) [172] and William Rivers [173, 174]. In these two wars, mental health casualties accumulated up to 10% of all soldiers. As a consequence, a division of "neuropsychiatry" was created in the US Army with William C. Menninger as its first director [1]. Interestingly, in 1943, when General George Patton slapped a soldier suffering from "combat exhaustion" at a frontline hospital during the Sicily campaign, and accused him of malingering, he was relieved of command and ordered to apologize by General Dwight D. Eisenhower.

In the middle of WWII, in 1942, Ludwig Binswanger (1881–1966) published his 1942 book "Grundformen und Erkenntnis menschlichen Daseins" (Basic Forms and the Realization of Human Being in the World) [175], and the next year, Jean-Paul Sartre (1905–1980; Fig. 10.49) published his book "L'Être et le néant: Essai d'ontologie phénoménologique" (Being and Nothingness: An Essay on Phenomenological Ontology) [176]. Sartre came to be one of the most prominent existential philosophers. In 1946, he published his book "Existentialism and Humanism" where he argued that existence precedes essence, and humans are

Fig. 10.49 Jean-Paul Sartre (1905–1980) c. 1967

defined by their actions, not by some predetermined purpose. This was both in accord with an atheistic perspective abut also by his belief that there is not any fixed constitution or nature. Character continuously changes as a consequence of experience (a notion strongly similar to the "blank slate"). He attributed "free will" a dominant role; humans are absolutely responsible for their actions and decisions, and internal subconscious forces play no role. His view was that persons could choose what to fully allow into consciousness, and no blame could be attributed to an uncontrollable unconscious. His line of thinking essentially considered mental illness to be the product of free will and choice, and later, he went on to support antipsychiatric movements by also writing the introduction to the 1960 book "The Divided Self" by Ronald David Laing (1927–1989). The ideas of Sartre imply a dualistic understanding of the mind-brain problem, which in many ways, it is similar to Descartes' "substance dualism." According to Sartre, there is "being for itself" (mind) and "being in itself" (matter) [85].

However, it was Ludwig Binswanger (Fig. 10.50) who developed the "theory of existential analysis" by elaborating on Heidegger's ideas. This "Daseinsanalytik" suggested the presence of three modes of experience: the Umwelt (the realm of biological and physical energies and entities), the Mitwelt (interpersonal and social

Fig. 10.50 Ludwig Binswanger (1881–1966)

realm), and the Eigenwelt (the realm of individual inner experience and meaning). According to Binswanger, mental disorder was the result of imbalance among these three modes of experience, with the self (Eigenwelt) being the dominant. He considered "self" to be the basic feature of "identity" which he thought was the product of the synthesis of the three modes of experience and was reflecting the "being in the world." He suggested that the basic element which determines the communication between human beings is not care (in German: "*Sorge*" according to Heidegger) but love (in German: "*liebe*") [104, 177–181].

Hans Georg Gadamer (1900–2002) rejected any need for a scientific methodology in human sciences and argued for the need to reconnect with concepts like education and self-cultivation (in German: *bildungsbegrieff*) and common sense (sensus communis). According to his ideas, it is exactly because of the existence of a priori cognitive expectations that we are able to seek truth, and as a consequence, the human mental life and existence could be conceptualized as an inner "discussion" [182, 183]. His works developed further the approach called "hermeneutics" (after the Greek word "*ερμηνεία*" meaning explanation, understanding) [85]. Other existential philosophers with similar ideas were Karl Jaspers (1883–1969), Eugene Minkowski (1885–1972), and Viktor Frankl (1905–1997). Especially, Frankl shaped the so-called Third Viennese School which proposed that the main driving force in the life of the individual human being was the "will to meaning," in other words the "reason to live." If such a meaning exists, the person can bear almost any adversity. Meaning is based on values, and Frankl described three types of values: creative, experiential, and attitudinal. During the second half of the 1940s and early 1950s, Erich Fromm (1900–1980) and David Riesman (1909–2002) argued that the contemporary European and American societies were exerting a dehumanizing and isolating effect on people (alienation), leading to discomfort and eventually to mental illness. Erich Fromm (Fig. 10.51) was a social philosopher and psychologist, and his approach to mental illness was focused on the role the society

Fig. 10.51 Erich Fromm (1900–1980)

could play in its development. Essentially, he tried to bridge psychoanalysis with Marxism and eventually distanced himself from the former for the sake of the later [184, 185].

The story of psychosurgery continued after WWII when, inspired by the work of Fiamberti, Freeman developed a new "transorbital" lobotomy in 1946 (Fig. 10.52) [186–193]. This new technique involved lifting the upper eyelid and placing the point of a thin surgical instrument (called orbitoclast or leukotome) under the eyelid and against the top of the eye socket. A hammer was used to drive the orbitoclast through the thin layer of bone and into the brain along the plane of the bridge of the nose. All cuts were designed to transect the connections of the prefrontal cortex to the thalamus. Its simplicity suggested the possibility of carrying it out in mental hospitals lacking the surgical facilities required for the earlier, more complex procedures [149]. The first patient to receive this new technique was Sallie Ellen Ionesco (1917–2007), a 29-year-old woman who was psychotic, severely depressed, and violently suicidal. This was one of the rare cases the method returned a miracle therapeutic result. According to the narration of her daughter, Angelene Forester, the operation restored her in her condition before the illness developed and "gave her her mother back" [194].

Eventually, in the United States, approximately 40,000 people were lobotomized and 17,000 in the UK, while in Scandinavian countries, the frequency was 2.5 higher in comparison to the United States [195–198].

During the 1930s and the 1940s, Robert King Merton (1910–2003; Fig. 10.53) made important contributions concerning the person's behavior in the frame of the social environment. He described a number of adaptational modes of behavior (conformity, innovation, ritualism, retreatism, and rebellion), but his most notable ideas concerned the concepts of "role model" and the "self-fulfilling prophecy," which corresponds to a behavior or process through which a false or negative belief or a similar expectation the person has influences the behavior to such an extent that makes the originally false conception come true [199–201]. The same period, Harry Stack Sullivan (1892–1949; Fig. 10.54) started as a psychoanalyst, but later, he focused on the interpersonal communication processes and interactions. He was influenced by the cultural and social ideas of Alfred Adler, Karen Horney, Erich Fromm, and others. His approach followed the idea of mental disorder as being the result of social influences and especially interpersonal relationships with loneliness being the most important causal factor. He derived the idea of "significant other(s)" and introduced the term "problems in living" to describe and interpret mental disorders. He also introduced the terms "participant observer" and "consensual validation" to describe the dynamics between therapist and patient. Overall, he paved the way for the antipsychiatric movements which appeared a few decades later. He published little during his lifetime, and his influence came mainly from his lectures. Most of his work was published post-humorously by Helen Swick Perry [202–204].

In the field of psychometrics, Raymond Cattell (1905–1997) proposed the presence of 16 primary factors as the latent structure of normal personality and was one of the first to utilize factor analysis in his research [205]. Henri Ey (1900–1977) developed his "organodynamic theory" concerning the hierarchical structure of the

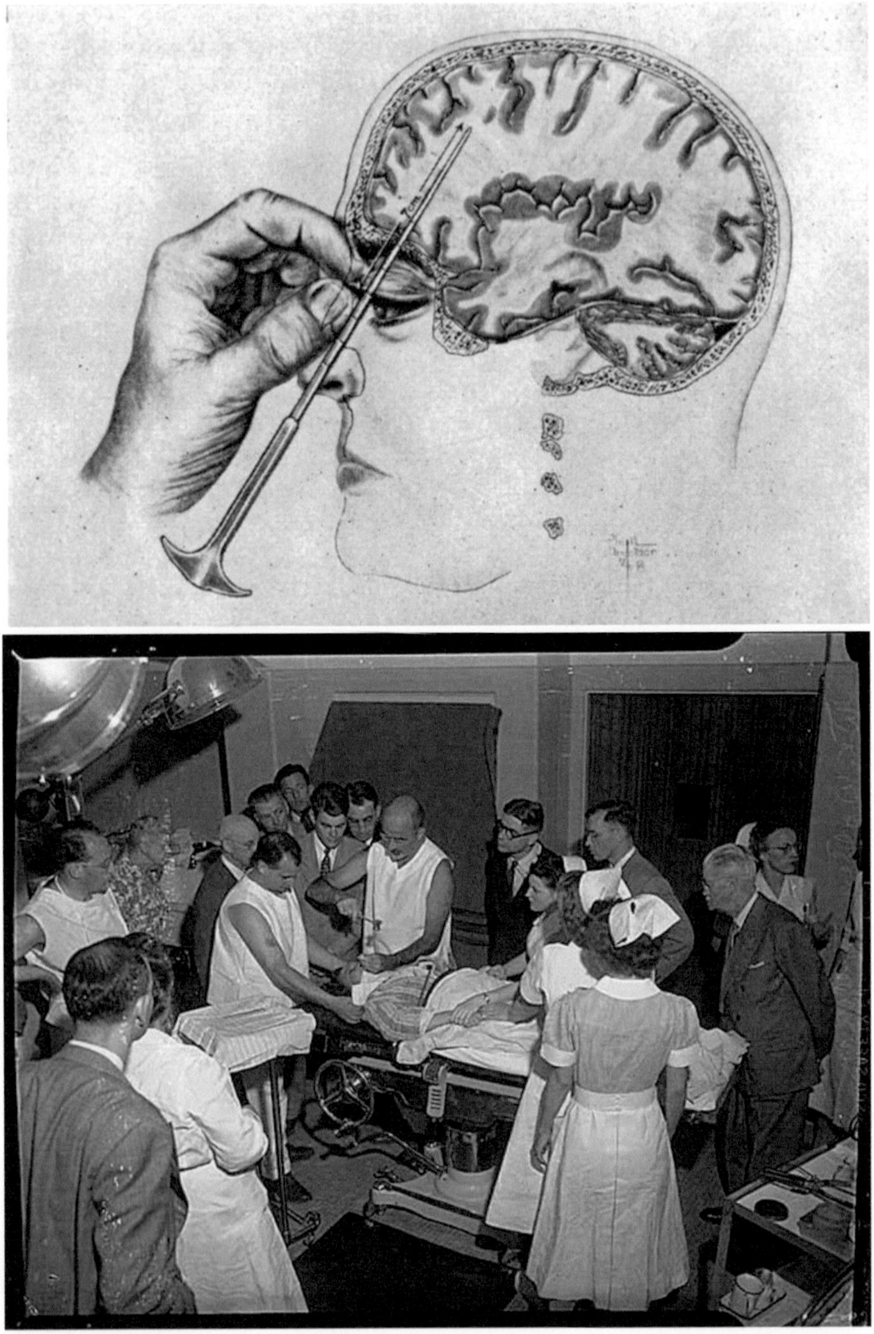

Fig. 10.52 Transorbital lobotomy by Walter Freeman (1895–1972) c. 1949 (Museum of History & Industry Photograph Collection, MOHAI, Seattle Post-Intelligencer Collection, 1986.5.25616)

Fig. 10.53 Robert King Merton (1910–2003)

states of consciousness, which included both biological as well as psychodynamic factors and was largely based on the ideas of Thomas Willis, Pierre Janet, and John Hughlings Jackson [206–208].

Albert Deutsch (1905–1975) was an impoverished journalist who in the mid-1940s wrote a number of articles in the lay press concerning the problematic condition of mental hospitals in the United States and supported his claims with a number of photographs he had taken [209]. In 1948, he wrote a book on the matter, "The Shame of the States" [210].

In 1947, Hugh Edward De Wardener (1915–2013; Fig. 10.55), on the basis of his experiences with a rice-only diet in a Japanese prison camp, suggested that Wernicke's encephalopathy was due to thiamine deficiency and treated these patients with thiamine. At that time, the Wernicke-Korsakoff syndrome was the cause behind the chronic admission of a significant number of mental patients because of alcohol abuse. As a result, thiamine treatments almost wiped out Wernicke's encephalopathy [211]. It is notable that De Wardener was arrested prisoner after the fall of Singapore like John Cade (1912–1980) who later reintroduced lithium therapy. They both spent years in Japanese concentration camps, and both made remarkable breakthrough observations during their captivity. It is also notable that Archibald Leman Cochrane (1909–1988), after whom the Cochrane

Fig. 10.54 Harry Stack Sullivan (1892–1949)

Collaboration took its name, was also taken prisoner during WWII after the battle of Crete in Greece and spent years in a concentration camp in Thessaloniki, Greece. During that time, he identified vitamin C deficiency as the cause of beriberi and applied successful treatment through a remarkable randomized placebo-controlled trial in the concentration camp [212, 213].

And then the storm hit. In September 1947, a bitter conflict between the somatic therapists and the analysts appeared for the first time openly and fiercefully. The previous year, at a meeting of the American Psychiatric Association in Chicago, the Group for the Advancement of Psychiatry (GAP) was found by the leading psychoanalyst William Menninger (1899–1966; Fig. 10.56) who was brother of Karl Menninger. In its first public bulletin, in 1947, the GAP launched a frontal attack on "shock therapy" especially on ECT suggesting it was "the object of promiscuous and indiscriminate use" [141]. This was not pure fiction since at that time, the ECT was widely used in asylums, often in punitive and untherapeutic ways in order to control patients, but the attack was also on its very scientific background and

Fig. 10.55 Hugh Edward De Wardener (1915–2013)

efficacy. Two years later, the film "Snake Pit" with Olivia de Havilland gave an unsympathetic portrayal of ECT (Fig. 10.57). The movie was based on the 1946 autobiographical book by Mary Jane Ward (1905–1981) which described her horrible experience in Rockland State Hospital, New York, after being erroneously diagnosed as suffering from schizophrenia.

In 1947 was the year the term "antidepressant" was coined for the first time in the commercial advertisement for the Benzedrine-analgesic combo "Edrisal."

Concerns on the use and consequences of psychosurgery were mounting already since 1944 [196], but Egas Moniz was awarded the Nobel Prize in 1949 "for his discovery of the therapeutic value of leukotomy in certain psychoses." The same year, he was shot by a paranoid patient (who did not have a leukotomy) and subsequently used a wheelchair [143]. The Soviet psychiatrist Vasily Gilyarovsky criticized lobotomy both in terms of efficacy and theoretical background. In his view, ECT was promoting an unacceptable mechanistic brain localization assumption [214]. Following his criticism, the Soviet Union was the first to ban this procedure on moral grounds in 1950 [215, 216].

The decades before the WWII, Melanie Klein (1882–1960) and Anna Freud (1895–1982) had formed two different schools of child development and psychoanalysis, and the debates between their followers during the 1940s (the so-called controversial discussions) split the British Psychoanalytical Society into the Kleinian, Anna Freudian, and independent training divisions, and this remains until today (Fig. 10.58).

What made Klein's "object relations theory" distinct from the rest of psychoanalytical thinking was the assumption that the final form of external relationships is

Fig. 10.56 William Menninger (1899–1966). On the right on the cover of Time magazine. (Time, October 25, 1948, © Time, Inc. Used under license)

already preformed in the brain of the infant. According to classical psychoanalysis, only object-seeking urges exist which eventually give birth to relationships which are not predetermined. Klein's theories are very close to those of Jung and bare many similarities with those of Ronald Dodds Fairbairn (1889–1964), Donald Woods Winnicott (1896–1971), and Henry Guntrip (1901–1975).

Donald Winnicott (1896–1971) was member of the Independent Group of British Psychoanalysts, and between 1930 and 1970, he developed his theories on transitional phenomena to which persons become deeply attached. The work of Carl Rogers (1902–1987) marked the beginning of a shift of psychotherapy from physicians and psychoanalysts to clinical psychologists and psychiatric social workers.

In 1948, the World Health Organization (WHO) was founded and immediately took over responsibility for the International Classification of Diseases (ICD). The sixth edition was published in 1949 and was the first to contain a section on mental disorders.

In 1949, in the Bundoora Repatriation Hospital, a veteran's hospital in a suburb of Melbourne, the Australian John Cade (1912–1980; Fig. 10.59) injected urine from patients with schizophrenia to guinea pigs to test the hypothesis that mania is caused by intoxication by a normal body element circulating in excess, while melancholia is the corresponding deprative condition. He used lithium urate as control and observed it caused the rodents to be tranquilized. In fact, it is possible that the animals were lethargic because of lithium toxicity, and this seems to be the case also with the patients of William Alexander Hammond, since the dosages reported by both researchers lead to lithium intoxication. Eventually, the same year, Cade reported positive results from the treatment of 10 acutely manic patients [217, 218];

Fig. 10.57 Poster of the 1948 film "Snake Pit" (up) and photo of a scene (down)

Fig. 10.58 Left: Melanie Klein (1882–1960). Right: Anna Freud (1895–1982)

however, 2 years later, he reported the first death because of lithium toxicity in a patient whose bipolar illness otherwise responded extremely well to treatment. Subsequent research with other ions (rubidium, cesium, lanthanum, neodymium, and strontium) gave no positive results. The story of John Cade shows how starting with a completely false theory (uric acid theory for bipolar disorder as developed in the nineteenth century) and probably through erroneous interpretation of results (the side effect of lethargy interpreted as therapeutic action) eventually led to the discovery of one of the most efficacious treatments.

Finally, two events characterized the decade of the 1940s. One was the theory of "organotherapy" developed by Wilhelm Reich (1897–1957). This was an outrageous and unscientific idea by a scientist with important contribution to psychoanalysis in the past but clearly almost psychotic during the last years of his life. This absurd theory gained much publicity, and surprisingly, it was embraced not only by large masses of laypeople but also by a large number of psychiatrists and psychologists worldwide for several years while followers of it still exist. On the contrary, an important but little known detail is the use of glutamic acid in the treatment of schizophrenia in the late years of the 1940s. Today, this seems like an early glutaminergic theory concerning the etiopathogenesis of schizophrenia [219].

Just before the end of the 1940s, US President Harry Truman established the National Institute of Mental Health (NIMH) in 1949 and appointed the psychoanalyst Robert Felix (1905–1990) as its first director. He served in this position for 15 years, and under his leadership, the NIMH refused to fund biological research and invested on the psychosocial approach.

Fig. 10.59 John Cade (1912–1980) from the Herald and& Weekly Times Portrait Collection, State Library of Victoria, Australia

References

1. Lieberman JA. Shrinks: the untold story of psychiatry. New York: Back Bay Books; 2016.
2. Torrey FE. Freudian fraud: the malignant effect of Freud's theory on American thought and culture. New York: Harper Collins; 1992.
3. Fulford KWM, Thornton T, Graham G. Phenomenology and psychopathology. In: Fulford KWM, Thornton T, Graham G, editors. The Oxford textbook of philosophy and psychiatry. Oxford: Oxford University Press; 2006b.
4. Tymieniecka A, editor. The human being in action: the irreducible element in man part II investigations at the intersection of philosophy and psychiatry (Analecta Husserliana). New York: Springer; 1978.
5. Globus GG. On Freud and Husserl. Am J Psychiatry. 1985;142(5):663–4. https://doi.org/10.1176/ajp.142.5.663b.
6. Gorner P. Reid, Husserl and phenomenology. Br J Hist Philos. 2001;9(3):545–55. https://doi.org/10.1080/09608780110072506.
7. Husserl E. Logical investigations. London: Routledge; 2001.

8. Husserl E. Ideas: general introduction to pure phenomenology. Eastford: Martino Fine Books; 2017.
9. Luft S, Schlimme JE. The phenomenology of intersubjectivity in Jaspers and Husserl: on the capacities and limits of empathy and communication in psychiatric praxis. Psychopathology. 2013;46(5):345–54. https://doi.org/10.1159/000353272.
10. Ricoeur P. Husserl: an analysis of his phenomenology. Northwestern University Press; 2007.
11. Evans RB. Titchener and American experimental psychology. Rev Hist Psicol. 1984;5(1–2):117–25.
12. Hindeland MJ. Edward Bradford Titchener: a pioneer in perception. J Hist Behav Sci. 1971;7:23–8. https://doi.org/10.1002/1520-6696(197101)7:1<23::aid-jhbs2300070104>3.0.co;2-v.
13. Krech D. Titchener on experimental psychology. Am Psychol. 1968;23(5):367–8. https://doi.org/10.1037//0003-066x.23.5.367.
14. Proctor RW, Evans R. E. B. Titchener, women psychologists, and the experimentalists. Am J Psychol. 2014;127(4):501–26. https://doi.org/10.5406/amerjpsyc.127.4.0501.
15. Rowe FB, Murray FS. A note on the Titchener influence on the first psychology laboratory in the south. J Hist Behav Sci. 1979;15(3):282–4. https://doi.org/10.1002/1520-6696(197907)15:3<282::aid-jhbs2300150308>3.0.co;2-b.
16. Foschi R, Cicciola E. Politics and naturalism in the 20th century psychology of Alfred Binet. Hist Psychol. 2006;9(4):267–89. https://doi.org/10.1037/1093-4510.9.4.267.
17. Nicolas S, Ferrand L. Alfred Binet and higher education. Hist Psychol. 2002;5(3):264–83. https://doi.org/10.1037/1093-4510.5.3.264.
18. Wolf TH. The emergence of Binet's conceptions and measurement of intelligence: a case history of the creative process. Part II. J Hist Behav Sci. 1969;5(3):207–37. https://doi.org/10.1002/1520-6696(196907)5:3<207::aid-jhbs2300050302>3.0.co;2-g.
19. Cuthbert AW. Seventh W.D.M. Paton Memorial Lecture. The man who never was--Walter Ernest Dixon FRS. Br J Pharmacol. 2001;133(7):945–50. https://doi.org/10.1038/sj.bjp.0704175.
20. Gray J. Ivan Petrovich Pavlov and the conditioned reflex. Brain Res Bull. 1999;50(5–6):433. https://doi.org/10.1016/s0361-9230(99)00180-x.
21. Pavlov IP. Conditioned reflexes: an investigation of the physiological activity of the cerebral cortex. London: Oxford University Press; 1927.
22. Pavlov IP. Conditioned reflexes (1969). JAMA. 1849–1936;209(13):2050–1.
23. Cambiaghi M, Sacchetti B. Ivan Petrovich Pavlov (1849-1936). J Neurol. 2015;262(6):1599–600. https://doi.org/10.1007/s00415-015-7743-2.
24. Haas LF. Ivan petrovich pavlov (1849-1936). J Neurol Neurosurg Psychiatry. 1999;67(3):299. https://doi.org/10.1136/jnnp.67.3.299.
25. Marks IM. The Nobel prize award in physiology to Ivan Petrovich Pavlov--1904. Aust N Z J Psychiatry. 2004;38(9):674–7. https://doi.org/10.1080/j.1440-1614.2004.01440.x.
26. Samoilov VO. Ivan Petrovich Pavlov (1849-1936). J Hist Neurosci. 2007;16(1–2):74–89. https://doi.org/10.1080/09647040600793232.
27. Kretschmer E. Physique and character. Oxford: Harcourt, Brace; 1925.
28. Sheldon W. The varieties of human physique: an introduction to constitutional psychology. New York: Harper; 1940.
29. Asprem E. A nice arrangement of heterodoxies: William McDougall and the professionalization of psychical research. J Hist Behav Sci. 2010;46(2):123–43. https://doi.org/10.1002/jhbs.20422.
30. Valentine ER. Spooks and spoofs. Hist Hum Sci. 2012;25(2):67–90. https://doi.org/10.1177/0952695111418745.
31. Beers CW. A mind that found itself: an autobiography. 1908. Am J Public Health. 2010;100(12):2354–6. https://doi.org/10.2105/ajph.100.12.2354.
32. Hierholzer R. A mind that found itself. BMJ. 2007;334(7601):1009.1002–9. https://doi.org/10.1136/bmj.39204.690995.59.

33. Scull A, Schulkin J. Psychobiology, psychiatry, and psychoanalysis: the intersecting careers of Adolf Meyer, Phyllis Greenacre, and Curt Richter. Med Hist. 2009;53(1):5–36. https://doi.org/10.1017/s002572730000329x.
34. Dewsbury DA. Psychobiology. Am Psychol. 1991;46(3):198–205. https://doi.org/10.1037//0003-066x.46.3.198.
35. Gelder MG. Aldolf Meyer. In: Berrios G, Freeman H, editors. 150 years of British psychiatry, vol. 1. London: Gaskell; 1991.
36. Kraepelin E. Manic-depressive insanity and paranoia. Edinburgh: Livingstone; 1921.
37. Bleuler E. Die Prognose der Dementia praecox (Schizophreniegruppe). Allgemeine Zeitschrift für Psychiatrie und psychischgerichtliche Medizin. 1908;65:436–64.
38. Fusar-Poli P, Politi P. Paul Eugen Bleuler and the birth of schizophrenia (1908). Am J Psychiatry. 2008;165(11):1407. https://doi.org/10.1176/appi.ajp.2008.08050714.
39. Weygandt W. Uber die mischzustande des manisch-depressiven irreseins. Munichs: JF Lechman; 1899.
40. Fulford KWM, Thornton T, Graham G. Karl Jaspers and general psychopathology. In: Fulford KWM, Thornton T, Graham G, editors. The Oxford textbook of philosophy and psychiatry. Oxford: Oxford University Press; 2006a.
41. Jaspers K. General psychopathology (trans: Hoenig J, Hamilton M, 1913). Chicago: The Johns Hopkins University Press; 1997.
42. Mundt C (2013) The philosophical roots of Karl Jasper's general psychopathology. In: Fulford KWM, Davies M, R G.T. Gipps et al. (eds) The Oxford handbook of philosophy and psychiatry. Oxford University Press Oxford. doi:https://doi.org/10.1093/oxfordhb/9780199579563.001.0001.
43. Fulford KWM, Thornton T, Graham G. Psychopathology and Methodenstreit. In: Fulford KWM, Thornton T, Graham G, editors. The Oxford textbook of philosophy and psychiatry. Oxford: Oxford University Press; 2006d.
44. Park SC. Karl Jaspers' general psychopathology (Allgemeine Psychopathologie) and its implication for the current psychiatry. Psychiatry Investig. 2019;16(2):99–108. https://doi.org/10.30773/pi.2018.12.19.2.
45. Wiggins O, Schwartz M. Edmund Husserl's influence on Karl Jaspers's phenomenology. Philos Psychiatry Psychol. 1997;4(1):15–36.
46. Wiggins OP, Schwartz MA, Spitzer M. Phenomenological/descriptive psychiatry: the methods of Edmund Husserl and Karl Jaspers. In: Spitzer M, Uehlein F, Schwartz M, Mundt C, editors. Phenomenology, language, and schizophrenia. New York: Springer; 1992. p. 46–69.
47. Kumazaki T. The theoretical root of Karl Jaspers' general psychopathology. Part 1: reconsidering the influence of phenomenology and hermeneutics. Hist Psychiatry. 2013;24(2):212–26. https://doi.org/10.1177/0957154X13476201.
48. Mundt C. Jaspers' concept of limit situation: extensions and clinical application. In: Breyer T, Fuchs T, Mundt C, editors. Karl Jaspers' psychopathology and philosophy. New York: Springer; 2014.
49. Potter N. Empathic foundations of clinical knowledge. In: Fulford KWM, Davies M, Gipps RGT, et al., editors. The Oxford handbook of philosophy and psychiatry. Oxford: Oxford University Press; 2013. https://doi.org/10.1093/oxfordhb/9780199579563.001.0001.
50. Maurer K, Volk S, Gerbaldo H. Auguste D and Alzheimer's disease. Lancet. 1997;349(9064):1546–9. https://doi.org/10.1016/S0140-6736(96)10203-8.
51. Lewis H. Boas, Franz. In: McGee RJ, Warms RL, editors. Theory in social and cultural anthropology: an encyclopedia. Thousand Oaks: SAGE; 2013.
52. Gossett T. Race: the history of an idea in America. New York: Oxford University Press; 1997.
53. Mead M. Coming of age in Samoa. Coming of age in Samoa. Oxford: Morrow; 1928.
54. Freeman D. Margaret Mead and Samoa. Cambridge: Harvard University Press; 1983.
55. Shankman P. Not even wrong: Margaret Mead, Derek Freeman, and the Samoans. Martin Orans. Am Anthropol. 1996;98(4):889. https://doi.org/10.1525/aa.1996.98.4.02a00320.
56. Shankman P. The trashing of Margaret Mead: anatomy of an anthropological controversy. Madison: University of Wisconsin Press; 2009.

57. Singer P. A Darwinian left. Yale: Yale University Press; 1999.
58. Bechterev V, Puusepp L. La chirugie des aliénes [Surgery of the insane]. Arch Int Neurol. 1912;34(1–17):69–89.
59. Feldman RP, Goodrich JT. Psychosurgery: a historical overview. Neurosurgery. 2001;48(3):647–57; discussion 657–9. https://doi.org/10.1097/00006123-200103000-00041.
60. O'Callaghan M, Carroll D. Psychosurgery: a scientific analysis. Lancaster: MTP Press Limited, Falcon House; 1982.
61. Puusepp L. Alcune considerazioni sugli interventi chirurgici nelle malattie mentali [Some Considerations about Surgery in Mental Illness]. G Acad Med Torino. 1937;100:3–16.
62. Barger G, Dale H. β-Lmidazolylethylamine, a depressor constituent of intestinal mucosa. J Physiol. 1910;41:499–503.
63. Loewi O. From the workshop of discoveries. Lawrence: University of Kansas Press; 1953.
64. Dale H. Accident and opportunism in medical research. Br Med J. 1948;2(4574):451–5. https://doi.org/10.1136/bmj.2.4574.451.
65. Jung C. Collected papers on analytical psychology. Tindall and Cox: Bailliere; 1916.
66. Jung C. Psychology and religion. Binghamton: The Vail-Ballou Press; 1960.
67. Jung C, Pauli W. The interpretation of nature and psyche. New York: Pantheon Books; 1955.
68. Cronk G. George Herbert Mead. In: Fieser J, Dowden B, editors. The internet encyclopedia of philosophy. 2005. https://www.iep.utm.edu/.
69. Ritzer G. Sociological theory. New York: McGraw-Hill; 2008.
70. Gillespie A. Becoming other: from social interaction to self-reflection. Charlotte: Information Age Publishing; 2006.
71. Mead G, Morris C. Mind, self, and society from the standpoint of a social behaviorist. Chicago: University of Chicago Press; 1967.
72. Shepherd B. A war of nerves: soldiers and psychiatrists 1914–1994. London: Pimlico; 2002.
73. Gibby RE, Zickar MJ. A history of the early days of personality testing in American industry: an obsession with adjustment. Hist Psychol. 2008;11(3):164–84. https://doi.org/10.1037/a0013041.
74. Lopez-Munoz F, Ucha-Udabe R, Alamo C. The history of barbiturates a century after their clinical introduction. Neuropsychiatr Dis Treat. 2005;1(4):329–43.
75. John D. Deep sleep: Australian doctors escape discipline. BMJ. 1993;306(6891):1501.
76. Kaplan RM. Deep sleep therapy in Australia. Australas Psychiatry. 2013;21(5):505–6. https://doi.org/10.1177/1039856213499621.
77. Stolerman IP, editor. Encyclopedia of psychopharmacology. 2nd ed. New York: Springer; 2015.
78. Swan N. Australian deep sleep report awakens anger. BMJ. 1991;302(6768):70–1.
79. Walton M. Deep sleep therapy and Chelmsford private hospital: have we learnt anything? Australas Psychiatry. 2013;21(3):206–12. https://doi.org/10.1177/1039856213486703.
80. Thomson G. Prof. Edward L. Thorndike. Nature. 1949;164(4168):474. https://doi.org/10.1038/164474a0.
81. Thorndike E. Individuality. Boston: Houghton Mifflin; 1911.
82. Baillie J. Introduction to Bertrand Russell. In: Baillie J, editor. Contemporary analytic philosophy. 2nd ed. Hoboken: Prentice Hall; 1997.
83. Glock H. What is analytic philosophy? Cambridge: Cambridge University Press; 2008.
84. Wittgenstein L. Philosophical investigations. 4th ed. New York: Wiley-Blackwell; 2009.
85. Fulford KWM, Thornton T, Graham G. Philosophical methods in mental health practice and research. In: Fulford KWM, Thornton T, Graham G, editors. The Oxford textbook of philosophy and psychiatry. Oxford: Oxford University Press; 2006c.
86. Watson JB, Rayner R. Conditioned emotional reactions. J Exp Psychol. 1920;3(1):1–14. https://doi.org/10.1037/h0069608.
87. Beck HP, Levinson S, Irons G. Finding little Albert: a journey to John B. Watson's infant laboratory. Am Psychol. 2009;64(7):605–14. https://doi.org/10.1037/a0017234.
88. Fridlund AJ, Beck HP, Goldie WD, Irons G. Little Albert: a neurologically impaired child. Hist Psychol. 2012;15(4):302–27. https://doi.org/10.1037/a0026720.

89. Cover Jones M. A laboratory study of fear: the case of Peter. Pedagog Semin. 1924;31:308–15.
90. Shaffer LS. From mirror self-recognition to the looking-glass self: exploring the justification hypothesis. J Clin Psychol. 2005;61(1):47–65. https://doi.org/10.1002/jclp.20090.
91. Krantz DL, Hall R, Allen D. William McDougall and the problem of purpose. J Hist Behav Sci. 1969;5(1):25–38. https://doi.org/10.1002/1520-6696(196901)5:1<25::aid-jhbs2300050104>3.0.co;2-s.
92. Searls D. The inkblots: Hermann Rorschach, his iconic test, and the power of seeing. New York: Crown; 2017.
93. Amann-Gainotti M. Contributions to the history of psychology: LXXXV. Jean Piaget, student of Pierre Janet (Paris 1919-1921). Percept Mot Skills. 1992;74(3 Pt 2):1011–5. https://doi.org/10.2466/pms.1992.74.3c.1011.
94. Broughton JM, Leadbeater B, Amsel E. Reflections on Piaget: proceedings of the Jean Piaget memorial conference. Teach Coll Rec. 1981;83(2):151–217.
95. Burman JT. Jean Piaget: images of a life and his factory. Hist Psychol. 2012;15(3):283–8. https://doi.org/10.1037/a0025930.
96. Schneider K. Die Schichtung des emotionalen Lebens und der Aufbau der Depressionszustände. Zeitschrift für die gesamte Neurologie und Psychiatrie. 1920;59(1):281–6. https://doi.org/10.1007/bf02901090.
97. Schneider K. Psychopathic personalities. London: Cassell; 1958.
98. Shorter E. A brief history of placebos and clinical trials in psychiatry. Can J Psychiatr. 2011;56(4):193–7. https://doi.org/10.1177/070674371105600402.
99. Kopeloff LM, Kopeloff N, Hinsie LE, Etchells JL. Tuberculous Bacillemia and dementia praecox. J Bacteriol. 1935;30(1):49–59.
100. Kopeloff N, Cheney C. Studies in focal infection: its presence and elimination in the functional psychoses. Am J Psychiatry. 1922;2:139–56.
101. Gordon F. French psychiatry and the new woman: the case of Dr Constance Pascal, 1877-1937. Hist Psychiatry. 2006;17(66 Pt 2):159–82. https://doi.org/10.1177/0957154X06056601.
102. Gordon F. Constance Pascal's Chagrins d'amour et psychoses (1935): a French psychiatrist's views on psychoanalysis. Hist Psychiatry. 2015;26(1):3–18. https://doi.org/10.1177/0957154X14542730.
103. Heidegger M. Being and time. New York: Harper Perennial; 2008.
104. Askay R, Farquhar-Lacewing J. Being unconscious: Heidegger and Freud. In: Fulford KWM, Davies M, Gipps RGT, et al., editors. The Oxford handbook of philosophy and psychiatry. Oxford University Press; 2013. https://doi.org/10.1093/oxfordhb/9780199579563.001.0001.
105. Cohen D. Looking Back: Freud and the British royal family. Psychologist. 2013;26:462–3.
106. Lindemann E. The psychopathological effect of sodium amytal. Proc Soc Exp Biol Med. 1931;28:864–6.
107. Lindemann E. Psychological changes in normal and abnormal individuals under the influence of sodium amytal. Am J Psychiatr. 1932;11(6):1083–91.
108. Moreno J. Impromptu man. 1st ed. New York: Bellevue Literary Press, NYU School of Medicine; 2014.
109. Marineau R. Jacob Levy Moreno: 1889-1974: father of psychodrama, sociometry, and group psychotherapy. London: Tavistock/Routledge; 1989.
110. Moreno J. Who shall survive?: Foundations of Sociometry, group psychotherapy, and Sociodrama. Beacon: Beacon House; 1977.
111. Nolte J. The philosophy, theory and methods of J. L. Moreno: the man who tried to become God. New York: Routledge; 2014.
112. Marmol F. Lithium: bipolar disorder and neurodegenerative diseases possible cellular mechanisms of the therapeutic effects of lithium. Prog Neuro-Psychopharmacol Biol Psychiatry. 2008;32(8):1761–71. https://doi.org/10.1016/j.pnpbp.2008.08.012.
113. Shorter E. The history of lithium therapy. Bipolar Disord. 2009b;11(Suppl 2):4–9. https://doi.org/10.1111/j.1399-5618.2009.00706.x.

114. Roy P. Global pharma and local science: the untold tale of reserpine. Indian J Psychiatry. 2018;60(Suppl 2):S277–83. https://doi.org/10.4103/psychiatry.IndianJPsychiatry_444_17.
115. Gananath S, Chandra Bose K. Rauwolfia serpentina: a new Indian drug for insanity and high blood pressure. Indian Med World. 1931;11:194–201.
116. Vakil RJ. A clinical trial of Rauwolfia serpentina in essential hypertension. Br Heart J. 1949;11(4):350–5. https://doi.org/10.1136/hrt.11.4.350.
117. Wortis J. On the response of schizophrenic subjects to hypoglycemic insulin shock. J Nerv Ment Dis. 1936;84:497–506.
118. Shorter E. Sakel versus Meduna: different strokes, different styles of scientific discovery. J ECT. 2009c;25(1):12–4. https://doi.org/10.1097/YCT.0b013e31818f5766.
119. Fink M. Induced seizures as psychiatric therapy: Ladislas Meduna's contributions in modern neuroscience. J ECT. 2004;20(3):133–6. https://doi.org/10.1097/00124509-200409000-00001.
120. Meduna LJ. The convulsive treatment: a reappraisal. J Clin Exp Psychopathol. 1954;15(3):219–33.
121. Nyirö J, Jablonszky A. Einige Daten zur Prognose der Epilepsie, mit besonderer Rücksicht auf die Konstitution. Psychiatrisch-Neurologische Wochenschrift. 1929;31:547–9.
122. Meduna LJ. Versuche über die biologische Beeinfl ussung des Ablaufes der. Schizophrenie: Campher und Cardiazol. Zeitschrift für die gesamte Neurologie und Psychiatrie. 1935;152:235–262.
123. Meduna LJ. Die Konvulsionstherapie der Schizophrenie. Halle: Marhold; 1937.
124. Polatin P, Friedman MM, Harris MM, Horwitz WA. Vertebral fractures produced by Metrazol-induced convulsions. J Am Med Assoc. 1939;112(17):1684–7. https://doi.org/10.1001/jama.1939.02800170030010.
125. Shorter E, Healy D. Shock therapy: a history of electroconvulsive treatment in mental illness. Rutgers University Press; 2007.
126. Sakel M. The classical Sakel shock treatment: a reappraisal. J Clin Exp Psychopathol. 1954b;15(3):255–316.
127. Shorter E. Before Prozac: the troubled history of mood disorders in psychiatry. Oxford: Oxford University Press; 2009a.
128. Ackner B, Harris A, Oldham AJ. Insulin treatment of schizophrenia; a controlled study. Lancet. 1957;272(6969):607–11. https://doi.org/10.1016/s0140-6736(57)91070-x.
129. Ackner B, Oldham AJ. Insulin treatment of schizophrenia. A three-year follow-up of a controlled study. Lancet. 1962;1(7228):504–6. https://doi.org/10.1016/s0140-6736(62)91477-0.
130. Fink M, Shaw R, Gross GE, Coleman FS. Comparative study of chlorpromazine and insulin coma in therapy of psychosis. J Am Med Assoc. 1958;166(15):1846–50. https://doi.org/10.1001/jama.1958.02990150042009.
131. Oldham AJ. Insulin treatment of schizophrenia--a controlled study. Anglo Ger Med Rev. 1965;2:689–701.
132. Prinzmetal M, Bloomberg W. The use of benzedrine for the treatment of narcolepsy. J Am Med Assoc. 1935;105:2051–4.
133. Moniz AE. Portuguese neurologist (1968). JAMA. 1874-1955;206(2):368–9. https://doi.org/10.1001/jama.1968.03150020084021.
134. Berrios GE. The origins of psychosurgery: Shaw, Burckhardt and Moniz. Hist Psychiatry. 1997;8(29 Pt 1):61–81. https://doi.org/10.1177/0957154X9700802905.
135. Haas LF. Neurological stamp. Egas Moniz (1874-1955). J Neurol Neurosurg Psychiatry. 2003;74(5):653. https://doi.org/10.1136/jnnp.74.5.653.
136. Valenstein E. The prefrontal area and psychosurgery. In: Uylings HBM, Van Eden CG, De Bruin JPC, Corner MA, Feenstra MGP, editors. The prefrontal cortex: its structure, function and pathology. Amsterdam: Elsevier; 1990. p. 539–54.
137. Jacobsen CF. A study of cerebral function in learning. The frontal lobes. J Comp Neurol. 1931;52(2):271–340. https://doi.org/10.1002/cne.900520205.

138. Heller AC, Amar AP, Liu CY, Apuzzo ML. Surgery of the mind and mood: a mosaic of issues in time and evolution. Neurosurgery. 2006;59(4):720–33; discussion 733–9. https://doi.org/10.1227/01.NEU.0000240227.72514.27.
139. Heller AC, Amar AP, Liu CY, Apuzzo ML. Surgery of the mind and mood: a mosaic of issues in time and evolution. Neurosurgery. 2008;62(6 Suppl 3):921–40. https://doi.org/10.1227/01.neu.0000333761.30877.8d.
140. Pressman J. Last resort: psychosurgery and the limits of medicine. Cambridge: Cambridge University Press; 2002.
141. Shorter E. A history of psychiatry: from the era of the asylum to the age of Prozac. New York: Wiley; 1997.
142. Shorter E. A historical dictionary of psychiatry. Oxford: Oxford University Press; 2005.
143. Finger S. Origins of neuroscience: a history of explorations into brain function. Oxford: Oxford University Press; 2001.
144. Freeman W. Prefrontal lobotomy: final report of 500 Freeman and Watts patients followed for 10 to 20 years. South Med J. 1958;51(6):739–45. https://doi.org/10.1097/00007611-195806000-00011.
145. Freeman W, Watts JW. Prefrontal lobotomy: the surgical relief of mental pain. Bull N Y Acad Med. 1942;18(12):794–812.
146. Freeman W, Watts JW. Prefrontal lobotomy; survey of 331 cases. Am J Med Sci. 1946;211:1–8. https://doi.org/10.1097/00000441-194601000-00001.
147. Golla FL, Freeman W, et al. Discussion: prefrontal leucotomy with reference to indications and results. Proc R Soc Med. 1946;39(8):443–58.
148. Armocida G. [A.M. Fiamberti and 'psycosurgery']. Med Secoli. 2007;19(2):457–74.
149. El-Hai J. The Lobotomist: a maverick medical genius and his tragic quest to rid the world of mental illness. Hoboken: Wiley; 2005.
150. Allport GW, Odbert HS. Trait-names: a psycho-lexical study. Psychol Monographs. 1936;47(1):i–171. https://doi.org/10.1037/h0093360.
151. John OP, Angleitner A, Ostendorf F. The lexical approach to personality: a historical review of trait taxonomic research. Eur J Personal. 1988;2(3):171–203. https://doi.org/10.1002/per.2410020302.
152. Bradley C. The behavior of children receiving Benzedrine. Am J Psychiatr. 1937;94(3):577–85. https://doi.org/10.1176/ajp.94.3.577.
153. Strohl MP. Bradley's Benzedrine studies on children with behavioral disorders. Yale J Biol Med. 2011;84(1):27–33.
154. Nathanson M. The central action of betaaminopropylbenzene (benzedrine) clinical observations. JAMA. 1937;108(7):528–31.
155. Wilbur D, MacLean A, Allen E. Clinical observations on the effect of benzedrine sulfate: a study of patients with states of chronic exhaustion, depression and psychoneurosis. JAMA. 1937;109(8):549–54.
156. Lewis A. States of depression: their clinical aetiological differentiation. Br Med J. 1938;4:875–8.
157. Gazdag G, Ungvari GS. Electroconvulsive therapy: 80 years old and still going strong. World J Psychiatry. 2019;9(1):1–6. https://doi.org/10.5498/wjp.v9.i1.1.
158. Wechsler D. The measurement of adult intelligence. Baltimore: Williams & Witkins; 1939.
159. Wechsler D. Wechsler intelligence scale for children. New York: The Psychological Corporation; 1949.
160. Rost KL. [Euthanasia films in Nazi Germany: the social history and film history background to a seduction toward death]. Zeitgesch. 2001;28(4):214–27.
161. Schmidt U. Medical films, ethics and euthanasia in Nazi Germany: the history of medical research and teaching films of the Reich Office for Educational Films, Reich Institute for Films in science and education, 1933-1945. Abh Gesch Med Naturwiss. 2002;92:1–394.
162. Makari G. Revolution in mind: the creation of psychoanalysis. New York: HarperCollins; 2008.

163. Benedict S, Shields L, O'Donnell AJ. Children's "euthanasia" in Nazi Germany. J Pediatr Nurs. 2009;24(6):506–16. https://doi.org/10.1016/j.pedn.2008.07.012.
164. Meusch M. Hadamar: a German psychiatric treatment center in WWII. Biomol Eng. 2001;17(2):65–9. https://doi.org/10.1016/s1389-0344(00)00067-8.
165. Strous RD. Nazi euthanasia of the mentally ill at Hadamar. Am J Psychiatry. 2006;163(1):27. https://doi.org/10.1176/appi.ajp.163.1.27.
166. Peters UW. German psychiatry. In: Freeman H, editor. A century of psychiatry. London: Mosby; 1999.
167. Bachrach S. In the name of public health--Nazi racial hygiene. N Engl J Med. 2004;351(5):417–20. https://doi.org/10.1056/NEJMp048136.
168. Lifton R. The Nazi doctors: medical killing and the psychology of genocide. New York: Basic Books; 1986.
169. Dub L, Lurie L. Use of Benzedrine in the depressed phase of the psychotic state. Ohio State Med J. 1939;35:39–45.
170. Salter A. Three techniques of autohypnosis. J Gen Psychol. 1941;24(2):423–38. https://doi.org/10.1080/00221309.1941.10544386.
171. Hathaway SR, McKinley JC. A multiphasic personality schedule (Minnesota): I. Construction of the schedule. J Psychol. 1940;10(2):249–54. https://doi.org/10.1080/00223980.1940.9917000.
172. Myers C. A contribution to the study of Shell shock. Lancet. 1915;185(4772):316–20. https://doi.org/10.1016/s0140-6736(00)52916-x.
173. Rivers WH. The repression of war experience. Proc R Soc Med. 1918;11(sect psych):1–20.
174. Rivers WH. Psychiatry and the war. Science (New York, NY). 1919;49(1268):367–9. https://doi.org/10.1126/science.49.1268.367.
175. Binswanger L. Grundformen und Erkenntnis menschlichen Daseins. Zürich: Niehans; 1942.
176. Sartre J. Being and nothingness: an essay on phenomenological ontology. New York: Washington Square Press; 1993.
177. Ghaemi SN. Rediscovering existential psychotherapy: the contribution of Ludwig Binswanger. Am J Psychother. 2001;55(1):51–64. https://doi.org/10.1176/appi.psychotherapy.2001.55.1.51.
178. Havens LL. The development of existential psychiatry (Karl Jaspers, E. Minkowski, and Otto Binswanger). J Nerv Ment Dis. 1972;154(5):309–31. https://doi.org/10.1097/00005053-197205000-00001.
179. Holt H. Ludwig binswanger (1881-1966): a tribute. J Existent. 1966;6(25):93–6.
180. Marazia C. Philosophical whitewashing. Ludwig Binswanger (1881-1966) and the sterilisation of manic-depressive patients. Med J. 2011;46(2):134–54.
181. Reppen J. Ludwig Binswanger and Sigmund Freud: portrait of a friendship. Psychoanal Rev. 2003;90(3):281–91. https://doi.org/10.1521/prev.90.3.281.23619.
182. Gadamer H. The relevance of the beautiful and other essays. Cambridge: Cambridge University Press; 1986.
183. Palmer R. Hermeneutics: interpretation theory in Schleiermacher, Dilthey, Heidegger, and Gadamer. Evanston: Northwestern University Press; 1969.
184. Fromm E. Escape from freedom. New York: Henry Holt; 1941.
185. Fromm E. The sane society. New York: Rinehart & Company; 1955.
186. Cheng S, Tait HS, Freeman W. Transorbital lobotomy versus electroconvulsive therapy in the treatment of mentally ill tuberculous patients. Am J Psychiatry. 1956;113(1):32–5. https://doi.org/10.1176/ajp.113.1.32.
187. Freeman W. Transorbital leucotomy. Lancet. 1948;2(6523):371–3. https://doi.org/10.1016/s0140-6736(48)90947-7.
188. Freeman W. Transorbital lobotomy; survey after from 1 to 3 years. Dis Nerv Syst. 1949;10(12):360–3.
189. Freeman W. Transorbital lobotomy; an extension of Fiamberti's method for use in severe schizophrenia. Sist Nerv. 1950;2(1):1–8.

190. Freeman W. Lobotomy; a comparison of prefrontal lobotomy with transorbital lobotomy. Surg Gynecol Obstet. 1951;92(5):603–6.
191. Freeman W. Transorbital lobotomy in state mental hospitals. J Med Soc N J. 1954;51(4): 148–50.
192. Freeman W, Williams JM. The lesions of transorbital lobotomy. Trans Am Neurol Assoc. 1951a;56:236–7.
193. Freeman W, Williams JM. Lesions of transorbital lobotomy. AMA Arch Neurol Psychiatry. 1951b;66(2):191–8. https://doi.org/10.1001/archneurpsyc.1951.02320080075007.
194. Forester A. 2008. http://www.pbs.org/wgbh/americanexperience/features/lobotomist-angelene-forester/.
195. Goldbeck-Wood S. Norway compensates lobotomy victims. BMJ. 1996;313(7059):708. https://doi.org/10.1136/bmj.313.7059.708a.
196. Ogren K, Sandlund M. Psychosurgery in Sweden 1944-1964. J Hist Neurosci. 2005;14(4):353–67. https://doi.org/10.1080/096470490897692.
197. Tranoy J. Lobotomy in Scandinavian psychiatry. J Mind Behav. 1996;17(1):1–20.
198. Tranoy J, Blomberg W. Lobotomy in Norwegian psychiatry. Hist Psychiatry. 2005;16(61 Pt 1):107–10. https://doi.org/10.1177/0957154X05052224.
199. Biggs M. Self-fulfilling prophecies. In: Hedström P, Bearman P, editors. The Oxford handbook of analytical sociology. Oxford: Oxford University Press; 2009.
200. Merton RK. The unanticipated consequences of purposive social action. Am Sociol Rev. 1936;1(6):894–904. https://doi.org/10.2307/2084615.
201. Merton RK. The self-fulfilling prophecy. Antioch Rev. 1948;8(2):193–210. https://doi.org/10.2307/4609267.
202. Evans Iii FB. Harry Stack Sullivan: interpersonal theory and psychotherapy. Florence: Taylor & Frances/Routledge; 1996.
203. Rioch DM. Recollections of Harry stack Sullivan and of the development of his interpersonal psychiatry. Psychiatry. 1985;48(2):141–58. https://doi.org/10.1080/00332747.1985.11024276.
204. Swick Perry H. Psychiatrist of America: the life of Henry stack Sullivan. Cambridge: Belknap Press; 1982.
205. Cattell RB. The description of personality: basic traits resolved into clusters. J Abnorm Soc Psychol. 1943;38(4):476–506. https://doi.org/10.1037/h0054116.
206. Brisset C, Bernard P. [Henri Ey (1900--1977)]. Nouv Presse Med. 1978;7(1):54.
207. Charlin A. [Obituary address in memory of Dr. Henri Ey]. Ann Med Psychol (Paris). 1978;136(1):121–125.
208. Kammerer T. [In memoriam Henri Ey]. Nervenarzt. 1978;49(6):313–315.
209. Weiss KJ. Albert Deutsch, 1905-1961. Am J Psychiatry. 2011;168(3):252. https://doi.org/10.1176/appi.ajp.2010.10101491.
210. Deutsch A. The shame of the states. New York: Harcourt, Brace; 1948.
211. De WH, Lennox B. Cerebral beriberi (Wernicke's encephalopathy); review of 52 cases in a Singapore prisoner-of-war hospital. Lancet. 1947;1(6436):11–7.
212. Cochrane AL. Sickness in Salonica: my first, worst, and most successful clinical trial. Br Med J (Clin Res Ed). 1984;289(6460):1726–7. https://doi.org/10.1136/bmj.289.6460.1726.
213. Stavrou A, Challoumas D, Dimitrakakis G. Archibald Cochrane (1909-1988): the father of evidence-based medicine. Interact Cardiovasc Thorac Surg. 2014;18(1):121–4. https://doi.org/10.1093/icvts/ivt451.
214. Gilyarovsky V. Pavlov's teaching is the basis of psychiatry (in Russian). Health Worker. 1950;37(14 Sep).
215. Diefenbach GJ, Diefenbach D, Baumeister A, West M. Portrayal of lobotomy in the popular press: 1935-1960. J Hist Neurosci. 1999;8(1):60–9. https://doi.org/10.1076/jhin.8.1.60.1766.
216. Lichterman BL. On the history of psychosurgery in Russia. Acta Neurochir. 1993;125(1–4): 1–4. https://doi.org/10.1007/bf01401819.
217. Cade JF. Lithium salts in the treatment of psychotic excitement. Med J Aust. 1949;2(10): 349–52. https://doi.org/10.1080/j.1440-1614.1999.06241.x.

218. Cade JF. Lithium salts in the treatment of psychotic excitement. 1949. Bull World Health Organ. 2000;78(4):518–20.
219. Kitzinger H, Arnold DG, et al. A preliminary study of the effects of glutamic acid on catatonic schizophrenics. Rorschach Res Exch J Proj Tech. 1949;13(2):210–8. https://doi.org/10.1080/10683402.1949.10381459.

Chapter 11
From the Second Half of the Twentieth Century to the Early Twenty-First: The Psychopharmacologic Era

The second half of the twentieth century started with a symbolic event. It was the establishment of the World Psychiatric Association (WPA) in 1950 (Fig. 11.1). The first president was Jean Delay, secretary was Henry Ey, and the first congress took place in Paris. The association was formally and legally established in 1961. In 1957 in Zurich, the International College of Neuropsychopharmacology (CINP) was established.

The first randomized controlled trial (RCT) in psychiatry took place in 1950 and was designed to test claims that cortisone was useful in treating psychotic disorders. The results demonstrated that it was not [1]. Psychoanalysts reacted by arguing that the processes being evaluated were too complex to be assessed by this method [2].

In 1951, Carl Ransom Rogers (1902–1987) developed the client-centered therapy. He rejected the term "patient" and introduced the term "client" instead of arguing that the term "patient" implied hierarchy, while the term "client" suggested equality. Rogers agreed with the ideas of Rousseau that people are naturally disposed to be kind, socially cooperative, and healthy and that restrictions in this potential were at the root of mental disorders [3].

The next year 1952, Hans Eysenck (1916–1997; Fig. 11.2) dismissed psychoanalysis as untestable and therefore inherently unscientific [4]. Eysenck was one of the first advocates of what later was called the "evidence-based approach." Unfortunately, there is strong evidence that his research career was characterized by problematic conflicts of interest as well as unscientific claims supported by problematic if not fraud data. For many, his ideas reflected an early version of an antipsychiatric approach [5, 6]. His ideas are discussed further in the part of the text concerning the 1970s in the current chapter. The same year the DSM-I was published, a blend of psychoanalytic and Meyerian views. Around the same time, existential psychotherapy appeared, based among others on the work of Karl Jaspers (1883–1969). It emphasized interactions and feelings here and now rather than past

K. N. Fountoulakis, *Psychiatry*, https://doi.org/10.1007/978-3-030-86541-2_11

Fig. 11.1 Establishment of the WPA during the first World Congress in Paris 1950

Fig. 11.2 Hans Eysenck (1916–1997)

events and interpretation [7]. The same period, an approach termed Familienpflege (family care) was popular in Central Europe. In England the concept of the "therapeutic community" was developed by Maxwell Jones (1907–1990). This type of treatment approach was based on social learning theories and concerned the creation of democratic-analytic groups where patients participated in equal terms with therapists and included discussion groups, educational films, psychodrama, and discussions of community life [8].

The active component reserpine was isolated in 1952 [9], and soon the term "tranquilizer" was introduced for the first time to describe its psychotropic properties. The same year the first proper RCT in psychiatry was conducted by Louis Lasagna (1923–2003; Fig. 11.3), who later was to become the dean of American pharmacology. In 1952 he started a trial to compare chloral hydrate and pentobarbital sodium with the recently introduced methylparafynol and with placebo. This was a properly designed double-blind, placebo-controlled, four-arm study with patients properly randomized [10].

In 2005, the Hungarian-born neurosurgeon George Udvarhelyi (1920–2010) publicly claimed that the famous Evita Peron underwent lobotomy for the treatment of her cancer pains, but later it was hypothesized that it was also organized by her husband General Juan Domingo Peron, President of Argentina, at that time, in order to calm her down, decrease her belligerence, and avoid a civil war. If it happened it was carried out in maximum secrecy by the distinguished American neurosurgeon James L. Poppen (1903–1978), and a recorded and videotaped interview with a former scrub nurse and confidante of him suggests that he also performed lobotomies on prisoners in the prisons of Buenos Aires probably to better test the procedure before operating on Evita [11, 12].

During the 1950s several researchers studied lithium and its usefulness in bipolar disorder [13]. However the important contribution that made the difference came from Denmark in 1952, when Erik Stromgren (1909–1993), Head at that time of the

Fig. 11.3 Louis Lasagna (1923–2003) Foto used with permission of Tufts Center for the Study of Drug Development

Aarhus University psychiatric clinic in Risskov, asked Mogens Schou (1918–2005; Figs. 11.4 and 11.25) to undertake a randomized controlled trial of lithium in mania [14]. Mogens Schou was not a psychiatrist but a physician specialized in clinical chemistry and had observed a dramatic therapeutic effect of long-term lithium treatment in his younger brother. Schou randomized acutely manic patients with a flip of a coin to lithium or placebo, and every 2 weeks, placebo and lithium were alternating. Some of the patients entered the study in a proper RCT way. In 1954 he published the results which made a significant impact [15]. However, lithium was still difficult to administer, and blood levels a matter of guesswork. The situation changed with the introduction of the Coleman flame photometer in 1958 which made the monitoring of plasma lithium levels much more precise in comparison to the previously used Beckman photometer.

While pharmaceutical research was ongoing since the eighteenth century, and in spite of the knowledge on the usefulness of lithium since the late nineteenth century, no efficacious treatment options were available until the 1950s. This decade is considered to be the time when the psychopharmacology revolution happened; however this is not entirely true. This beginning is linked to the introduction of

Fig. 11.4 Mogens Schou (1918–2005)

antipsychotics; however their synthesis is the end result of a long research journey which originated in the dye industry and initially concerned an endeavor to develop anti-malaria treatments [16, 17]. In the UK in 1856, Sir William Henry Perkin (1838–1907) synthesized the first aniline dye, called mauve. This was followed by other aniline-based dyes, and in 1876 the German chemist Heinrich Caro (1834–1910) developed the first fully artificially synthesized medication, which could also be used as a dye, the methylene blue, which was a phenothiazine derivative. In 1883, the German August Bernthsen (1855–1931; Fig. 9.28) identified the phenothiazine core in the methylene blue molecule [18].

The methylene blue is an extraordinary molecule. In 1899 in Italy, Pietro Bodoni used it to treat psychotic agitation, but probably because of the development of barbiturates, its use never spread. It is interesting that in 1986 compelling data emerged on its usefulness as a prophylactic treatment in bipolar disorder [19]. Biochemically it is closely related to iminodibenzyl, or summer blue, which is the origin for tricyclic antidepressants as well as many anticonvulsants including carbamazepine.

Because during WWI Germany was cut off from quinine supplies, an effort was made to develop alternative anti-malaria treatments. During WWII it was the Allies that cut off from quinine supplies and experimented with the phenothiazine ring to produce synthetic anti-malaria agents. Previously, Paul Ehrlich (1854–1915) had reported in 1891 that methylene blue was effective in treating the symptoms of malaria, while in 1937 Daniel Bovet (1907–1992; Fig. 11.5) synthesized thymoxyethyldiethylamine. This was the first antihistaminergic, and he received the Noble Prize in 1957 among others also for his work on antihistaminergics and drugs affecting the brain. In 1944 Henry Gilman (1893–1986) produced a number of phenothiazine derivatives which however were not efficacious as anti-malaria drugs. Eventually at the Rhone-Poulenc laboratories, promethazine was synthesized in 1947, as a potent antihistamine agent, and soon it was obvious that it possesses strong sedative properties. Later, in 1988, James Black (1924–2010; Fig. 11.5) would receive also the Nobel Prize for his work on histamine receptors, during the 1950s and 1960s.

In 1945, the French surgeon Pierre Huguenard (1924–2006) used promethazine together with pethidine as part of a cocktail to induce relaxation and indifference in surgical patients. In 1950, another surgeon, Henri Laborit (1914–1995; Fig. 11.6), suggested that promethazine could stabilize the central nervous system by causing "artificial hibernation" and described this state as "sedation without narcosis." The same year again in Paris, Paul Guiraud (1882–1974) tried promethazine in patients with schizophrenia and agitated psychosis.

Laborit and Huguenard were surgeons at an era when surgeons were often responsible for anesthesia also. Laborit suggested to Rhône-Poulenc that they develop a compound with better stabilizing properties according to the concept of the "lytic cocktail" as this was conceptualized by both Laborit and Huguenard. Laborit was trying "hibernotherapy," that is, he was trying to reduce body metabolism during the operation through hypothermia. In his mind, antihistaminergics could stabilize the body reaction to stress because of hypothermia [20].

Fig. 11.5 Left: Daniel Bovet (1907–1992), Nobel laureate 1957 for his work on antihistaminergics and drugs affecting the brain. Right: James Black (1924–2010) Nobel laureate 1988 for his work on histamine receptors

Fig. 11.6 Top: Henri Laborit (1914–1995). Middle: Pierre Deniker (1917–1998). Bottom: Jean Delay (1907–1987)

In response to this demand, the chemist Paul Charpentier (Fig. 11.7) produced a series of compounds in 1950 and selected chlorpromazine as the one with the least peripheral activity. The testing showed that chlorpromazine had a plethora of pharmacological effects since it exerted an effect on the sympathetic system by reversing the effects of epinephrine, it manifested anticholinergic and anti-emetic effects, it

Fig. 11.7 Left: Paul Charpentier. Right: Simone Courvoisier

potentiated barbiturates, and in animals it had marked cardiovascular effects. Simone Courvoisier (Fig. 11.7) found that chlorpromazine produced indifference to aversive stimuli in rats. In perhaps the first industrial use of a behavioral test to screen for pharmacological properties, Courvoisier used the rope climbing test to explore the properties of chlorpromazine [21]. This test was developed in 1915 by David Macht (1882–1961) for his studies on opioid salts. The testing showed that the lack of response on the rope test following administration of chlorpromazine was more clear-cut than with any of the other compounds. Failure to respond was not because of sedation or impaired coordination but because of some sort of indifference [18].

Chlorpromazine was first used by Laborit at the Val-de-Grâce military hospital in Paris, as an anesthetic booster in intravenous doses of 50–100 mg on surgery patients. The trials confirmed that this "lytic cocktail" of chlorpromazine, promethazine, and an analgesic was superior to the agents usually used at that time, it reduced the shock, and the patients were feeling psychologically better after the operation ([22]). Laborit also noted a hypothermic effect and suggested it may induce artificial hibernation. He thought this might allow the body to better tolerate major surgery by reducing shock, which was a novel idea at the time. Rhône-Poulenc released chlorpromazine onto the market in 1953 under the trade name Largactil, which was derived from "large" and "activity." It is interesting that another trade name was "Hibernal" probably because of the wide acceptance of the method of sleep therapy, while the third widely known trade name was "Thorazine" and was derived from a rearrangement of various elements in the full chemical name.

Laborit went further, introduced the term "ataractic" from Greek "αταραξία" (calmness, placidity), and recommended that chlorpromazine could be used to manage combat stress reactions. In Korea some US soldiers injured in battlefield became indifferent after taking it and probably lost rescue opportunities, and subsequently they died. After these incidents, chlorpromazine use in combat situations stopped. The French military also used it for various purposes from stress alleviation to prevention of sea-sickness often with devastating results [18].

Apart from his role in the establishing chlorpromazine as an anesthetic drug, the role of Laborit in its use in mental patients is unclear, and the question, which was the first psychiatrist to use chlorpromazine, is difficult to answer [23]. This confusion and dispute is the main reason behind the fact that the Nobel Prize Committee never recognized any of the discoverers of chlorpromazine despite its revolutionary clinical significance. Records suggest that it was Léon Chertok (1911–1993), a psychiatrist interested in hypnosis and which was awarded with the "Croix de Guerre" for his actions during WWII, who suggested for the first time its use in mental patients to Marcel Montassut who at that time was Médecin-Chef at the Psychiatrique de Villejuif hospital. Together in 1951 they administered an intravenous dose to the fellow psychiatrist Cornelia Quarti (1923–1984) who acted as a volunteer [24]. Cornelia "Mimma" Quarti was also an active member of the Italian resistance during WWII in Bergamo, Italy, and from her side, this was probably an act of continuous dedication in the fight for a better world. The first trial on mental patients was carried out in 1952 by Pierre Hamon, Jean Paraire, and Jean Velluz at the Val-de-Grâce hospital, probably after suggestions by Laborit to use it along with artificial hibernation to stabilize the nervous system. Essentially these three psychiatrists were the first to start working with chlorpromazine but not the first to publish, and probably they used it in combination with barbiturates. The first reported chlorpromazine-treated patient was a 57-year-old laborer who was admitted to the Val-de-Grace because of erratic, uncontrollable behavior probably in the frame of acute mania. Response was evident already at day 1, and complete remission was achieved after 3 weeks of treatment [25].

Officially the breakthrough development came when Pierre Deniker (1917–1998; Fig. 11.6) started using chlorpromazine together with hibernation ("hibernotherapie") [26, 27] in psychotic patients at the Hôpital Sainte-Anne in Paris. However, soon it was obvious that the hospital pharmacy was unable to provide with the quantities of ice necessary for the hibernation treatment of such a great number of patients. Therefore, the only option was to give chlorpromazine alone, and subsequently Jean Delay (1907–1987; Fig. 11.6) and Pierre Deniker concluded that it was the net biochemical effect of the substance in the human brain alone that was causing the therapeutic effect [20]. The first publication of strong evidence on the efficacy of chlorpromazine monotherapy with daily injections at dosages of 75–100 mg daily in 38 psychotic patients without the use of other sedating agents or any hibernation technique came in 1952 by Delay and Deniker which since are considered the fathers of modern psychopharmacology. Their paper was published shortly before that of Hamon et al. [28].

In March 1953, *The New York Times* reported that Dr. Hakim of Ahmedabad in India had received a gold medal for reporting the efficacy of a medication named "Siledin" which contained Rauwolfia among other natural ingredients, in the treatment of patients with schizophrenia during a local medical congress. These findings reported response rates of 80% to the combination "Siledin" and ECT in 146 patients with manic depression or schizophrenia [18, 29, 30]. The first placebo-controlled trial with reserpine in psychotic patients was carried in 1953 by Leo Hollister (1920–2000) [31]. However, reserpine soon fell out of use partially because it presented no commercial interest.

In the meantime, the announcement of the therapeutic results with chlorpromazine was received with reservation. However subsequent experiments in 1954 and 1955 with chlorpromazine [32–36] and reserpine [31, 37] indicated that disinterested or hostile observers could not in good conscience claim that these drugs had no effect, whatever they might think about their overall benefits. In 1956 Delay and Deniker published the results from the use of chlorpromazine in 300 patients [38]. The same year the first case of what later would be called "neuroleptic malignant syndrome" was reported [39]. It received its name by Delay and Deniker in 1960 (syndrome malin des neuroleptiques) [40, 41]. The efficacy of chlorpromazine was eventually solidly established in 1960 with the conduction of properly designed studies [42, 43].

It is interesting that allergic reactions were reported even by nurses as a result of their contact with the drug, and this probably embedded the view of chlorpromazine as a potent agent. Such reactions never occurred since [18]. It is also interesting that when parkinsonian side effects were observed for the first time [44], most neurologists which were asked of their opinion considered the cases as "hysterical," and to conclude their embarrassment from the clinical picture, they were seeing some suggested that these agents made patients prone to hysteria ("hysterizing" agents). However, by 1955 the case of extrapyramidal side effects was widely accepted, and anticholinergic agents were administered.

In the meantime, Paul MacLean (1913–1990; Fig. 10.33) coined the term "limbic system" in 1952 as well as the term "triune brain" to describe a circuit more or less similar to that of Papez. According to MacLean there are three brains, not one (the "triune brain") [45]. Each represents a different stratum of evolution and is formed on top of the evolutionary older layer. He named them reptilian or primitive brain, limbic system or Paleomammalian brain, and neocortex or Neomammalian brain. He suggested that they operate in close cooperation but conflicts between them are not rare. Essentially this was an advanced and more elaborated rephrase of the ideas of John Hughlings Jackson and remotely of Thomas Willis.

Earlier Frieda Fromm-Reichmann (1889–1957) had introduced the concept of the "schizophrenogenic mother" as the cause of psychosis. She argued that mother could induce psychosis to her child through a toxic pattern of behavior [46]. Building on this, in the 1950s, Gregory Bateson (1904–1980), husband of Margaret Mead and prominent anthropologist, developed the "double bind" theory of schizophrenia. According to him, not only the mother is the sickest member of the family, but the core problem responsible for the emergence of psychosis is conflicting demands (the double bind) [47]. These ideas had been widely accepted for decades and caused many severe problems within families [48–50]. Parents and especially mothers had to endure the tragedy of having a kid with mental illness and additionally to be blamed for this. It is important to note that these theories emerged without any empirical support at all, only of the basis of pure intellectual exercise. At that time, theorizing was almost out of control, and also the wording of the constructing of theories was often ridiculous as in the case of Melanie Klein, which suggested that "as the child spits out the mother's milk, fearing the mother will revenge herself because of his hatred of her" [51].

In terms of nosology and classification, for most of the twentieth century, psychiatry in Europe continued to follow the biomedical model, while in the USA, the psychosocial model and psychoanalytical approaches were dominant. Sir Aubrey Lewis (1900–1975; Fig. 11.8) proposed a continuum model from anxiety to psychosis [52], while on the contrary, Sir Martin Roth (1917–2006; Fig. 11.8) suggested categorization of these disorders. In 1957 Karl Leonhard (1902–1988; Fig. 11.8) proposed that the term "bipolar disorder" should replace "manic-depression," and he also made a distinction between "monopolar" (unipolar depression) and "bipolar" illness. Eventually his suggestion was officially and widely adopted and included in the DSM-III. He also developed a classification of bipolar disorder and distinguished it from cycloid psychoses. His work along with that of Carlo Perris (1928–2000) and George Winokur (1925–1996; Fig. 11.8) helped establishing the distinction between unipolar and bipolar depression [53–55].

It seems that the first thoughts on the need of a reliable classification of mental disorders emerged during the efforts of the World Health Organization (WHO) to develop ICD-8 in New York in 1959. Official historiography credits the philosopher Carl Gustav Hempel (1905–1997) with this idea [56, 57], but more recent research into the records suggests that Hempel although lectured on the advantages and disadvantages of a causal-theoretical vs. a purely descriptive approach, clearly supported the former. It was Sir Aubrey Lewis who advocated for the descriptive approach, but his suggestions were rejected by the majority of the time [58, 59].

Starting in 1953 and until 1981, Jacques Marie Émile Lacan (1901–1981; Fig. 11.9) developed an approach to psychoanalysis with an emphasis on language and the need to make contact with the pre-language period in the unconscious. He established a whole school of psychoanalytic psychotherapy bearing his name.

Claude Levi-Strauss (1908–1990; Fig. 11.10) is best known as the father of "structural anthropology." He worked on many aspects of human society from an anthropological standpoint and suggested that the alliance between members of the society, which provides solidarity, is achieved through marriage rather than ancestral descent. Concerning mental processes and mechanisms, he suggested that they are identical across cultures although their external manifestations might grossly differ. As he wrote in his seminal book *Tristes Tropiques*, the minds of "savage" people are identical to those of "civilized" people [60–62].

Fig. 11.8 Left: Sir Aubrey Lewis (1900–1975) c. 1969, Center left: Sir Martin Roth (1917–2006), Center right: Karl Leonhard (1902–1988), Right: George Winokur (1925–1996)

Fig. 11.9 Jacques Marie Émile Lacan (1901–1981)

Fig. 11.10 Claude Levi-Strauss (1908–1990)

The discovery of chlorpromazine led to the development of other antipsychotics and also paved the road for the development of antidepressants. This revolution in psychopharmacology had also profound implications for the diagnostic approach and classification, since it encouraged the categorical classification of patients into discrete nosological entities rather than describing them according to a dimensional approach. This process, which reflected a return to the biomedical model and keeping distance from psychoanalytical approaches, had had already started years ago.

The discovery of psychopharmacotherapy simply provided an external validating criterion to the nosological concepts summarized by Kraepelin half a century before.

But the psychopharmacology revolution had also social implications too. For the first time, a large number of mental patients could be released to the community with their symptoms under control [63]. This spread concern to society. There are reports that petitions were drawn up protesting the discharge of these patients. Also mayors and curates visited Vinatier and Bassens hospitals to persuade the doctors not to release patients while patients' relatives and neighbors issued warnings that they knew better concerning the patient's true nature than the doctors did [18]. In turn this led to the introduction of the sectorizing of psychiatric services, with teams of mental health professionals being responsible for patients within specific areas and the emergence of community psychiatry. In the USA in 1955, the Joint Commission on Mental Illness and Health was established, and in 1961 it issued a report entitled Action for Mental Health. This led to the Community Mental Health Act in 1963 which mandated the establishment of community mental health centers as an alternative to hospitals. However, this initiative led to the predominance of the public health model over the clinical model and to the creation of services focusing on emotional distress and psychosocial problems faced by people in disadvantaged communities. Thus, it failed to effectively cover severe and enduring mental illness and eventually led to the creation of new homeless and imprisoned populations comprised by mental patients that would previously live in asylums (Fig. 11.11) [64, 65].

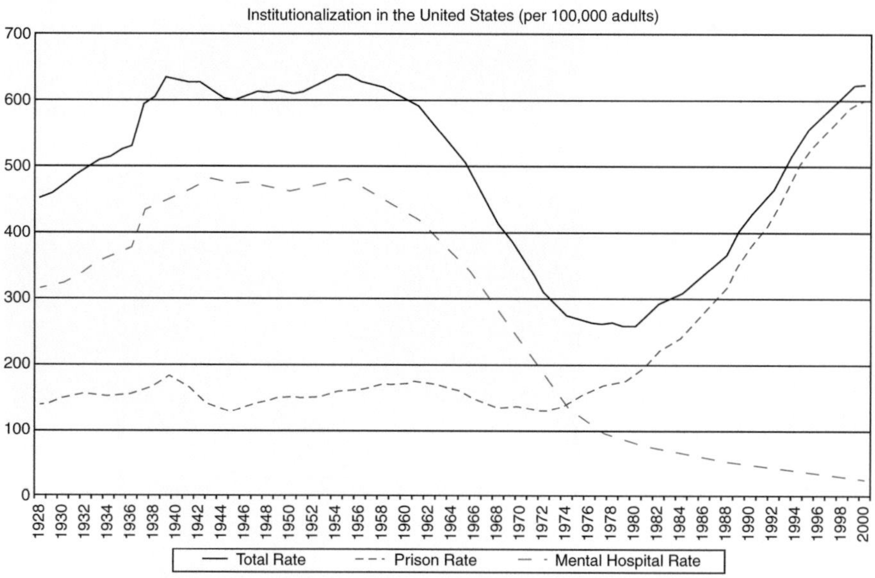

Fig. 11.11 Creation of new homeless and imprisoned populations comprised by mental patients that would previously live in asylums. Chart from [64, 65]

The Social Psychiatric Unit at the Maudsley Hospital in London began to investigate the live in the asylums and in 1962 proposed the concept of the "institutional neurosis" [66]. Beginning in 1967, at Hope Hospital and Salford Royal, Hugh Freeman (1930–2011) attempted to set up a comprehensive system of care that was integrated into the community.

Overall, as a result of the development of antipsychotics, the number of patients treated in asylums declined dramatically, at least in the developed world, and subsequently there was an up to 80% decrease in asylum beds, in spite of the fact that a truly adequate network of community psychiatry services was never developed anywhere in the world. The Italian story is interesting on this matter, since in 1978 after the pressure by the radical psychiatrist Franco Basaglia (1924–1980; Fig. 11.12), further admission of people to mental hospitals was made illegal.

However, psychopharmacology came to conflict with psychoanalysis which was taking US psychiatry by storm. By 1960, almost every major psychiatry position in the USA was occupied by a psychoanalyst. Psychoanalysis expanded to include also treatment of psychosis, and it was the Menninger Clinic in Topeka, Kansas, that pushed for a combination of psychoanalysis and asylum psychiatry. In this frame, an extreme concept concerning the definition of mental illness came in 1963 from

Fig. 11.12 Franco Basaglia (1924–1980)

Karl Menninger (1893–1990; Fig. 11.13) who suggested that mental illness is not the exception; on the contrary most people suffer from some kind of it, and many suffer from a severe form for most of their lives [67]. Karl Menninger declared that it was possible to achieve "a state of being weller than well" essentially giving psychoanalysis the status of a lifestyle rather than treatment option, while his brother William was at the cover of the *Time* magazine in 1948 posing as the psychiatrist who holds the key to the brain secrets (Fig. 10.56). To some extent, in the decades that followed and until the arrival of DSM-III in the 1970s, a political dimension was also prominent and obvious in psychiatry, with psychoanalysis being adopted by capitalism, the old-fashioned European tradition by Marxism, while the evolving European New Left (nouvelle gauche) embraced existentialism.

In 1948, Marthe Louise Vogt (1903–2003) and Sir John Carew Eccles (1903–1997) demonstrated the existence of acetylcholine in the brain [68–70] and in 1954 the existence of norepinephrine [71]. In 1953 Betty Twarog (1927–2013) and Irvine Heinly Page (1901–1991) as well as T.B. Crawford and Sir John Henry Gaddum (1900–1965) demonstrated that also serotonin could be found in the brain [72, 73]. Still, there was doubt whether these substances were playing a role in brain function or were simply carried there incidentally by the blood. In spite of these doubts and disputes, biological models of mental disorders and proposed biochemical pathways of treatment started to appear [74–79]. It was in 1955 that the appearance of spectrophotofluorimeter made possible the first correlation of changes in levels of a substance in the brain (serotonin) with the effects of a drug (reserpine) [80].

Fig. 11.13 Karl Menninger (1893–1990)

In 1952 the Church of Scientology was founded, and right from the beginning, it was hostile to the disciplines of psychiatry and psychology. As a religion, scientology preaches that humans are immortal beings who have forgotten their true nature and past lives. Thus, it recommends that they should undergo a process through which they will re-experience painful or traumatic events from their past, so as to free themselves. In this frame it rejects psychiatric diagnosis as well as treatment, especially with medication.

In 1953 Paul Janssen (1926–2003; Fig. 11.14) started working on the improvement of the properties of pethidine. Gradually he transformed the pethidine molecule and turned it from opioid to a butyrophenone, and eventually in 1958 he synthesized haloperidol. Surprisingly the early trials of haloperidol in the USA were discouraging [81]. Eventually it was approved in 1965, 10 years later than in Europe. Similarly it was not before 1964 that a NIMH-funded placebo-controlled study proved the efficacy of chlorpromazine beyond doubt [82, 83].

In 1955 Alfred Pletscher (1916–2006) reported that reserpine caused the neurons to dump their stores of serotonin [80]. The same year studies from Japan concluded that amphetamines cause a psychotic-like state and they could be used to induce a model of schizophrenia [84]. This was the result of the flooding of Japanese society with amphetamines after WWII.

Also in 1955 the term "neuroleptic" was proposed to the Académie Nationale de Médecine by Jean Delay and Pierre Deniker to describe the specific neuropharmacologic effects of chlorpromazine [85, 86]. The term comes from the Greek words

Fig. 11.14 Paul Janssen (1926–2003)

νεύρο ("neuro" which originally meant "tendon" but later came to refer to nerves only) and *λήψις* (lipsis, meaning seizing something). The second word is often used in neurology and psychiatry in various terms like "thymoleptic" (controlling of affect), "epilepsy" (complete seizure of the body), "catalepsy" (lack of movement seizes the body), "narcolepsy" (sleep seizes the body), etc. The term "neuroleptic" was meaning that the drug was taking control of the nervous system of the patient with desired (antipsychotic) and undesired (extrapyramidal) effects.

One year later, in 1956, benzodiazepines were introduced by Leo Henryk Steinbach (1908–2005). The next year, Nathan Kline (1916–1983; Fig. 11.15) and Robert Noce (1914–1995) shared the 1957 Lasker Prize for their work in the

Fig. 11.15 Left to right: Nathan Schellenberg Kline (1916–1983) and John C. Saunders, at the Rockland State Hospital, Orangeburg, NY (from the Smithsonian Institution Archives, Accession 90–105, Science Service Records, Image No. SIA2008–4868)

discovery of reserpine, and the same year, the Lasker Prize was also given to Pierre Deniker, Henri Laborit, and Heinz Lehmann, for their work on chlorpromazine. Interestingly, as a recognition of the contribution of Indian medicine, the Lasker award was also given to the cardiologist Rustom Jal Vakil (1911–1974) whose work on reserpine was based on the reports of Kaviraj Gananath Sen and Kartik Chandra Bose in the 1930s [87].

While the usefulness of antipsychotic medications was gradually established and they were rapidly being accepted as treatment options by the psychiatric community, their precise mode of action remained elusive. The concept that receptors were behind the effects of medication on the human body was still a matter of debate. At the root of this debate was a resistance to accept a materialistic view of the human body since electrophysiology was leaving space for spirituality, but this was not the case with chemistry. There are authors suggesting that the biological aspect of psychiatry has finally been accepted in part because of the "spiritual" and cultural effect of psychedelic drugs. This debate had profound implications on the essence of psychiatry itself [88–91].

A few years earlier, it had been reported that the lysergic acid diethylamide (LSD) blocked peripheral serotonin receptors [78]. This was combined with the observation that the antihypertensive agent reserpine caused depression in many hypertensive patients [92]. It has been also noted that reserpine depleted brain monoamines available for release from the synapse [93] and that it caused sedation and motor retardation in animals.

In parallel with antipsychotics, the history of antidepressants begins with the recognition of the efficacy of iproniazid against depression. During WWII, Hoffmann-La Roche had procured large supplies of the chemical hydrazine to be used as rocket fuel by the German army. After the war on the basis of this research and with the use of that supply, it went on in 1951 to synthesize two antituberculosis drugs, isoniazid (Rimifon) and iproniazid (Marsilid).

The same year (1951), Irving Selikoff (1912–1992) and Edward Robitzek (1913–1984), at the Seaview Hospital on Staten Island, started studying isoniazid and iproniazid, and additionally to the anti-tuberculosis efficacy, they reported that these two agents exerted some kind of stimulating effect which often led to disciplinary problems, with some patients becoming mildly euphoric and dancing in the halls [94–101]. Today, having the experience of how rare such events might be, these reports seem kind of peculiar; however they led to a breakthrough discovery. Max Lurie (born 1920) was a psychiatrist in Cincinnati at that time, and together with Harry Salzer (born 1906), they were probably the first to try these substances on depressed patients with a clear aim to treat their mental disorder and reported that two-thirds of patients improved [102, 103]. They also coined the term "antidepressant" for the first time in 1952 [18]. Iproniazid was considered to be mainly a potent monoamine oxidase inhibitor (MAOI). This was the result of the work of the Swiss chemist Ernst Albert Zeller (1907–1987), at Northwestern University in Chicago, which had discovered in 1952 that iproniazid, but not isoniazid, inhibited the action of the enzyme monoamine oxidase [104–110].

Following this, several authors discussed on the possible psychiatric applications of iproniazid ([111], 1956; [112]). In 1953 Gordon Kamman (1898–1961), at the Fergus Falls Historic State Hospital in Minnesota, showed that iproniazid was not working in schizophrenia [113]. Jackson Smith at Huston Texas showed that it could be efficacious in depression [114, 115]. However, George Crane (1901–1995) at the Westchester Division of New York's Montefiore Hospital reported on unwanted behavioral effects of iproniazid, including hyperactivity, euphoria, and hostility [96].

In parallel, during the 1950s Julius Axelrod (1912–2004; Fig. 11.16) described the mechanism of the reuptake of monoamines and discovered the catechol-O-methyl transferase (COMT) [116] and in 1961 that imipramine acted by blocking the reuptake of norepinephrine in peripheral tissues and brain slices [117]. For his work he won the Nobel Prize in 1970 [118–121]. In this frame, in 1959 the pharmacologist John C. Saunders (Fig. 11.15), then a research associate at Ciba's offices, claimed that at a conference in 1955 he was the first to discuss the monoamine theory of depression on the basis of his work on reserpine [122], but the transcript of the 1955 conference made no note of Saunders's participation [123]. At that time he had joined the team of Nathan Kline, and in 1956, Kline together with Saunders

Fig. 11.16 Julius Axelrod (1912–2004), Nobel laureate 1970

and Harry P. Loomer (1909–1995), who was head of the clinical service at Rockland State, administered iproniazid to a mixed sample of 17 chronic female mental patients characterized by depressive or negative symptoms. They reported a 47% response rate after 5 weeks and an impressive 70% after 5 months, and these researchers attributed iproniazid the properties of a "psychic energizer" [97, 124–127].

In 1964 Nathan Kline received a second Lasker Prize for his work with iproniazid, but Loomer and Saunders claimed that they should be credited too [128]. Loomer did not insist, but eventually, in 1981, Saunders collected a third of the financial part of the prize after a long debate at court.

In the middle of the psychopharmacological era, the ideas of Karl Popper (1902–1994; Fig. 11.17) concerning the philosophy of science came to radically change the balance of scientific debates. Popper proposed that falsifiability should be the "sine qua non" of a sustainable scientific theory. In plain words, scientific theories are only those who could be tested under experimental conditions which do not protect them from being rejected and leave no place for debate after the experimental data are produced. According to this approach, the prime examples of this approach are the non-scientific nature of creativist theories (where an all-mighty God could serve to explain any failures, peculiarities, and inconsistencies) vs. the

Fig. 11.17 Karl Popper (1902–1994)

scientific nature of the Darwinian theory which could be tested under falsifiable conditions. Popper's approach triggered concerns on the scientific nature of psychological theories and psychotherapies but also for a large variety of social, economic, and political theories.

In the meantime, other research groups have also studied iproniazid. Frank Ayd (1920–2008) reported in 1957 that approximately half of 39 psychiatric patients with mixed diagnosis at Franklin Square Hospital in Baltimore improved [129]. The first large study which confirmed the efficacy of iproniazid was carried out in 1959, by C.M.B. Pare and Merton Sandler (1926–2014) at the Bethlem Royal and Maudsley Hospitals in London. They reported a 52% response rate in a study sample of 50 depressed patients suitable for ECT treatment [130]. Theodore Robie reported that the use of iproniazid reduced the need for ECT by 70% at New Jersey's Orange Memorial Hospital [131, 132]. Samuel Joel was the first to use iproniazid in six OCD patients and reported positive results [133]. Nathan Kline went on and used a combination of iproniazid and amphetamine and reported that often response came within 24–48 hours. An interesting observation came from the St. Thomas's Hospital in London, where iproniazid was the choice of treatment in a specific group of depressed patients with what later would be labeled "atypical depression" [134].

Unfortunately, Kline's career did not end well. On the one hand, he started prescribing risky and extravagant combinations at high dosages and was experimenting without taking into consideration the changing legislation. Investigation by the FDA revealed that he had not acquired neither the necessary board approvals nor consent from the patients who participated in one of his famous studies concerning beta-endorphin [135]. Eventually in 1982 he was compelled to sign a federal consent decree swearing to never conduct research on drugs again [51].

Of course iproniazid was also tested in Paris, where Jean Delay heard of this effect from his pulmonology colleagues at Cochin Hospital. In 1952 (before Lurie and Salzer), Delay, with the resident Jean-Francois Buisson, reported the positive effect of isoniazid on depressed patients [136, 137].

Unhappily, iproniazid was dogged by liver toxicity. According to the FDA, the risk of liver toxicity was about 1 per 3–4000 patients with a 20% lethality rate. In 1958 Hoffmann-La Roche sought to corral the problem by changing the recommended dose on the label, but eventually iproniazid was recalled in 1961, due to reports of lethal hepatotoxicity [138, 139]. Other MAOIs followed including phenelzine, nialamide, and pheniprazine. A significant problem with the MAOIs was that they cause episodes of severe hypertensive crisis related to eating certain foods (e.g., cheese containing tyramine), and soon it was clear that patients under treatment with MAOIs should avoid them [140].

In the meantime, during the 1940s and 1950s, Friedrich (Frederick) Salomon Perls (1893–1970; Fig. 11.18) developed the gestalt therapy which holds that a neurotic patient is repressing one part of the whole (gestalt) and that, by therapeutically reliving a painful experience, the patient can restore wholeness. The core of the gestalt therapy process is based on enhanced awareness of sensation, perception, bodily feelings, emotion, and behavior, in the present moment. Also, emphasis is

Fig. 11.18 Friedrich (Frederick) Salomon Perls (1893–1970)

put on relationship and the contact between the self, its environment, and the others [141]. In 1958 appeared the first example of what would later be called cognitive therapy. It was called rational emotive therapy and was developed by Albert Ellis (1913–2007) on the basis of the assumption that highly aroused emotions greatly impair cognitive functions [142].

In 1959 was the first description of tardive dyskinesia [143], while the first report in the English language that made the syndrome widely known came the next year [144]. It took some years to confirm that it was a drug-induced side effect [145–148].

In spite of the fact that iproniazid existed years before, the era of antidepressants started when the Ciba-Geigy Pharmaceutical Company was working on the improvement of the chlorpromazine molecule, and it synthesized imipramine (G 22355) by adding the side chain of chlorpromazine onto the iminodibenzyl nucleus (derived from the dye "summer blue"), a nucleus similar to the phenothiazine nucleus. It seems that initially Robert Domenjoz (1939–1993), head of pharmacology at Geigy, insisted in studying this new compound in schizophrenia in spite of repeated failures even when the dosage reached 1500 mg/day. One of the researchers who were trying imipramine was Roland Kuhn (1912–2005; Fig. 11.19) in the psychiatric hospital at M*ü*nsterlingen in the canton of Thurgau, situated on the southern border of Lake Constance [149]. In spite of the wish of the industry, Kuhn started administering the agent to patients with depression, and in the morning of the sixth day of treatment,

Fig. 11.19 Roland Kuhn (1912–2005)

January 18, 1956, a patient named Paula manifested a significant improvement in her condition. Unfortunately, neither the industry nor his fellow psychiatrists showed any interest in these results. In a rare twist of luck, the businessman, scholar, and influential Geigy stockholder Robert Boehringer (1884–1974) heard of Kuhn's results and asked him to treat his wife, which was suffering from depression. With his wife responding within a week, Boehringer persuaded Geigy to develop the drug as an antidepressant [150, 151]. Thus imipramine was the first tricyclic agent, and its antidepressant effect was first discovered by Kuhn [152] who described imipramine as a "thymoleptic" after the Greek words *θυμικόν* (thymikon, meaning emotions) and *λήψης* (lipsis meaning holding), thus suggesting that the new compound was "controlling the emotions" [153, 154]. Later, this agent appeared to act by inhibiting the reuptake of norepinephrine and serotonin, in the brain [155–159].

Delay and Deniker preferred instead the terms "psycho-analeptic" or "mood stimulants," while in 1959 they proposed the term "antidepressant." In 1958, at the Ciba-Geigy laboratories, chlorimipramine was synthesized with the chlorination of imipramine.

A number of controlled trials were conducted in the next few years, and authors were enthousiastic concerning the efficacy of imipramine. However a pooled analysis of these trials came out in 1965 and revealed that efficacy was clear but not that impressive, since two-thirds of patients under imipramine responded in comparison to one-third under placebo [160, 161]. At that time, discussions concerning the most appropriate patient group to treat with imipramine proliferated. There was a broad consensus that melancholic patients responded best, thus setting a long-lasting opinion which would be dominant among clinicians until today that endogenous/melancholic depression should be treated with medication but neurotic/characterological should not.

At the same period, Joseph Wolpe (1915–1997; Fig. 11.20) found that reintroducing a frightened animal to the fear-conditioning stimulus reversed the neurosis. This discovery laid the basis for somatic desensitization and subsequently reciprocal inhibition [162, 163]. While British behaviorists accepted that patients could be mentally ill and treatment could correct disabilities, in parallel Burrhus Frederic Skinner (1904–1990; Fig. 11.20) developed a radical behaviorism approach which suggested that all deviant behavior, including psychosis, was a result of faulty learning. He rejected the concept of free will and argued that human behavior depends on the consequences of previous actions (principle of reinforcement) [164–170]. Along with the conceptualization of "operant conditioning" on the basis of a three-term contingency (stimulus-response-reinforcer), he created the operant conditioning chamber known as a "Skinner Box" for the experimental analysis of animal behavior (Fig. 11.21).

Fig. 11.20 Left: Joseph Wolpe (1915–1997). Right: Burrhus Frederic Skinner (1904–1990)

Fig. 11.21 Skinner working with an operant conditioning chamber also known as a "Skinner Box". Permission asked from info@brain.harvard.edu

In 1958, the Germans Fritz Hünziker with J. Schmutz and E. Eichenberger at the Wander Pharmaceutical Company in Bern, Switzerland, among other compounds synthesized clozapine, and Günther Stille put it through preclinical testing. During clinical testing, clozapine was proved to be a difficult and even dangerous compound. In the trial run by Deniker and Pierre Simon, 4 out of 19 patients died, one from an irreversible agranulocytosis and another from malignant hyperthermia. The trials confirmed its efficacy and the low rate of extrapyramidal side effects, but it was not proven superior to the already existing antipsychotic agents [171–174].

It has been discussed that one of the research targets of the group which synthesized clozapine was to refute the belief that extrapyramidal side effects were utterly linked with the antipsychotic effect. Thus, it is said, but never confirmed, that the name "Leponex" came from the combination of the words *lepus* (means hare in Latin) and *ex* (former or out). Thus name Leponex was widely taken as a rebuff to the ideas advocated by Paul Janssen, Hans-Joachim Haase (1922–1997, whose name means hare), and Pier Deniker [18]. Haase had argued that there could be no antipsychotic efficacy without extrapyramidal side effects [175], and this statement was so strong in the psychiatric community that the manufacturer of clozapine was unwilling to market it because of the lack of them [173].

In 1959 an initiative was launched to standardize clinical assessment and treatment outcome to eventually create the Early Clinical Drug Evaluation Unit (ECDEU) [176]. The same year, Kurt Schneider (1887–1967; Fig. 10.31) proposed the first rank or Schneiderian criteria for schizophrenia [177]. In 1960 Jean Guyotat (born 1920) reported for the first time on the treatment efficacy of clomipramine in obsessive-convulsive disorder [178]. In 1961 the works of Sir John Carew Eccles (1903–1997), David Curtis (born 1927), and Jeff Watkins (born 1929) led to the recognition of glutamate as a neurotransmitter for the first time and essentially changed the whole paradigm of brain function [179, 180]. The same year, the Canadian psychiatrist Heinz Lehmann (1911–1999) suggested the term "antipsychotics" for neuroleptics [181]. An important observation led to the conclusion that several agents had an abuse and dependence potential, and this had a clear biological basis [182]. In 1962, Delay and Deniker described a syndrome they called "syndrome malin des neuroleptiques" [40, 41, 183, 184] whose first case had been reported in 1956 [39].

From 1960 to 1963, Hoffman-La Roche introduced the first benzodiazepines in the market. These were chlordiazepoxide (Librium) and diazepam (Valium) for the treatment of anxiety. Benzodiazepines became a class of very popular agents but at the cost of high potential for abuse and dependence.

This decade also hosted the first complete biological theory on the mechanisms underlying schizophrenia and its treatment. It was formulated in the early 1960s mainly by Arvid Carlsson (1923–2018; Fig. 11.22) who later received the Nobel Prize in 2000. Previously Carlsson had identified dopamine mainly on the basis that the administration of norepinephrine or serotonin alone could not reverse the sedative effects of reserpine. However L-dopa could reverse it and simultaneously alleviated parkinsonian symptoms caused by reserpine [185–188]. Additionally, the catecholamine hypothesis of depression, developed by Joseph J. Schildkraut (1934–2006; Fig. 11.23) in 1965, was also to a large extend based on Carlsson's work [189, 190].

In 1959, Nikita Khrushchev, a then leader of the USSR, pointed out that "… to those who might start calling for opposition to communism on this 'basis,' we say that now, too, there are people who fight against communism, but clearly the mental state of such people is not normal." After this speech, in the 1960s, Andrei Vladimirovich Snezhnevsky (1904–1987; Fig. 11.24) and his colleagues introduced the diagnosis of "sluggish schizophrenia" for people who dissented politically [192]. Dissent was considered to be a form of schizophrenia with delusions of reform. This diagnosis was used exclusively in the Soviet Union as well as in several Eastern Bloc countries, and it was applied also to persons without any psychotic symptoms on the assumption they will develop in the future. This is considered the prime example of psychiatric abuse for political reasons in the USSR [191, 193–206]. Snezhnevsky was accused that he personally signed a commission decision on legal insanity of mentally healthy persons, which however were considered to be enemies of the Soviet state, including Vladimir Bukovsky, Natalya Gorbanevskaya,

Fig. 11.22 Arvid Carlsson (1923–2018)

Leonid Plyushch, Mikola Plakhotnyuk, and Pyotr Grigorenko [191]. It is said that he attributed the diagnosis of sluggish schizophrenia to Andrei Sakharov and Joseph Brodsky in absentia. Interestingly, the American Psychiatric Association at its annual meeting held in San Francisco in 1970 honored Snezhnevsky as a "distinguished fellow" for his "outstanding contribution to psychiatry and related sciences" [203], but in 1980, the Special Committee on the Political Abuse of Psychiatry, established by the Royal College of Psychiatrists in 1978, charged Snezhnevsky as responsible for the compulsory detention of the celebrated dissident, Leonid Plyushch. The RCPsych concluded that there is sufficient evidence to suggest that "Professor Snezhnevsky has acted unethically and no longer warrants a place of honor in the Royal College of Psychiatrists" [207, 208]. On the other hand, Snezhnevsky was honored with the title of a "Hero of Socialist Labour," two "Orders of Lenin," four "Orders of the Red Star," and the "USSR State Prize." In an interesting intervention, Yuri Fyodorovich Orlov (1924–2020), a nuclear physicist and human rights activist, former Soviet dissident, and founder of the Moscow Helsinki Group and the Soviet Amnesty International group, argued that all happened against Snezhnevsky's will and in reality it was the work of Georgy Vasilyevich Morozov (1920–2012; Fig. 11.24), the director of the Serbsky Institute, who has collaborated with the KGB since his student's days.

Fig. 11.23 Joseph J. Schildkraut (1934–2006)

Additionally, during the Khrushchev-Brezhnev era in the former Soviet Union, the KGB had the control of forensic psychiatric institutions and used them in an arbitrary way to label political opponents of the regime as suffering from psychosis and incarcerated them for long periods in "special psychiatric hospitals" [209]. Because of this, in 1983 the Soviet Psychiatric Association voluntarily withdrew from the World Psychiatric Association (WPA) when facing almost certain expulsion [210]. It was readmitted back to the WPA only after the Gorbachev reforms, when many of these political prisoners were released. In his article of 2002, a former president of the American Psychiatric Association Alan Abraham Stone (1929–), who as a member of team had examined Soviet dissident Pyotr Grigorenko and found him mentally healthy in 1979, put forward the academically revisionist theory that there was no political abuse of psychiatry in the Soviet Union [211].

China has been accused of manifesting similar practices but not to the extent it happened in the Soviet Union. These accusations concerned the way members of the Falun Gong movement (a meditative movement drawing on several Eastern religious doctrines) were treated [212, 213].

A critical event happened in the early 1960s and dramatically changed health care world-wide. Thalidomide was first marketed in 1957 in West Germany as a

Fig. 11.24 Left: Andrei Vladimirovich Snezhnevsky (1904–1987), Right: Georgy Vasilyevich Morozov (1920–2012), From the book *Cold War in Psychiatry: Human Factors, Secret Actors* by Robert [191]

sedative or hypnotic and also for the treatment of nausea, anxiety, insomnia, gastritis, and tension. It was widely used to treat nausea and morning sickness in pregnant women. Soon it became an over-the-counter drug in West Germany. However soon throughout the world, about 10,000 cases of newborn with phocomelia due to thalidomide were reported. Half of them died. Other effects included deformed eyes, hearts, alimentary and urinary tracts, blindness, and deafness. Thalidomide never made it to the US market because of the persistence of Frances Kathleen Oldham Kelsey (1914–2015) who asked for more data on safety. In the aftermath tougher rules for the testing and licensing of drugs were introduced, such as the Kefauver Harris Amendment in the USA and the Directive 65/65/EEC1 in the EU. The Kefauver Harris Amendment was passed unanimously by congress in October 1962. According to it the pharma industry was required to demonstrate both the efficacy and safety of new drugs and also to report all side effects and adverse reactions to the FDA. It also demanded that informed consent should be obtained of patients before participating in clinical trials. The FDA was strengthened, and the Drug Efficacy Study Implementation was initiated to reclassify drugs already on the market.

One of the thalidomide case consequences was that from that point on, medication treatment was considered "risky" in principle. This was a new idea and the concept of treatment changed radically. Drug development and drug availability started following strict rules, and medications were required to target specific indications; they would be available only by prescription, and their efficacy and safety should have been proven through randomized controlled trials (RCTs). Especially for psychiatric medication, these meant the use of operationalized diagnostic criteria and psychometric rating scales, the emphasis was put on categorical rather than dimensional models of disease and diagnosis, and the cost for the development of new medications exploded.

In the USA in 1960, Samuel Gershon joined the Schizophrenia and Psychopharmacology Joint Research Project of the University of Michigan at the mental hospital in Ypsilanti, Michigan, and the same year along with Arthur Yuwiler, also at Ypsilanti, they published the first North American paper on lithium [214]. In 1964 the Danish psychiatrist Poul Christian Baastrup (1918–2002; Fig. 11.25) demonstrated the efficacy of lithium for the maintenance phase [215]. For the next few years, there was a significant academic opposition to the use of lithium as the standard treatment for BD, and much emphasis was given to its toxicity. Aubrey Lewis (1900–1975; Fig. 11.8), professor of psychiatry and head of the Maudsley, considered lithium treatment "dangerous nonsense," and Michael Shepherd (1923–1995) was also extremely negative towards it and suggested that lithium was toxic in mania and that claims of efficacy for it in preventing depression rested on "dubious scientific methodology" [216–221]. However more studies established lithium and robustly linked it to the treatment of all phases of Bipolar disorder [14, 222–231]. Latter Fred Goodwin (1936–2020; Fig. 11.26) suggested it could be also useful in the treatment of depression as add-on to antidepressants [232–237]. The

Fig. 11.25 Left: Poul Christian Baastrup (1918–2002), Center: John Cade (1912–1980), Right: Mogens Schou (1918–2005)

Fig. 11.26 Fred Goodwin (1936–2020)

recommended serum lithium levels were determined with certainty in 1976 [238]. The term "mood normalizer" was proposed by Mogens Schou for lithium [239] after the term "mood stabilizer," which was used during the 1950s to refer to a combination of amphetamine and barbiturate to treat patients with neurotic instability but not patients with bipolar disorder.

Lithium treatment for bipolar disorder was approved in 1961 in France, in 1966 in the UK, in 1967 in Germany, and in 1970 in Italy and the USA. In 1974, this application was extended to its use as a preventive agent for manic-depressive illness [227]. Following the success with lithium treatment, valproate was introduced in 1966 [240] and latter carbamazepine [241].

The first neuroleptic depot appeared in 1963 and was fluphenazine enanthate (Moditen). It could give steady blood levels for up to 2 weeks. Fluphenazine decanoate (Modecate) followed in 1965 and could last up to 4 weeks.

Eric Berne (born Eric Lennard Bernstein, 1910–1970; Fig. 11.27) introduced the transactional analysis and put the emphasis on the analysis of patients' social interactions by postulating that a life script created by parents in early childhood shapes and inhibits their lives. The deciphering of that script could lead to treatment. In 1964 he published his best-selling book *Games People Play* [242].

Noam Chomsky (born 1928; Fig. 11.28) is widely known more for his political ideas and activity, but his contribution to neuroscience is also significant. Since the

Fig. 11.27 Eric Berne (1910–1970)

1960s he formulated a number of novel and radical theories of language and its development, including the universal grammar theory, the generative grammar theory, the Chomsky hierarchy, and the minimalist program [243–246]. According to him, there is an intrinsic element concerning the grammatical structure, and this is biologically embedded in humans. Therefore, children need only minimal input from the environment so that to fully develop complex language skills ("poverty of the stimulus" argument) [247–255]. As a consequence he also played a major role in the decline of behaviorism, and he was strongly critical of the theories of Skinner [256, 257].

In the 1960s and 1970s, Norman Geschwind (1926–1984; Fig. 11.29) followed the work of Sperry and further shed light on the differential function and collaboration between the two brain hemispheres. He unveiled structural and functional asymmetries and showed that the planum temporale was three times larger in the left hemisphere. He was able to show aspects of the complex exchange of information and collaboration between the two hemispheres [258]. However, the efforts to find some kind of a biological index for the diagnosis of mental disorders failed in a dramatically disappointing way [259, 260].

Timothy Leary (1920–1996; Fig. 11.30) was one of the most fascinating public figures during the 1960s and the 1970s and is considered to be the guru of the counterculture of that period [261–263]. In 1961 and after spending some days on

Fig. 11.28 Noam Chomsky (born 1928)

vacation in Mexico, he became convinced that psychedelic drugs provided a means to explore the mind [264] and started his controversial experiments with psilocybin and LSD in Harvard University as well as at the Marsh Chapel with divinity students and in the Concord prison [265]. Eventually he lost his position at the Harvard in 1963 and went on to create a mansion near New York known as "Millbrook" to continue with his experimentations. He was arrested several times and was held in more than 30 different prisons around the world. After a prison escape and in the middle of unrest throughout the USA, he was labeled as "the most dangerous man in America" by then President Nixon.

One of the developments during the 1960s was the rise of the civil rights movement, feminism, and labor movement. These movements were of extreme importance for society since they marked the transition from the more authoritarian socio-political environment of the pre-war period to the modern democratic environment, characterized by more justice, wide participation in decision-making, criticism of authority, and global transparency. Their contribution to the promotion and establishing of universal human and civil rights and the improvement of the overall quality of life was enormous. The change that occurred in human society within a few decades was unpresidented. This was an important development for psychiatry also since all these movements were focusing on oppression, and since the beginning, they borrowed rhetoric from psychiatry and especially from

Fig. 11.29 Norman Geschwind (1926–1984). Photo curtesy of MM Mesulam

Fig. 11.30 Left: Allen Ginsberg (1926–1997) poet, philosopher, and author. Center: Timothy Leary (1920–1996). Right: John C. Lilly (1915–2001) neuroscientist, psychoanalyst, psychonaut, and philosopher. Polaroid photograph taken Easter Sunday 1991, at the home of Dr. Oscar Janiger, photo by Philip H. Bailey

psychoanalysis and psychosocial theories. Maybe inevitably, this transitory phase had a dark side also. Many of the intellectuals involved in this movement were primary hostile against all institutions of western society and considered them to be oppressive rather than serving. A prevailing idea utilized psychiatry as a model and example for society. Soon many activists came to consider psychiatry in the frame of the struggle of liberation vs. oppression. According to this concept, psychiatric patients were considered to be an oppressed minority. At that time there was a hope that the oppressed throughout the world could unite and support and draw strength from each other and mental patients were seen as such an oppressed minority group in need for support by civil rights activists [266].

The most important and astonishing development within this frame was the emergence of anti-psychiatry. It was a product of the interaction between social psychiatry and social activism and begun as social activism within the boundaries of mainstream psychiatry. In 1961, Thomas Szasz (1920–1912; Fig. 11.31) published *The Myth of Mental Illness* and argued that the very concept of mental illness was a logical absurdity [267]. His target was mainly the psychoanalytical industry in the USA and its problematic model and practice. But since psychoanalysis was identified with mainstream psychiatry of the time in the USA, this approach was perceived as an attack on psychiatry itself [268] and eventually became one.

Michel Foucault (1926–1984; Fig. 11.31) argued that since the eighteenth century and because the Enlightenment pushed civilization to embrace reason, society lost its tolerance towards madness, and as a result there was a social confinement and isolation of mental patients [269]. According to him and to Erving Goffman (1922–1982), psychiatric treatment in hospitals is essentially evil [269, 270]. Foucault was essentially an anarchist, who worked on a wide variety of intellectual topics of its time, and his approach was to synthesize history, philosophy,

Fig. 11.31 Left: Thomas Szasz (1920–1912), Center: Michel Foucault (1926–1984), Right: Herbert Marcuse (1898–1979)

epistemology of the human sciences, and genealogy of the main social, political, and economic institutions of modernity [271] in an admixture which was considered to be chaotic by many authors. His style was to focus on the other (hidden) side and to try to explain the whole phenomenon on the basis of this concealed side. Though appealing as an approach, the validity of such a method is questionable. His basic idea was that during the twentieth century there is the rise of "biopower," that is, of the technocratic sciences and organized societies with all aspects of human life being subject to it [272]. He considered all the institutions of society as oppressive, and in this frame, the individual person has become conceivable and manageable only as a "human capital." Concerning psychiatry and mental health, Foucault argued that there was no liberation of the mentally ill during the nineteenth century; only an objectification of the concept of their freedom occurred. This was because according to this theory, the concept of mental illness is considered to be the product of this "biopower environment" with technocratic-economic organization of the whole society, and "patients" are in reality human capital waiting to be restored and to be optimized as a biological capital with the use of medicine [271]. Thus, psychiatry is part of the disciplinary system, and psychiatrists together with other mental health professionals are simply the instrument of public order enforcement. In the frame of Foucault's historical interpretation, mental illness is included under the umbrella of medicine only because in the struggle among different spheres of power, including religion, philosophy, sociology, and medicine, the latter prevailed. An oxymoron in his theory is that essentially, he considers this to have been inevitable and maybe irreversible. The way Foucault uses the historical events surrounding the madness of King George III of England is interesting. While this case could be used as proof that mental illness is not a social construct but instead it reflects true illness associated with disability, for Foucault it is simply the proof of the overwhelming power of psychiatry, which is able to remove even Kings from power.

Herbert Marcuse (1898–1979; Fig. 11.31) suggested that psychoanalysis perpetuates and complicates rather than solves the problems and supported the use of psychedelic drugs as a way to raise consciousness [273]. In 1967, David Cooper (1931–1986; Fig. 11.32) published *Psychiatry and Anti-psychiatry*. In this book he proposed that the major obstacle to the recovery of patients with schizophrenia was the mainstream psychiatry and accused contemporary mental health care for alienating physicians and debilitating milieu. Cooper was the first to refuse the existence of mental illness [274]. This was later supported by the theory that schizophrenia is a recently appeared disorder in the history of mankind [275, 276]. A number of books followed which criticized psychiatry in terms of civil rights and with the antipsychiatric approach [270, 277, 278]. Eventually the core nosology of psychiatry was challenged.

Several patient advocacy societies were established during this period, including the Mental Patients' Union in the UK and the Insane Liberation Front in 1970, the Mental Patients' Liberation Front in 1971, and the Network Against Psychiatric Assault in 1972 in the USA [279, 280].

Fig. 11.32 Left: David Cooper (1931–1986). Right: Ronald David Laing (1927–1989)

In this environment under turmoil came the Helsinki Declaration of 1964. It elevated to a principle of absolute moral certainty that one must not lie to patients neither in research studies nor concerning their diagnosis and treatment in everyday clinical practice. For psychiatry this was a challenge, to inform and educate patients concerning their mental health status and diagnosis, and additionally it made a number of frequently used "placebos" no longer an acceptable practice since it was considered to be something between lying and fraud [281, 282].

The rise of existentialist philosophy in Europe probably also played a role. For existentialist philosophers it was clear that mental disorder was a matter of free choice. Persons were free to choose whether between mental disorder and health and essentially mental disorder was considered to be the ultimate act of freedom; the person was choosing to reject an unbearable reality. Existentialism built on the Aristotelian ideas that everything comes from the senses and the "blank slate" of Lock and Hobbes. If freedom was the issue, humans could be free only if they are governed by reason without any divine control or natural restrains or limitations (like inherent cognitive biases, innate ideas, subconscious or unconscious processes, previous trauma, or genetic determinants). In such a pure idealistic frame of understanding the mind, mental disorder cannot exist in the form of a medical condition, since its existence implies objective, natural, and non-moral limitations to human freedom. Therefore, as the father of irrationalism, Saint Augustine reversed the question why now that Christianity prevailed the Roman Empire collapses, so did existentialists with mental disorder. Their answer was that insanity does not prove the existence of natural limits in human freedom, but on the contrary, it is itself the proof of the presence of unlimited freedom since it denies the acceptance of an unwanted reality.

Jean-Paul Sartre (1905–1980; Fig. 10.49 and Fig. 11.33) in his foreword to the book *The Divided Self* by Ronald David Laing (1927–1989; Fig. 11.32) noted that "mental illness is the revolt that the free organism in its total entity invents in order to live in an unbearable situation" [283]. Existentialism had a profound effect on scientific thinking itself. Various theories concerning the essence and nature of "being" were developed. One of them was skepticism which questioned reality itself; how do we know what we perceive as the environment is not merely a dream? A version of skepticism put forward the possibility that all others are no more than person-imitating automata or mindless zombies [284–289]. Skepticism should be considered as an extreme version of the internalistic doubts originally phrased by Rene Descartes in his *Meditations*. Wittgenstein was clear-cut and rejected these approaches as making no sense [290].

Fig. 11.33 Standing from left to right: Jacques Lacan, Cécile Éluard, Pierre Reverdy, Louise Leiris, Pablo Picasso, Zanie de Campan, Valentine Hugo, Simone de Beauvoir, Brassaï Sitting, from left to right: Jean-Paul Sartre, Albert Camus, Michel Leiris, Jean Aubier. Photo taken in March 19, 1944 by Brassaï (Gyula Halász)

These approaches and concepts put all treatments available for mental patients under dispute. Psychosurgery and ECT were the most attacked, but also medications were considered as harmful especially with the recognition of serious and visually impressive adverse effects like tardive dyskinesia. Addictions were considered to be a result of the medicopharmaceutical model and approach. Opposition to ECT was growing around the world. In 1962 Ken Kesey (1935–2001; Fig. 11.34) published the book *One Flew Over the Cuckoo's Nest* in which he analogized society's oppressors with psychiatric ward staff and the means of social and political oppression with physical treatments of mental disorders, especially ECT and psychosurgery [291].

Unfortunately, later it was proved that the US government was involved in cases of abuse of psychiatry, during the period when such abuses were widespread in the USSR. Donald Ewen Cameron (1901–1967) who served as President of the Canadian, American, and World Psychiatric Associations, the American Psychopathological Association, and the Society of Biological Psychiatry developed the theory that it could be possible to cure schizophrenia by erasing existing memories and reprogramming the mind [292]. Although many documents are now declassified and are available online, the full history and details are not known yet, but it has been reported (but not confirmed) that from 1957 to 1964 he participated in the MKUltra experiments (sometimes referred to a CIA's mind control program). Of course, it is impossible to verify the truth for programs like this, but reports suggest it involved administering LSD, various paralytic drugs, and ECT at 30–40 times the normal power to patients without their consent. The idea behind this was to put the patient into a drug-induced coma for weeks at a time (and up to 3 months in one case) while playing tape loops of noise or simple statements. One disturbing detail was that these experiments were allegedly carried out in patients suffering from mild symptomatology, including anxiety disorders and postpartum depression

Fig. 11.34 Left: Ken Kesey (1935–2001) photo from Oregon Historical Society Research Library. Center left: The poster of the film. Center right: Michael Douglas, Milos Forman, Louise Fletcher, Jack Nicholson, and Saul Zaentz, posing with the five Oscar awards (1976). Right: Dr. Dean Brooks (1916–2013) is shown with the hydrotherapy console that a patient uses to break a window and escape from the mental hospital in the film *One Flew Over the Cuckoo's Nest*. Brooks also appeared in the movie as a psychiatrist

without their informed consent [293]. It was reported that many of them suffered permanent debilitation after these treatments, including incontinence, amnesia, aphasia, and behavioral problems [294, 295]. Much of these might fall into the realm conspiracy theory, but the documents pertaining to the project were destroyed so no definite answer can be given.

As a result of the overall changing of the landscape in mental health, during the 1960s and 1970s, there was an exponential increase in the number of psychologists in Europe and North America. Also psychologists started pushing away from testing and started embracing and also demanding to be involved in psychotherapies especially since many psychiatrists started abandoning psychoanalysis for the sake of medication treatment.

Albert Ellis (1913–2007; Fig. 11.35) had already introduced rational emotive therapy in 1958 [142]. Ellis believed that emotions rather than logic dominate behavior [296]. In the 1970s he went on to describe 12 irrational beliefs pertaining to an ideal self-perception and expectations of achievements and acceptance/love by others, perfectionism, justice, and locus of control [297]. In this changing environment came the work of Aaron Temkin Beck (born 1921; Fig. 11.35), who developed a pragmatic form of psychoanalysis which lay somewhere between a behavioral and a psychodynamic approach [298, 299]. Initially Beck was a devoted psychoanalyst who however wanted to find scientific support for his ideas. He designed a research effort to test whether the psychoanalytic theory that depression was "anger or hostility turned inwards" and that depressed patients wanted to punish themselves. However, his research results showed that depressed patients felt much better when succeeding in a task and much worse when failing, thus rejecting the anger and hostility hypothesis. He went on and developed a theory which argued that the patient's problems emerged because of the presence of developmentally formulated

Fig. 11.35 Left: Albert Ellis (1913–2007). Right: Aaron Temkin Beck (born 1921)

cognitive distortions. Thus, the therapist helps the patient identify them and learn alternative and more realistic cognitive approaches. In an expansion of Kant's theories, Beck suggested that schemata, that is, groups of cognitive rules, dictate behavior [300, 301].

During 1950–1970, Erik Homburger Erikson (born Erik Salomonsen; 1902–1994) developed concepts of adult identity, identity diffusion, and crisis, while in the 1960s and 1970s, Milton Erickson (1901–1980) developed hypnotic psychotherapy.

In 1954 Abraham Harold Maslow (1908–1970; Fig. 11.36) published his book *Motivation and Personality* [302] and in 1962 the book *Toward a Psychology of Being* [303]. His theories were based on observations in healthy normal individuals and focused on the normal function of psychological mechanisms including the unconscious. He specifically focused on motivation, and he suggested the presence of two basic forms of human motives: growth (B) and deficiency (D). He went further to develop the "Hierarchy of Human Needs" which placed human needs and motives from the more deficiency-oriented constructs (basic biological needs, safety, etc.) at the bottom to the more growth-oriented constructs (meta-needs or

Fig. 11.36 Abraham Harold Maslow (1908–1970)

self-actualization) at the top of the pyramid. The healthier the individual, the less the unresolved issues he/she has with lower needs. The presence of unresolved issues lower in the pyramid leads to inability to pursue higher-order needs, and the individual is stuck with the effort to fulfil basic needs, leading to psychopathology.

A surprise came from the Nobel Prize laureate Linus Pauling (1901–1994), who in 1968 suggested (without any evidence) that the cause of schizophrenia lays in a deficiency of the enzyme nicotine adenine deaminase (NAD). Therefore schizophrenia, according to Pauling, was an analog of brain pellagra, and a large amount of niacins or nicotinic acids could be the appropriate treatment. This approach was in accord with the ideas of Abram Hoffer (1917–2009), and this signaled the rise of orthomolecular psychiatry which heralded that the appropriate treatment for mental disorders was vitamins (often in giga-dosages), various metals involved in biological processes, or other naturally occurring elements [304]. Linus Pauling was advocating for the usefulness of vitamin C to treat a variety of diseases from common cold to cancer, although he had no data available to support his claims. This unscientific approach by a Nobel laureate seems unusual, but in fact it is not; a significant number of such people have embraced metaphysical, spiritual, or other paranormal, supernatural, and unscientific theories and ideas, including psychic mediums, psychokinesis, AIDS denial, anti-vaccine, and conspiracy theories.

In the 1960s and 1970s, Roger Sperry (1913–1994; Fig. 11.37) carried out his work on split-brain which eventually will win him the 1981 Nobel Prize in Physiology and Medicine [305]. Until then the role of the corpus callosum was unknown, and most researchers were following the ideas of Andrew Akelaitis (1904–1955) who in the 1940s suggested no significant role for it [306]. However, Sperry used advanced procedures and proved the pivotal role and function of the corpus callosum for the integration of cognitive processes and described the peculiar clinical pictures of patients with split-brain. Additionally he argued that the left hemisphere was the site of the analytical, verbal, and more rational processes, while the right was the site of emotions, creativity, imagination, and impulses [307, 308].

After the thalidomide scandal, the next hallmark in the shaping of the pharmaceutical research and market came in 1966, when Henry Beecher (1904–1976) published an article where he described 22 research studies in detail. These studies took place in the USA, and it was clearly obvious that the patients which participated in them had not been informed that they were participating in a trial; on the contrary they thought they were receiving standard medical care [309]. This revelation changed the landscape overnight. Doctors and researchers were turned from heroic and almost divine figures to oppressors, and the patient-doctor relationship would never again be the same. As a result of growing distrust, variations in medical practice were no longer seen as interesting and useful opinions and approaches and became the target of scrutiny and regulation. Medical research started following a number of rules and regulations making it very expensive; as a result, it started following more and more the ways of the economic market. In this increasingly negativistic landscape, a complication for psychiatry was that it was the only medical specialty with some of its treatments having political consequences.

Fig. 11.37 Roger Sperry (1913–1994), Nobel laureate 1981

In May 1968 during the Paris uprising, the office of Jean Delay in the Department of Psychiatry at the University of Paris was ransacked by French students. By virtue of his position, Jean Delay was a symbol of the order the students were attacking. Additionally, his role in the evolution of psychopharmacology made him a special target as the proud father of the dreaded "camisole chimique" (chemical strait-jacket). The next year, in Japan, following visits by Szasz and others, Japanese students revolted and occupied the Department of Psychiatry at Tokyo University, at that time chaired by Hiroshi Utena, and hold it for 10 years, thus bringing research to a halt.

Since the mid-1970s, the number of patients treated in asylums declined dramatically. An up to 80% decrease in asylum beds is reported in many countries. Psychiatric departments in general hospitals appeared, and community services were developed. The policy was that chronically ill patients are cared for in nursing homes, in board-and-care homes, halfway houses, supervised or semi-supervised apartments, and other facilities. The problem was and still is that this model was never developed to cover the entire population, and the system is frequently problematic, fragmented, and in need of integration. A strong critique is that it led to a dramatic increase in the homeless population and in imprisoned mental health patients (Fig. 11.11). During the same time period and since the late 1960s, important reforms in mental health laws took place around the world. The main idea was to emphasize the rights of the mental patients to contest involuntary hospitalization and refuse treatments. Also, the shift of the insanity defense from "not guilty by reason of insanity" to "guilty but mentally ill" gave courts the right to ask for and monitor psychiatric treatment. Especially the Tarasoff case in 1976 and the Ramona case in 1994, but also other legal cases, put psychotherapy on a new basis since therapists realized that apart from the moral obligation to treat and protect the patient, there is also a legal obligation to protect others from the consequences of the patients' behavior.

In 1969 sulpiride was synthesized and proved to be an antipsychotic with fewer extrapyramidal adverse events and lower risk for tardive dyskinesia.

Already with the early years of the 1970s, a large discussion and controversy emerged concerning psychiatry as a science, and most of the debate was around the reliability and validity of psychiatric diagnosis. In 1971 a comparison of diagnostic habits between London and New York was published and suggested that American psychiatrists tended to diagnose schizophrenia more often than their British counterparts [310]. Following this, in 1972, the St. Louis group, which was led by Eli Robins (1921–1994; Fig. 11.38) and included the then resident John Feighner, Sam Guze (1923–2000; Fig. 11.38), George Winokur (Figs. 11.8 and 11.38), Robert Woodruff, and Rod Muñoz, published a full set of its operational criteria, and this was the first effort to operationalize psychiatric diagnosis and classification [311, 312], a line of work that eventually led to the DSM operationalized diagnostic criteria [313]. And then, the next year, the David Rosenhan (1929–2012; Fig. 11.39) study (or "Thud" study) was published reporting that 19 normal persons presented to psychiatric hospitals claiming that they had heard the word "thud"; all of them were hospitalized for 19 days on average, all but one received the diagnosis of schizophrenia in spite of behaving normally during hospitalization and were treated with antipsychotics [314]. It is highly likely that Rosenhan himself was one of these pseudo-patients under the alias "David Lurie" and was hospitalized for 9 days at Haverford State Hospital in Haverford, Pa., in February 1969. The experience as it

Fig. 11.38 From left to right: Eli Robins (1921–1994; from Becker Medical Library, Washington University School of Medicine), Sam Guze (1923–2000; from Psychiatric Bulletin, 25(2), 77–77), Rod Muñoz, George Winokur (1925–1996), and John Feighner (1937–2006), the St. Louis group

Fig. 11.39 David Rosenhan (1929–2012). Image courtesy of Stanford Law School

is described in his dairies was horrible, and colleagues described a shaken, changed man after this. Following this publication Rosenhan was challenged by one hospital to send pseudo-patients; during the following weeks, the hospital staff identified 41 pseudo-patients out of 250 new admissions, but Rosenhan had not sent anyone. The true value of this experiment remains controversial until today, since in general the medical (not only the psychiatric) diagnosis and care rely extensively on the sincerity of the patient and deliberate faking by normal persons is not anticipated. It has been also argued that the paper published by Rosenhan was not accurate in many ways. For example, during his own hospitalization, he reported more symptoms besides hearing the word "thud," including suicidal ideation. If this was so and the pseudo-patients in fact utilized an elaborated deception, this could explain the reason of prolonged hospitalizations and treatments. It is possible that the study was cherry-picking and reported selectively only the negative and suppressed the positive experiences of pseudo-patients. It is also possible that many of these pseudo-patients might never existed at all [315]. However the case pointed the lack of objective and independent laboratory diagnostic indices in psychiatry [314, 316–322]. The issues raised with that study were raised again when Lauren Slater (born

1963) claimed she repeated the same experiment by presenting herself at eight different psychiatric emergency rooms pretending to be a patient. She received a diagnosis of depression with psychotic features but was not hospitalized. Her conclusion was that although circumstances have changed, the diagnosis of mental disorder is still unreliable [323]. Since then other authors have failed to confirm these findings [324], she has been accused of fraud [325], and in response she claimed that she is under attack from "a gang of psychologists and psychiatrists" [326]. Additionally, psychiatry was accused, that through psychoanalysis, it had focused on the worried well and neglected those in true need [327].

Back to the 1970s, the classification of homosexuality as a mental disorder also emerged as another important issue for psychiatry. This practice had emerged both from psychoanalytic theories [328] but also as an adaptation of psychiatric nosology to social norms and the legislation. It is important to note that decriminalization of homosexual acts did not occur in any European or North American country (except France) before the dawn of twentieth century; in most of them occurred after WWII and in several US states (e.g., Texas, Florida, Alabama, Louisiana, Idaho, etc.), decriminalization of homosexual acts was abolished only in the first years of the twenty-first century. In England and Wales, it was abolished in 1967, and one of the most famous victims was the brilliant mathematician Alan Turing (1912–1954, the so-called father of computers and modern information science. After an invaluable contribution to allied victory during WWII, he was convicted to chemical (hormonal) castration and eventually died probably by suicide at the age of 41.

In *DSM-I* homosexuality was considered a "sociopathic personality disturbance," while in DSM-II it was recognized as a separate mental disorder. Even therapeutic methods had been developed with Charles Socarides (1922–2005; Fig. 11.40), being a leading figure of this approach. It is interesting that his son Richard Socarides (born 1954; Fig. 11.40) came out openly as gay and became advisor to President Clinton and an advocate for the rights of the LGBT community. In spite of his son, Charles Socarides remained convinced that homosexuality is a mental disorder in need of specific treatment until his death [329, 330].

For years there were activist groups and demonstrations challenging the views of society and demanding decriminalization of homosexuality and non-discrimination policies by the state. One such demonstration occurred in 1970 at the annual congress of the American Psychiatric Association in San Francisco, and a key feature was that gay activists sided with the anti-psychiatric movement during that demonstration. One sensational event happened in 1972 when the psychiatrist and gay rights activist John Fryer (1937–2003), with a courageous act, delivered a speech at the APA annual meeting under the name Dr. H. Anonymous (Fig. 11.41), wearing a tuxedo, wig, and fright mask. He started his speech with the words: "I am a homosexual. I am a psychiatrist" [51].

At that particular period, Robert Spitzer (1932–2015; Fig. 11.42) had undertaken the leadership of the workgroup for the next DSM edition and was determined to make it to be a game-changer [331]. It was during the February 1973

Fig. 11.40 Charles Socarides (left) and Richard Socarides (right)

Fig. 11.41 Left: Barbara Gittings, Frank Kameny, and John Fryer in disguise as "Dr. H. Anonymous". Right: John Fryer (1937–2003), curtesy of the AGLP

meeting of the APA Board of Trustees that the decision to endorse empirical science rather than Freudian dogma was taken and Spitzer seemed to be the right man to carry this extremely difficult task [332]. The basic idea was to shift the basis of the definition of mental disorder from the "cause/etiology" which was advocated by psychoanalysis, to a combination of persistence of symptoms, distress, and disability [333].

Fig. 11.42 Robert Spitzer (1932–2015) posing with DSM-III (curtesy of the New York State Psychiatric Institute, University of Columbia, USA)

Among the other difficult issues, Spitzer had to deal with the question of homosexuality. Among his actions was to have a secret meeting with a group of gay psychiatrists, and to his surprise, this included chairs of prominent psychiatry departments and a former APA president. All were hiding their sexual orientation and were living double lives. Eventually it was decided that the principle of subjective distress should be applied and that homosexuality on its own was not a mental disorder.

Two were the main forces hostile to the new DSM concept. One was the psychoanalytically oriented members of the APA as well as the American Psychoanalytical Association, who considered the new approach superficial and stupid and argued in a poetic (but not scientific) way that "the elimination of the psychiatric past by the *DSM-III* Task Force can be compared to the director of a national museum destroying his Rembrandts, Goyas, Utrillos, van Goghs, etc. because he believes his collection of Comic-Strip Type Warhols has greater relevance" [334]. But as previously said, at that time the variability of psychiatric diagnoses had become an infamous symbol of lost credibility of psychiatry, and there was much scientific and public criticism on the topic [335]. Definitions were vague and understanding was idiosyncratic [336–340].

The second force of resistance was the American Psychological Association who held an occupational interest to keep mental disorder a psychosocial issue rather than a medical condition and was benefited by the anti-psychiatric movement. However, when the American Psychological Association threatened to develop its own classification system, Spitzer encouraged them to do so; in his polite words, this would enrich the scientific literature. They never did.

The development of DSM-III which was published in 1980, was coordinated with the development of the night edition of the International Classification of Disease published by the World Health Organization (ICD-9), which was published in 1975 and implemented in 1978. However, ICD did not eliminate "homosexuality disorder" until the tenth edition in 1990. Enduring differences between the two systems were that the DSM-III introduced explicit diagnostic criteria, a multiaxial diagnostic assessment system, and an atheoretical approach. However compromise was inevitable, and much of decisions were made not based on data but on consensus [341]. In spite of these shortcomings, this did not stopped DSM-III from constituting a clear step forward [333].

In parallel, psychodynamic psychiatry was also advancing. In the 1970s, Heinz Kohut (1913–1981) developed the theory of self-psychology, and James Masterson (1926–2010) and Otto Kernberg (born 1928) developed two different concepts for diagnosing and treating borderline and narcissistic disorders (Fig. 11.43). According to Kernberg, there are higher, intermediate, and lower levels of character pathology. He considered the intermediate and the lower levels to correspond to borderline personality organization (not personality disorder per se). A new type of psychotherapeutic intervention appeared in the 1970s when Myrna Weissman (born 1935) and Gerald Klerman (1928–1992) introduced interpersonal therapy (IPT) as a brief

Fig. 11.43 Left: Heinz Kohut (1913–1981), Center: James Masterson (1926–2010). Right: Otto Kernberg (born 1928)

therapeutic intervention to be delivered by mental health professionals, psychiatrists, psychologists, or social workers and focusing on the relationship between the onset of depression and current interpersonal problems: grief, disputes, transitions, and loneliness (Fig. 11.44). A distinct feature of this psychotherapy was that it was the first to define the procedures in a manual so that the treatment could be tested in clinical trials. The developers insisted that psychotherapy could be tested in clinical trials like medication and that this evidence was required before disseminating its use. As a result, there are now over 140 clinical trials testing its efficacy. It has been adapted for other disorders and formats such as in groups, and it has been recently used to train community health workers with no mental health background in the first clinical trial in Africa [342–346]. The same decade, Theodore Millon (1928–2014; Fig. 11.45) developed his theory of personality which is largely behind the personality categories of the DSM. His approach is based on a Skinnerian model of coping patterns which reflect ways to receive positive or avoid negative reinforcements (a bipole) in the frame of the bipoles of pleasure vs. pain and self vs. others [347].

The first "psychiatric survivor" groups had already emerged by this time, often united by their opposition to ECT. The big blow to ECT was given in 1975 by the film *One Flew over the Cuckoo's Nest* which was based on the 1962 novel by Ken Kesey. The film was produced by Michael Douglas and was directed by Milos Forman with Jack Nicholson in the leading role and psychiatrist Dean Kent Brooks (1916–2013; Fig. 11.34) in a minor role. It won all five main Oscars at that year and embedded a negative image of ECT and psychiatry in the attitude of the general public (Fig. 11.34). Michael Douglas' father, Kirk Douglas, a civil rights advocate himself, had purchased the rights of the novel for theater and film production and

Fig. 11.44 Left: Myrna Weissman (1935–). Right: Gerald Klerman (1928–1992)

Fig. 11.45 Theodore Millon (1928–2014)

adapted it for the theater in 1963 starring himself together with Gene Wilder. For an analysis of the plot from a political-ideological perspective, please see Chap. 22.

In the USA, legislatures drafted motions to ban the use of ECT. Around the world it started becoming effectively impossible to use ECT. It was essentially banned in cantons of Switzerland and in 1999 in private hospitals in Italy. Many books critical of ECT were published, and Thomas Szasz commented on Leonard Roy Frank's (1932–2015) *History of Shock Treatment* [348] "what the rock and the stake were to the inquisition, what the concentration camp and the gas chamber were to national socialism, the mental hospital and electroshock are to institutional psychiatry." To be noted there is a new TV show under the title *Ratched* which constitutes a prequel series revolving around the character of Nurse Ratched from *One Flew Over the Cuckoo's Nest* and how she evolved from a nurse to a full-fledged monster. The series tracks her murderous progression through the mental healthcare system. The show appeared in September 2020 in Netflix, again with Michael Douglas as the executive producer. It was a crude and cheap reflection of the Ratched of the film and destroyed any allegory of the original character. Interestingly, Michael Douglas' wife, the Catherine Zeta Jones, has announced that she suffers from bipolar disorder and underwent a number of hospitalizations in rehabilitation clinics. In spite of repeated interviews to fight the stigma and to discuss her inner experience, no information on the type of treatment and care she has received was ever released.

The "social revolution" within psychiatry went even further, and even some large and reputable hospitals started holding therapeutic meetings that often included not only patients and mental health professionals but also everyone whose work got him with a direct or indirect contact with the ward, e.g., administration staff, gardeners, and rest of hospital staff as well as the patients. Many participants in these meetings were against all forms of coercion. It is of course reasonable to expect that in such meetings lay opinion and attitude and contra-rationalism would prevail over scientific and professional knowledge. Often, the prevailing attitude was that any kind of restriction of patient is unacceptable and basing on experimental data was correlated with experimenting in concentration camps.

In 1975, in Finland 16 patients out of 2260 exposed (0.7%) to clozapine developed agranulocytosis, and 8 of them subsequently died from secondary infections. As a consequence, clozapine was withdrawn from Finland, and all applications in other countries were suspended. There was never a satisfactory answer as to what exactly had caused the Finnish problems [349–351], but this was not the first time clozapine was related with deaths; as in the trial run by Deniker and Pierre Simon several years before, 20% of patients had died; and Deniker expressed a very negative opinion on the usefulness and safety of the drug.

Building on the works of Adolf Meyer, George Engel (1913–1999; Fig. 11.46) proposed the biopsychosocial model in an attempt to bring together the biological, social, and psychological elements of etiology and treatment of mental disorders

Fig. 11.46 George Engel (1913–1999)

in a frame analogous to general systems theory [352–355]. Although this model enjoyed much popularity, eventually it was abandoned simply because the effect of specific factors could not be proven or, instead, they were proven to be irrelevant in specific disorders. It constituted a politically correct compromise among different approaches and schools of psychiatry; it was also in accord with lay beliefs and common reason but lacked and essentially never obtained scientific data to support it.

In the late 1970s, Arvid Carlsson developed the first selective serotonin reuptake inhibitor (SSRI) which was zimelidine for the Swedish company Astra AB. It was a pyridylallylamine, and it is structurally different from other antidepressants. Zimelidine was marketed in 1982, but soon its use was linked to rare case reports of Guillain-Barré syndrome, and subsequently it was withdrawn from the market [356].

Fluoxetine was synthesized in 1972 for Eli Lilly and Company by Bryan Barnet Molloy (1939–2004; Fig. 11.47) and Robert Rathbun and entered the market in 1986 under the trade name "Prozac." It was the first blockbuster SSRI and had broad therapeutic indications including depression, OCD, bulimia, panic disorders, and premenstrual dysphoric disorder. It is on the World Health Organization's List of Essential Medicines.

In 1975 appeared the first attempts to test the efficacy of psychotherapeutic methods through the conduction of properly designed clinical trials [357, 358].

Fig. 11.47 Bryan Barnet Molloy (1939–2004) courtesy of Elli Lilly Co Archives. All rights reserved

In the 1960s and 70s, there was a trend towards ever larger doses, and the approach of megadose therapy emerged. Patients were to be rapidly neurolepticized as soon as they were hospitalized, and "haloperidol narcosis" became frequent, corresponding to some 25 gr of intravenous chlorpromazine per day. This practice led to deaths from neuroleptic malignant syndrome, asphyxiation due to dystonia, and cardiovascular collapse [359–362]. In 1988 a series of articles argued that there is no benefit with the use of doses higher than 600 mg of chlorpromazine or 30 mg of haloperidol per day [363]. Other studies supported the need for low dose of antipsychotics [364–369].

In the USA, during the presidency of Jimmy Carter in 1977, the congress created the National Committee for the Protection of Human Subjects of Biomedical and Behavioral Research. This committee investigated the practice of psychosurgery and arrived at an overall negative conclusion. It reported that only some extremely limited and properly performed psychosurgery could have some positive effects. After this overall rejection of psychosurgery practices, recently there was some movement for the Nobel Foundation to rescind the prize it awarded to Moniz, but the Foundation declined [370].

These were times when claims concerning the existence of oppressed minorities were often extreme. The possibility they were dictated by personal interests or acquaintances cannot be ruled out. It is important to note that nobody ever gave a fight against the ridiculous concept of "schizophrenogenic mother," developed by Frieda Fromm-Reichmann (1889–1957) and Gregory Bateson (1904–1980) [46, 47] which put the blame on families and especially on mothers in a tragic and unfair way and caused severe problems and burden without having any scientific support at all [48–50]. These mothers of patients with schizophrenia were literally accused by psychoanalysts that they were the direct cause of the mental illness of their children and were indeed oppressed, but nobody stood in their defense. Instead, several prominent civil rights activists engaged in movements for the legalization of pedophilia and the decriminalization of child abuse, especially in France. Pedophilia is included among DSM-V and ICD-10 diagnostic categories, with the arbitrary age threshold for minors to be set to 13 years (while in most countries, the legal age is 14–16), but in 1977 a petition calling for the decriminalization of all consenting relations between adults and minors was signed by among others Jean-Paul Sartre, Simone de Beauvoir, Michel Foucault, Jacques Derrida, Roland Barthes, and the child psychoanalyst Françoise Dolto (1908–1988) [371]. As Michael Foucault argued, the central idea was that prohibition of such relationships was a direct violation of children's civil rights and an abuse by itself [372, 373]. In this frame, the notion of child protection was considered to be a kind of patriarchal oppression and denounced as a danger to the child's personal integrity [374]. Foucault also argued that the whole idea of "consent" was a trap because nobody really gives his/her consensus until the act is performed and thus the act itself is a measure of consent [372, 373]. These arguments were hosted by prestigious newspapers and journals [375, 376] and generally were setting the threshold age at 12 [375], but in a 1979 letter to the newspaper *Libération* in support of Gérard R., an accused child sex criminal awaiting his trial, they defended sexual encounters with minors as young as 6 years old [376]. Opponents of this movement were often accused as "reactionary," and pedophiles were considered as victims.

In September 1979, the National Alliance for the Mentally Illness (NAMI) was founded, and it became one of the most potent consumer lobbying force in mental health. Its mission was to promote research and the provision of care to mental health patients and their families. It also functions as a mutual-help organization.

In the meantime, also in the Astra AB laboratories, Lennart Florvall developed remoxipride in 1978, as a selective D-2 antagonist. It was marketed in 1990, but because of cases of potentially lethal aplastic anemia, it was also withdrawn from the market in 1993 [377].

Herman van Praag (born 1929; Fig. 11.48) proposed a new approach in 1981. He named it sociobiological psychiatry, and according to it, there are significant limitations of pharmacotherapy and psychotherapy; therefore their combination is essential [378]. This theory tried to bridge once more the divergent and often colliding psychiatric schools by allowing an eclectic and synthesizing approach which would leave space for both. Van Praag later went on to explore the role as well as the possible therapeutic utility of various social factors, including religiosity [379].

In the late 1980s, in the Netherlands, the psychiatrist Marius Romme (born 1934) together with the mental health advocate Sandra Escher (born 1945) and Patsy Hague who used to be one of Romme's patients, established one of the most

Fig. 11.48 Herman van Praag (born 1929)

enduring and hostile to all orthodox therapies group under the name "Hearing Voices Self-Help Movement" (Fig. 11.49) [380, 381].

In 1987 the DSM-III-R was published by the APA. This version was radically different from DSM-I and DSM-II since operationalized criteria were introduced and new classifications and new names were utilized. DSM-IV followed in 1994, DMS-IV-TR in 2000, and DSM-V in 2013.

Having started in 1947, Hans Eysenck (1916–1997; Fig. 11.2) formulated a basic proposal of an empirically derived model of personality on the basis of two personality dimensions, extraversion and neuroticism [382]. His beginning has been discussed earlier in the current chapter. He was the first to analyze personality differences using an empirical/statistical method. In the late 1970s, he completed the model with the addition of a third dimension, psychoticism [383, 384]. Overall, he proposed that the basic factors were neuroticism (tendency to experience negative emotions), extraversion (tendency to enjoy positive events), and psychoticism (cognitive style). He also proposed a neurobiological basis for these personality dimensions mainly in terms of arousal and limbic system function. Eysenck's

Fig. 11.49 Screenshot of the International Hearing Voices Network (taken on April 12, 2021)

theory and all the theories that derived from it concern approach/reward, inhibition/punishment, and aggression/flight [385–394]. In parallel, Eysenck developed a psychological and personality theory of politics by arguing that the traditional left-right distinction could be psychologically interpreted in the frame of tender vs. tough mindedness and subsequently in the frame of introversion vs. extraversion [395, 396].

Robert Roger McCrae (born 1949) and Paul Costa (born 1942; Fig. 11.50) proposed the five-factor model (Big Five) [397] which includes neuroticism, extroversion, agreeableness, openness, and conscientiousness and constitutes an further development of Eysenck's theory. The older concept of "psychoticism" was substituted by agreeableness and conscientiousness, while openness has some degree of overlap with extroversion [398].

The work of Robert Cloninger (born 1944; Fig. 11.51) is characterized by an attempt to intimately connect temperamental characteristics with individual differences in genetics, neurotransmitter systems, and behavioral conditioning. He described novelty seeking (anger), harm avoidance (fear), reward dependence (attachment), and persistence (ambition) [399, 400]. His research suggests that temperament components can be assessed as early as preschool age [401] and remain moderately stable throughout a person's lifespan except for changes from behavioral conditioning [402].

Hagop S. Akiskal (1944–2021; Fig. 11.52) has conceived temperament as the affective predisposition or reactivity, based on the original descriptions of fundamental states (manic or hyperthymic, irritable, cyclothymic, anxious, and depressive) by Kraepelin in 1921. Empirical research has confirmed the hypothesized four-dimensional factor structure, and it is in agreement with those previously

Fig. 11.50 Left: Robert Roger McCrae (born 1949). Right: Paul T. Costa

Fig. 11.51 Robert Cloninger (born 1944)

proposed on clinical populations. In an additional exploratory analysis, a depressive type constituted the antipode of hyperthymia and was distinguished from cyclothymia; the irritable temperament appeared somewhat independent. These are close to the classic descriptions of Kraepelin, Kretschmer, and Schneider [403].

Maybe the most important development of the 1980s was the starting of the retreat of the welfare model, at least partially as a consequence of the combination of the exploding increase in health costs and economic shortages. It is interesting that mental health and related welfare came under fire from both sides of the political spectrum; left-winged politics adopted an approach with many anti-psychiatric features, while the right-winged were in principle hostile to welfarism and were inclined to push mental health out of medicine and into the cheaper domain of social care, and this is why they accepted a similar approach. Another related development was that national government agencies and bodies started recognizing the various mental advocate groups mainly in the frame of adopting the idea that the patient is essentially a health services consumer (or user) and of a shift towards market approaches to healthcare provision [404].

In 1993, the APA raised concerns about the possibility of false accusations of abuse based on memories recovered in therapy [405].

Fig. 11.52 Hagop S. Akiskal (born 1944)

In the meantime the enthusiasm for clozapine had not disappeared, and in the late 1980s, a pivotal study showed that it was superior to chlorpromazine in refractory patients [406, 407]. On the basis of this study, clozapine gains the reputation of a superior efficacy in comparison to other antipsychotics and especially in refractory patients. This reputation holds until today in spite of the fact that the hard data do not support it [408]. Also on the basis of this study, clozapine returned to the market starting with USA in 1990, but close laboratory monitoring of the patients was mandatory.

Since 1991 detailed operational treatment guidelines were developed by the American Psychiatric Association, and after 2000 there was a wave of "pragmatic" trials [409–422]. Also, during this period, the first meta-analytic studies emerged [423], and the evidence-based medicine principles gained ground in the treatment recommendations. In 1993 the Cochrane Collaboration was found and so far has published a number of systematic reviews in mental health.

During the last three decades of the twentieth century, Eric Kandel (born 1930; Fig. 11.53) experimented with the *Aplysia* sea slug, which has only 20,000 nerve cells, in order to clarify biochemical processes pertaining to memory formation. He

Fig. 11.53 Eric Kandel (born 1930), Nobel laureate 2000

managed to explain why short-term memory involves only minor modulation of the synapses, while on the contrary, the long-term memory requires new synaptic linkages with the utilization of the CREB protein. Kandel earned the 2000 Nobel Prize in Medicine and Physiology.

In the 1990s a new class of antipsychotics appeared. They were called Second Generation Antipsychotics (SGAs) but became more widely known as "atypical antipsychotics." The term "atypical" implied that they lacked some of the side effects the older medication had, and for many authors, these side effects were utterly correlated with the antipsychotic therapeutic effect. Thus atypical antipsychotics had a significantly lower frequency of extrapyramidal symptoms and signs and related conditions (e.g., tardive dyskinesia or neuroleptic malignant syndrome) [424, 425].

Risperidone was approved in 1994. Sertindole entered the market in 1996 in the UK, but it was not approved in the USA because it produced QT prolongation. As a consequence, it was withdrawn from the rest of the world also. Olanzapine and quetiapine followed in 1996 and 1997, respectively. Essentially, risperidone is an improvement of the chemical structure of haloperidol, while olanzapine and quetiapine derived from clozapine.

Though widely prescribed, antipsychotics were "academically" accepted for the treatment of bipolar disorder only after 2000. Until that time antipsychotics were not considered to be the treatment of choice for bipolar patients, mostly because of fears for the induction of tardive dyskinesia. Second generation agents carry a low incidence of tardive dyskinesia, and this softened the defenses to conduct clinical trials with them. For the first time, it was olanzapine which was recommended by the Texas algorithm in 2002 [426].

One of the major issues of the 1990s was the rising use of amphetamines and SSRIs in the treatment of younger and younger children [427]. In 2007 the FDA required all antidepressants to carry a black box warning stating that antidepressants may increase the risk of suicide in people younger than 25 years old, on the basis of a meta-analysis of all trial data submitted to the FDA [428]. In January 2008, the FDA issued an alert (but not black box) about an increased risk for suicidality with treatment with antiepileptic drugs [429–431].

Beginning in 2008, a number of meta-analytic studies questioned the clinical usefulness of antidepressants and enjoyed much attention by the mass media

Fig. 11.54 Mass media headlines on the non-efficacy of antidepressants after the publication of Kirsch in 2008

(Fig. 11.54). It has been shown that there is a significant bias in the publication of antidepressant trials [432] and that the effect size of the medication group in comparison to placebo is rather small [433–440]. Under-reporting of data and especially of negative trials as well as misleading ways of publication proved to be a widespread phenomenon [441–443].

All the meta-analytic studies mentioned above were based on five "datasets." The datasets are the Khan et al. set [437, 444], the Turner et al. set [432], the Kirsch et al. set [438], the Fournier et al. set [435], and the Undurraga et al. set [440].

Following this conclusion, several authors and agencies like the NICE suggested the utilization of "alternative" treatment options (e.g., exercise and psychotherapy) in mildly depressed patients and pharmacotherapy only for the most severe cases. Among other things, these authors and authorities did not take into consideration that, peculiarly, similar findings were reported concerning psychotherapy [445–448].

On the basis of these results, a "conspiracy theory" involving the FDA was proposed [449, 450]. Furthermore, by "overstretching" the interpretation of the data, it has been suggested that because they do not incur drug risks, alternative therapies (e.g., exercise and psychotherapy) may be a better treatment choice for depression [449]. These triggered much interest from the mass media and from intellects outside the mental health area, often with a biased and ideologically loaded approach [451]. However, the most important suggestion was that initial severity plays a major role and antidepressants might not have any effect at all in mildly depressed patients [435, 437, 438].

However further meta-analyses suggested that antidepressants are clearly superior to placebo and initial severity cannot and should not serve as a clinical cue for treatment choice [452–457]. Unfortunately, the mass media showed an impressive indifference concerning the final conclusion of the antidepressants debate. There was not a single headline with them.

Recently a similar debate emerged concerning the usefulness of antipsychotics especially in the long term. A series of publications argued that long-term antipsychotic treatment could not only be useless but also harmful because it might worsen the long-term outcome of patients [458–461]. In this frame, a number of issues have been raised, including the return of the dopamine sensitization hypothesis and the possibility antipsychotics cause brain atrophy. These emerged after the publication of several studies recently, which revived an old debate and suggested a more favorable outcome for those patients who discontinued antipsychotic medication soon after the resolution of the acute phase [462–464]. Additionally many authors insist on possible harmful effects of long-term antipsychotic treatment [460, 465, 466].

It is a fact that the literature lacks properly designed and conducted studies concerning the long-term effect of antipsychotics. Long-term studies of more than 2–3 years are naturalistic, and our knowledge beyond a 3-year follow-up is limited [467, 468]. On the other hand, there is a wide agreement on the efficacy of antipsychotics during the acute phase, and this is based on an abundance of hard data and among others a bulk of placebo-controlled studies, accompanied by a series of reliable meta-analyses [380, 469–472]. This certainty covers also the early maintenance phase and up to 3 years after the acute episode [468, 473]. Careful analysis suggests that antipsychotics do not cause any kind of supersensitivity psychosis [474, 475], and it is highly unlikely they cause any additional brain atrophy; although definite meta-analyses are lacking, the strongest data suggest a better outcome of patients under long-term treatment.

One novel, interesting, and potentially dangerous feature of this ongoing debate on the usefulness or potential dangers of antipsychotic treatment is that lay persons publish in scientific medical journals [461], while on the other hand, letters by scholars, not accepted for publication by scientific journals, are celebrated and published with an accusational attitude in lay web sites [476, 477]. This is partially but not entirely new since it had happened in the 1940s and 1950s in the frame of the attack against ECT by William Menninger (1899–1966; Fig. 10.56) and later critics of ECT [2].

An important trend since the 1990s was the increasing involvement of economic and market/commercial factors in medical and thus in psychiatric research. Academic bias was a problem even before, but now the admixture was even more complex, since success or failure of a product to make it to the market demanded the resources of a multinational company, and often it determined its survival or its economic default. In turn, clinical freedom became constrained by algorithms and guidelines, while much of mental health care is now provided by general practitioners (GPs).

The increasing costs of mental health care and especially of the psychosocial services led to the development of managed care and to other organized efforts to keep the cost under control and often to reduce it. There is not anymore such thing as the confidential psychiatrist-patient relationship which excluded all third parties. The decisions are not entirely up to them, but third parties have a say also. One of the direct consequences was that psychological treatments tend to pass into the hands of psychologists, social workers, and non-medical healthcare providers, while the discipline of psychiatry faces an existential identity dilemma.

Of all medical disciplines, psychiatry is the one which has benefited from high technology the least. It is unknown whether and when this might change in the future. The first truly digital psychiatry application for mobile phones to assist treatment exclusively on the basis of hard evidence was launched in mid-2019 by the International College of Neuropsychopharmacology (CINP; Fig. 11.55) and was based on its newly developed guidelines on the treatment of bipolar disorder [478–481].

However, in a wider scale, currently, the only real contribution of high-tech in mental health is keeping of records and more sophisticated psychometric and neuropsychological tools of uncertain clinical utility. Cellphone applications showed some usefulness in the management of patients and their everyday routine and experience. Virtual reality is expected to open new horizons by providing a more "naturalistic" simulation of life events and experiences and thus assisting psychotherapeutic and rehabilitation efforts. Probably the horizons of the future are impossible to predict.

The list of people with eminent contribution to psychiatry and mental health includes many scientists who marked research as well as teaching. It is impossible to mention all of them, but it is important to cite Norman Sartorius (born 1935;

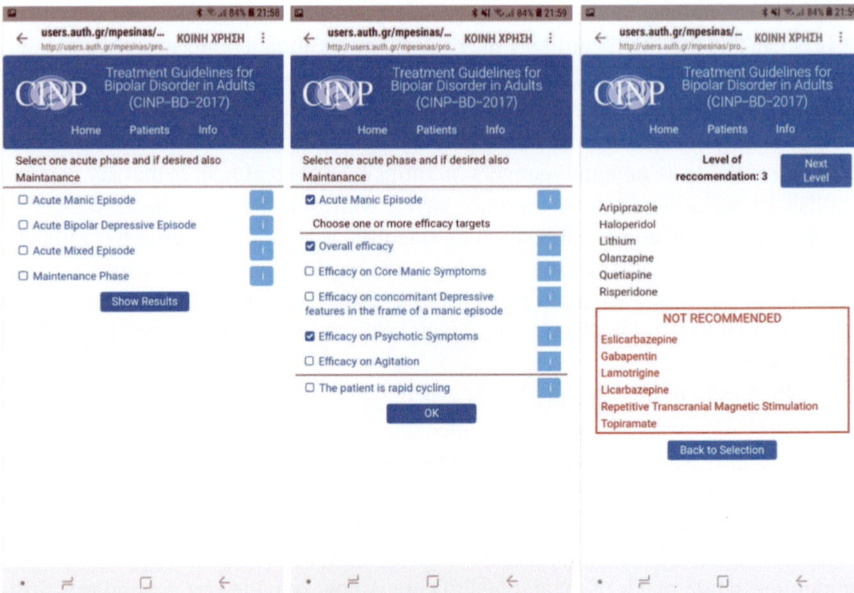

Fig. 11.55 Screenshots from the CINP mobile cell application for the treatment guidelines for bipolar disorder

Fig. 11.56 Norman Sartorius (born 1935)

Fig. 11.57 Hans Jurgen Moeller (born 1945)

Fig. 11.56), Hans Jurgen Moeller (born 1945; Fig. 11.57), Siegfried Kasper (born 1950; Fig. 11.58), and Hagop S. Akiskal (born 1944; Fig. 11.52) who acted as mentors of the author of the current book and guided me among the wandering rocks of psychiatry and especially of psychopharmacology. It is also important to cite the works of Jeffrey Lieberman (born 1948; Fig. 11.59), leader of the CATIE pragmatic trial as well as Nancy Andreasen (born 1938; Fig. 11.60). A whole generation of researchers, most of them today in their 50s, are pushing towards new developments

Fig. 11.58 Siegfried Kasper (born 1950)

Fig. 11.59 Jeffrey Lieberman (born 1948)

Fig. 11.60 Nancy Andreasen (born 1938)

and new breakthrough findings to further improve the lives of mental patients and public mental health.

References

1. Rees WL, Healy D. The place of clinical trials in the development of psychopharmacology. Hist Psychiatry. 1997;8(29 pt 1):1–20. https://doi.org/10.1177/0957154X9700802901.
2. Shorter E. A history of psychiatry: from the era of the asylum to the age of Prozac. New York: Wiley; 1997.
3. Rogers C. Client-centered therapy: its current practice, implications and theory. London: Constable; 1951.
4. Eysenck HJ. The effects of psychotherapy: an evaluation. J Consult Psychol. 1952;16(5):319–24. https://doi.org/10.1037/h0063633.
5. Pelosi AJ. Personality and fatal diseases: revisiting a scientific scandal. J Health Psychol. 2019;24(4):421–39. https://doi.org/10.1177/1359105318822045.

6. Pelosi AJ, Appleby L. Psychological influences on cancer and ischaemic heart disease. BMJ. 1992;304(6837):1295–8. https://doi.org/10.1136/bmj.304.6837.1295.
7. Fulford KWM, Thornton T, Graham G. Karl Jaspers and general psychopathology. In: Fulford KWM, Thornton T, Graham G, editors. The Oxford textbook of philosophy and psychiatry. Oxford: Oxford University Press; 2006.
8. Jones M. The therapeutic community: a new treatment method in psychiatry. New York: Basic Books; 1953.
9. Mueller J, Schlittler E, Bein H. Reserpine, the sedative principle of Rauwolfia serpentina benth. Experientia. 1952;8:338.
10. Lasagna L. A comparison of hypnotic agents. J Pharmacol Exp Ther. 1954;111(1):9–20.
11. Nijensohn DE. Prefrontal lobotomy on Evita was done for behavior/personality modification, not just for pain control. Neurosurg Focus. 2015;39(1):E12. https://doi.org/10.3171/2015.3.FOCUS14843.
12. Nijensohn DE, Savastano LE, Kaplan AD, Laws ER Jr. New evidence of prefrontal lobotomy in the last months of the illness of Eva Peron. World Neurosurg. 2012;77(3–4):583–90. https://doi.org/10.1016/j.wneu.2011.02.036.
13. Noack CH, Trautner EM. The lithium treatment of maniacal psychosis. Med J Aust. 1951;2(7):219–22.
14. Bech P. The full story of lithium. A tribute to Mogens Schou (1918-2005). Psychother Psychosom. 2006;75(5):265–9. https://doi.org/10.1159/000093947.
15. Schou M, Juel-Nielsen N, Stromgren E, Voldby H. The treatment of manic psychoses by the administration of lithium salts. J Neurol Neurosurg Psychiatry. 1954;17(4):250–60. https://doi.org/10.1136/jnnp.17.4.250.
16. Ayd FJ Jr. Invited ACNP lecture. The early history of modern psychopharmacology. Neuropsychopharmacology. 1991;5(2):71–84.
17. Shen WW. A history of antipsychotic drug development. Compr Psychiatry. 1999;40(6):407–14. https://doi.org/10.1016/s0010-440x(99)90082-2.
18. Healy D. The creation of psychopharmacology. Cambridge: Harvard University Press; 2002.
19. Naylor GJ, Martin B, Hopwood SE, Watson Y. A two-year double-blind crossover trial of the prophylactic effect of methylene blue in manic-depressive psychosis. Biol Psychiatry. 1986;21(10):915–20. https://doi.org/10.1016/0006-3223(86)90265-9.
20. Thuillier J. The ten years which changed the face of mental illness. London: Martin Dunitz; 1999.
21. Ban TA. Fifty years chlorpromazine: a historical perspective. Neuropsychiatr Dis Treat. 2007;3(4):495–500.
22. Laborit H, Huguenard P. L'hibernation artificielle par moyen pharmacodynamiques et physiques. Presse Med. 1951;59:1329.
23. Shen WW, Giesler MC. The discoverers of the therapeutic effect of chlorpromazine in psychiatry: qui etaient les vrais premiers praticiens? Can J Psychiatry. 1998;43(4):423–4.
24. Stip E. Who pioneered the use of antipsychotics in North America? Can J Psychiatry. 2015;60(3 Suppl 2):S5–13.
25. Hamon J, Paraire J, Velluz J. Rémarques sur l'action du 4560R.P. sur l'agitation maniaque. Ann Med Psychol. 1952;110:331–5.
26. Delay J, Deniker P, Harl JM. Therapeutic method derived from hiberno-therapy in excitation and agitation states. Ann Med Psychol (Paris). 1952a;110(2 2):267–73.
27. Delay J, Deniker P, Tardieu Y. Hibernotherapy and sleep therapy in psychiatric and psychosomatic therapeutics. Presse Med. 1953;61(58):1165–6.
28. Delay J, Deniker P, Harl JM. Therapeutic use in psychiatry of phenothiazine of central elective action (4560 RP). Ann Med Psychol (Paris). 1952b;110(2 1):112–7.
29. Hakim R. Indigenous drugs in the treatment of mental diseases. In: VI Gujrat and Saurashtra provincial medical conference, Marzo, India; 1953. p. 14–6.
30. Lopez-Munoz F, Bhatara VS, Alamo C, Cuenca E. Historical approach to reserpine discovery and its introduction in psychiatry. Actas espanolas de psiquiatria. 2004;32(6):387–95.

31. Hollister LE, Krieger GE, Kringel A, Roberts RH. Treatment of chronic schizophrenic reactions with reserpine. Ann N Y Acad Sci. 1955;61(1):92–100. https://doi.org/10.1111/j.1749-6632.1955.tb42455.x.
32. Elkes J, Elkes C. Effect of chlorpromazine on the behavior of chronically overactive psychotic patients. Br Med J. 1954;2(4887):560–5. https://doi.org/10.1136/bmj.2.4887.560.
33. Lehmann HE, Hanrahan GE. Chlorpromazine; new inhibiting agent for psychomotor excitement and manic states. AMA Arch Neurol Psychiatry. 1954;71(2):227–37.
34. Miksztal MW. Chlorpromazine (thorazine) and reserpine in residential treatment of neuropsychiatric disorders in children. J Nerv Ment Dis. 1956;123(5):477–9. https://doi.org/10.1097/00005053-195605000-00009.
35. Rees WL, Lambert C. The value and limitations of chlorpromazine in the treatment of anxiety states. J Ment Sci. 1955;101(425):834–40. https://doi.org/10.1192/bjp.101.425.834.
36. Winkelman NW Jr. Chlorpromazine in the treatment of neuropsychiatric disorders. J Am Med Assoc. 1954;155(1):18–21. https://doi.org/10.1001/jama.1954.03690190024007.
37. Davies DL, Shepherd M. Reserpine in the treatment of anxious and depressed patients. Lancet. 1955;269(6881):117–20. https://doi.org/10.1016/s0140-6736(55)92118-8.
38. Delay J, Deniker P, Ropert R. Study of 300 case histories of psychotic patients treated with chlorpromazine in closed wards since 1952. L'Encephale. 1956;45(4):528–35.
39. Ayd FJ Jr. Fatal hyperpyrexia during chlorpromazine therapy. J Clin Exp Psychopathol. 1956;17(2):189–92.
40. Berman BD. Neuroleptic malignant syndrome: a review for neurohospitalists. The Neurohospitalist. 2011;1(1):41–7. https://doi.org/10.1177/1941875210386491.
41. Delay J, Pichot P, Lemperiere T, Elissalde B, Peigne F. A non-phenothiazine and non-reserpine major neuroleptic, haloperidol, in the treatment of psychoses. Ann Med Psychol (Paris). 1960;118(1):145–52.
42. Casey JF, Bennett IF, Lindley CJ, Hollister LE, Gordon MH, Springer NN. Drug therapy in schizophrenia. A controlled study of the relative effectiveness of chlorpromazine, promazine, phenobarbital, and placebo. AMA Arch Gen Psychiatry. 1960a;2:210–20. https://doi.org/10.1001/archpsyc.1960.03590080086012.
43. Casey JF, Lasky JJ, Klett CJ, Hollister LE. Treatment of schizophrenic reactions with phenothiazine derivatives. A comparative study of chlorpromazine, triflupromazine, mepazine, prochlorperazine, perphenazine, and phenobarbital. Am J Psychiatry. 1960b;117:97–105. https://doi.org/10.1176/ajp.117.2.97.
44. Steck H. Le syndrome extrapyramidal et diencephalique au Largactil et au Serpasil. Ann Med Psychol (Paris). 1954;112:737–43.
45. MacLean PD. A triune concept of the brain and behaviour. Hincks memorial lectures. Toronto: University of Toronto Press; 1973.
46. Fromm-Reichmann F. Notes on the development of treatment of schizophrenics by psychoanalytic psychotherapy. Psychiatry. 1948;11(3):263–73. https://doi.org/10.1080/00332747.1948.11022688.
47. Bateson G, Jackson DD, Haley J, Weakland J. Toward a theory of schizophrenia. Behav Sci. 2007;1(4):251–64. https://doi.org/10.1002/bs.3830010402.
48. Harrington A. The fall of the schizophrenogenic mother. Lancet. 2012;379(9823):1292–3. https://doi.org/10.1016/s0140-6736(12)60546-7.
49. Morris JA. The fall of the schizophrenogenic mother. Lancet. 2012;380(9837):110. https://doi.org/10.1016/S0140-6736(12)61180-5.
50. Neill J. Whatever became of the schizophrenogenic mother? Am J Psychother. 1990;44(4):499–505. https://doi.org/10.1176/appi.psychotherapy.1990.44.4.499.
51. Lieberman JA. Shrinks: the untold story of psychiatry. New York: Back Bay Books; 2016.
52. Lewis A. States of depression: their clinical aetiological differentiation. Br Med J. 1938;4:875–8.
53. Leonhard K. Cycloid psychoses, often erroneously considered as schizophrenia. Psychiatr Neurol Med Psychol (Leipz). 1957a;9(12):359–65.
54. Leonhard K. Pathogenesis of manic-depressive disease. Nervenarzt. 1957b;28(6):271–2.
55. Leonhard K. The classification of endogenous psychoses. 5th ed. New York: Irvington; 1979.

56. Kendell R. The role of diagnosis in psychiatry. Oxford: Blackwell Scientific Publications; 1975.
57. Sadler JZ, Wiggins OP, Schwartz MA. Philosophical perspectives on psychiatric diagnostic classification. Baltimore: Johns Hopkins University Press; 1994.
58. Fulford K, Sartorius N. A secret history of ICD and the hidden future of DSM. In: Broome M, Bortolotti L, editors. Psychiatry as cognitive neuroscience: philosophical perspectives. Oxford: Oxford University Press; 2009. p. 29–48.
59. Fulford KWM, Davies M, Richard GTG, Graham G, Sadler J, Stanghellini G, Thornton T. The next hundred years: watching our Ps and Q. In: Fulford KWM, Davies M, Richard GTG, et al., editors. The Oxford handbook of philosophy and psychiatry. Oxford: Oxford University Press; 2013. https://doi.org/10.1093/oxfordhb/9780199579563.001.0001.
60. Leach E. Claude Levi-Straus. Revised ed. New York: Viking Press; 1974.
61. Moore J. Visions of culture: an introduction to anthropological theories and theorists. 3rd ed. Walnut Creek: Altamira; 2009.
62. Strauss C. Tristes Tropiques (1973 English translation by John and Doreen Weightman). New York: Atheneum; 1955.
63. Swazey J. Chlorpromazine in psychiatry: a study in therapeutic innovation. Cambridge: Massachusetts Institute of Technology Press; 1974.
64. Harcourt B. From the asylum to the prison: rethinking the incarceration revolution. Tex L Rev. 2006;84:1751.
65. Harcourt BE. An institutionalization effect: the impact of mental hospitalization and imprisonment on homicide in the United States, 1934–2001. J Leg Stud. 2011;40(1):39–83. https://doi.org/10.1086/658404.
66. Wing J, Brown G. Institutionalism and schizophrenia. London: Cambridge University Press; 1970.
67. Menninger K. The vital balance: the life process in mental health and illness. New York: Viking Penguin; 1963.
68. Eccles JC. Acetylcholine and synaptic transmission in the spinal cord. J Neurophysiol. 1947;10(3):197–204. https://doi.org/10.1152/jn.1947.10.3.197.
69. Eccles JC, Eccles DM, Fatt P. Pharmacological investigations on a central synapse operated by acetylcholine. J Physiol. 1956;131(1):154–69. https://doi.org/10.1113/jphysiol.1956.sp005452.
70. Feldberg W, Vogt M. Acetylcholine synthesis in different regions of the central nervous system. J Physiol. 1948;107(3):372–81. https://doi.org/10.1113/jphysiol.1948.sp004282.
71. Vogt M. Norepinephrine and epinephrine in the central nervous system. Pharmacol Rev. 1954;6(1):31–2.
72. Amin AH, Crawford TB, Gaddum JH. The distribution of substance P and 5-hydroxytryptamine in the central nervous system of the dog. J Physiol. 1954;126(3):596–618. https://doi.org/10.1113/jphysiol.1954.sp005229.
73. Twarog BM, Page IH. Serotonin content of some mammalian tissues and urine and a method for its determination. Am J Phys. 1953;175(1):157–61. https://doi.org/10.1152/ajplegacy.1953.175.1.157.
74. Bradley PB, Elkes J. The effects of some drugs on the electrical activity of the brain. Brain J Neurol. 1957;80(1):77–117. https://doi.org/10.1093/brain/80.1.77.
75. Elkes J, Elkes C, Bradley PB. The effect of some drugs on the electrical activity of the brain, and on behaviour. J Ment Sci. 1954;100(418):125–8. https://doi.org/10.1192/bjp.100.418.125.
76. Gaddum JH. Antagonism between lysergic acid diethylamide and 5-hydroxytryptamine. J Physiol. 1953;121(1):15P.
77. Shaw E, Woolley DW. Yohimbine and ergot alkaloids as naturally occurring antimetabolites of serotonin. J Biol Chem. 1953;203(2):979–89.
78. Woolley DW, Shaw E. A biochemical and pharmacological suggestion about certain mental disorders. Proc Natl Acad Sci U S A. 1954a;40(4):228–31. https://doi.org/10.1073/pnas.40.4.228.
79. Woolley DW, Shaw E. Some neurophysiological aspects of serotonin. Br Med J. 1954b;2(4880):122–6. https://doi.org/10.1136/bmj.2.4880.122.

80. Pletscher A, Shore PA, Brodie BB. Serotonin release as a possible mechanism of reserpine action. Science. 1955;122(3165):374–5. https://doi.org/10.1126/science.122.3165.374.
81. Denber HC, Rajotte P, Kauffman D. Problems in evaluation of R-1625. Am J Psychiatry. 1959;116:356–7. https://doi.org/10.1176/ajp.116.4.356.
82. Goldberg SC, Klerman GL, Cole JO. Changes in schizophrenic psychopathology and Ward behaviour as a function of phenothiazine treatment. Br J Psychiatry J Ment Sci. 1965;111:120–33. https://doi.org/10.1192/bjp.111.471.120.
83. Guttmacher MS. Phenothiazine treatment in acute schizophrenia; effectiveness: the National Institute of Mental Health psychopharmacology service center collaborative study group. Arch Gen Psychiatry. 1964;10:246–61.
84. Tatetsu S, Goto A, Fujiwara T. Kakuseizai chuudoku (Psychostimulant Toxicosis). Tokyo: Igaku-shoin; 1956.
85. Deniker P. From chlorpromazine to tardive dyskinesia (brief history of the neuroleptics). Psychiatric J Univ Ottawa. 1989;14(1):253–9.
86. King C, Voruganti LN. What's in a name? The evolution of the nomenclature of antipsychotic drugs. J Psychiatry Neurosci. 2002;27(3):168–75.
87. Vakil RJ. A clinical trial of Rauwolfia serpentina in essential hypertension. Br Heart J. 1949;11(4):350–5. https://doi.org/10.1136/hrt.11.4.350.
88. Gerard RW. Physiology and psychiatry. Am J Psychiatry. 1949;106(3):161–73. https://doi.org/10.1176/ajp.106.3.161.
89. Gerard RW. The biological roots of psychiatry. Am J Psychiatry. 1955a;112(2):81–90. https://doi.org/10.1176/ajp.112.2.81.
90. Gerard RW. Biological roots of psychiatry. Science. 1955b;122(3162):225–30. https://doi.org/10.1126/science.122.3162.225.
91. Gordon WW. Cerebral physiology and psychiatry. J Ment Sci. 1948;94(394):118–32. https://doi.org/10.1192/bjp.94.394.118.
92. Muller JC, Pryor WW, Gibbons JE, Orgain ES. Depression and anxiety occurring during Rauwolfia therapy. J Am Med Assoc. 1955;159(9):836–9. https://doi.org/10.1001/jama.1955.02960260006002.
93. Shore PA, Silver SL, Brodie BB. Interaction of reserpine, serotonin, and lysergic acid diethylamide in brain. Science. 1955;122(3163):284–5. https://doi.org/10.1126/science.122.3163.284-a.
94. Bloch RG, Dooneief AS, Buchberg AS, Spellman S. The clinical effect of isoniazid and iproniazid in the treatment of pulmonary tuberculosis. Trans Annu Meet Natl Tuberc Assoc. 1953;49:177–8.
95. Bloch RG, Dooneief AS, Buchberg AS, Spellman S. The clinical effect of isoniazid and iproniazid in the treatment of pulmonary tuberculosis. Ann Intern Med. 1954;40(5):881–900. https://doi.org/10.7326/0003-4819-40-5-881.
96. Crane GE. The psychiatric side-effects of iproniazid. Am J Psychiatry. 1956;112(7):494–501. https://doi.org/10.1176/ajp.112.7.494.
97. Kline NS. Clinical experience with iproniazid (marsilid). J Clin Exp Psychopathol. 1958;19(2, Suppl. 1):72–8. discussion 78-79
98. Pleasure H. Psychiatric and neurological side-effects of isoniazid and iproniazid. AMA Arch Neurol Psychiatry. 1954;72(3):313–20. https://doi.org/10.1001/archneurpsyc.1954.02330030047004.
99. Robitzek EH, Selikoff IJ, Ornstein GG. Chemotherapy of human tuberculosis with hydrazine derivatives of isonicotinic acid; preliminary report of representative cases. Q Bull Sea View Hosp. 1952;13(1):27–51.
100. Selifoff IJ, Robitzek EH, Ornstein GG. Toxicity of hydrazine derivatives of isonicotinic acid in the chemotherapy of human tuberculosis; a preliminary report. Q Bull Sea View Hosp. 1952;13(1):17–26.
101. Selikoff IJ, Robitzek EH. Tuberculosis chemotherapy with hydrazine derivatives of isonicotinic acid. Dis Chest. 1952;21(4):385–438. https://doi.org/10.1378/chest.21.4.385.

102. Salzer HM, Lurie ML. Anxiety and depressive states treated with isonicotinyl hydrazide (isoniazid). AMA Arch Neurol Psychiatry. 1953;70(3):317–24. https://doi.org/10.1001/archneurpsyc.1953.02320330042005.
103. Salzer HM, Lurie ML. Depressive states treated with isonicotinyl hydrazide (isoniazid); a follow-up study. Ohio Med. 1955;51(5):437–41.
104. Bailey SD, Bucci L, Gosline E, Kline NS, Park IH, Rochlin D, Saunders JC, Vaisberg M. Comparison of iproniazid with other amine oxidase inhibitors, including W-1544, JB-516, RO 4-1018, and RO 5-0700. Ann N Y Acad Sci. 1959;80:652–68. https://doi.org/10.1111/j.1749-6632.1959.tb49243.x.
105. Griesemer EC, Barsky J, Dragstedt CA, Wells JA, Zeller EA. Potentiating effect of iproniazid on the pharmacological action of sympathomimetic amines. Proc Soc Exper Biol Med Soc Exp Biol Med. 1953;84(3):699–701. https://doi.org/10.3181/00379727-84-20757.
106. Griesemer EC, Dragstedt CA, Wells JA, Zeller EA. Adrenergic blockade by iproniazid. Experientia. 1955;11(5):182–3. https://doi.org/10.1007/BF02161306.
107. Lauer JW, Inskip WM, Bernsohn J, Zeller EA. Observations of schizophrenic patients after iproniazid and tryptophan. AMA Arch Neurol Psychiatry. 1958;80(1):122–30. https://doi.org/10.1001/archneurpsyc.1958.02340070140021.
108. Rebhun J, Feinberg SM, Zeller EA. Potentiating effect of iproniazid on action of some sympathicomimetic amines. Proc Soc Exper Biol Med Soc Exp Biol Med. 1954;87(1):218–20. https://doi.org/10.3181/00379727-87-21339.
109. Zeller EA, Barsky J, Fouts JR, Kirchheimer WF, Vanorden LS. Influence of Isonicotinic acid hydrazide (Inh) and 1-Isonicotinyl-2-isopropyl hydrazide (Iih) on bacterial and mammalian enzymes. Experientia. 1952;8(9):349–50. https://doi.org/10.1007/Bf02174413.
110. Zeller EA, Sarkar S. Amine oxidases. XIX. Inhibition of monoamine oxidase by phenylcyclopropylamines and iproniazid. J Biol Chem. 1962;237:2333–6.
111. Crane GE. Iproniazid (marsilid) phosphate, a therapeutic agent for mental disorders and debilitating diseases. Psychiatric Res Rep. 1957;8:142–52.
112. Ayd FJ Jr. Chemical treatment of depression. Ann N Y Acad Sci. 1959;80:734–44. https://doi.org/10.1111/j.1749-6632.1959.tb49250.x.
113. Kamman GR, Freeman JG, Lucero RJ. The effect of 1-isonicotynl 2-isopropyl hydrazide (IIH) on the behavior of long-term mental patients. J Nerv Ment Dis. 1953;118(5):391–407. https://doi.org/10.1097/00005053-195311000-00002.
114. Smith J. The use of the isopropyl derivative of Isonicotinylhydrazine (Marsilid) in the treatment of mental disease. Am Practitioner Dig Treat. 1953a;4:519–20.
115. Smith JA. The use of the isopropyl derivative of isonicotinylhydrazine (marsilid) in the treatment of mental disease; a preliminary report. Am Pract Dig Treat. 1953b;4(8):519–20.
116. Axelrod J, Tomchick R. Enzymatic O-methylation of epinephrine and other catechols. J Biol Chem. 1958;233(3):702–5.
117. Axelrod J, Whitby LG, Hertting G. Effect of psychotropic drugs on the uptake of H3-norepinephrine by tissues. Science. 1961;133(3450):383–4. https://doi.org/10.1126/science.133.3450.383.
118. Coyle JT. Julius Axelrod (1912-2004). Mol Psychiatry. 2005;10(3):225–6. https://doi.org/10.1038/sj.mp.4001650.
119. Pincock S. Julius Axelrod Lancet. 2005;365(9457):380. https://doi.org/10.1016/S0140-6736(05)17814-3.
120. Shampo MA, Kyle RA. Julius Axelrod--American biochemist and Nobel prize winner. Mayo Clin Proc. 1994;69(2):136. https://doi.org/10.1016/s0025-6196(12)61039-8.
121. Snyder SH. Obituary: Julius Axelrod (1912-2004). Nature. 2005;433(7026):593. https://doi.org/10.1038/433593a.
122. Zeller EA, Blanksma LA, Burkard WP, Pacha WL, Lazanas JC. In vitro and inhibition of amine oxidases. Ann N Y Acad Sci. 1959;80:583–9. https://doi.org/10.1111/j.1749-6632.1959.tb49236.x.
123. APA Psychiatric Research Reports 1 In, Washington, DC, 1955.

124. Kline NS. Uses of reserpine, the newer phenothiazines, and iproniazid. Res Publ Assoc Res Nerv Ment Dis. 1959;37:218–44.
125. Loomer H, Saunders J, Kline N. A clinical and psychodynamic evaluation of iproniazid as a psychic energizer. In: Hollender M, editor. Research in affects, American Psychiatric Association, psychiatric research reports 8. Washington, DC: APA; 1957a.
126. Loomer HP, Saunders JC, Kline NS. A clinical and pharmacodynamic evaluation of iproniazid as a psychic energizer. Psychiatric Res Rep. 1957b;8:129–41.
127. Saunders JC, Radinger N, Rochlin D, Kline NS. Treatment of depressed and regressed patients with iproniazid and reserpine. Dis Nerv Syst. 1959;20(1):31–9.
128. Saunders JC, Kline NS. Lasker award - priority claim. J Am Med Assoc. 1965;191(10):865. https://doi.org/10.1001/jama.1965.03080100083031.
129. Ayd FJ Jr. A preliminary report on marsilid. Am J Psychiatry. 1957;114(5):459. https://doi.org/10.1176/ajp.114.5.459.
130. Pare CM, Sandler M. A clinical and biochemical study of a trial of iproniazid in the treatment of depression. J Neurol Neurosurg Psychiatry. 1959;22:247–51. https://doi.org/10.1136/jnnp.22.3.247.
131. Robie TR. An effective oral chemotherapy for melancholia; iproniazid. Dis Nerv Syst. 1958a;19(7, Part 2):49–52.
132. Robie TR. Iproniazid chemotherapy in melancholia. Am J Psychiatry. 1958b;115(5):402–9. https://doi.org/10.1176/ajp.115.5.402.
133. Joel SW. Twenty month study of iproniazid therapy. Dis Nerv Syst. 1959;20:521–4.
134. West ED, Dally PJ. Effects of iproniazid in depressive syndromes. Br Med J. 1959;1(5136):1491–4. https://doi.org/10.1136/bmj.1.5136.1491.
135. Kline NS, Li CH, Lehmann HE, Lajtha A, Laski E, Cooper T. Beta-endorphin--induced changes in schizophrenic and depressed patients. Arch Gen Psychiatry. 1977;34(9):1111–3. https://doi.org/10.1001/archpsyc.1977.01770210125012.
136. Delay J, Buisson JF. Psychic action of isoniazid in the treatment of depressive states. J Clin Exp Psychopathol. 1958;19(2, Suppl. 1):51–5.
137. Delay J, Deniker P, Buisson JF, Haim A. Treatment of depressive states by the isonicotinic acid derivatives isoniazid & iproniazid. Ann Med Psychol (Paris). 1959;117(1):125–32.
138. Nelson SD, Mitchell JR, Snodgrass WR, Timbrell JA. Hepatotoxicity and metabolism of iproniazid and isopropylhydrazine. J Pharmacol Exp Ther. 1978;206(3):574–85.
139. Timbrell JA. The role of metabolism in the hepatotoxicity of isoniazid and iproniazid. Drug Metab Rev. 1979;10(1):125–47. https://doi.org/10.3109/03602537908993904.
140. Blackwell B. Hypertensive crisis due to monoamine-oxidase inhibitors. Lancet. 1963;2(7313):849–50. https://doi.org/10.1016/s0140-6736(63)92743-0.
141. Perls F. Gestalt therapy: excitement and growth in the human personality. new ed. Highland: The Gestalt Journal Press; 1977.
142. Ellis A. Rational psychotherapy. J Gen Psychol. 1958;59(1):35–49. https://doi.org/10.1080/00221309.1958.9710170.
143. Sigwald J, Bouttier D, Raymond C, Piot C. Quatre cas de dykinésie facio-bucco-linguo-masticatrice a évolution prolongée secondaire a un traitement par les neuroleptiques. Rev Neurol. 1959;100:751–5.
144. Uhrbrand L, Faurbye A. Reversible and irreversible dyskinesia after treatment with perphenazine, chlorpromazine, reserpine, and ECT. Psychopharmacologia. 1960;1:408–18.
145. Ayd F. Persistent dyskinesia. Med Sci. 1967;18:32–40.
146. Crane GE. Dyskinesia and neuroleptics. Arch Gen Psychiatry. 1968a;19(6):700–3. https://doi.org/10.1001/archpsyc.1968.01740120060009.
147. Crane GE. Tardive dyskinesia in patients treated with major neuroleptics: a review of the literature. Am J Psychiatry. 1968b;124(8):40–8. https://doi.org/10.1176/ajp.124.8S.40.
148. Schelkunov EL. Adrenergic effect of chronic administration of neuroleptics. Nature. 1967;214(5094):1210–2. https://doi.org/10.1038/2141210a0.

149. Cahn C. Roland Kuhn, 1912–2005. Neuropsychopharmacology. 2006;31(5):1096. https://doi.org/10.1038/sj.npp.1301026.
150. Kirsh D, Ogas O. The drug hunters: the improbable quest to discover new medicines. New York: Arcade; 2017.
151. Shorter E. A historical dictionary of psychiatry. Oxford: Oxford University Press; 2005.
152. Kuhn R. Über die Behandlung depressiver Zustände mit einem Iminodibenzylderivat (G 22355). Schweiz Med Wochenschr. 1957;87:1135–40.
153. Bossong F. In memory of Roland Kuhn (1912-2005) and 50 years of imipramine. Nervenarzt. 2008;79(9):1080–6. https://doi.org/10.1007/s00115-008-2492-4.
154. Kuhn R. The treatment of depressive states with G 22355 (imipramine hydrochloride). Am J Psychiatry. 1958;115(5):459–64. https://doi.org/10.1176/ajp.115.5.459.
155. Carlsson A, Fuxe K, Ungerstedt U. The effect of imipramine on central 5-hydroxytryptamine neurons. J Pharm Pharmacol. 1968;20(2):150–1. https://doi.org/10.1111/j.2042-7158.1968.tb09706.x.
156. Gershon S, Holmberg G, Mattsson E, Mattsson N, Marshall A. Imipramine hydrochloride. Its effects on clinical, autonomic, and psychological functions. Arch Gen Psychiatry. 1962;6:96–101. https://doi.org/10.1001/archpsyc.1962.01710190098011.
157. Glowinski J, Axelrod J. Effect of drugs on the uptake, release, and metabolism of H3-norepinephrine in the rat brain. J Pharmacol Exp Ther. 1965;149:43–9.
158. Herting G, Axelrod J, Whitby LG. Effect of drugs on the uptake and metabolism of H3-norepinephrine. J Pharmacol Exp Ther. 1961;134:146–53.
159. Lidbrink P, Jonsson G, Fuxe K. The effect of imipramine-like drugs and antihistamine drugs on uptake mechanisms in the central noradrenaline and 5-hydroxytryptamine neurons. Neuropharmacology. 1971;10(5):521–36. https://doi.org/10.1016/0028-3908(71)90018-9.
160. Klerman GL, Cole JO. Clinical pharmacology of imipramine and related antidepressant compounds. Pharmacol Rev. 1965;17:101–41.
161. Klerman GL, Cole JO. Clinical pharmacology of imipramine and related antidepressant compounds. Int J Psychiatry. 1967;3(4):267–304.
162. Eysenck H. Behaviour therapy and the neuroses. Oxford: Pergamon Press; 1960.
163. Wolpe J. Psychotherapy by reciprocal inhibition. Stanford: Stanford University Press; 1958.
164. Morse WH, Skinner BF. Some factors involved in the stimulus control of operant behavior. J Exp Anal Behav. 1958;1:103–7. https://doi.org/10.1901/jeab.1958.1-103.
165. Skinner BF. Are theories of learning necessary? Psychol Rev. 1950;57(4):193–216. https://doi.org/10.1037/h0054367.
166. Skinner BF. The control of human behavior. Trans N Y Acad Sci. 1955;17(7):547–51. https://doi.org/10.1111/j.2164-0947.1955.tb02820.x.
167. Skinner BF. Diagramming schedules of reinforcement. J Exp Anal Behav. 1958;1:67–8. https://doi.org/10.1901/jeab.1958.1-67.
168. Skinner BF. Contingencies of reinforcement in the design of a culture. Behav Sci. 1966a;11(3):159–66. https://doi.org/10.1002/bs.3830110302.
169. Skinner BF. What is the experimental analysis of behavior? J Exp Anal Behav. 1966b;9(3):213–8. https://doi.org/10.1901/jeab.1966.9-213.
170. Skinner BF, Morse WH. Concurrent activity under fixed-interval reinforcement. J Comp Physiol Psychol. 1957;50(3):279–81. https://doi.org/10.1037/h0047021.
171. Baldessarini RJ, Frankenburg FR. Clozapine. A novel antipsychotic agent. N Engl J Med. 1991;324(11):746–54. https://doi.org/10.1056/NEJM199103143241107.
172. Healy D. Psychopharmacology and the ethics of resource allocation. Br J Psychiatry J Ment Sci. 1993;162:23–9.; discussion 29-37. https://doi.org/10.1192/bjp.162.1.23.
173. Hippius H. The history of clozapine. Psychopharmacology. 1989;99(Suppl):S3–5. https://doi.org/10.1007/BF00442551.
174. Simon P. Twenty-first century drug development. In: Healy D, editor. The Psychopharmacologists III. Boca Raton: CRC Press; 2000. p. 526–7.

175. Haase H, Janssen P. The action of neuroleptic drugs: a psychiatric, neurologic, and pharmacological investigation. Amsterdam: North Holland; 1958.
176. Schooler NR. Jonathan O. Cole, MD (1925-2009): innovator in clinical psychopharmacology and of the ECDEU/NCDEU tradition. J Clin Psychiatry. 2011;72(3):286–7. https://doi.org/10.4088/JCP.10com06726.
177. Schneider K. Clinical psychopathology. New York: Grune and Stratton; 1959.
178. Guyotat J, Marin D, Bonhomme R. Imipramine treatment in other than depressive states. J de medecine de Lyon. 1960;41:367–75.
179. Curtis DR, Phillis JW, Watkins JC. Actions of amino-acids on the isolated hemisected spinal cord of the toad. 1961. Br J Pharmacol. 1997;120(4 Suppl):204–25.; discussion 202-203. https://doi.org/10.1111/j.1476-5381.1997.tb06801.x.
180. Watkins JC, Jane DE. The glutamate story. Br J Pharmacol. 2006;147(Suppl 1):S100–8. https://doi.org/10.1038/sj.bjp.0706444.
181. Lehmann HE. New drugs in psychiatric therapy. Can Med Assoc J. 1961;85:1145–51.
182. Hollister LE, Motzenbecker FP, Degan RO. Withdrawal reactions from chlordiazepoxide ("Librium"). Psychopharmacologia. 1961;2:63–8. https://doi.org/10.1007/BF00429621.
183. Caroff SN. The neuroleptic malignant syndrome. J Clin Psychiatry. 1980;41(3):79–83.
184. Morris HH 3rd, McCormick WF, Reinarz JA. Neuroleptic malignant syndrome. Arch Neurol. 1980;37(7):462–3. https://doi.org/10.1001/archneur.1980.00500560092019.
185. Carlsson A. The occurrence, distribution and physiological role of catecholamines in the nervous system. Pharmacol Rev. 1959;11(2, Part 2):490–3.
186. Carlsson A. Arvid Carlsson. In: Squire L (ed) he history of neuroscience in autobiography, vol. 2. San Diego: Academic Press; 1998. p. 28–66.
187. Carlsson A, Lindqvist M. Effect of chlorpromazine or haloperidol on formation of 3methoxytyramine and Normetanephrine in mouse brain. Acta Pharmacol Toxicol. 1963;20:140–4. https://doi.org/10.1111/j.1600-0773.1963.tb01730.x.
188. van Rossum JM. The significance of dopamine-receptor blockade for the mechanism of action of neuroleptic drugs. Archives internationales de pharmacodynamie et de therapie. 1966;160(2):492–4.
189. Schildkraut JJ. The catecholamine hypothesis of affective disorders: a review of supporting evidence. Am J Psychiatry. 1965;122(5):509–22. https://doi.org/10.1176/ajp.122.5.509.
190. Schildkraut JJ. The catecholamine hypothesis of affective disorders. A review of supporting evidence. Int J Psychiatry. 1967;4(3):203–17.
191. Van Voren R. Cold war in psychiatry: human factors. Rodopi: Secret Actors; 2009.
192. Lavretsky H. The Russian concept of schizophrenia: a review of the literature. Schizophr Bull. 1998;24(4):537–57. https://doi.org/10.1093/oxfordjournals.schbul.a033348.
193. Aliev N. “Sluggish schizophrenia” (concerning the article by A.B. Smulevich and P.V. Mikhalev) (S.S. Korsakov zhurnal nevropatologii i psikhiatrii), no. 12, 1987, pp. 186-187. Zh Nevropatol Psikhiatr Im S S Korsakova. 1992;92(2):132.
194. Andreyev H. Political dissent and "sluggish" schizophrenia in the Soviet Union. Br Med J (Clin Res Ed). 1986;293(6550):822. https://doi.org/10.1136/bmj.293.6550.822-a.
195. Bloch S, Reddaway P. Soviet psychiatric abuse: the shadow over world psychiatry. Boulder: Westview Press; 1985.
196. Boiarshinova TN. Psychological diagnosis of sluggish schizophrenia (according to thematic apperception test findings). Zh Nevropatol Psikhiatr Im S S Korsakova. 1975;75(11):1680–2.
197. Gladkova KI. Basic patterns in the formation of sluggish schizophrenia according to epidemiologic findings. Zh Nevropatol Psikhiatr Im S S Korsakova. 1977;77(2):253–6.
198. Makhova TA. Features of the course of sluggish psychopathic-like schizophrenia, its diagnosis and forensic psychiatric significance. Zh Nevropatol Psikhiatr Im S S Korsakova. 1982;82(5):113–9.
199. Merskey H, Shafran B. Political hazards in the diagnosis of ‘sluggish schizophrenia’. Br J Psychiatry J Ment Sci. 1986;148:247–56. https://doi.org/10.1192/bjp.148.3.247.

200. Polubinskaya SV. Reform in psychiatry in post-soviet countries. Acta Psychiatr Scand Suppl. 2000;399(399):106–8. https://doi.org/10.1111/j.0902-4441.2000.007s020[dash]24.x.
201. Sarsembaev KT. Social-work adaptation of patients with sluggish schizophrenia in the late stages of the course of the disease (according to the results of a clinico-epidemiologic study). Zh Nevropatol Psikhiatr Im S S Korsakova. 1985;85(8):1217–21.
202. Smulevich AB. Sluggish schizophrenia in the modern classification of mental illness. Schizophr Bull. 1989;15(4):533–9. https://doi.org/10.1093/schbul/15.4.533.
203. Szasz T. Therapeutic state soviet psychiatry: its supporters in the West. Inquiry. 1978; (January 2):4–5.
204. Vilenskii OG, Kolomiichenko LN. Clinical features and work capacity of patients with sluggish schizophrenia and a hypochondriacal syndrome. Zh Nevropatol Psikhiatr Im S S Korsakova. 1983;83(5):728–32.
205. Wilkinson G. Political dissent and "sluggish" schizophrenia in the Soviet Union. Br Med J (Clin Res Ed). 1986;293(6548):641–2. https://doi.org/10.1136/bmj.293.6548.641.
206. Zairov GK, Gurdorzh T, Bavuuzhav O. (Clinico-catamnestic study) clinical picture, treatment and social-occupational adaptation of patients suffering schizophrenia with a sluggish, uninterrupted course. Zh Nevropatol Psikhiatr Im S S Korsakova. 1977;77(5):722–5.
207. Adler N, Gluzman S. Soviet special psychiatric hospitals. Where the system was criminal and the inmates were sane. Br J Psychiatry J Ment Sci. 1993;163(6):713–20. https://doi.org/10.1192/bjp.163.6.713.
208. Levine S. The special Committee on the political abuse of psychiatry. Bulletin of the Royal College of Psychiatrists. 2018;5(5):94–5. https://doi.org/10.1192/s0140078900011445.
209. Finlayson J. Political abuse of psychiatry with a special focus on the USSR. Bull Royal College Psychiatr. 2018;11(4):144–5. https://doi.org/10.1192/pb.11.4.144.
210. Wynn A. The Soviet Union and the world psychiatric association. Lancet. 1983;1(8321):406–8. https://doi.org/10.1016/s0140-6736(83)91514-3.
211. Stone AA. Psychiatrists on the side of the angels: the Falun gong and soviet Jewry. J Am Acad Psychiatry Law. 2002;30(1):107–11.
212. Lyons D, O'Malley A. The labelling of dissent: politics and psychiatry behind the Great Wall. Psychiatr Bull. 2002;26:434.
213. van Voren R. The WPA world congress in Yokohama and the issue of political abuse of psychiatry in China. Psychiatr Bull. 2018;26(12):441–2. https://doi.org/10.1192/pb.26.12.441.
214. Gershon S, Yuwiler A. Lithium ion: a specific psychopharmacological approach to the treatment of mania. J Neuropsychiatr. 1960;1:229–41.
215. Baastrup PC. The use of lithium in manic-depressive psychosis. Compr Psychiatry. 1964;5(6):396–408. https://doi.org/10.1016/s0010-440x(64)80049-3.
216. Blackwell B. Need for careful evaluation of lithium. Am J Psychiatry. 1969;125(8):1131. https://doi.org/10.1176/ajp.125.8.1131.
217. Blackwell B. Lithium. Lancet. 1970;2(7678):875–6. https://doi.org/10.1016/s0140-6736(70)92041-6.
218. Blackwell B. Lithium prophylaxis in recurrent affective disorders. Br J Psychiatry J Ment Sci. 1971;118(542):131–2. https://doi.org/10.1192/bjp.118.542.131.
219. Blackwell B. Prophylactic lithium: science or science fiction? Am Heart J. 1972;83(1):139–41. https://doi.org/10.1016/0002-8703(72)90118-4.
220. Blackwell B, Shepherd M. Prophylactic lithium: another therapeutic myth? An examination of the evidence to date. Lancet. 1968;1(7549):968–71. https://doi.org/10.1016/s0140-6736(68)90917-3.
221. Shepherd M. A prophylactic myth. Int J Psychiatry. 1970;9:423–5.
222. Angst J, Grof P, Schou M. Lithium Lancet. 1969;1(7605):1097.
223. Angst J, Weis P, Grof P, Baastrup PC, Schou M. Lithium prophylaxis in recurrent affective disorders. Br J Psychiatry J Ment Sci. 1970;116(535):604–14. https://doi.org/10.1192/bjp.116.535.604.

224. Baastrup PC, Poulsen JC, Schou M, Thomsen K, Amdisen A. Prophylactic lithium: double blind discontinuation in manic-depressive and recurrent-depressive disorders. Lancet. 1970;2(7668):326–30. https://doi.org/10.1016/s0140-6736(70)92870-9.
225. Baastrup PC, Schou M. Lithium as a prophylactic agent - its effect against recurrent depressions and manic-depressive psychosis. Arch Gen Psychiatry. 1967;16(2):162.
226. Johnstone EC, Crow TJ, Frith CD, Owens DG. The Northwick Park "functional" psychosis study: diagnosis and treatment response. Lancet. 1988;2(8603):119–25. https://doi.org/10.1016/s0140-6736(88)90682-4.
227. Mitchell PB, Hadzi-Pavlovic D. Lithium treatment for bipolar disorder. Bull World Health Organ. 2000;78(4):515–7.
228. Schioldann J. John Cade's seminal lithium paper turns fifty. Acta Psychiatr Scand. 1999;100(6):403–5. https://doi.org/10.1111/j.1600-0447.1999.tb10889.x.
229. Schioldann J. 'On periodical depressions and their pathogenesis' by Carl Lange (1886). Hist Psychiatry. 2011;22(85 Pt 1):108–30. https://doi.org/10.1177/0957154X10396807.
230. Schioldann J. Obituary. Hist Psychiatry. 2016;17(2):247–52. https://doi.org/10.1177/0957154x06061602.
231. Schou M, Baastrup PC, Grof P, Weis P, Angst J. Pharmacological and clinical problems of lithium prophylaxis. Br J Psychiatry J Ment Sci. 1970;116(535):615–9. https://doi.org/10.1192/bjp.116.535.615.
232. Goodwin FK. Rationale for long-term treatment of bipolar disorder and evidence for long-term lithium treatment. J Clin Psychiatry. 2002;63(Suppl 10):5–12.
233. Goodwin FK, Fireman B, Simon GE, Hunkeler EM, Lee J, Revicki D. Suicide risk in bipolar disorder during treatment with lithium and divalproex. JAMA. 2003;290(11):1467–73. https://doi.org/10.1001/jama.290.11.1467.
234. Goodwin FK, Murphy DL, Bunney WE Jr. Lithium. Lancet. 1969a;2(7613):212–3. https://doi.org/10.1016/s0140-6736(69)91446-9.
235. Goodwin FK, Murphy DL, Bunney WE Jr. Lithium-carbonate treatment in depression and mania. A longitudinal double-blind study. Arch Gen Psychiatry. 1969b;21(4):486–96. https://doi.org/10.1001/archpsyc.1969.01740220102012.
236. Goodwin FK, Murphy DL, Dunner DL, Bunney WE Jr. Lithium response in unipolar versus bipolar depression. Am J Psychiatry. 1972;129(1):44–7. https://doi.org/10.1176/ajp.129.1.44.
237. Goodwin FK, Zis AP. Lithium in the treatment of mania: comparisons with neuroleptics. Arch Gen Psychiatry. 1979;36(8):840–4. https://doi.org/10.1001/archpsyc.1979.01780080014003.
238. Bech P, Vendsborg PB, Rafaelsen OJ. Lithium maintenance treatment of manic-melancholic patients: its role in the daily routine. Acta Psychiatr Scand. 1976;53(1):70–81. https://doi.org/10.1111/j.1600-0447.1976.tb00060.x.
239. Schou M. Normothymotics, "mood-normalizers": are lithium and the imipramine drugs specific for affective disorders? Br J Psychiatry J Ment Sci. 1963;109:803–9. https://doi.org/10.1192/bjp.109.463.803.
240. Lambert PA, Carraz G, Borselli S, Carbel S. Neuropsychotropic action of a new anti-epileptic agent: depamide. Ann Med Psychol (Paris). 1966;124(5):707–10.
241. Okuma T, Inanaga K, Otsuki S, Sarai K, Takahashi R, Hazama H, Mori A, Watanabe M. Comparison of the antimanic efficacy of carbamazepine and chlorpromazine: a double-blind controlled study. Psychopharmacology. 1979;66(3):211–7. https://doi.org/10.1007/BF00428308.
242. Berne E. Games people play - the basic hand book of transactional analysis. New York: Ballantine Books; 1964.
243. Jager G, Rogers J. Formal language theory: refining the Chomsky hierarchy. Philos Trans R Soc Lond Ser B Biol Sci. 2012;367(1598):1956–70. https://doi.org/10.1098/rstb.2012.0077.
244. Ottl B, Jager G, Kaup B. Does formal complexity reflect cognitive complexity? Investigating aspects of the Chomsky hierarchy in an artificial language learning study. PLoS One. 2015;10(4):e0123059. https://doi.org/10.1371/journal.pone.0123059.

245. Piattelli-Palmarini M. Ever since language and learning: afterthoughts on the Piaget-Chomsky debate. Cognition. 1994;50(1–3):315–46. https://doi.org/10.1016/0010-0277(94)90034-5.
246. Wang L, Bastiaansen M, Yang Y, Hagoort P. Information structure influences depth of syntactic processing: event-related potential evidence for the Chomsky illusion. PLoS One. 2012;7(10):e47917. https://doi.org/10.1371/journal.pone.0047917.
247. Berwick RC, Friederici AD, Chomsky N, Bolhuis JJ. Evolution, brain, and the nature of language. Trends Cogn Sci. 2013;17(2):89–98. https://doi.org/10.1016/j.tics.2012.12.002.
248. Berwick RC, Pietroski P, Yankama B, Chomsky N. Poverty of the stimulus revisited. Cogn Sci. 2011;35(7):1207–42. https://doi.org/10.1111/j.1551-6709.2011.01189.x.
249. Chomsky N. The development of grammar in child language: formal discussion. Monogr Soc Res Child Dev. 1964;29:35–9.
250. Chomsky N. The United States and the universality of human rights. Int J Health Serv. 1999;29(3):623–6. https://doi.org/10.2190/V3QN-ELK5-LF26-46KF.
251. Chomsky N. Universals of human nature. Psychother Psychosom. 2005;74(5):263–8. https://doi.org/10.1159/000086316.
252. Chomsky N. Language architecture and its import for evolution. Neurosci Biobehav Rev. 2017;81(Pt B):295–300. https://doi.org/10.1016/j.neubiorev.2017.01.053.
253. Everaert MBH, Huybregts MAC, Berwick RC, Chomsky N, Tattersall I, Moro A, Bolhuis JJ. What is language and how could it have evolved? Trends Cogn Sci. 2017;21(8):569–71. https://doi.org/10.1016/j.tics.2017.05.007.
254. Friederici AD, Chomsky N, Berwick RC, Moro A, Bolhuis JJ. Language, mind and brain. Nat Hum Behav. 2017;1(10):713–22. https://doi.org/10.1038/s41562-017-0184-4.
255. Yang C, Crain S, Berwick RC, Chomsky N, Bolhuis JJ. The growth of language: universal grammar, experience, and principles of computation. Neurosci Biobehav Rev. 2017;81(Pt B):103–19. https://doi.org/10.1016/j.neubiorev.2016.12.023.
256. Andresen J. Skinner and Chomsky 30 years later. Or: the return of the repressed. Behav Anal. 1991;14(1):49–60. https://doi.org/10.1007/BF03392552.
257. Virues-Ortega J. The case against B. F. Skinner 45 years later: an encounter with N. Chomsky. Behav Anal. 2006;29(2):243–51. https://doi.org/10.1007/BF03392133.
258. Sandrone S. Norman Geschwind (1926-1984). J Neurol. 2013;260(12):3197–8. https://doi.org/10.1007/s00415-013-6871-9.
259. Clarke CA, Sheppard PM. Lessons of the "pink spot". BMJ. 1967;1(5542):757. https://doi.org/10.1136/bmj.1.5542.757-b.
260. Ellman GL, Jones RT, Rychert RC. Mauve spot and schizophrenia. Am J Psychiatry. 1968;125(6):849–51. https://doi.org/10.1176/ajp.125.6.849.
261. Goffman K, Joy D. Counterculture through the ages: from Abraham to acid house. New York: Villard; 2004.
262. Greenfield R. Timothy Leary: a biography. Orlando: Harcourt, Incorporated; 2006.
263. Leary T, Ginsberg A. High priest. New York: Ronin Publishing; 1995.
264. Lee M, Shlain B. Acid dreams: the complete social history of LSD: the CIA, the sixties, and beyond. New York: Grove Press; 1992.
265. Leary T, Metzner R, Presnell M, Weil G, Schwitzgebel R, Kinne S. A new behavior-change program using psilocybin. Psychother-Theor Res. 1965;2(2):61–72.
266. Fanon F. The wretched of the earth. Harmondsworth: Penguin Books; 1967.
267. Szasz T. The myth of mental illness: foundations of a theory of personal conduct. New York: Hober-Harper; 1961.
268. Healy D, Savage M, Michael P, Harris M, Hirst D, Carter M, Cattell D, McMonagle T, Sohler N, Susser E. Psychiatric bed utilization: 1896 and 1996 compared. Psychol Med. 2001;31(5):779–90. https://doi.org/10.1017/s0033291701003968.
269. Foucault M. Histoire de la folie à l'âge classique. Paris: Plon; 1961.
270. Goffman E. Asylums: essays on the social situation of mental patients and other inmates. New York: Anchor Books; 1961.
271. Leoni F. From madness to mental illness: psychiatry and biopolitics in Michel Faoucault. In: Fulford KWM, Davies M, Richard GTG, et al., editors. The Oxford handbook of philoso-

phy and psychiatry. Oxford: Oxford University Press; 2013. https://doi.org/10.1093/oxfordhb/9780199579563.001.0001.
272. Foucault M. Madness and civilization: a history of insanity in the age of reason. New York: Random House; 1988a.
273. Marcuse H. One-Dimensional Man. Boston: Beacon Press; 1964.
274. Cooper D. Psychiatry and anti-psychiatry Tavistock. London: Routledge; 1967.
275. Hare E. Was insanity on the increase? The fifty-sixth Maudsley lecture. Br J Psychiatry J Ment Sci. 1983;142:439–55. https://doi.org/10.1192/bjp.142.5.439.
276. Turner TH. Schizophrenia as a permanent problem. Some aspects of historical evidence in the recency (new disease) hypothesis. Hist Psychiatry. 1992;3(12):413–29. https://doi.org/10.1177/0957154X9200301202.
277. Scheff T. Being mentally ill: a sociological theory. London: Weidenfield & Nicolson; 1966.
278. Scheff T. Labeling Madness. New York: Prentice-Hall; 1975.
279. Chamberlin J. On our own: patient-controlled alternatives to the mental health system. In: On our own: patient-controlled alternatives to the mental health system. New York: McGraw-Hill; 1978.
280. Chamberlin J. The ex-Patients' movement: where We've been and where We're going. J Mind Behavior. 1990;11(3/4):323–36.
281. Bok S. The ethics of giving placebos. Sci Am. 1974;231(5):17–23. https://doi.org/10.1038/scientificamerican1174-17.
282. Brody H. The lie that heals: the ethics of giving placebos. Ann Intern Med. 1982;97(1):112–8. https://doi.org/10.7326/0003-4819-97-1-112.
283. Laing R. The politics of experience. Harmondsworth: Penguin Books; 1967.
284. Avramides A. Skepticism about knowledge of other minds. In: The Routledge companion to epistemology. London: Routledge; 2011. https://doi.org/10.4324/9780203839065.ch40.
285. DeRose K. Contextualism and skepticism. Oxford Scholarship Online. 2018; https://doi.org/10.1093/oso/9780199564477.003.0004.
286. Dretske FI. Observational terms. Philos Rev. 1964;73(1):25–42. https://doi.org/10.2307/2183200.
287. Hyslop A. The analogical inference to other minds. Amsterdam: Springer; 1995a. https://doi.org/10.1007/978-94-015-8510-1_5.
288. Hyslop A. Other minds and scientific inference. Amsterdam: Springer; 1995b. https://doi.org/10.1007/978-94-015-8510-1_4.
289. Stine GC. Skepticism, relevant alternatives, and deductive closure. Philos Stud. 1976;29(4):249–61. https://doi.org/10.1007/Bf00411885.
290. Wittgenstein L. On certainty. New York: Harper; 1969.
291. Kesey K. One flew over the Cuckoo's Nest Methuen. London: Penguin; 1962.
292. Cameron DE, Lohrenz JG, Handcock KA. The depatterning treatment of schizophrenia. Compr Psychiatry. 1962;3(2):65–76. https://doi.org/10.1016/s0010-440x(62)80015-7.
293. Rauh J, Turner J. Anatomy of a public interest case against the CIA. Hamline J Pub L & Pol'y. 1990;11(2):307.
294. Collins A. In the sleep room: the story of CIA brainwashing experiments in Canada. Toronto: Key Porter Books; 1988.
295. Marks J. The search for the Manchurian candidate. New York: Times Books; 1979.
296. Dryden W, Bond FW. Reason and emotion in psychotherapy: Albert Ellis. Br J Psychiatry J Ment Sci. 1994;165(1):131–5. https://doi.org/10.1192/bjp.165.1.131.
297. Ellis A. Reason and emotion in psychotherapy. Secaucus: Carol Publishing Group; 1994. Revised and Updated
298. Beck A. Depression. New York: Harper & Row; 1967.
299. Beck A. Cognitive therapy and the emotional disorders international. New York: Universities Press; 1976.
300. Beck AT. Depression: causes and treatment. Philadelphia: University of Pennsylvania Press; 1972.

301. Beck AT, Haigh EA. Advances in cognitive theory and therapy: the generic cognitive model. Annu Rev Clin Psychol. 2014;10(1):1–24. https://doi.org/10.1146/annurev-clinpsy-032813-153734.
302. Maslow AH. Motivation and personality. Oxford: Harpers; 1954.
303. Maslow A. Toward a psychology of being. Toward a psychology of being. Princeton: D Van Nostrand; 1962. https://doi.org/10.1037/10793-000.
304. Pauling L. Orthomolecular psychiatry. Varying the concentrations of substances normally present in the human body may control mental disease. Science. 1968;160(3825):265–71. https://doi.org/10.1126/science.160.3825.265.
305. Sperry RW. Cerebral organization and behavior: the split brain behaves in many respects like two separate brains, providing new research possibilities. Science. 1961;133(3466):1749–57. https://doi.org/10.1126/science.133.3466.1749.
306. Akelaitis AJ. A study of gnosis, praxis and language following section of the corpus callosum and anterior commissure. J Neurosurg. 1944;1(2):94–102. https://doi.org/10.3171/jns.1944.1.2.0094.
307. Gazzaniga MS, Bogen JE, Sperry RW. Some functional effects of sectioning the cerebral commissures in man. Proc Natl Acad Sci U S A. 1962;48:1765–9. https://doi.org/10.1073/pnas.48.10.1765.
308. Sperry R. Some effects of disconnecting the cerebral hemispheres. Science. 1982;217(4566):1223–6. https://doi.org/10.1126/science.7112125.
309. Beecher HK. Ethics and clinical research. N Engl J Med. 1966;274(24):1354–60. https://doi.org/10.1056/NEJM196606162742405.
310. Kendell RE, Cooper JE, Gourlay AJ, Copeland JR, Sharpe L, Gurland BJ. Diagnostic criteria of American and British psychiatrists. Arch Gen Psychiatry. 1971;25(2):123–30. https://doi.org/10.1001/archpsyc.1971.01750140027006.
311. Feighner JP, Robins E, Guze SB, Woodruff RA Jr, Winokur G, Munoz R. Diagnostic criteria for use in psychiatric research. Arch Gen Psychiatry. 1972;26(1):57–63. https://doi.org/10.1001/archpsyc.1972.01750190059011.
312. Robins E, Guze SB. Establishment of diagnostic validity in psychiatric illness: its application to schizophrenia. Am J Psychiatry. 1970;126(7):983–7. https://doi.org/10.1176/ajp.126.7.983.
313. Kendler KS, Munoz RA, Murphy G. The development of the Feighner criteria: a historical perspective. Am J Psychiatry. 2010;167(2):134–42. https://doi.org/10.1176/appi.ajp.2009.09081155.
314. Rosenhan DL. On being sane in insane places. Science. 1973;179(4070):250–8. https://doi.org/10.1126/science.179.4070.250.
315. Cahalan S. The great pretender: the undercover Mission that changed our understanding of madness. New York: Grand Central Publishing; 2019.
316. Millon T. Reflections on Rosenhan's "on being sane in insane places". J Abnorm Psychol. 1975;84(5):456–61. https://doi.org/10.1037/h0077120.
317. Spitzer RL. On pseudoscience in science, logic in remission, and psychiatric diagnosis: a critique of Rosenhan's "on being sane in insane places". J Abnorm Psychol. 1975;84(5):442–52. https://doi.org/10.1037/h0077124.
318. Spitzer RL. More on pseudoscience in science and the case for psychiatric diagnosis. A critique of D.L. Rosenhan's "on being sane in insane places" and "the Contestual nature of psychiatric diagnosis". Arch Gen Psychiatry. 1976;33(4):459–70. https://doi.org/10.1001/archpsyc.1976.01770040029007.
319. Thomison JB. Editorial: on being sane in insane places. J Tenn Med Assoc. 1974;67(3):247–5.
320. Thomison JB. On being sane in insane places-reprise. J Tenn Med Assoc. 1979;72(5):382–3.
321. Wallerstein RS. Discussion of Rosenhan's "on being sane in insane places". Bull Menn Clin. 1973;37(5):526–30.
322. Weiner B. "on being sane in insane places": a process (attributional) analysis and critique. J Abnorm Psychol. 1975;84(5):433–41. https://doi.org/10.1037/h0077126.
323. Slater L. Opening Skinner's Box. London: Bloomsbury; 2004.

324. Spitzer RL, Lilienfeld SO, Miller MB. Rosenhan revisited: the scientific credibility of Lauren Slater's pseudopatient diagnosis study. J Nerv Ment Dis. 2005;193(11):734–9. https://doi.org/10.1097/01.nmd.0000185992.16053.5c.
325. Zimmerman M. Pseudopatient or pseudoscience: a reviewer's perspective. J Nerv Ment Dis. 2005;193(11):740–2. https://doi.org/10.1097/01.nmd.0000186016.78212.d8.
326. Slater L. Reply to Spitzer and colleagues. J Nerv Ment Dis. 2005;193(11):743–4. https://doi.org/10.1097/01.nmd.0000185883.10503.98.
327. Mayes R, Horwitz AV. DSM-III and the revolution in the classification of mental illness. J Hist Behav Sci. 2005;41(3):249–67. https://doi.org/10.1002/jhbs.20103.
328. Friedman RC, Downey JI. Psychoanalysis and the model of homosexuality as psychopathology: a historical overview. Am J Psychoanal. 1998;58(3):249–70. https://doi.org/10.1023/a:1022583223820.
329. Socarides CW. Homosexuality: psychoanalytic therapy. Northvale: Jason Aronson, Inc.; 1989.
330. Socarides CW. Homosexuality: a freedom too far. Phoenix: Adam Margrave Books; 1995.
331. Lieberman J, First M. Robert Spitzer: a psychiatrist for the ages. Neuropsychopharmacology. 2016;41(13):3124. https://doi.org/10.1038/npp.2016.30.
332. Decker HS. The making of DSM-III: a diagnostic Manual's conquest of American psychiatry. New York: Oxford University Press; 2013.
333. Bayer R, Spitzer RL. Neurosis, psychodynamics, and DSM-III. A history of the controversy. Arch Gen Psychiatry. 1985;42(2):187–96. https://doi.org/10.1001/archpsyc.1985.01790250081011.
334. Berk H, Jaso H. Memorantum; 1976.
335. Kirk S, Kutchins H. The selling of DSM: the rhetoric of science in psychiatry. New York: Aldine de Gruyter; 1992.
336. Jaspers K. General Psychopathology (trans: Hoenig J, Hamilton M). The Johns Hopkins University Press (1997), Chicago; 1913.
337. Othmer E, Othmer S. The clinical interview using DSM-IV-TR: Vol 1: fundamentals. In: The clinical interview using DSM-IV-TR: Vol 1: fundamentals. Arlington: American Psychiatric Publishing, Inc.; 2002.
338. Oyebode F. Sims' symptoms in the mind: textbook of descriptive psychopathology. Edinburgh: Elsevier; 2018.
339. Turner SM, Hersen M. The interviewing process. In: Hersen M, Turner S, editors. Diagnostic interviewing. 3rd ed. New York: Spring Street; 2003. p. 3–11.
340. Ventura J, Liberman RP, Green MF, Shaner A, Mintz J. Training and quality assurance with the structured clinical interview for DSM-IV (SCID-I/P). Psychiatry Res. 1998;79(2):163–73. https://doi.org/10.1016/s0165-1781(98)00038-9.
341. Davies J. How voting and consensus created the diagnostic and statistical manual of mental disorders (DSM-III). Anthropol Med. 2017;24(1):32–46. https://doi.org/10.1080/13648470.2016.1226684.
342. Klerman GL, Weissman MM. Social causes of depression. Lancet. 1978;2(8091):680–1. https://doi.org/10.1016/s0140-6736(78)92792-7.
343. Weissman M, Klerman G, Markowitz J. The guide to interpersonal psychotherapy. Oxford: Oxford University Press; 2018.
344. Weissman MM, Klerman GL. Psychotherapy with depressed women: an empirical study of content themes and reflection. Br J Psychiatry J Ment Sci. 1973;123(572):55–61. https://doi.org/10.1192/bjp.123.1.55.
345. Weissman MM, Klerman GL, Paykel ES, Prusoff B, Hanson B. Treatment effects on the social adjustment of depressed patients. Arch Gen Psychiatry. 1974;30(6):771–8. https://doi.org/10.1001/archpsyc.1974.01760120033006.
346. Weissman MM, Paykel ES, Siegel R, Klerman GL. The social role performance of depressed women: comparisons with a normal group. Am J Orthopsychiatry. 1971;41(3):390–405. https://doi.org/10.1111/j.1939-0025.1971.tb01126.x.

347. Millon T, Grossman S, Millon C, Meagher S, Ramnath R. Personality disorders in modern life, 2nd ed. personality disorders in modern life. 2nd ed. Hoboken: Wiley; 2004.
348. Frank L. The history of shock treatment. San Francisco: Leonard Roy Frank; 1978.
349. Amsler HA, Teerenhovi L, Barth E, Harjula K, Vuopio P. Agranulocytosis in patients treated with clozapine. A study of the Finnish epidemic. Acta Psychiatr Scand. 1977;56(4):241–8. https://doi.org/10.1111/j.1600-0447.1977.tb00224.x.
350. de la Chapelle A, Kari C, Nurminen M, Hernberg S. Clozapine-induced agranulocytosis. A genetic and epidemiologic study. Hum Genet. 1977;37(2):183–94. https://doi.org/10.1007/BF00393581.
351. Idanpaan-Heikkila J, Alhava E, Olkinuora M, Palva I. Letter: clozapine and agranulocytosis. Lancet. 1975;2(7935):611. https://doi.org/10.1016/s0140-6736(75)90206-8.
352. Engel GL. The need for a new medical model: a challenge for biomedicine. Science. 1977;196(4286):129–36. https://doi.org/10.1126/science.847460.
353. Engel GL. The biopsychosocial model and the education of health professionals. Ann N Y Acad Sci. 1978;310:169–87. https://doi.org/10.1111/j.1749-6632.1978.tb22070.x.
354. Engel GL. The clinical application of the biopsychosocial model. Am J Psychiatry. 1980;137(5):535–44. https://doi.org/10.1176/ajp.137.5.535.
355. Engel GL. Sounding board. The biopsychosocial model and medical education. Who are to be the teachers? N Engl J Med. 1982;306(13):802–5. https://doi.org/10.1056/NEJM198204013061311.
356. Fagius J, Osterman PO, Siden A, Wiholm BE. Guillain-Barre syndrome following zimeldine treatment. J Neurol Neurosurg Psychiatry. 1985;48(1):65–9. https://doi.org/10.1136/jnnp.48.1.65.
357. Luborsky L, Singer B, Luborsky L. Comparative studies of psychotherapies. Is it true that "everywon has one and all must have prizes"? Arch Gen Psychiatry. 1975;32(8):995–1008. https://doi.org/10.1001/archpsyc.1975.01760260059004.
358. Sloane RB, Staples FR, Cristol AH, Yorkston NJ. Short-term analytically oriented psychotherapy versus behavior therapy. Am J Psychiatry. 1975;132(4):373–7. https://doi.org/10.1176/ajp.132.4.373.
359. Abrahamson D. High-dose antipsychotic medication. Br J Psychiatry J Ment Sci. 1994;165(2):269. https://doi.org/10.1192/bjp.165.2.269a.
360. Bennie EH. High-dose antipsychotic medication. Br J Psychiatry J Ment Sci. 1994;165(4):553. https://doi.org/10.1192/bjp.165.4.553a.
361. Thompson C. The use of high-dose antipsychotic medication. Br J Psychiatry J Ment Sci. 1994;164(4):448–58. https://doi.org/10.1192/bjp.164.4.448.
362. Will D, Wrate RM, Bhate S, Taylor P, James T, Rothery D, Clark A. High-dose antipsychotic medication. Br J Psychiatry J Ment Sci. 1994;165(2):269–70. https://doi.org/10.1192/bjp.165.2.269b.
363. Baldessarini RJ, Cohen BM, Teicher MH. Significance of neuroleptic dose and plasma level in the pharmacological treatment of psychoses. Arch Gen Psychiatry. 1988;45(1):79–91. https://doi.org/10.1001/archpsyc.1988.01800250095013.
364. Coppens HJ, Slooff CJ, Paans AM, Wiegman T, Vaalburg W, Korf J. High central D2-dopamine receptor occupancy as assessed with positron emission tomography in medicated but therapy-resistant schizophrenic patients. Biol Psychiatry. 1991;29(7):629–34. https://doi.org/10.1016/0006-3223(91)90132-6.
365. Farde L, Wiesel FA, Halldin C, Sedvall G. Central D2-dopamine receptor occupancy in schizophrenic patients treated with antipsychotic drugs. Arch Gen Psychiatry. 1988;45(1):71–6. https://doi.org/10.1001/archpsyc.1988.01800250087012.
366. McCreadie RG, MacDonald IM. High dosage haloperidol in chronic schizophrenia. Br J Psychiatry J Ment Sci. 1977;131:310–6. https://doi.org/10.1192/bjp.131.3.310.
367. Rifkin A, Doddi S, Karajgi B, Borenstein M, Wachspress M. Dosage of haloperidol for schizophrenia. Arch Gen Psychiatry. 1991;48(2):166–70. https://doi.org/10.1001/archpsyc.1991.01810260074011.

368. Van Putten T, Marder SR, Mintz J. The therapeutic index of haloperidol in newly admitted schizophrenic patients. Psychopharmacol Bull. 1987;23(1):201–5.
369. Van Putten T, Marder SR, Mintz J. A controlled dose comparison of haloperidol in newly admitted schizophrenic patients. Arch Gen Psychiatry. 1990;47(8):754–8. https://doi.org/10.1001/archpsyc.1990.01810200062009.
370. Sutherland J. Should they de-Nobel Moniz? The Guardian, 5 August; 2004.
371. Mottier V. Sexuality: a very short introduction. Oxford: Oxford University Press; 2008.
372. Foucault M. Sexuality morality and the law. New York: Routledge; 1988b.
373. Foucault M, Hocquenghem G, Danet J. Semiotext(e) special intervention series 2: loving boys/loving children Semiotext(e) magazine. New York: Semiotext(e); 1980.
374. Déchaux J-H. Le sacre de l'enfant. Regards Sur Une passion contemporaine (consecrating the child). Rev Fr Sociol. 2014;55(3):537–61. https://doi.org/10.3917/rfs.553.0537.
375. Aragon L, Ponge F, Barthes R, de Beauvoir S, Belladona J, Bon M, Sartre J, et al. Letter. Le Monde. 1977;26:1977.
376. Elles P, Bruckner P, Moustaki G, et al. Letter Liberasion; 1979.
377. unknown. Development of a new antipsychotic Remoxipride. Proceedings of an international symposium. Monte Carlo, Monaco, June 19-20, 1989. Acta Psychiatr Scand Suppl. 1990;358:1–188.
378. Van Praag HM. Socio-biological psychiatry. Compr Psychiatry. 1981;22(5):441–50. https://doi.org/10.1016/0010-440x(81)90032-8.
379. van Praag HM. Religiosity, a personality trait to be reckoned within psychiatry. World Psychiatry. 2013;12(1):33–4. https://doi.org/10.1002/wps.20006.
380. Leucht S, Pitschel-Walz G, Abraham D, Kissling W. Efficacy and extrapyramidal side-effects of the new antipsychotics olanzapine, quetiapine, risperidone, and sertindole compared to conventional antipsychotics and placebo. A meta-analysis of randomized controlled trials. Schizophr Res. 1999;35(1):51–68. https://doi.org/10.1016/s0920-9964(98)00105-4.
381. Romme MA, Escher AD. Hearing voices. Schizophr Bull. 1989;15(2):209–16. https://doi.org/10.1093/schbul/15.2.209.
382. Eysenck HJ. Dimensions of personality. New York: Routledge; 1947.
383. Eysenck H, Eysenck S. Psychoticism as a dimension of personality. London: Hodder & Stoughton; 1976.
384. Eysenck HJ. Psychosis and psychoticism: a reply to bishop. J Abnorm Psychol. 1977;86(4):427–30.
385. Eysenck HJ, Eysenck SB. A factorial study of an interview-questionnaire. J Clin Psychol. 1962;18:286–90. https://doi.org/10.1002/1097-4679(196207)18:3<286::aid-jclp2270180311>3.0.co;2-u.
386. Eysenck SB, Adelaja O, Eysenck HJ. A comparative study of personality in Nigerian and English subjects. J Soc Psychol. 1977;102(Second Half):171–8. https://doi.org/10.1080/00224545.1977.9713261.
387. Eysenck SB, Eysenck HJ. An improved short questionnaire for the measurement of extraversion and neuroticism. Life Sci (1962). 1964;3:1103–9. https://doi.org/10.1016/0024-3205(64)90125-0.
388. Eysenck SB, Eysenck HJ. Physiological reactivity to sensory stimulation as a measure of personality. Psychol Rep. 1967;20(1):45–6. https://doi.org/10.2466/pr0.1967.20.1.45.
389. Eysenck SB, Eysenck HJ. The measurement of psychoticism: a study of factor stability and reliability. Br J Soc Clin Psychol. 1968;7(4):286–94. https://doi.org/10.1111/j.2044-8260.1968.tb00571.x.
390. Eysenck SB, Eysenck HJ. Scores on three personality variables as a function of age, sex and social class. Br J Soc Clin Psychol. 1969;8(1):69–76. https://doi.org/10.1111/j.2044-8260.1969.tb00588.x.
391. Eysenck SB, Eysenck HJ. The questionnaire measurement of psychoticism. Psychol Med. 1972;2(1):50–5. https://doi.org/10.1017/s0033291700045608.

392. Eysenck SB, Eysenck HJ. The place of impulsiveness in a dimensional system of personality description. Br J Soc Clin Psychol. 1977;16(1):57–68. https://doi.org/10.1111/j.2044-8260.1977.tb01003.x.
393. Eysenck SB, Eysenck HJ, Shaw L. The modification of personality and lie scale scores by special 'honesty' instructions. Br J Soc Clin Psychol. 1974;13(1):41–50. https://doi.org/10.1111/j.2044-8260.1974.tb00876.x.
394. Eysenck SB, White O, Eysenck HJ. Personality and mental illness. Psychol Rep. 1976;39(3):1011–22. https://doi.org/10.2466/pr0.1976.39.3.1011.
395. Eysenck HJ. The psychology of politics and the personality similarities between fascists and communists. Psychol Bull. 1956;53(6):431–8. https://doi.org/10.1037/h0043987.
396. Eysenck HJ. Political-ideology and science. Am Psychol. 1982;37(11):1288–9. https://doi.org/10.1037/0003-066x.37.11.1288.
397. McCrae RR, Costa PT Jr. Validation of the five-factor model of personality across instruments and observers. J Pers Soc Psychol. 1987;52(1):81–90. https://doi.org/10.1037//0022-3514.52.1.81.
398. Markon KE, Krueger RF, Watson D. Delineating the structure of normal and abnormal personality: an integrative hierarchical approach. J Pers Soc Psychol. 2005;88(1):139–57. https://doi.org/10.1037/0022-3514.88.1.139.
399. Cloninger CR. A systematic method for clinical description and classification of personality variants. Propos Arch Gen Psychiat. 1987;44(6):573–88. https://doi.org/10.1001/archpsyc.1987.01800180093014.
400. Cloninger CR, Svrakic DM, Przybeck TR. A psychobiological model of temperament and character. Arch Gen Psychiatry. 1993;50(12):975–90. https://doi.org/10.1001/archpsyc.1993.01820240059008.
401. Constantino JN, Cloninger CR, Clarke AR, Hashemi B, Przybeck T. Application of the seven-factor model of personality to early childhood. Psychiatry Res. 2002;109(3):229–43. https://doi.org/10.1016/s0165-1781(02)00008-2.
402. Josefsson K, Jokela M, Cloninger CR, Hintsanen M, Salo J, Hintsa T, Pulkki-Raback L, Keltikangas-Jarvinen L. Maturity and change in personality: developmental trends of temperament and character in adulthood. Dev Psychopathol. 2012;25(3):713–27.
403. Akiskal HS, Placidi GF, Maremmani I, Signoretta S, Liguori A, Gervasi R, Mallya G, Puzantian VR. TEMPS-I: delineating the most discriminant traits of the cyclothymic, depressive, hyperthymic and irritable temperaments in a nonpatient population. J Affect Disord. 1998;51(1):7–19. https://doi.org/10.1016/s0165-0327(98)00152-9.
404. Campbell P, Roberts A. Survivors' history. Life Day. 2009;13(3):33–9.
405. American Psychiatric Association. American psychiatric association, statement on memories of sexual abuse. Washington, D.C: APA Press; 1993.
406. Kane J, Honigfeld G, Singer J, Meltzer H. Clozapine for the treatment-resistant schizophrenic. A double-blind comparison with chlorpromazine. Arch Gen Psychiatry. 1988;45(9):789–96. https://doi.org/10.1001/archpsyc.1988.01800330013001.
407. Meltzer HY, Bastani B, Kwon KY, Ramirez LF, Burnett S, Sharpe J. A prospective study of clozapine in treatment-resistant schizophrenic patients. I Preliminary Rep Psychopharmacol. 1989a;99(Suppl):S68–72. https://doi.org/10.1007/BF00442563.
408. Samara MT, Dold M, Gianatsi M, Nikolakopoulou A, Helfer B, Salanti G, Leucht S. Efficacy, acceptability, and tolerability of antipsychotics in treatment-resistant schizophrenia: a network meta-analysis. JAMA Psychiat. 2016;73(3):199–210. https://doi.org/10.1001/jamapsychiatry.2015.2955.
409. Boter H, Peuskens J, Libiger J, Fleischhacker WW, Davidson M, Galderisi S, Kahn RS, group Es. Effectiveness of antipsychotics in first-episode schizophrenia and schizophreniform disorder on response and remission: an open randomized clinical trial (EUFEST). Schizophr Res. 2009;115(2–3):97–103. https://doi.org/10.1016/j.schres.2009.09.019.
410. Fava M, Rush AJ, Trivedi MH, Nierenberg AA, Thase ME, Sackeim HA, Quitkin FM, Wisniewski S, Lavori PW, Rosenbaum JF, Kupfer DJ. Background and rationale for the

sequenced treatment alternatives to relieve depression (STAR*D) study. Psychiatr Clin North Am. 2003;26(2):457–94. https://doi.org/10.1016/s0193-953x(02)00107-7.
411. Fleischhacker WW, Keet IP, Kahn RS, Committee ES. The European First episode schizophrenia trial (EUFEST): rationale and design of the trial. Schizophr Res. 2005;78(2–3):147–56. https://doi.org/10.1016/j.schres.2005.06.004.
412. Hierholzer R. Remission rates for depression in STAR*D study. Am J Psychiatry. 2006;163(7):1293.; author reply 1293-1294. https://doi.org/10.1176/appi.ajp.163.7.1293.
413. Jones PB, Barnes TR, Davies L, Dunn G, Lloyd H, Hayhurst KP, Murray RM, Markwick A, Lewis SW. Randomized controlled trial of the effect on quality of life of second- vs first-generation antipsychotic drugs in schizophrenia: cost utility of the latest antipsychotic drugs in schizophrenia study (CUtLASS 1). Arch Gen Psychiatry. 2006;63(10):1079–87. https://doi.org/10.1001/archpsyc.63.10.1079.
414. Lieberman JA. Effectiveness of antipsychotic drugs in patients with chronic schizophrenia: efficacy, safety and cost outcomes of CATIE and other trials. J Clin Psychiatry. 2007;68(2):e04. https://doi.org/10.4088/jcp.0207e04.
415. Rush AJ. STAR*D: what have we learned? Am J Psychiatry. 2007;164(2):201–4. https://doi.org/10.1176/ajp.2007.164.2.201.
416. Rush AJ, Fava M, Wisniewski SR, Lavori PW, Trivedi MH, Sackeim HA, Thase ME, Nierenberg AA, Quitkin FM, Kashner TM, Kupfer DJ, Rosenbaum JF, Alpert J, Stewart JW, PJ MG, Biggs MM, Shores-Wilson K, Lebowitz BD, Ritz L, Niederehe G, Group SDI. Sequenced treatment alternatives to relieve depression (STAR*D): rationale and design. Control Clin Trials. 2004;25(1):119–42. https://doi.org/10.1016/s0197-2456(03)00112-0.
417. Sachs GS, Thase ME, Otto MW, Bauer M, Miklowitz D, Wisniewski SR, Lavori P, Lebowitz B, Rudorfer M, Frank E, Nierenberg AA, Fava M, Bowden C, Ketter T, Marangell L, Calabrese J, Kupfer D, Rosenbaum JF. Rationale, design, and methods of the systematic treatment enhancement program for bipolar disorder (STEP-BD). Biol Psychiatry. 2003;53(11):1028–42. https://doi.org/10.1016/s0006-3223(03)00165-3.
418. Schneider LS, Ismail MS, Dagerman K, Davis S, Olin J, McManus D, Pfeiffer E, Ryan JM, Sultzer DL, Tariot PN. Clinical antipsychotic trials of intervention effectiveness (CATIE): Alzheimer's disease trial. Schizophr Bull. 2003;29(1):57–72. https://doi.org/10.1093/oxfordjournals.schbul.a006991.
419. Schneider LS, Tariot PN, Lyketsos CG, Dagerman KS, Davis KL, Davis S, Hsiao JK, Jeste DV, Katz IR, Olin JT, Pollock BG, Rabins PV, Rosenheck RA, Small GW, Lebowitz B, Lieberman JA. National Institute of Mental Health clinical antipsychotic trials of intervention effectiveness (CATIE): Alzheimer disease trial methodology. Am J Geriat Psychiat. 2001;9(4):346–60.
420. Simon NM, Otto MW, Weiss RD, Bauer MS, Miyahara S, Wisniewski SR, Thase ME, Kogan J, Frank E, Nierenberg AA, Calabrese JR, Sachs GS, Pollack MH, Investigators S-B. Pharmacotherapy for bipolar disorder and comorbid conditions: baseline data from STEP-BD. J Clin Psychopharmacol. 2004;24(5):512–20. https://doi.org/10.1097/01.jcp.0000138772.40515.70.
421. Stroup TS, McEvoy JP, Swartz MS, Byerly MJ, Glick ID, Canive JM, McGee MF, Simpson GM, Stevens MC, Lieberman JA. The National Institute of Mental Health clinical antipsychotic trials of intervention effectiveness (CATIE) project: schizophrenia trial design and protocol development. Schizophr Bull. 2003;29(1):15–31. https://doi.org/10.1093/oxfordjournals.schbul.a006986.
422. Thase ME. STEP-BD and bipolar depression: what have we learned? Curr Psychiatry Rep. 2007;9(6):497–503. https://doi.org/10.1007/s11920-007-0068-9.
423. Geddes J, Freemantle N, Harrison P, Bebbington P. Atypical antipsychotics in the treatment of schizophrenia: systematic overview and meta-regression analysis. BMJ. 2000;321(7273):1371–6. https://doi.org/10.1136/bmj.321.7273.1371.
424. Copolov D. New name for atypical antipsychotics? Am J Psychiatry. 1997;154(3):439. https://doi.org/10.1176/ajp.154.3.439a.

425. Meltzer HY, Matsubara S, Lee JC. Classification of typical and atypical antipsychotic drugs on the basis of dopamine D-1, D-2 and serotonin2 pKi values. J Pharmacol Exp Ther. 1989b;251(1):238–46.
426. Suppes T, Dennehy EB, Swann AC, Bowden CL, Calabrese JR, Hirschfeld RM, Keck PE Jr, Sachs GS, Crismon ML, Toprac MG, Shon SP, Texas Consensus Conference Panel on Medication Treatment of Bipolar D. Report of the Texas consensus conference panel on medication treatment of bipolar disorder 2000. J Clin Psychiatry. 2002;63(4):288–99. https://doi.org/10.4088/jcp.v63n0404.
427. Zito JM, Safer DJ, dosReis S, Gardner JF, Boles M, Lynch F. Trends in the prescribing of psychotropic medications to preschoolers. JAMA. 2000;283(8):1025–30. https://doi.org/10.1001/jama.283.8.1025.
428. Friedman RA. Antidepressants' black-box warning--10 years later. N Engl J Med. 2014;371(18):1666–8. https://doi.org/10.1056/NEJMp1408480.
429. Fountoulakis KN, Gonda X, Baghai TC, Baldwin DS, Bauer M, Blier P, Gattaz W, Hasler G, Moller HJ, Tandon R, Vieta E, Kasper S. Report of the WPA section of pharmacopsychiatry on the relationship of antiepileptic drugs with suicidality in epilepsy. Int J Psychiatry Clin Pract. 2015;19(3):158–67. https://doi.org/10.3109/13651501.2014.1000930.
430. Fountoulakis KN, Gonda X, Samara M, Siapera M, Karavelas V, Ristic DI, Iacovides A. Antiepileptic drugs and suicidality. J Psychopharmacol. 2012;26(11):1401–7. https://doi.org/10.1177/0269881112440514.
431. Siamouli M, Samara M, Fountoulakis KN. Is antiepileptic-induced suicidality a data-based class effect or an exaggeration? A comment on the literature. Harv Rev Psychiatry. 2014;22(6):379–81. https://doi.org/10.1097/HRP.0000000000000039.
432. Turner EH, Matthews AM, Linardatos E, Tell RA, Rosenthal R. Selective publication of antidepressant trials and its influence on apparent efficacy. N Engl J Med. 2008;358(3):252–60. https://doi.org/10.1056/NEJMsa065779.
433. Barbui C, Furukawa TA, Cipriani A. Effectiveness of paroxetine in the treatment of acute major depression in adults: a systematic re-examination of published and unpublished data from randomized trials. CMAJ. 2008;178(3):296–305. https://doi.org/10.1503/cmaj.070693.
434. Bech P, Cialdella P, Haugh MC, Birkett MA, Hours A, Boissel JP, Tollefson GD. Meta-analysis of randomised controlled trials of fluoxetine v. placebo and tricyclic antidepressants in the short-term treatment of major depression. Br J Psychiatry J Ment Sci. 2000;176:421–8. https://doi.org/10.1192/bjp.176.5.421.
435. Fournier JC, DeRubeis RJ, Hollon SD, Dimidjian S, Amsterdam JD, Shelton RC, Fawcett J. Antidepressant drug effects and depression severity: a patient-level meta-analysis. JAMA. 2010;303(1):47–53. https://doi.org/10.1001/jama.2009.1943.
436. Ghaemi SN. Why antidepressants are not antidepressants: STEP-BD, STAR*D, and the return of neurotic depression. Bipolar Disord. 2008;10(8):957–68. https://doi.org/10.1111/j.1399-5618.2008.00639.x.
437. Khan A, Leventhal RM, Khan SR, Brown WA. Severity of depression and response to antidepressants and placebo: an analysis of the Food and Drug Administration database. J Clin Psychopharmacol. 2002;22(1):40–5. https://doi.org/10.1097/00004714-200202000-00007.
438. Kirsch I, Deacon BJ, Huedo-Medina TB, Scoboria A, Moore TJ, Johnson BT. Initial severity and antidepressant benefits: a meta-analysis of data submitted to the Food and Drug Administration. PLoS Med. 2008;5(2):e45. https://doi.org/10.1371/journal.pmed.0050045.
439. Moncrieff J, Wessely S, Hardy R. Active placebos versus antidepressants for depression. Cochrane Database Syst Rev. 2004;1:CD003012. https://doi.org/10.1002/14651858.CD003012.pub2.
440. Undurraga J, Baldessarini RJ. Randomized, placebo-controlled trials of antidepressants for acute major depression: thirty-year meta-analytic review. Neuropsychopharmacology. 2012;37(4):851–64. https://doi.org/10.1038/npp.2011.306.
441. Gilbody SM, Song F. Publication bias and the integrity of psychiatry research. Psychol Med. 2000;30(2):253–8. https://doi.org/10.1017/s0033291700001732.

442. Huston P, Moher D. Redundancy, disaggregation, and the integrity of medical research. Lancet. 1996;347(9007):1024–6. https://doi.org/10.1016/s0140-6736(96)90153-1.
443. Rennie D. Fair conduct and fair reporting of clinical trials. JAMA. 1999;282(18):1766–8. https://doi.org/10.1001/jama.282.18.1766.
444. Khan A, Warner HA, Brown WA. Symptom reduction and suicide risk in patients treated with placebo in antidepressant clinical trials: an analysis of the Food and Drug Administration database. Arch Gen Psychiatry. 2000;57(4):311–7. https://doi.org/10.1001/archpsyc.57.4.311.
445. Cuijpers P, Clignet F, van Meijel B, van Straten A, Li J, Andersson G. Psychological treatment of depression in inpatients: a systematic review and meta-analysis. Clin Psychol Rev. 2011;31(3):353–60. https://doi.org/10.1016/j.cpr.2011.01.002.
446. Cuijpers P, Smit F, Bohlmeijer E, Hollon SD, Andersson G. Efficacy of cognitive-behavioural therapy and other psychological treatments for adult depression: meta-analytic study of publication bias. Br J Psychiatry J Ment Sci. 2010a;196(3):173–8. https://doi.org/10.1192/bjp.bp.109.066001.
447. Cuijpers P, van Straten A, Bohlmeijer E, Hollon SD, Andersson G. The effects of psychotherapy for adult depression are overestimated: a meta-analysis of study quality and effect size. Psychol Med. 2010b;40(2):211–23. https://doi.org/10.1017/S0033291709006114.
448. Driessen E, Cuijpers P, Hollon SD, Dekker JJ. Does pretreatment severity moderate the efficacy of psychological treatment of adult outpatient depression? A meta-analysis. J Consult Clin Psychol. 2010;78(5):668–80. https://doi.org/10.1037/a0020570.
449. Kirsch I. Antidepressants and the placebo response. Epidemiol Psichiatr Soc. 2009a;18(4):318–22. https://doi.org/10.1017/s1121189x00000282.
450. Kirsch I. The Emperor's new drugs: exploding the antidepressant myth. London: The Bodley Head; 2009b.
451. Fountoulakis KN, Hoschl C, Kasper S, Lopez-Ibor J, Moller HJ. The media and intellectuals' response to medical publications: the antidepressants' case. Ann General Psychiatry. 2013a;12(1):11. https://doi.org/10.1186/1744-859X-12-11.
452. Cipriani A, Furukawa TA, Salanti G, Chaimani A, Atkinson LZ, Ogawa Y, Leucht S, Ruhe HG, Turner EH, Higgins JPT, Egger M, Takeshima N, Hayasaka Y, Imai H, Shinohara K, Tajika A, Ioannidis JPA, Geddes JR. Comparative efficacy and acceptability of 21 antidepressant drugs for the acute treatment of adults with major depressive disorder: a systematic review and network meta-analysis. Lancet. 2018;391(10128):1357–66. https://doi.org/10.1016/S0140-6736(17)32802-7.
453. Eriksson E, Hieronymus F. The alleged lack of efficacy of antidepressants in non-severe depression: a myth debunked. Acta Psychiatr Scand. 2018;137(6):447–9. https://doi.org/10.1111/acps.12902.
454. Fountoulakis KN, Moller HJ. Efficacy of antidepressants: a re-analysis and re-interpretation of the Kirsch data. Int J Neuropsychopharmacol. 2011;14(3):405–12. https://doi.org/10.1017/S1461145710000957.
455. Fountoulakis KN, Samara MT, Siamouli M. Burning issues in the meta-analysis of pharmaceutical trials for depression. J Psychopharmacol. 2014;28(2):106–17. https://doi.org/10.1177/0269881113504014.
456. Fountoulakis KN, Veroniki AA, Siamouli M, Moller HJ. No role for initial severity on the efficacy of antidepressants: results of a multi-meta-analysis. Ann General Psychiatry. 2013b;12(1):26. https://doi.org/10.1186/1744-859X-12-26.
457. Horder J, Matthews P, Waldmann R. Placebo, prozac and PLoS: significant lessons for psychopharmacology. J Psychopharmacol. 2011;25(10):1277–88. https://doi.org/10.1177/0269881110372544.
458. Buchanan RW, Kreyenbuhl J, Kelly DL, Noel JM, Boggs DL, Fischer BA, Himelhoch S, Fang B, Peterson E, Aquino PR, Keller W, Schizophrenia Patient Outcomes Research T. The

2009 schizophrenia PORT psychopharmacological treatment recommendations and summary statements. Schizophr Bull. 2010;36(1):71–93. https://doi.org/10.1093/schbul/sbp116.
459. Gotzsche PC, Young AH, Crace J. Does long term use of psychiatric drugs cause more harm than good? BMJ. 2015;350(May 12):h2435. https://doi.org/10.1136/bmj.h2435.
460. Moncrieff J. Antipsychotic maintenance treatment: time to rethink? PLoS Med. 2015;12(8):e1001861. https://doi.org/10.1371/journal.pmed.1001861.
461. Whitaker R. The case against antipsychotic drugs: a 50-year record of doing more harm than good. Med Hypotheses. 2004;62(1):5–13. https://doi.org/10.1016/S0306-9877(03)00293-7.
462. Calton T, Ferriter M, Huband N, Spandler H. A systematic review of the Soteria paradigm for the treatment of people diagnosed with schizophrenia. Schizophr Bull. 2008;34(1):181–92. https://doi.org/10.1093/schbul/sbm047.
463. Rappaport M, Hopkins HK, Hall K, Belleza T, Silverman J. Are there schizophrenics for whom drugs may be unnecessary or contraindicated? Int Pharmacopsychiat. 1978;13(2):100–11. https://doi.org/10.1159/000468327.
464. Wunderink L, Nieboer RM, Wiersma D, Sytema S, Nienhuis FJ. Recovery in remitted first-episode psychosis at 7 years of follow-up of an early dose reduction/discontinuation or maintenance treatment strategy: long-term follow-up of a 2-year randomized clinical trial. JAMA Psychiat. 2013;70(9):913–20. https://doi.org/10.1001/jamapsychiatry.2013.19.
465. Moncrieff J. Inconvenient truths about antipsychotics: a response to Goff et al. Mad in America; 2017. https://www.madinamerica.com/2017/05/inconvenient-truths-about-antipsychotics-a-response-to-goff-et-al/.2018.
466. Murray RM, Quattrone D, Natesan S, van Os J, Nordentoft M, Howes O, Di Forti M, Taylor D. Should psychiatrists be more cautious about the long-term prophylactic use of antipsychotics? Br J Psychiatry J Ment Sci. 2016;209(5):361–5. https://doi.org/10.1192/bjp.bp.116.182683.
467. Leucht S, Heres S, Hamann J, Kane JM. Methodological issues in current antipsychotic drug trials. Schizophr Bull. 2008;34(2):275–85. https://doi.org/10.1093/schbul/sbm159.
468. Leucht S, Tardy M, Komossa K, Heres S, Kissling W, Davis JM. Maintenance treatment with antipsychotic drugs for schizophrenia. Cochrane Database Syst Rev. 2012b;5:CD008016. https://doi.org/10.1002/14651858.CD008016.pub2.
469. Haddad PM, Correll CU. The acute efficacy of antipsychotics in schizophrenia: a review of recent meta-analyses. Ther Adv Psychopharmacol. 2018;8(11):303–18. https://doi.org/10.1177/2045125318781475.
470. Leucht S, Arbter D, Engel RR, Kissling W, Davis JM. How effective are second-generation antipsychotic drugs? A meta-analysis of placebo-controlled trials. Mol Psychiatry. 2009a;14(4):429–47. https://doi.org/10.1038/sj.mp.4002136.
471. Leucht S, Cipriani A, Spineli L, Mavridis D, Orey D, Richter F, Samara M, Barbui C, Engel RR, Geddes JR, Kissling W, Stapf MP, Lassig B, Salanti G, Davis JM. Comparative efficacy and tolerability of 15 antipsychotic drugs in schizophrenia: a multiple-treatments meta-analysis. Lancet. 2013;382(9896):951–62. https://doi.org/10.1016/S0140-6736(13)60733-3.
472. Leucht S, Kissling W, Davis JM. Second-generation antipsychotics for schizophrenia: can we resolve the conflict? Psychol Med. 2009b;39(10):1591–602. https://doi.org/10.1017/S0033291709005455.
473. Leucht S, Hierl S, Kissling W, Dold M, Davis JM. Putting the efficacy of psychiatric and general medicine medication into perspective: review of meta-analyses. Br J Psychiatry J Ment Sci. 2012a;200(2):97–106. https://doi.org/10.1192/bjp.bp.111.096594.
474. Leucht S, Tardy M, Komossa K, Heres S, Kissling W, Salanti G, Davis JM. Antipsychotic drugs versus placebo for relapse prevention in schizophrenia: a systematic review and meta-analysis. Lancet. 2012c;379(9831):2063–71. https://doi.org/10.1016/S0140-6736(12)60239-6.

475. Tiihonen J, Tanskanen A, Taipale H. 20-year Nationwide follow-up study on discontinuation of antipsychotic treatment in First-episode schizophrenia. Am J Psychiatry. 2018;175(8):765–73. https://doi.org/10.1176/appi.ajp.2018.17091001.
476. Gøtzsche P. Psychiatry ignores an Elephant in the room. Mad in America; 2017. https://www.madinamerica.com/2017/09/psychiatry-ignores-elephant-room/.2018.
477. Whitaker R. Thou shall not criticize our drugs; 2017. https://www.madinamerica.com/2017/09/thou-shall-not-criticize-our-drugs/. Accessed 5 July 2018.
478. Fountoulakis KN, Grunze H, Vieta E, Young A, Yatham L, Blier P, Kasper S, Moeller HJ. The International College of Neuro-Psychopharmacology (CINP) treatment guidelines for bipolar disorder in adults (CINP-BD-2017), part 3: the clinical guidelines. Int J Neuropsychopharmacol. 2017;20(2):180–95. https://doi.org/10.1093/ijnp/pyw109.
479. Fountoulakis KN, Vieta E, Young A, Yatham L, Grunze H, Blier P, Moeller HJ, Kasper S. The International College of Neuropsychopharmacology (CINP) treatment guidelines for bipolar disorder in adults (CINP-BD-2017), part 4: unmet needs in the treatment of bipolar disorder and recommendations for future research. Int J Neuropsychopharmacol. 2017;20(2):196–205. https://doi.org/10.1093/ijnp/pyw072.
480. Fountoulakis KN, Yatham L, Grunze H, Vieta E, Young A, Blier P, Kasper S, Moeller HJ. The International College of Neuro-Psychopharmacology (CINP) treatment guidelines for bipolar disorder in adults (CINP-BD-2017), part 2: review, grading of the evidence, and a precise algorithm. Int J Neuropsychopharmacol. 2017c;20(2):121–79. https://doi.org/10.1093/ijnp/pyw100.
481. Fountoulakis KN, Young A, Yatham L, Grunze H, Vieta E, Blier P, Moeller HJ, Kasper S. The International College of Neuropsychopharmacology (CINP) treatment guidelines for bipolar disorder in adults (CINP-BD-2017), part 1: background and methods of the development of guidelines. Int J Neuropsychopharmacol. 2017d;20(2):98–120. https://doi.org/10.1093/ijnp/pyw091.

Part II
The Philosophical and Scientific Roots

Chapter 12
Introduction to the Relation Between Psychiatry and Philosophy

Why is it important, in a book about a medical specialty, to discuss philosophical matters? Why is it important to educate ourselves in matters pertaining to spiritual, existential, religious, and other non-medical and more importantly non-scientific issues? Why this is important for a psychiatrist, but it is indifferent for a surgeon or a dermatologist? And what exactly is important?

Two are the main reasons. The first is that the psychiatrist himself needs to have a clear picture of what his discipline is about: what is important and what is not, what is valid, and what is pure and useless verbal eloquence especially concerning diagnosis [1–3]. Essentially, psychiatry cannot function without philosophical presupposition and reflection. Psychiatry and philosophy share many concerns including the mind-body problem, free will, ethics, and epistemology, among others [4]. The second is that he/she should be able to engage in hard discussions with patients, their families but also with lay but sensitized persons concerning issues pertaining to the human experience, existence, and feelings, but also what mental illness is and why treatment is efficacious, as well as pertaining political and often religious aspects. Most philosophical theorists are skeptical if not negative about the very existence of psychiatric diagnoses, and the greatest challenge would be when discussants are highly educated in those topics and have a negativistic or even hostile attitude towards psychiatry.

The major problem here is that most psychiatrists study philosophy not to clarify these issues and to arm themselves, but on the contrary, they jump on the horse of philosophy and fine arts only to escape clinical psychiatry and the medical model. Most of them think that an anarchy of philosophies about mental illness in their mind increases their freedom of choice concerning their practice style; it enriches the field and widens horizons. This is a completely unscientific stance and suggests that medical theory and practice are a matter of taste, like gastronomy, not the result of rigorous research.

On the other hand, cross-disciplinary work between philosophy and psychiatry has taken off in the 1990s, the "decade of the brain" [5]. The question is: how

K. N. Fountoulakis, *Psychiatry*, https://doi.org/10.1007/978-3-030-86541-2_12

fruitful this was for clinical psychiatry? The obvious answer could be "not at all" since questions remain unanswered, while attacks on the essence of psychiatry as a scientific discipline and a medical specialty are ever increasing, especially among scholars.

These issues are even more troublesome when they arise in the form of disagreements within a multidisciplinary team whose members might hold different backgrounds and philosophical views [6]. Such difficulties in multidisciplinary teams arise from the fact that although almost all mental health professionals claim they accept some version of a biopsychosocial model, in reality the discrepancies are huge, and often personality conflicts are disguised into scientific conflicts [7]. Thus, the so-called new philosophy of psychiatry manifests many problems of this magnitude [8].

On the other hand, the advances in neurosciences during the last few decades caused dramatic tectonic changes in philosophical thinking. In reality, philosophy finds it difficult to keep up with neuroscientific developments, which clearly point to a pure (eliminative) materialism, although there is a significant gap of knowledge. Neuroscience is the strongest factor that demolishes traditional philosophical thinking [9]. It has been predicted that the combined pressure from the advances in neurosciences and the ever-increasing intensity of the service user voice would lead to a golden era in scientific psychiatry [4]. So far nothing like this happened; on the contrary, unscientific views and policies supported by strong lobbies seem to gain ground and official acceptance [10].

The philosophical approach with the strongest endurance in psychiatry is phenomenology but in a very practically oriented and adjusted form. This approach was the basis for the psychopathology as conceptualized by Karl Jaspers [11], which was perhaps the first philosopher-psychiatrist. At the opposite end, the phenomenological-existential philosophers including Jean-Paul Sartre paved the way for anti-psychiatry and the non-rational theories which tantalize mental health even today. The fundamental problem for philosophy is that the existence of mental illness poses unacceptable limits to the idea of human freedom and free will; as a consequence, there are two basic options: either mental illness exists and human freedom is a limited concept without universal value, or on the contrary, mental disorder is in fact a matter of free choice.

The prevention of abuses of psychiatry for purposes of political or social control needs excellent knowledge and utilization of philosophical thinking and mobilization of cross-cultural, historical, and political-ideological and often religious reserves. It seems that the vulnerability of psychiatry and failure to prevent its abuse is usually the end result of the accumulation of negative events and situations in all the above domains. This was the case of psychiatric abuse in the Soviet Union [12], and it falls into what John Locke called "enthusiasms" [13], which make rational thinking to fail. A similar enthusiasm led psychiatry to adopt psychoanalysis to an extreme manner.

The big question is whether philosophy could make any difference in psychiatry. If no difference is expected eventually for clinical practice, then philosophy is unimportant the same way it is for endocrinology and other medical disciplines. Some philosophers argue that this is the case and philosophy is justified to remain primarily intellectual rather than practical [4]. According to this line of thinking, in principle and in essence, philosophy and psychiatry cannot translate fully one into the other and will never achieve this. Tricky opinions that the "highest common factor" rather than the "lowest common denominator" approach should be sought to bridge the gap and in this way "keeping the best of both" [4] is simply not logical and definitely unscientific.

This idea of "keeping the best of both" often accompanied by phrases like "widening the horizons" or "enriching our discipline" is catastrophic because it dissociates philosophical thinking from scientific advances and reduces everything to meaningless elegant eloquence. On the contrary, it is clinically meaningful as well as necessary to go as deep as possible with philosophical thinking and face all discrepancies and contradiction in the theory and practice of mental health and disease and also their socio-political implications. But the laws of science should never be violated.

The close relationship of psychiatry with philosophy, and the absolute need for the average clinician to have at least some knowledge of philosophical and socio-political issues, reflects the fact that our discipline is more difficult, more complex, and more demanding than the other specialties of medicine. It is more demanding, not only in terms of interpersonal skills, but also in terms of a wider education which transcends the boundaries of medicine, and not only borrows, but also influences other disciplines and schools of thought, and its essence is appealing even to the average lay person. There are many films around psychiatry and mental illness; probably none around cardiosurgery or most other medical specialties.

Philosophy is important in psychiatry for much the same reason that it is important in theoretical physics. Both disciplines demand clear thinking about concepts as well as sophisticated scientific instruments for gathering data [14, 15]. But another way of conceptualizing things could be that such philosophy is no philosophy at all; it is merely methodology elaboration and subsequently "technology development" rather than creative thinking. Well, a scientist has no problem with such an approach.

In the first quarter of the twenty-first century, philosophy is not where it was a century ago but still faces the temptation to misuse or misinterpret scientific discoveries as this was made evident with the famous Sokal case [16–18]. So, is really psychiatry going to continue working within multiple and in some respects mutually inconsistent theoretical models? There are authors suggesting that psychiatry should step out of the mainstream medical model and embrace a model similar to theoretical physics which utilizes several often inconsistent theoretical models [19]. The problem with this approach is that psychiatry currently operates in this inconsistent way, while theoretical physics is in search for a unifying theory. Thus, this "environment of inconsistency" in physics should be considered a temporary necessity not the ultimate goal.

References

1. Aronowitz RA. When do symptoms become a disease? Ann Intern Med. 2001;134(9 Pt 2):803–8. https://doi.org/10.7326/0003-4819-134-9_part_2-200105011-00002.
2. First MB, Regier DA. Separation of anxiety and depressive disorders: new tools will lead to more valid classification system. BMJ. 2003;327(7419):869–70. https://doi.org/10.1136/bmj.327.7419.869-b; author reply 870.
3. Frances A. Preface. In: Sadler J, Wiggins O, Schwartz M, editors. Philosophical perspectives on psychiatric diagnostic classification. Baltimore: Johns Hopkins University Press; 1994. p. vii–ix.
4. Fulford KWM, Davies M, Gipps GT, Graham G, Sadler J, Stanghellini G, Thornton T. The next hundred years: watching our Ps and Q. In: Fulford KWM, Davies M, Gipps GT, et al., editors. The oxford handbook of philosophy and psychiatry. Oxford: Oxford University Press; 2013. https://doi.org/10.1093/oxfordhb/9780199579563.001.0001.
5. Fulford KWM, Davies M, Gipps GT, Graham G, Sadler J, Stanghellini G, Thornton T, editors. The Oxford handbook of philosophy and psychiatry. Oxford: Oxford University Press; 2013. https://doi.org/10.1093/oxfordhb/9780199579563.001.0001.
6. Fulford K, Morris K, Sadler J. Past improbable, future possible: the renaissance in philosophy and psychiatry. In: Fulford K, Morris K, Sadler J, editors. Nature and narrative: an introduction to the new philosophy of psychiatry. Oxford: Oxford University Press; 2003. p. 1–41.
7. Colombo A, Bendelow G, Fulford B, Williams S. Evaluating the influence of implicit models of mental disorder on processes of shared decision making within community-based multi-disciplinary teams. Soc Sci Med. 2003;56(7):1557–70. https://doi.org/10.1016/s0277-9536(02)00156-9.
8. Graham G, Lynn Stephens G. An introduction to philosophical psychopathology: its nature, scope, and emergence. In: Graham G, Lynn Stephens G, editors. Philosophical psychopathology. Cambridge: MIT Press; 1994. p. 1–23.
9. Andreasen N. Brave new brain: conquering mental illness in the era of the genome. Oxford: Oxford University Press; 2001.
10. Fountoulakis KN, Souliotis K. The right to choose treatment-without-treatment: respecting civil rights or an unprecedented manifestation of 'reverse stigma'? Acta Neuropsychiatr. 2019;31(1):56–8. https://doi.org/10.1017/neu.2018.24.
11. Jaspers K. Allgemeine Psychopathologie. Berlin: Springer; 1913.
12. Fulford KW, Smirnov AY, Snow E. Concepts of disease and the abuse of psychiatry in the USSR. Br J Psychiatry. 1993;162:801–10. https://doi.org/10.1192/bjp.162.6.801.
13. Locke J, Nidditch PH. An essay concerning human understanding. Oxford: Clarendon; 1979.
14. Fulford K. Teleology without tears: naturalism, neo-naturalism and evaluationism in the analysis of function statements in biology (and a bet on the twenty-first century). Philos Psychiatry Psychol. 2000;7:77–94.
15. Fulford KW, Stanghellini G, Broome M. What can philosophy do for psychiatry? World Psychiatry. 2004;3(3):130–5.
16. Epstein WM. Confirmational response bias among social work journals. Sci Technol Hum Values. 2016;15(1):9–38. https://doi.org/10.1177/016224399001500102.
17. Hilgartner S. The Sokal affair in context. Sci Technol Hum Values. 1997;22(4):506–22.
18. Sokal A. Transgressing the boundaries: Towards a transformative hermeneutics of quantum gravity; 1996.
19. Meares R. Towards a psyche for psychiatry. In: Fulford KWM, Morris KJ, Sadler JZ, Stanghellini G, editors. Nature and narrative: an introduction to the new philosophy of psychiatry. Oxford: Oxford University Press; 2003. p. 43–56.

Chapter 13
The Concept and Definition of Mental Illness

13.1 Does Mental Illness Really Exist?

The prime question concerns the reason psychiatry exists as a medical specialty. This could only be justified with the existence of mental conditions that need diagnosis and treatment in the way medicine conceptualizes and treats. If there is no such thing as mental disorder or disease and essentially we are dealing with a sociocultural or psychological phenomenon, psychiatry is not only mistaken, but indeed it could be harmful. So, the core question is "do mental disorders exist as medical issues that require diagnosis and treatment?" To answer this question, we need first to clarify a couple of things. We are dealing with two overlapping but separate issues.

First it is the method of intellectual process and investigation we are going to follow. We will follow the scientific method, which follows strict rules and arrives at documented conclusions. If we were to discuss the issue in the philosophical, at least the analytical philosophy method should be followed, and this definitely eventually leads to the scientific method. A second characteristic of science which delineates it from philosophy is the utilization of data that come from the systematic execution of structured research. Thus, when we embark on the ship of science, theorizing is not the main way to arrive at answers. Theories are useful to focus on how to gather data, and also they could be extremely valuable in data interpretation. But data or evidence are the king of science.

Second, not all science is medicine, but medicine is science. So while the first step would be to adopt the scientific method in the study of mental phenomena and the so-called disorders, the result of the scientific method might not place them in the realm of medicine and the neurosciences. This is neither mandatory nor inevitable. Again, it will depend on the evidence that is on systematic empirical observation and mathematical analysis and strictly logical interpretation of the gathered data. This process could prove that indeed there are mental disorders, or on the contrary, it could prove that they are nothing more than sociocultural reflections. It

K. N. Fountoulakis, *Psychiatry*, https://doi.org/10.1007/978-3-030-86541-2_13

could prove that neuroscience explains mental functions, or on the contrary it could show that mental functioning is unrelated to brain function, the same way it is unrelated to kidney function (in healthy people). Thus, the scientific method is essential because it is the only one which can provide with reliable and valid answers. Which these answers will be is a totally different issue.

If the scientific method suggests or definitely proves that mental functions are based on brain function and mental disorders do exist, then the only reasonable pathway to follow is the medical/neuroscientific model. What the author needs to clarify beyond doubts and misunderstandings is that he does not consider the scientific and the non-scientific method in equal terms, and this might confuse the reader which is not a physician or neuroscientist. The author is sharp in this; the scientific method is the only way to go on, and this might eventually permit for intellectual exercises into the metaphysical or even the supernatural but only when the limits of science are exhausted and on condition their continuous expansion is constantly taken under consideration. On the contrary the author takes equal distances from the medical and the non-medical model and allows the scientific method to give the answer in a way as objective and evidence-based as possible. Now again, this, for the non-physician and non-neuroscientist reader could be considered as an unfair bias; this is because all the non-medical models on ‘mental’ (let’s call it disorder) are non- or un-scientific. Well it is true vice versa. If we reject science, then all sorts of irrational proposals find room to flourish and be justified.

Having clarified the ground upon which we are going to elaborate and argue, we can now proceed with the working on the questions and to seek answers, without poetic, aesthetic, or ideological inclinations.

The main issue is the argument made by many scholars, anti-psychiatrists, and recently supporters of post-psychiatry that the concept of “mental disorder” is simply a social construct and as such it is the product of modern civilization and the commercialization of medicine and more particularly of psychiatry, especially by the pharmaceutical industry. Additionally, mental illness as a concept and practice is considered to be the product of the attempt by the society to control behavior.

These arguments are absolute, and they suggest there is absolutely nothing more in mental disorder apart from social control plus the deleterious effect of modern civilization and commercial interests. However, if these were true, then there would not have been reports on mental illness which can be dated back to antiquity and even at prehistoric times. Many scholars trace the formal conceptualization of mental illness as back as to Plato and his work *Πολιτεία* (*Republic*) [1, 2], while descriptions of mental patients and their behavior can be found in the Iliad and Odyssey, the Bible, or the sacred scripts of ancient Egypt and Mesopotamia. Such an assumption would imply also that mental illness is related to any form of civilization, and even at the level civilization was during the prehistoric times. The fact is striking; at least as early as the Roman era, special treatment in front of the law was reserved specifically for insane people. However, there are no clues to suggest that things were better (or worse) during the pre-civilization very early times of humankind on earth. What can be argued with certainty is that mental illness is at least not exclusively the product of modern civilization and neither is at least nor exclusively the product of commercialization of medicine.

There are only a few things which characterize human societies no matter the geographical location, level of technological development, or sociocultural characteristics. Among these few elements are spoken language and mental illness. The latter is accepted and characterized as a form of human behavior so different and so incomprehensible that it is attributed to divine or daemonic intervention and there is a need for special measures for its treatment. Throughout the ages, these persons were not considered as having a different opinion or a different view of the world, but as persons who have lost their contact with reality or the control over themselves. This is why, already since antiquity, the legislation tried to separate the assessment of the actions of "patients" from those of "normal" people. If social suppression would be the case, then such a different consideration would not have developed. Suppression not protection or justice would be the case. There is no civilization at any historical era or geographical location which considered mental patients in a similar way with other people; they were either considered as "patients" or as "victims or collaborators of supernatural forces" (or something in between or both) but never as revolutionaries or dissidents. The reverse was often true, that is, revolutionaries or dissidents were labeled as suffering from a mental disorder, as was the case in the former USSR, but this is a completely different issue. This is a universal historical fact which is often neglected or distorted in order to serve ideological or political aims.

The above mean that any argument against the existence of mental disorders, based on social, political, or historical grounds, is erroneous. Thus, mental disorder exists beyond any doubt, and this is very important to be clearly stated. Since antiquity, and long before modern sociocultural and political establishment existed, there are reports and comments on "mental illness," suggesting that something exists that has to do with "abnormal behavior" and it is perceived as in need for treatment/correction. Whether social, political, economic, spiritual, or other non-biological factors influence the triggering and/or the course of mental disorder is a completely different matter, open for debate.

To rephrase it and conclude, what stands for the general concept of mental disorder exists in spontaneous records since antiquity and throughout the centuries, long before modern socio-political and economic establishments were in place, and irrespective of the socio-political, cultural, and economic environment of the time. Any suggestion that mental illness does not exist or it is the product of modern social factors is simply not based neither on data nor common sense.

This chapter will discuss the various conceptual as well as technical issues pertaining to the definition of mental disorder. Starting from the point that there is at least some extreme form of behavior which should be considered to be a mental disorder, the next question is whether it is possible to distinguish between normal and abnormal behavior in medical terms without the influence of other factors. If the answer to this is more or less "yes," then the next question is whether this abnormal behavior can be reduced to more elementally and medically utilized concepts, that is, neurobiological processes. Eventually the ultimate question is whether we can synthesize the above into an operationalized definition for mental disorder, with measurable reliability and validity. The sub-chapters that follow will deal with each of these questions separately.

13.2 Is It Possible to Distinguish Between Normal Behavior and Mental Disorder?

The first comment should be that there are definitely "culturally sanctioned abnormal behaviors" and as such they do not qualify to be considered as a mental disorder [3]. These include bereavement, anger, grief, or even violent or self-destructive and suicidal behaviors in the frame of extreme situations (e.g., war) and under the influence of the specific culture and society.

The second comment concerns "right" and "wrong," in other words how behavior is perceived and validated by society and according to societal norms and demands. Plato was the first to bind vices with mental health and suggested that wrongdoing is related to some kind of mental disorder: a view shared to some extent also by Aristotle [4, 5]. In the next few centuries, the concept of metaphysical sin emerged in all the monotheistic Mediterranean religions [4, 5], and interestingly the "cardinal seven deadly sins" are reflected in a number of mental disorders in both contemporary classification systems, e.g., narcissistic and antisocial personalities, bulimia/binge eating, etc. [6]. During the same period, vices were separated from mental disorder on the basis of "free choice"; while healthy people were considered to enjoy the ability for free choice and free will, mental patients were not, and subsequently mental disorder and the accompanying behaviors were not the result of such a free choice by the individual.

Denial of the existence of mental illness on the basis of sociocultural preferences is not the only problem. With the domination by psychoanalysis on psychiatric thinking, the boundaries between mental illness and mental health disappeared because psychoanalytic theory argues that everybody has some type of neurotic conflict [7–11]. Because of that, and because of a wide disagreement concerning what exactly mental disorders are, psychiatry celebrated more exotic and unsubstantiated treatment methods through the ages than any other medical specialty (Fig. 13.1). Even worse, with a metaphysical background (often close to the supernatural) always present, most mental health professionals view this as an important wealth for the field rather than a fundamental problem. When in the 1970s the American Psychiatric Association embarked on an endeavor to medicalize psychiatry and to develop an as-reliable-as-possible nomenclature and nosology, opposition was fierce both from the American Psychoanalytic Association and from the American Psychological Association. The arguments psychoanalysts made concerning the development of DSM-III are characteristic of this attitude: "the elimination of the psychiatric past, by the DSM-III Task Force can be compared to the

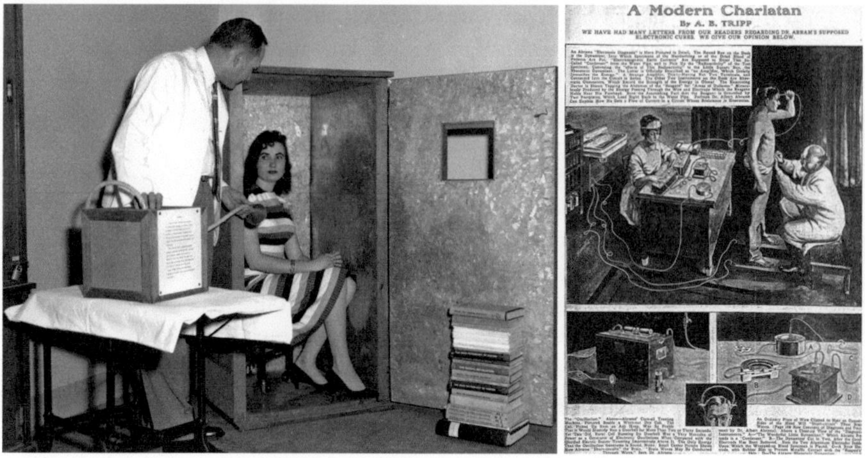

Fig. 13.1 Left: A patient in an orgone chamber. Right: Radionic therapeutics, exposed as fraud in an illustration from Science & Invention Magazine, October 1923

director of a national museum destroying his Rembrandts, Goyas, Utrillos, van Goghs, etc. because he believes his collection of Comic-Strip Type Warhols has greater relevance" [12, 13].

With the development of a more rigorous approach to the above questions which are "existential" for psychiatry, a number of other issues emerged:

How culturally bound is our concept and definition of mental disorder [14–16]?

What should be done with syndromes like bereavement which could be considered to be a "culturally sanctioned response" to an adverse event [17]?

The most astonishing development was the collapse of the distinction between what is biological and what is social, since it has been shown that socialization processes are deeply influenced by biology, while on the other hand, the expression of phenotypes often depends significantly on the social environment. The so-called gene-environment interaction is currently a hot topic and attracts increasing attention [18, 19].

13.3 Reductionism, Mental Illness, and Psychiatry

Reductionism is a philosophical approach which suggests that observed phenomena can be described in terms of much simpler underlying processes [20]. It is based on three models which include translation, derivation, and explanation [21–23]. There are several main facets of this theory [24]:

- Ontological (sweeping) reductionism, suggesting that the entire reality is reducible to a finite number of simple elements [25]. Dualism poses a limit to

reductionism, while in principle monism permits its absolute application, providing no supernatural explanations are accepted. Concerning mental phenomena, Thomas Nagel (born 1937) speaks of "psychophysical reductionism" as the attempt to reduce them to simple events that belong to the fields of physics and chemistry [26]. Such attempts eventually led to "token ontological reductionism" which argues that a system is nothing more than the sum of its simpler parts [27], which reflects "extinctive materialism" or even "eliminationism" (the phenomena do not exist, are illusions) [28]. At this point there is an important debate concerning whether the existence of "emergentism" and "property dualism" would restrict the application of reductionism, especially when there is historicity [29, 30] and complex systems with properties impossible to be reduced to simpler structures [31, 32]; therefore ontological reductionism rejects emergentism [33]. Such radical views are considered by many as mistaken [34]. However, rejection of reductionism on the basis of non-linear relationships [35] or high complexity [36, 37] as emergentism and property dualism do seems a convenient ideological rather than a solid theoretical approach and a way to embrace supernatural explanations.
- Methodological reductionism, which refers to the scientific attempts to utilize reductionism in order to explain nature.
- Theory reductionism, which demands that a newer theory brings progress by reducing the observed phenomena into more fundamental elements in comparison to older theories.
- Epistemological reductionism refers to the use of levels of reduction in an effort to explain phenomena with subsequent levels leading to deeper understanding. This means that all scientific disciplines are based on another discipline that studies more elementary phenomena, e.g., biology is based on chemistry which in turn is based on physics. This is in accord with the hierarchical classification of scientific disciplines by the physicist and Nobel laureate Philip Warren Anderson (born 1923), from particle physics to sociology. Anderson was cautious to note that at each level different laws might be valid without any direct reference to previous levels [38]. The question pertaining to psychiatry and mental disorders is whether they could be positioned within this hierarchical classification and whether they are based on biology, and as a result there is an hierarchical organization with levels of abstraction [39, 40]. It is possible that if such a hierarchical organization exists, it is tree-like or network-like rather than linear.
- Creeping reductionism constitutes a compromise, according to which there is partially applied reductionism to explain partial aspects but not the entire phenomenon, while the possibility of emergentism as the ultimate combining mechanism remains a valid option [25]. It keeps the door open to supernatural explanations.

Opponents of reductionism often label it as "fragmentalism" suggesting that it implies an inappropriate fragmentation and destruction of things which exist only as wholes and their parts are irrelevant to the conceptualizing of the whole [41, 42].

The opposite of reductionism is "downward causation" and was introduced by Peter Bøgh Andersen (1945–2010), Claus Emmeche (born 1956), Niels Ole Finnemann (born 1945), and Peder Voetmann Christiansen (born 1939), among others. It suggests that lower-level phenomena could be explained or determined by higher-level ones [43–45], but this occurs indirectly with the environment as a mediator. For example, failure of behavior to response adequately to environmental stimuli produces an effect of the environment on lower structures so as new and more adequate behaviors emerge. This is the case with systems theory and cybernetics. The approach is based on the assumption that the whole is more than the sum of the parts (a theory originally proposed by Aristotle, also called "holism"), and essentially it adopts the emergentism theory to explain how this surplus appears [46, 47]. It also confuses "feedback" with "causation" [48]. The problem with emergentism when suggesting that the properties of the whole cannot be reduced to the properties of the parts is that there will always be some scientific pathway to explain how this surplus of properties emerged; the opposite assumption would be that these properties appeared from nothing and their appearance does not obey to natural laws. The example of water is helpful here. Although for the naked eye and the naïve observer, water cannot be reduced to atoms of oxygen and hydrogen, and when these two simpler structures are combined in a complex system, new and radically different properties emerge concerning the product (water). Today this can be explained and understood in terms of physics and chemistry. Conclusively, water and its properties are reducible to hydrogen and oxygen atoms and their properties. Also in the case of complex interactions like those described in systems theory and cybernetics, the presence of feedback loops which transcend levels of hierarchical reductionism does not reject reductionism per se. It merely questions the separation between levels. On the other hand, expecting that these levels would be sharply distinct should be considered a naïve and too simplistic approach.

More specifically for the realm of brain and mind, there is a long tradition of scholars who consider the phenomena pertaining to this realm to be distinct from those in the natural sciences, and as a consequence, mental phenomena could not be subjective to reduction, especially reduction to biology. For psychiatry, the prospect of reductionism identifies with the following of the scientific method and subsequently with the place of psychiatry among the other specialties of medicine and disciplines of neurobiology and neurosciences [25, 49–51]. It is a prerequisite or a consequence of the extinguishing of supernatural and metaphysical from psychiatric theory and practice.

A sweeping approach to mental illness would demand that every disorder should be reduced to matter and to the processes at the molecular level or even deeper [51–53]. Although sophisticated models of overlapping prototypes and complex interactions could be proposed [54–57], the basic demands and consequences of the reductionistic method cannot be avoided.

For some, the organization of mind and brain is extremely important for the conceptualization of mental disorder. Many authors suggest that the human mind is a concrete and deep modular organization which might not correspond to biological structure. If such a very specific modular structure is present, it has been argued that any lack of social functioning or false belief becomes a disorder since dysfunction can easily be identified with the dysfunctioning module and can be studied according to the rules of science [58]. On the other hand, if a lack of a modular organization of the mind is accepted, then according to some authors, the naturalistic approach to mental disorder becomes problematic because of the large complexity of the human mind, which makes its study impossible. This could be partially true, since complexity poses a barrier to our ability to understand the mind, but the approach itself simply reveals insufficient knowledge of biology. Modules do exist, both in terms of function and in structure, but they are not entirely separate, and they exist in interplay with other modules in an extremely complex network. In the brain, separation of functioning is not a yes/no or black and white phenomenon, but still local phenomena are often isolated as it is shown with specific limited brain lesions (e.g., aphasias). This is indeed an issue, but it does not pause problems in the principle of the naturalistic approach; it poses a limit to our ability to understand nature, which is a quite different problem (see also Chap. 16).

At the end of the day, the core scientific problem is that brain organization and the mind are not only difficult to approach in neurobiological terms, but also their complexity is beyond our current capacity to mathematically analyze and model. Will we expect ever to have such a mathematical and analytic capacity? The magnitude of the problem is so big that the obvious answer is a risky, difficult, and reserved "no." This difficult problem is coupled with another basic problem which is that many authors tried to define mental disorder in an obsessive way and took too rigidly and as eternal truth the scientific knowledge of their time or their own limited education. Their concepts of medical condition or disease are clearly antiquated. It is to be noted, that although Karl Jaspers started from the viewpoint of existentialism, he is more in accord with pragmatism in his suggestion that mental conditions could be conceptualized as "understandable" vs. "un-understandable" with normality identifying with the "understandable," neurotic conditions being "partially understandable" and psychotic illness as "un-understandable" [59–61].

As a consequence of the above, and as a practical compromise, a creeping reductionism is already in place today, concerning our dealing with the problem of mental phenomena and mental disorders, and the question is how far this method could practically go. In practice, all models manifest high variability and uncertainty; at least partially this is because of the way life has developed in the frame of evolution, and this is what the biologist and Nobel laureate Francois Jacob (1920–2013) called "tinkering" [62].

13.4 The Difficult Road Towards Reliability and Validity

With many fundamental questions on the nature and essence of mental illness and psychiatry being incompletely answered, most often assessment, diagnosis, and treatment are carried out on the basis of the therapist's personal inclination of preference to a certain "school" within the psychiatric tradition. In most cases, this "preference" was based on aesthetics or personal opinion rather than scientific thought, and it seemed legitimate for mental health professionals to declare that they follow a certain school rather than another, without feeling any need to scientifically support their choice. It seems as if this choice is a matter of taste rather than knowledge. As a late 1960 study which compared the diagnostic preferences between US and UK psychiatrists has shown, there was a big discrepancy in the way each national group perceived and was diagnosing schizophrenia [63].

The reliability and especially the validity of psychiatric diagnosis were specifically questioned with the famous Rosenhan's study for the first time [64] and were raised again when Lauren Slater (born 1963) claimed she repeated the same experiment by presenting herself at eight different psychiatric emergency rooms pretending to be a patient (see also Chap. 11 for details). She received a diagnosis of depression with psychotic features but was not hospitalized. Her conclusion was that although circumstances have changed, the diagnosis of mental disorder is still unreliable [65]. Since then other authors have failed to confirm these findings [66], she has been accused of fraud [67], and in response she claimed that she is under attack from "a gang of psychologists and psychiatrists" [68].

No matter how fair or not these critiques were, the fact is that as a medical specialty, and in order to function within the discipline of medicine, psychiatry needs to define its rules, methods, and targets, in accord with the enduring tradition of a prestigious profession in a difficult and demanding environment. The sole existence of a supposed condition is not sufficient; it should satisfy a number of criteria in order to be the focus of medicine. This is an important element of the scientific way medicine follows. Medicine is not (yet) a strict and complete positive science, but it tends to become one, and at any case it is oriented towards science and its rules and absolutely rejects supernatural explanations. According to these rules, the psychiatric literature (as all medical literature) should satisfy (among other things) the fundamental characteristics of scientific knowledge. This type of knowledge is:

- Abstractive, i.e., it focuses on common denominators across observations rather than trying to describe each observation separately and independently in detail. Observation of individuals should be reducible to conclusions concerning groups.
- Reliable, i.e., observations are replicable across different time points and with different observers agreeing on what they observe.
- Valid, that is, there is a specific underlying biological lesion or dysfunction which is reflected in the manifestation of clinical symptoms [69]. This criterion demands the observed behavior to be directly linked to some kind of neurobiological substrate.
- Have predictive value, i.e., knowledge can be used to predict future events.
- Is additive across observations and observers, meaning that what is learned from the observation of one group by one observer can be transferred and used in the management of another group by another (or the same) observer.
- Ideally it should conform with the falsification criterion of Karl Raimund Popper (1902–1994; Fig. 11.17) [70]. According to this, a statement, hypothesis, or theory has falsifiability if it could be contradicted by a basic statement, which could

be tested through an observation no matter whether this would be theoretical or impossible to conduct for practical but not for logical reasons. This means that, for example, such an experiment could be carried out if sufficient resources are used, but practically it could be too expensive, but not methodologically impossible. Passing the falsification criterion does not mean the theory is correct; it w5ll be proven wrong. It simply means that it is a scientific theory, free of logical quantum leaps and distortions and, more importantly, free of supernatural elements.

If these qualities are not present in psychiatric knowledge, then psychiatry cannot be considered to be part of medicine, but rather of literacy, philosophy, or culture. It is interesting that even these fields are slowly adopting the scientific way, with an increasing frequency, although currently this can be seen only in isolated paradigms.

Reliability and validity are core features for any scientific concept and process, and they are strongly bound together although they are conceptually independent from each other. The reason is that when reliability increases, it also drags validity. In psychiatry, it is evident that all the controversial historical changes can be understood as outcomes of a misguided attempt to pursue validity at the expense of reliability. One should have in mind that epistemologically reliability is primary and validity follows; reliability is a necessary but not sufficient condition for validity. To say it in plain words, it is necessary for us to first agree on the object of your study, and after that, we will be able to study it in depth. And this is exactly what was achieved in the last half century. By avoiding theory-saturated and highly inferential approaches, there was a dramatic reduction in the diagnostic disagreement between psychiatrists, and the more accurate clinical conceptualization of specific mental disorders was developed.

Increase in reliability improves validity almost by definition, although some authors argue that in some cases improving reliability comes at the cost of validity and they claim that this is the case for psychiatry [71]. This is in sharp contrast to the definitions and the scientific method as we currently understand it; these claims simply reflect mourning for the loss of beautiful ideas and theories which however were unscientific and a matter of belief rather than empirical knowledge and data.

It is true that in the pursuit of higher reliability, contemporary classification systems and especially the DSM adopted an atheoretical approach which led to the impoverishment of the description of mental disorders by eliminating psychodynamics. According to some authors, this in turn led to distorted understanding of psychopathology and the proliferation of the diagnostic codes.

What many scholars tend to forget is that operationalized diagnostic criteria tend to standardize diagnosis not to fully describe mental disorders. Therefore, the picture they draw is small and narrow and lacks details, not to mention the lack of essential elements. Their biggest problem currently is not the omission of psychodynamic insight but the lack of the utilization of biological markers. On the other hand, it is unscientific to argue that the inclusion of unreliable elements would really add to the validity of the concepts. It will surely make them more impressive and appealing, but this has nothing to do with their measurable validity.

An example would be important here, and a very good example could be the reliability and validity of the observation of a glass of milk by different observers. Let's suppose that all of them wear red eyeglasses; in this case all of them would agree they see a glass filled with a kind of red liquid. They all agree in this observation, so it is reliable. This does not mean that it is also valid; we know that milk is white. Additionally, there is no single waveform length to characterize white light since it is the admixture of all visible colors. So, in this sense there is no external independent laboratory criterion to define validity. If observers use different color eyeglasses, they will report different colors for milk, and the identification of milk among glasses filled with various different liquids with different colors will be difficult if impossible. If you do not know the nature of the liquid, you simply see its color, and this could be correct providing you do not wear colored eyeglasses. Reliability means that we make all observers wear the same color of eyeglasses (or preferably no eyeglasses at all). In this way they are able to agree on how to categorize liquids of the same color, and eventually they can identify glasses of milk, although this will be restricted by the color of liquids. Taste could give a final identification, but instead of tasting a great number of glasses, we limit ourselves to taste only those containing a white liquid. If in this case milk can be separated from the other liquids, there is a hope that a laboratory test (e.g., biochemical composition) will unveil the nature of milk. An additional obvious limitation here is that not all white liquids are milk and wearing, for example, red glasses will likely identify as same color all white and red liquids alike. This is exactly what happens with clinical assessment. The problem is that if every observer is left free to choose the color of the eyeglasses by himself without any a priori rules, then several other colored liquids will be also chosen, and the sample will be too heterogeneous to study. This is why reliability proceeds validity, and this is why adding unreliable elements we think increases validity; in reality they reduce or destroy it. An unreliable method produces results depending on who makes the assessment (in the example what color of eyeglasses are used) not what the target of the assessment is (white color). As already implied, there is no perfectly reliable method; in our example, with the highly reliable condition, all observers wear red glasses, and they will gather all white and red liquids together, which is a much better outcome in comparison to gathering all sorts of colors but far from perfect if we want to gather only glasses of milk. Not even a perfect identification of colors would do; not all white liquids are milk. Input from another sense (e.g., taste) would be absolutely necessary, since input from this sense is independent from vision. Similarly, descriptive psychopathology is unlikely to be sufficient to solve the problem of identifying mental disorders; input from an independent method is necessary, and such a method is only biological markers.

On the other hand, validity includes content, criterion-related, and construct validity [72]. These three facets do not always run in parallel and are to somewhat independent of one another. Essentially, the idea is that a valid process, instrument, or tool assesses the content of what it is intended to assess [73–75]. According to our color-and-glass-of-milk example, a laboratory examination of the content of the glass in terms of proteins, lipids, and other nutrients would be valid since it will give a definite answer as to whether the respective liquid is milk or not.

As described in detail in Chap. 11, the first thoughts on the need of a reliable classification of mental disorders emerged during the efforts of the World Health Organization (WHO) to develop ICD-8 in New York in 1959. While the official historiography [76, 77] credits the philosopher Carl Gustav Hempel (1905–1997) with this idea [78], more recent research into the records suggests that Hempel although lectured on the advantages and disadvantages of a causal-theoretical vs. a purely descriptive approach, he clearly supported the former. It was Sir Aubrey Lewis who advocated for the descriptive approach, but his suggestions were rejected by the majority of the time [72, 79, 80]. No matter the historical details, it was DSM-III more than a decade later who adopted a pragmatic approach, in an attempt to increase reliability. This approach was based on the philosophy of empiricism and on the science and measurement called "operationalism," and it was exactly what Hempel had described but had not supported in 1959 [78]. Robert Spitzer and the DSM-III workgroup took the operationalized approach further and introduced a strong commitment to the biomedical model [81].

On the other hand, it is obvious that diagnosis and diagnostic classification shapes medical practice but also gives feedback and reciprocally shapes medical theory, eventually leading to a number of treatments but also to social responses. Inevitably the diagnostic and classification procedures include cultural elements in addition to the pure scientific ones [82]. While the early attempts to define "disease" stressed the functional orientation of biological systems (as well as of behavior) towards survival and better adjustment in a Darwinian frame, humanity has progressed far beyond such elementary and naïve approaches.

One important critique on contemporary nomenclature, diagnostic approach, and the classification systems is that by passing the years and with the development of newer versions of classification texts, mental disorders and their subtypes are multiplied. This was sometimes called "atomization of psychopathology" [83, 84]. Some argue that this is due to pressure by commercial interests and more specifically by the pharmaceutical industry. This is largely unreasonable; the industry has no benefit, for example, in the sub-classification of psychotic disorders or in the separation between anxiety and depressions. It would commercially be more favorable to have a limited number of broad diagnostic categories, for whom it would have been easier to obtain labeling for medications. With the emerging restrictions in prescribing of psychotropics, the pool of patients a specific medication targets becomes smaller with the introduction of new diagnostic categories, and in case prescription labeling is precisely enforced in combination with an accurate diagnosis, the vast majority of mental patients will be deprived from access to pharmacotherapy.

At the center of this whole debate is the role and the usefulness of the psychiatric interview, which currently is at least semi-structured, and it is guided by the targeted assessment of symptoms and subsequently of diagnostic criteria. This method was developed so that the therapist is able to objectively and reliably arrive at a nosographic diagnosis by reducing variability [85–87]. The core epistemological assumption is that the features of mental disorders are in principle accessible to question-and-answer techniques [88] and observable behaviors could be reliably interpreted as reflecting inner states of the mind [89]. However many authors insist that no matter the structure or the method, the interview remains an interpersonal exploration of inner experience, and it is fundamentally subjective [90]. According to this, an interview is a linguistic event, and not a behavioral-verbal interchange simply mediated by language, and that the modern way of interviewing is problematic because it is dominated by de-narratization [91–93]. Thus, during the interview a laborious process of interpretation is necessary that focuses on the personal level of analysis and permits the understanding of psychopathology and its peculiar correlation with feelings, meanings, and values of the subjects [69, 94–98]. Unfortunately, no matter how appealing these opinions could be, research has proven that such procedures are totally unreliable and subsequently they lack validity. These claims dispute the value of positioning psychiatry within the field of medical specialties [85, 98–100].

Another critique to guided or structured interviews is the possible presence of "procrustean errors" (adjusting the findings to fit criteria) [101] or "tunnel vision" [102]. These both sources of bias are supposed to reduce validity [103], but this is wishful conclusions not based on real data. Such assumptions expose lack of knowledge of research and scientific methods and put the emphasis on authority, not research data, as the source of knowledge.

13.5 Defining Illness in Medicine: The Medical Model

Historically, as already said, mental disorders were recognized since antiquity, millennia before psychiatry emerged as a separate field, and thus it emerged out of a real need. However, when it comes to definitions, things become quite complex.

Psychiatry faces not only all the problems and confusions medicine faces when it tries to define illness and treatment, but additionally many more, and also at a qualitatively distinct and often extreme level. Therefore before attempting to conclude with a definition of what mental illness is, one needs to briefly review what illness in general is and which are the contemporary nomenclature and definitions (Fig. 13.2) [97, 104–106].

The concept of somatic (bodily) disorder is also still the topic of debate in spite of advances in physiology and the understanding of the mechanisms of disease. The not-so-current nineteenth-century understanding of illness is based on a specific and identifiable physical lesion and is not plausible in psychiatry, but also it is outdated and inadequate for most fields in the rest of medicine [107]. Defining the so-called medical model in a satisfactory way is difficult, and a trivial conceptualization is based on another definition, that of health. According to this, the "medical model" is a system eventually aiming to the preservation or the

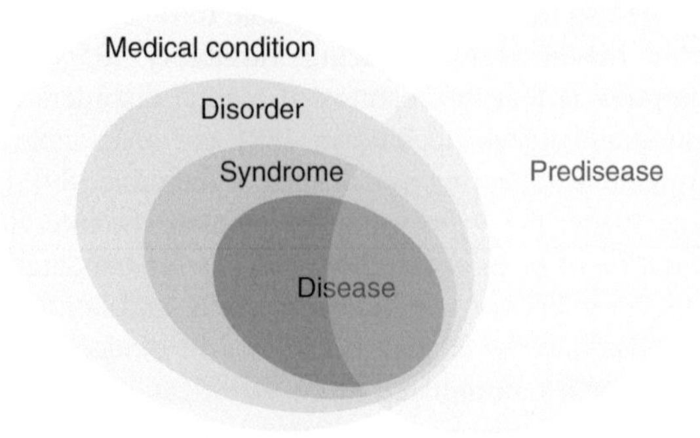

Fig. 13.2 Overlapping of medical condition, disorder, syndrome, disease and predisease

restoration of "health" [108]. Another approach suggests that many of those conditions which are the focus of modern medicine as well as their etiology seem to be simply extremes of conditions which can be understood in the frame of evolutionary biology [89, 109–111]. But since many medical conditions are poorly understood, a minimal conceptualization of the medical model would include no commitments about the underlying physical structure illness. But on the other hand, the most fruitful areas for philosophical debate are those who pose with questions difficult to answer; these questions help uncover the way the human body and mind work [112].

A basic and widely accepted assumption in the philosophy of medicine is that somatic disorders are value-free, while mental disorders are extremely complex in terms of value. This is somewhat in accord with Plato who identified morality with mental health and immorality with mental disease. Subsequently, to the extent mental disorders are tailored to the medical model, their value context will be decreased. This is not straightforward anymore because it has happened that advances in medicine and the diagnosis and treatment of medical conditions often led to the increasing of the value content, as, for example, in reproductive medicine [95, 111]. Additionally, progress in ethology made the study of behavior less value-oriented and more neutral-scientific [89]. The problem is not as deeply rooted as in psychiatry but still exists and even more seems to increase in magnitude for the whole of medicine.

Medical Condition

The broader term used to denote the focus of interest of medicine is "medical condition." It includes not only all diseases, lesions, and disorders but also all those non-pathological conditions which require and receive some kind of medical intervention.

Examples of non-pathological medical conditions are pregnancy and childbirth, myopia (most cases are variations of normal sight), as well as obesity and all the conditions requiring cosmetic plastic surgery. These conditions, though not abnormal per se, could be the cause of significant dysphoria or even disability for a variety of reasons, both personal and societal. Additionally, aging and death are not abnormal per se, since they inevitably occur for all people; on the contrary there is nothing more natural. However, the changes that come with age as well as increasing the duration of life and postponing death are central in the scope of medicine and health care.

Because of its vague meaning, as a label, "medical condition" is better accepted by patients who do not consider their condition to be deleterious, but on the contrary, it is also denounced by proponents of various rights movements. The attitude of insurance agencies towards general medical conditions varies and is determined by their wish to cut costs; therefore, those medical conditions considered to be non-pathological might not be covered by health insurance, and there is a tendency for "naturally occurring events" like delivery not to be taken care of doctors but to be carried out at home by a midwife, the way it was done a century ago.

The term "general medical condition" is used in the *Diagnostic and Statistical Manual of Mental Disorders (DSM)* to denote the presence of a "somatic" medical condition. This does not suggest that mental disorders are not "medical conditions"; they should be considered to be "specific medical conditions." In the *Diagnostic and Statistical Manual of Mental Disorders (DSM)*, the term was selected in order to avoid terms like "somatic condition" which would point to some kind of body-mind dichotomy and Cartesian dualism.

Disease

The term "disease" corresponds to any medical condition which disrupts the structure of the physiological function of the body, through a known way. Typically, it is associated with specific symptoms and signs, and it could be the result of the influence of external factors such as pathogens or by internal homeostatic dysfunctions. What distinguishes "disease" from all the other terms used to denote various medical conditions is that diseases are well-defined conditions with pathognomonic features which put the diagnosis with high reliability and validity. The terms "illness" and "sickness" are more or less similar to "disease."

Such an example is vascular brain disease, which is characterized by ischemic lesions in the brain, visible through brain MRI, and can be verified with pathological examination usually postmortem. Vascular brain disease is an interesting medical condition because the mere presence of vascular lesions might not cause any kind of symptoms and additionally a degree of vascular encephalopathy is considered to be "natural" or "expected" after a certain age. However, although they do not cause symptoms, nevertheless such lesions have strong predictive validity since they suggest neurocognitive impairment will probably develop after some years.

Recently the "Osler's Rule" (after Sir William Osler, 1849–1919) has been proposed. Its original form suggests that "a neurological defect has to be related to a specific lesion." This is in contrast to "Hickam's dictum," which states that the neurological defect can be due to several lesions. This fits perfectly with vascular brain disease where most often, in order for symptoms to emerge, multiple focal lesions or even extended diffuse vascular encephalopathy are necessary.

The modified "Osler's Rule" according to Ghaemi states that we should "treat diseases not symptoms" [113–115], meaning that we need to treat the underlying pathology. This suggestion clearly stresses the fact that there is no known biological substrate for mental disorders and therefore they cannot be considered as formal diseases, and as a consequence the biomedical model is under dispute.

Syndrome

When the medical condition has a specific clinical picture including several medical signs, symptoms, or other characteristics which tend to occur together, but often the picture manifests a significant degree of heterogeneity and there are no pathognomonic features to put the diagnosis with high reliability and validity, then the condition is called "syndrome."

There are exceptions to this rule probably because of traditional reasons, e.g., Down syndrome. Other medical conditions are labeled as "disease" or "syndrome" depending on the specific etiopathogenesis in the specific patient, but without following any specific rule. For example, "Parkinsonian syndrome" is the preferred term when the signs are the consequence of treatment with antipsychotic medication, while "Parkinson's disease" is the preferred term when the signs are manifestations of a degenerative brain process. In both cases the pathophysiology is well recognized so, but in the case of treatment-induced parkinsonism, the word "syndrome" is used in order to denote that no irreversible tissue damage exists. Often even after the discovery of the underlying etiopathogenetic cause, the condition keeps the "syndrome" name for traditional reasons as physicians are already familiar with it.

Disorder

The term "disorder" is broader than the syndrome and less specific. On the one hand, it includes monosymptomatic conditions which do not fulfill the complexity criteria of the "syndrome" and, on the other hand, as a term is more acceptable by many groups of patients since it is considered to be more value-neutral and less stigmatizing than the terms disease or illness.

In many areas of medicine, the term has been used in order to stress the "state" and "physiological" rather than "structural" nature of the condition (e.g., in metabolic disorders), and according to many authors, this is the core meaning of the term.

Predisease

Predisease is a term used to denote the presence of a subclinical or prodromal vanguard of a medical condition. Predisease might be completely asymptomatic and could be uncovered only after sophisticated testing, or it could be characterized by an attenuated and non-specific clinical picture which shares similarities with the full-blown condition.

After some time, a significant number of persons with predisease will evolve into the full-blown medical condition. Common examples are prediabetes and prehypertension.

Predisease is a vague and problematic area of research where the legitimate concern for the future medical status of the person is perplexed by the presence of various conflicts of interest including disease mongering, medicalization, and overtreatment.

At the core of these problems is our lack of knowledge on how many of these cases constitute the first stage of a disease and how many are simply non-pathological variations of normal functioning without carrying the risk to develop into a disease.

13.6 Mental Illness

Since antiquity, the concept of mental illness indisputably followed that of medical disease or illness. Of course, a question is what the place of medicine was at those times, but this is a completely different story, because neither religion had exactly its current meaning, place, and role.

It is important that the term "therapy," a term closely linked with disease and medicine, is used in ancient texts when referring to the alleviation of disordered behavior in such cases. Mania and melancholia are among the earliest terms in medical nomenclature. While diagnosis was always a common practice, limited only by the available knowledge of the time, diagnostic classification and a rigorous approach to patient assessment and diagnosis emerged relatively recently.

The Nature of Mental Illness

Some authors argue that the application of the biomedical model in psychiatry absolutely suggests that mental disorders are natural kinds [116–118]; however this is not necessary. The concept of "natural kinds" was first developed by Plato and suggests that in the universe there are ideal categories which serve as prototypes. For Plato this concerned both objects and abstract concepts like the ideally good human being. But today when we are talking about natural kinds, a prerequisite is that they obey to natural laws. For example, natural kinds are natural elements (e.g., $K+$, $Ca++$, etc.) or substances like water (H_2O). In medicine, several conditions could be considered to be natural kinds, including insulin-dependent diabetes mellitus or thalassemia. Others are probably not, including various autoimmune disorders and other conditions with overlapping manifestations, laboratory findings, and difficult differential diagnosis. Similarly, mental disorders could be conceptualized in a broad homeostatic dysregulation and dysfunction without clear categories or boundaries and definitions and with elusive etiopathogenesis and underlying mechanisms [119–124].

In this frame, contemporary classification systems treat mental disorders as "practical" not "natural" kinds [122]. An interesting approach has been proposed by some authors, which consider mental disorder to fall into a type of a "Roschian concept" [56], in accord with the Wittgensteinian family resemblance concept. This concept suggests that a network of similarities rather than precise definitions is more adequate to describe a group of similar objects [125, 126]. Wittgenstein used games as an example and argued that there are a number of human activities which are considered to be "games," but the list is quite heterogeneous since it includes sports, cards, board games, etc., and there is not a core group of features that characterizes them all. There are many subgroups, and there seems to be a hierarchical subgrouping in the form of a tree; patterns of characteristic features manifest only partial overlapping. Under such a conceptualization, there could be some kind of prototypical diagnostic categories, but individual patients may vary a lot from each other and are diagnosed not on the basis of the similarities among them, but on the basis of the similarities they manifest with the prototypical diagnosis. Deep in essence, even the Roschian concept demands the existence of some kind of a "prototype" or "kind," the difference is that it considers this prototype to be polymorphic, i.e., multiple forms of it could exist almost independently of each other. While a natural kind resembles a single fruit, a Roschian concept resembles a bunch of grapes. Additionally, such a concept is top-down, not bottom-up. This means that it is not based on the analysis of raw data, but instead it is a theoretical and to some extent arbitrary construction, which serves as an umbrella to host a heterogeneous group of patients. However, in all these families of concepts, a key common feature always exists and bridges all concepts and brings them together, the same way the stem brings grapes together to form a bunch. This is the philosophical principle which is practically reflected in the adoption of polythetic diagnostic criteria sets by modern classification systems.

Theoretical Approaches to the Definition of Mental Disorder

The developments, in the wider medical and healthcare environment, accompanied by radical changes in the socio-political environment, made necessary the precise definition of what "mental disorder" is, and if this is not possible, at least what a "mental health condition" could be. It is evident that the term "mental disorder" currently seems to lack a single consistent operational definition able to cover all situations; therefore different approaches to definition might be necessary, depending on the case [3, 127]. There are radically different approaches to the nature of mental disorders [128, 129], varying from considering them as maladaptive learned behaviors [130], merely a social construct [131, 132], suppressive mechanisms of political control [133–135], spiritual experiences [136], the result of family dynamics [137], or simple life problems [138, 139]. Often, the use of language is of prime importance in the formation of these theories [112], while the frequent use of eloquent extensive metaphors exposes lack of deep knowledge. Currently the concept of mental disorder itself seems vague, and most authors believe this is a consequence of the scientific immaturity of our field. The big question is whether this vagueness is an inherent property of mental disorder itself rather than an epistemic problem [140].

There is wide consensus that mental health conditions do not satisfy the criteria to be considered as "illnesses" or "diseases," and since sometimes there is only a limited number of symptoms present, the term "disorder" is widely accepted as the appropriate term.

During the last century or so, at a theoretical and philosophical level, five are the major approaches when someone tries to define mental disorders: realism, pragmatism, constructivism, naturalism, and normativism [72]:

- Realism assumes that the content of science is real and independent of human activities. According to the medical model, a neurobiological dysfunction of some kind, also in small subsystems, underlies every mental disorder [141–143], and on the basis of this dysfunction, it should be possible to distinguish between dysfunctional brains from merely unpopular behaviors which is still not practically possible [144].
- The problem is that the biological nature of mental disorders is yet unknown, which is a problem not uncommon with other medical conditions in other fields of medicine also.
- Pragmatism argues for a sensible approach to mental disorders by simply seeking categories that perform well in the world in an otherwise atheoretical way. This is currently the basis of contemporary classification systems, which combines reason with observation in a practical way.
- Constructivism suggests that social forces shape our conceptualization of mental disorders. According to it, mental disorders do not really exist, but instead they were invented in order to serve socio-political aims and demands. This approach has been associated with anti-psychiatry activists. This is in sharp contrast to the empirical finding that several psychological functions differ qualitatively from their normal counterparts, e.g., delusions from beliefs [145]. The existentialist approach should be considered as part of this third approach. A softer version suggests that although mental disorders do exist, the social component is by far the dominant.
- Naturalism claims that disorder can be defined in terms of biological dysfunction (either with a causal or an etiological role), and it is a value-free concept [72, 95, 118, 146–152].
- Normativism contrasts the previous approaches [153–158] and suggests that shifting values and social norms caused changes in mental disorder. These thoughts are based on the statistical theory of disease (rare somatic or mental conditions are abnormal, and if dysfunction accompanies them, then there is a need for treatment) and the evolutionary approach to health and disease.

While the above principles are theoretically strong in their pure form, in reality their elements seem to intermix. For example, although it is true that truly socially constructed mental disorders are rare if exist at all, there are socio-politically defined aspects of mental disorders as these were shaped historically [159]. Even mental function could be culturally determined. For example, while oral language is universal and thus probably hardwired in the brain, because it exists in every group of humans around the globe during all eras of history, written language is not. There seems to be some king of universality in the use of icons or hieroglyphics, but alphabetical writing was developed only once in human history in the east Mediterranean basin. However, writing and reading is an important function in modern life, and it is a standard component in neuropsychological testing. But whether it is a true function remains a matter of debate [58, 155]. Dyslexia on the other hand is recognized as a mental disorder although its impairment seems to be largely restricted to a culturally defined and developed ability.

In the frame of the above principles, a number of conceptual theories emerged and fueled intense debates. It is of outmost importance to see how the theoretical principles are combined or utilized in real-world theories and debates. Maybe the most interesting debate was Kendell vs. Szasz in the mid-1970s [107, 139]. This debate seems to continue until recently [160]. In essence, the pivotal point of disagreement between them concerned the meaning of somatic disorder [111]. Szasz defined it in a very narrow and antiquitating way as "the structural and functional integrity of the human body," while Kendell used broadly evolutionary terms involving a "biological disadvantage." Additionally, Kendell argued the main reason for the difficulty in defining mental disorders is that advances in neurosciences are far behind the progress seen in other areas of medicine. On the contrary, Szasz argued that in contrast to somatic diseases, mental disorders are exclusively moral concepts, and with this he adopted a pure constructivist approach [107, 111, 139]. The approach of Kendell was based on realism and naturalism with a compromise on the basis of pragmatism. A third approach (as mentioned earlier) is the concept of "harmful dysfunctions," defined in evolutionary terms, with "harm" referring to social contexts and social value judgments, while "dysfunction" refers to problematic biological mechanisms [76, 77, 79, 89, 109, 110, 146, 150, 161–163]. This approach is based mainly on realism and naturalism but also utilizes a minimum input from constructivism. A fourth one argues for the need for an ahistorical approach, the "systemic capacity view of function," in which functions are considered as components of broader mechanisms specifically designed to keep a system in homeostasis, without the need to define what "dysfunction" stands for [164]. This later idea is based on naturalism. All these approaches are adequate to a certain extent, but they also manifest deep deficits which make them insufficient as universal approaches. For example, although there are evidence to support the usefulness of systems or subsystems dysfunction in accord with Boorse's ideas in many disorders [141, 142], there are cases with some kind of either supposed dysfunction or alternative ways of function which are not considered to be disorders. Such is the case of homosexuality, unless it can be proven that there is some kind of group selective advantage carried by differences in sexual orientation [165]. On the other hand, the functionalistic approach in the frame of Wakefield's ideas is also problematic because it does not take into consideration the so-called biological bad states.

These are not disorders, but they are considered unwanted and put the individual at a disadvantage when it comes to compete for resources or mating. Ugly face is such a "biological bad state" as are some minor dysmorphic features including big nose or prominent ears. Interestingly many of these states are the focus of other fields of medicine, mostly of cosmetic plastic surgery. It is not unusual for these bad states to cause much more harm to the individual than minor diseases or disorders. And what about aging? All human beings are expected to age, and as a result, there will suffer from a progressive failure of all their bodily systems and eventually they will die. Thus, aging and its consequences are part of the normal cycle of life, but still they are the focus of medical intervention. Here, it is rather the operationally convenient rule that all reversible or treatable conditions, e.g., diabetes mellitus, have to be considered as medical diseases, while all the rest as natural decay of the human body. Many changes fall in the area between and often do not constitute a dysfunction per se, but still they are the focus of medical intervention, often purely for aesthetic reasons (e.g., wrinkles). Again, psychiatry is not alone among the other specialties of medicine to carry the burden of value-loaded concepts and practice.

The utilization of an example with vision is again suitable here in order to understand better Wakefield's ideas. Humans as well as many other species of animals possess color vision, that is, the ability to see colors in a significant variety and resolution. This gives an evolutionary advantage (apart to modern-day advantages) to those individuals who possess this ability, simply because among other things, they can differentiate between ripe and non-ripe fruits. Ripe fruits have higher content in carbohydrates and other nutrients and tend to have vivid colors, often reddish, while non-ripe tend to be green. Thus, lacking color vision either completely or partially puts the individual at a disadvantageous position when it comes to competition for food resources. Problematic color vision could not be considered a somatic illness per se, and its effect on the life of the individual depends on the environment, but it could lead even to death under certain circumstances. In most societies this dysfunction results in some kind of restrictions concerning the work the individual can carry out and the responsibilities he/she can undertake.

However, Wakefield's theory cannot explain how it is possible that certain mental disorders including sociopathy or behaviors of substance abuse and dependence could reflect specific traits which used to enhance adjustment in the past and in environments with radically different demands (e.g., low in resources), but they constitute "mismatches" in modern environments [110, 166].

Finally, a more practical but rather too simplistic approach has been developed by Bill Fulford and suggests that someone is ill if they experience a failure of ordinary action [167]. This is maybe the most representative of the pragmatic approaches, but it is also a poor compromise between the conflicting theories since it tends to bypass rather than to solve the existing problems and discrepancies.

Modern psychiatric nosology capitalizes on the abovementioned theoretical principles and theories and has managed to produce a fair classification of mental disorders as this is reflected in the two major contemporary classification systems, the ICD-10 and the DSM-V. Classification as a scientific adventure is not new; it was the cornerstone of Aristotle's philosophy who developed the idea of "Ιεραρχική

Κλίμακα" (Scala Natura; Fig. 13.3) to classify all beings and objects on earth, but in modern times, probably the first pioneer in the field was Thomas Sydenham (1624–1689) who preceded the botanic classification work "Systema Naturae" (1735; Fig. 13.4) by Carl Linnaeus (Carl von Linné; 1707–1778). His botanic work inspired others in various fields of science to use classification as a scientific tool to uncover the underlying nature of things. This was the case also in medicine, and in psychiatry it was eventually celebrated by Emil Kraepelin (1856–1926) who in 1912 developed the nosological concepts of dementia praecox (later to be called schizophrenia) and manic depression insanity (later to be called bipolar disorder), and in this way, he put the foundations of modern diagnostic approaches and systems in psychiatry. During the second half of the twentieth century, modern classification systems with operationalized diagnostic criteria appeared partially as a consequence of the failure of contemporary psychoanalytic approaches and because of the pressure from the society, administrative institutions, and needs [13, 168–171].

But before this, during the late nineteenth and the first half of the twentieth century, there was a great schism in psychiatry with two opposing approaches. The first was heralded by Emil Kraepelin and concerned an emphasis on the systematic approach to psychiatric diagnosis and classification, and the second by Sigmund Freud (1856–1939; Fig. 9.38), who developed psychoanalysis on the basis of

Fig. 13.3 Ramon Lull's Ladder of Ascent and Descent of the Mind, c. 1305, based on Aristotle's Scala Naturae

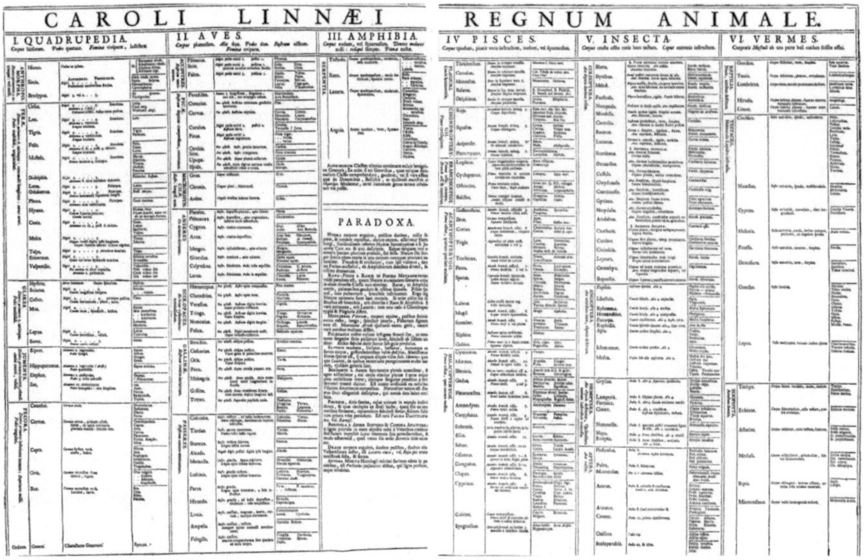

Fig. 13.4 Table of the Animal Kingdom (Regnum Animale) from Carolus Linnaeus's first edition (1735) of Systema Naturae

interpretation rather than observation and rejected the concept of psychiatric diagnosis. In line with this, Jerry Fodor (1935–2017) argued that "explanation" is necessary in order to define the abnormal but not the normal [172, 173].

However, this was a break that lasted only a few decades, and soon afterwards psychiatry returned to the nosological approach, with Robert Spitzer (1932–2015; Fig. 11.42) being the champion of this return to the roots. The introduction of classification systems based on empirical data made psychiatric diagnoses more comparable to the diagnoses in the other medical specialties and in close link with the scientific way and the rules of science [107, 174]. Eventually, psychiatry should be seen as a kind of cognitive neuroscience [143].

The way human brain functions and mental processes are structured is of prime importance. For example, there were some attempts to interpret them in terms of hardware vs. software and in this way to host both biological and psychodynamic psychiatry under the same roof [175]. This approach is considered today to be naïve and is abandoned.

Can the Biomedical Model Be Rejected?

A crucial question would be, is it possible to reject the medical model for mental illness? Where would this lead to? There are authors suggesting that such a rejection "does not commit you to rejecting materialism about human beings, but it does commit you to rejecting a fundamentally mechanistic view about what people are like" [143]. At the end of the day, mental disorders are definitely value-complex rather than value-simple, as somatic disorders usually are [176]. While advances in neurosciences could reduce this value load [141, 142, 146, 148, 177–179], some authors warn about the risks of emerging abusively dehumanizing consequences [111, 180]. Such an assumption is difficult to follow, for reasons discussed in detail in Chap. 16. The real problem is that many scholars feel that the medical model restricts human freedom and implies a sense of weakness, pessimism, or lack of control. Especially the "locus of control" seems to exist outside the will of the person, and his/her fate is controlled by others, who are "professionals in the field." This latter "professional relationship" between therapist and patient triggers more feelings of lost independence and freedom.

This is mainly why the anti-medical approach focuses not on empirical data but on subjective experience and interpretation and rejects abstraction in favor of individualized understanding. It favors generic non-scientific interventions close to lay persons' beliefs, including diet, exercise, or societal change. Such approaches are very difficult to tackle because they a priori reject the scientific way; they do not engage in discussions that follow the rules of science or even at least those of analytical philosophy. They retreat only when the everyday evidence is observable with the naked eye and are impossible to neglect; even then, they seem extraordinarily resilient.

Last but not least, the apologies of the authors in the introduction of DSM-IV because of the use of the term "mental disorder" as it could imply an anachronistic distinction between "mental" and "somatic" disorders are at least partially irrelevant; "mental" is an adjective for medical disorders, as are "gastroenterological," "neurological," or "endocrinological."

Definitions of Mental Disorder by Official Organizations

The World Health Organization (WHO) creator of the classification system International Classification of Diseases, Tenth Revision (ICD-10) avoids giving any direct definition of mental disorder; instead it shifts the discussion towards mental health and disability. Mental health is not merely the absence of mental illness, and additionally it does not identify with "well-being" [181, 182]. According to the WHO, the basic concept is that mental health could be defined as "a state of complete physical, mental and social well-being and not merely the absence of disease or infirmity" [183]. The definition becomes more sophisticated by suggesting that "mental health is defined as a state of well-being in which every individual realizes his or her own potential, can cope with the normal stresses of life, can work productively and fruitfully, and is able to make a contribution to her or his community" [182]. In this frame, the true nature of "mental health" is the ability to respond appropriately and to adjust to common social, occupational, and personal challenges [181, 182]. This is often but not always or necessary accompanied by positive feelings and well-being or professional success, especially during specific phases of life like adolescence or when facing difficult circumstances like migration or catastrophic situations [184].

However this attempt to avoid defining the grey zone between mental health and disease is loaded with a different kind of problems, including ethical (e.g., hedonism) [184], and maybe with even more and unexpected cultural issues [185].

On the contrary, the American Psychiatric Association (APA), creator of the second major classification system the *Diagnostic and Statistical Manual of Mental Disorders (DSM)* tackles the issue in a direct way and with partial success. While in the early editions of DSM such an effort to define mental disorder was not present [186] and the definitions were derived exclusively from medical practice and taken as granted, gradually the need to specifically define mental disorder emerged. The emphasis was put on the presence of distress/impairment which constitutes the cardinal characteristic of individuals seeking professional help [186–190]. In the previous APA classification system, the DSM-IV, a mental disorder was defined as a psychological syndrome or pattern which is associated with distress, disability, increased risk, of death or causes a significant loss of autonomy. Normal responses (e.g., grief) were excluded as well as "deviant" behavior for political, religious, or societal reasons, not arising from a dysfunction within the individual [191]. In the last edition, the DSM-V, the approach became more elaborated, and in addition to the DSM-IV definition, a mental disorder is supposed to be "characterized by clinically significant disturbance in an individual's cognition, emotion regulation, or behavior that reflects a dysfunction in the psychological, biological, or developmental processes underlying mental functioning" [192]. But to put things in their right value and size, the DSM conceptualization is a "position statement" rather than an operationalized definition per se. It discusses the key features a mental disorder has: distress/impairment, and it is distinct from normal distress and a cultural/subcultural way of behaving and is not determined in principle by the presence of conflict with society [187]. However, when this statement is utilized as an operationalized definition, many practical shortcomings emerge especially in the "grey zones."

The European Psychiatric Association (EPA) attempted to define mental health as "a dynamic state of internal equilibrium which enables individuals to use their abilities in harmony with the universal values of society. Basic cognitive and social skills; ability to recognize, express and modulate one's own emotions, as well as empathize with others; flexibility and ability to cope with adverse life events and function in social roles; and harmonious relationship between body and mind represent important components of mental health which contribute, to varying degrees, to the state of internal equilibrium" [193]. This is of course an excellent description but too far away from an operationalized definition.

It is important to note that according to all the above definitions, the mere presence of symptoms alone does not generally justify by itself for the diagnosis of a mental disorder, at least officially. Even concerning psychotic disorders, it has been reported that approximately 3–5% of the general population seems to have transitory psychotic experiences with approximately 75–90% of them disappearing over time [194–196]. Only a minority of 25–30% evolves into a full-blown psychotic disorder within two to three decades [197, 198]. Most of them do not seek professional help although it seems that they were likely to have felt the need for or to have been encouraged by others to seek treatment [194]. Those seeking treatment are characterized by distress and personalizing appraisals, mostly threatening [199–201] and more severe global impairment [197]. Whether those who seek care differ from those who do not in terms of severity or quality of the psychotic experience is a matter of debate [199, 202, 203], while they do not seem to differ in terms of family history of psychotic disorders [197].

At this point again, one should consider the possibility for mental disorders to fall into a type of a "Roschian concept" [56, 125, 126], and individual patients may vary a lot from one other and are diagnosed not on the basis of similarities among patients but on the basis of the similarities the patients manifest with the prototypical diagnosis, in a top-down approach.

All these approaches share common elements, but also they manifest significant differences in terms of their basic philosophy. Eventually they chart the spectrum from severe mental illness in one pole to the positive mental health in the other. It is rather easy to define the two opposite poles. Severe mental disorder is always characterized by significant distress or impairment of personal functioning [3, 127, 204] accompanied by a pervasive inability of the person to react appropriately and to adjust to common social, occupational, and personal challenges. In a philosophical-cognitive view, there is some kind of severe disruption of "intentionality" as a core feature of expressed behavior [205].

On the contrary, the "health pole" is characterized by the opposite features, but the crucial element is not the presence of a state of unchallenged happiness. This is a misunderstanding, emerging from happiness as the core meaning and aim of life according to Aristotle and the philosophers of the Enlightenment [206]. According to them, "happiness" is quite different from "hedonism" (from the Greek word *ηδονή* meaning the lust or pleasure). In fact, happiness was considered to be the aim of life, mainly in order to downgrade the role of religious-based meanings in life. Happiness was a more pragmatic as well as humanistic proposal, but the meaning of happiness remained open to debate, leaving space for religious as well as secular interpretations. For many, happiness constituted a basic human right, and society started changing towards a structure that could allow pursuit of happiness. Change was radical; not only the seeking of happiness became perfectly legitimate, but on the other hand not being happy was now considered to be a problem in need for a solution. Further on this path, the concept of "happiness" would be later utilized in several efforts to define mental disorder.

The Special Case of Personality Disorders

A particularly problematic diagnostic group is personality disorders which by definition carry a large socially determined element in their definition and criteria and, from its very inception, has been infused with philosophical assumptions and controversies [207]. Nowhere else in psychiatry and the study of human behavior, the dualistic approaches and the debate of moral vs. illness are so strong either overtly or covertly. Nowhere else scholars are more reluctant to bridge and unify the theoretical approaches, even when something like this seems utterly possible. The reason for this seems to be that the conflict is mainly a conflict of principles (dualism vs. monism) rather than an informed disagreement on topics. On the other hand, from all mental disorders, personality disorders are those more loaded with values and closely related to beliefs, desires, and action in a normative way which is currently impossible to be reduced to neurobiology [208].

The modern concept of personality disorders was developed in accord with the work and proposals of Theodore Millon (1928–2014; Fig. 11.45) [209–211]. Their diagnosis is based on the disrupted social and interpersonal behavior and functioning of the individual, rather than his/her internal experience, but often they do not satisfy any formal criteria for the presence of a mental disorder [212, 213]. Patients with personality disorders tend to seek professional help only when their problems are complicated, and the condition starts resembling a mental disorder with the appearance of dysphoria, social impairment, or some other kind of dysfunction. Even so, psychiatry already includes several diagnostic categories that could be viewed as primarily a conflict between the individual and the society, called "vice-laden" and include pedophilia, antisocial personality, and others [214–216].

There is a bulk of data from empirical research of more than half a century suggesting that "normal" personality is based on the existence of five factors and two superfactors. The two superfactors were named as "alpha" and "beta" since their psychological meaning was unclear [217], but others considered them to reflect "virtue" and "dynamism" [218]. In this way, "virtue" reflects the moral aspect (kindness, reliability, and decency), while "dynamism" reflects enthusiasm and energy. Many authors argue that the above suggest that in spite of the long attempts to eliminate value from psychiatric diagnosis, this is still strongly present in personality disorders [207] and psychodynamic practice, especially psychoanalytic [96]. However, a more technocratic and value-free interpretation of these findings is that one of the factors represents the internal reference system (self), while the other the way, these rules are applied during the interaction with the environment (self-environment interaction) [219]. There are several other interpretations, also in line with a value-free approach, that conceptualize these two superfactors as "stability" and "plasticity" [220] or "ego control" and "ego resiliency" [221, 222].

On the one hand, it seems that half of the total variation of personality traits could be attributed to genes [223, 224], but on the other hand, several authors argue that many diagnostic criteria used in personality disorders clearly reflect values rather than symptoms and mix the "dysfunctional" with the "disliked" [215, 225, 226]. In contrast, others suggest that there is a creative but delicate admixture of values/vice and health/disorder both conceptually and practically, with the ability to adapt being the prime criterion, thus rendering any theoretical disputes obsolete [227, 228]. Others conceptualize personality traits as being structured like virtues, with the extreme ends of each dimension being considered pathological [229]. These arguments, of course, do not solve the theoretical problem. They constitute a pragmatic compromise and simply redirect the debate to more fundamental questions concerning mental health.

The distilling of the literature concerning personality disorders has led to the formulation of a limited number of conceptual models. These models are directly relevant to the question whether personality disorders can be considered to be mental disorders per se. They are as follows [207, 228, 230–232]:

- The vulnerability model which suggests personality disorders are clinically relevant conditions in a similar way hypertension and hypercholesterolemia are; thus they constitute risk factors for the development of less controversial disorders including psychoses or mood disorders [233, 234]. This clearly rejects the place of personality disorders in the diagnostic nomenclature as stand-alone mental disorders by themselves.
- The pathoplasticity model considers personality disorders to be clinically relevant because they modify the course and outcome of other mental disorders [235]. Again, this clearly rejects the place of personality disorders in the diagnostic nomenclature as stand-alone mental disorders by themselves.
- The spectrum model conceptualizes personality disorders as being milder or attenuated expressions of the same genetic predispositions that underlie major disorders especially psychotic and mood disorders [236, 237].
- The capacity failure model suggests personality disorders are essentially developmental disorders characterized by the failure to develop one or more psychological basic necessary capacities [69, 207, 228, 232].

Delinquency and violence are distinct and more complex problems in relationship to the diagnosis of personality disorders. Psychopathic individuals fail to respond with behaviors suggesting that they are able to discriminate conventional rules from moral rules, which is a cognitive skill possessed even by preschool children. They are also less likely to learn from negative feedback while they manifest attenuated autonomic responses to stress and punishment [216, 238–243]. As characteristically is said, "they know, but don't care" concerning moral rules [216]. There is a great debate with some authors suggesting that psychopathic individuals have such an impairment that they should not be considered legally responsive for their deeds [244–246], while on the contrary, other authors consider them to be fully responsible before the law [247, 248]. This is not captured by contemporary classification systems [249]. One of the consequences is that in the UK the compulsory admission to the hospital of people under the UK Mental Health Act exposes the magnitude of the problem because for reasons of public safety, and without the presence of guilt or even a committed crime, persons suffering from a mental disorder and high risk for committing acts of violence or crimes are detained [187, 250].

A significant minority of persons with personality disorders, especially those with psychopathic or narcissistic features, might be more than average successful professionally and with higher intelligence. It is not the rule, but the "exception that proves the rule." The grey field of personality disorders constitutes a trap for psychiatry; it could well attract the discipline to deal with psychosocial issues and extremes of normal behavior, rather than mental disorders per se.

The example of psychopathy exposes the problem of the vice-mental disorder relationship and the problems it poses, both theoretical and practical, concerning the boundaries and the nature of mental disorders. Currently there is no reliable solution to this problem.

The Concept of "Mental Disorder" in a Wider Sociocultural and Politico-Legal Frame

The diagnosis of mental disorder or, on the contrary, the certification of mental health has serious consequences affecting all aspects of the life of the individual. Among other things, diagnosis gives access to expensive specialized treatment and care within the health system, as well as to benefits. It also protects from condemnation; the individual suffers from some kind of medical condition, and he/she is not lazy, rude, bad, or immoral. However, on the other hand, it could result in stigma, social exclusion, and infringement of rights, and a double burden could be imposed on the individual: that of both the moral condemnation and the stigma of mental illness [58, 251, 252].

In spite of the efforts, the standardization of procedures, and the use of operationalized criteria, the variability of psychiatric diagnoses is still too big. This, and in combination with the lack of diagnostic biomedical laboratory examinations, made it the main focus of public criticism [85]. In the lay culture but also for many scholars, mental states are subjective, and therefore it is impossible to capture their essence though mere observation especially structured. Following this line of thinking, subjective experiences are more specific phenomena to establish diagnosis, and thus therapy should be individualized. Subsequently, it has been argued that nosographic diagnostic categorization is inadequate in principle [102], but there is no data to support such an assumption. In spite of claims that "understanding" is essential for the diagnosis and treatment [90], all data are against such an assumption. All the existing data are based on the nosographic categorization for a number of reasons which are discussed in Chaps. 13, 17, and 18.

Although several authors argue concerning the presence of culturally specific elements in mental phenomena [97, 253, 254], this has not been proven empirically, and the literature is rather in support of the universality of mental disorder manifestations, with only a few culturally bound syndromes [253–256]. Even the celebrated better outcome of schizophrenia in developing countries in comparison to developed ones [257–259] has not been eventually confirmed [260–263].

Many psychiatrists believe that the underlying biological processes determine the "form" of symptoms which are common between cultures, and the personal and cultural factors determine the "content" of symptoms. Although many authors are in favor of this compromise, such an approach does not support a significant role for sociocultural factors in the etiopathogenesis of mental disorder or its definition, let alone its treatment.

However, along this line of thought, George Engel (1913–1999) proposed the biopsychosocial model (Fig. 13.5) in an attempt to bring together the biological, social, and psychological elements of etiology and treatment of mental disorders in a frame analogous to general systems theory [264–267]. Although this model enjoyed much popularity, the effect of specific factors cannot be proven, or, instead, they were proven to be irrelevant in specific disorders. It constitutes a politically correct attempt to reconcile opposing and often conflicting approaches to psychiatry, and it is reasonable and in accord with lay opinion concerning mental illness but lacks scientific support. It is also incapable to incorporate or to resolve moral dilemma [268]. It is both surprising and also disappointing that this model is still widely discussed and taught in medical students and residents in its original formulation. Even worse, in spite of its poor performance in scientific terms, recently it has been expanded to include also spirituality in a four-element model [269–271]. On second thought, this "expansion" might reflect the decay of the model instead of its flourishing.

The core elaboration of this model [272] proposes three ways to conceive mental disorder:

- The first is deficit-based and argues that psychopathology emerges as the result of selective and minor neurobiological dysfunction. This dysfunction is not an all-or-nothing phenomenon and probably includes various degrees of spreading as well as of severity [273]. The "disruption of intentionality in behavior" as a characteristic of mental disorders [205] should also be considered in the frame of this deficit model.
- The second is on the basis of input-based models according to whom psychopathology is the result of problematic input to otherwise normal neurobiological functioning of the brain. These models include behavioral and learning theories. On the basis of this, and with widening the model to include not only over-

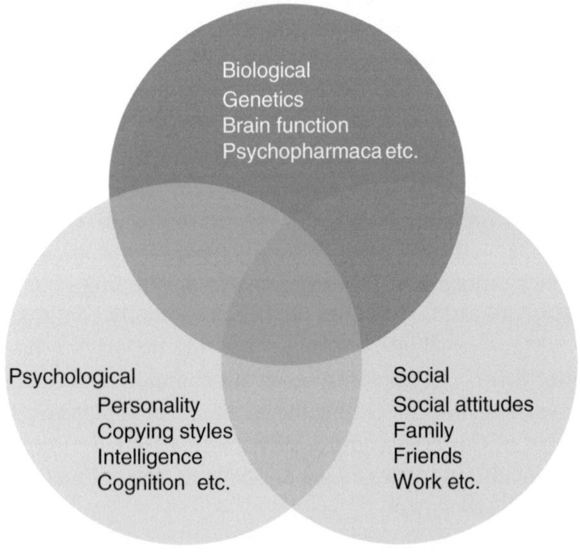

Fig. 13.5 Graphic representation of the biopsychosocial model proposed by George Engel (1913–1999)

whelming input but also deficient filters at the first pass (in the periphery or at early stages of processing), several theories have been proposed concerning psychotic symptoms as the consequence of the effort of the brain to make sense out of bizarre or chaotic sensory input [274–277]. The problem with all those theories is that they are neither in accord with clinical pharmaceutical research [278] nor with the finding that somehow neurocognitive decline is inversely correlated with positive symptoms [279].

- The third approach includes the so-called direct pathology models according to which research focuses on the modeling of the psychopathological condition itself rather than seeking the answers on the basis of knowledge from normal functioning or environmental input.

The above three approaches are rarely seen in isolation; the most usual picture is a hybrid of them in various admixtures.

The manifestation of mental disorder might depend on the environment, although this is the exception rather than the rule, with rates of mental disorders being similar across countries and cultures. One important example of a mental disorder which manifests itself depending on the circumstances and the environment are specific phobias. Phobias are very frequent in the general population but are perceived as problematic only when they interfere with the well-being or the functioning of the individual. In many occasions the person might not have the slightest idea that he/she carries such a phobia. For example, while a farmer in an agricultural village might not even imagine he/she might suffer from claustrophobia or fear of heights, such phobias could be extremely disabling for people living in large cities and need to take the elevator to go to their work in a skyscraper. Vice versa, an urban professional might not know that he/she is afraid of snakes or specific animals, but a farmer with such a phobia will find it difficult to work. Many people are afraid of driving (vehophobia); this is no real problem when one lives in the center of a big city with convenient mass transportation, but it could be a significant issue when living in rural areas.

It is important to briefly discuss a number of issues pertaining to the concepts of lay people as well as of intellectuals outside the fields of medicine, psychiatry, neuroscience, and psychology. A prevailing lay belief is that mental illness is subject to personal choice, and this is supported directly or indirectly by various theories including theories within psychology and dynamic psychiatry. In the realm of philosophy and in various politico-ideological groups, this belief is reflected in discussions pertaining to the relationship of mental illness and human free will, while in religious discussions, the issue is discussed with reference to morality. For overt psychotic disorders with grossly disordered behavior, there is little discussion on whether the person has any control over his/her own behavior. The issue is not a purely theoretical, since it has at least one significant practical consequence: it determines the way the law and the state will deal with the behavior of persons suffering from a mental disorder.

Almost two centuries ago, a new diagnostic category appeared in the American south; it was "drapetomania," and it was supposedly a mental disorder from which escaping slaves were specifically suffering [280, 281]. This clearly pseudo-scientific opinion was simply an expression of racism and was ridiculed in the American north. Samuel Cartwright (1793–1863) went further and argued that the cause of this "mental illness" was that some slave owners were treating slaves almost as equals.

In the past, social norms determined much of psychiatric nosology. This is no longer a fact concerning academic psychiatry, although social biases might influence the diagnostic approach of the individual psychiatrist. One should always have in mind that only recently with DSM-III, homosexuality stopped being considered to be a mental disorder [282]. Previously, the place of homosexuality among mental disorders had been dictated by erroneous psychoanalytical theories, social pressure, and lay beliefs and attitudes. On the other hand, the change reflects changes in policy and public opinion produced by the gay movement and the improvement in civil and personal rights rather than changes in the psychiatric literature which has never documented any relationship of sexual orientation or preferences per se and mental disorder, in spite of some psychoanalytic ideas [283]. Interestingly, the ridiculous concept of "schizophrenogenic mother," developed by Frieda Fromm-Reichmann (1889–1957) and Gregory Bateson (1904–1980) [284, 285] which put the blame on families and especially on mothers in a tragic and unfair way and caused severe problems and burden without having any scientific support at all [286–288], never came under the fire of any activist group.

A very important and unique feature of mental disorder is that often it is accompanied by complete lack of insight, and this results in profoundly negative consequences concerning the health but also the life of the individual as well as of others. In such cases, involuntary treatment might deem necessary, but there is much controversy. Involuntary treatment is ordered by the court according to the recommendation by the treating physicians, but it prerequisites the "presence of a mental disorder" as well as the establishment of the presence of a dangerous to self or others "lack of insight." In this frame, delinquency and violence in persons with personality disorders pose many ethical and legal dilemmas. In the UK the compulsory admission to the hospital of people under the UK Mental Health Act exposes the magnitude of the problem because for reasons of public safety, and without the presence of guilt or even a committed crime, persons suffering from a mental disorder and high risk for committing acts of violence or crimes are detained [187, 250].

As with most medical questions, the answer is neither simple nor straightforward, and it is definitely impossible for the lay person to understand completely. Most of intellectuals without a fair background in neuroscience or psychiatry also find it difficult to understand. Often the discussion becomes polarized with leaders and followers, instead for scholars and arguments. In the absence of a psychotic disorder for overt criminal behaviors, there is little disagreement that they are not the focus primarily of psychiatry [214–216]. On the other hand, through the ages, society has developed various institutions in order to control deviant behavior and to ensure safety and security, and asylums were among them [289–294].

At the end of the day, it is clear that the presence of mental disorder is not a matter of choice although its etiopathogenesis remains elusive. However, in cases other than psychosis, the behavior and especially those behaviors punished by law should be considered as being under the control of the individual, although in specific instances (e.g., impulse control disorders), resisting a specific behavior could be the source of significant distress. Intellectuals discussing the free will issue should have in mind that neurobiological research so far suggests that human free will is not an "all or none" or "black and white" phenomenon; instead it could manifest itself with various degrees of freedom and restrictions, often depending on the environment, the psychological state of the mind, or the physiological status of the body. Depending on the specific diagnosis, mental disorders are characterized by fewer freedom degrees and more restrictions in comparison to mental health (see also Chap. 16).

It is important to note that diagnosis of any kind both somatic and mental is not part of the person's self or identity. Identity politics play a role since mental illness tends to give a core feature in the patients' sense of identity [295, 296]. However, there are no "diabetic," "obese," or "schizophrenic" people or patients but rather patients or persons suffering from diabetes, obesity, or schizophrenia. This is called the "people-first language," and it was first appeared in 1988 as recommended by advocacy groups in the USA. The rationale behind this is that language use significantly shapes perceptions of the world and forms ideological preconceptions. Critics of this practice argue that this is simply a euphemism which inappropriately separates the person from one of its major features, and in some instances, it might stress the disability instead of diminishing it [297].

Last but not least, and having said a lot concerning the difficult problem of personality disorder, one should at least make a comment on President Trump's behavior and the criticism from scholars, from the media, and also by psychiatrists. The question here is very delicate and implications deeper than one might think. Disagreeing and not voting for Donald Trump is one thing. Criticizing its political attitude and actions, especially in relation to the storming of the Capitol on January 6, 2021, is not only legitimate but also absolutely necessary in order to protect democracy. However, putting him under a specific clinical diagnosis without clinically examining him is quite different. Interestingly, narcissistic personality seems to be the diagnosis put forward by many, while neurocognitive deterioration does not seem to have many supporters [298]. Then the gross methodological question is, how can you put such a diagnosis of personality disorder, which demands lifelong history and very early age at onset, accompanied by dysfunction or subjective dysphoria, to a person which by standard norms has superior performance and even managed to become President of the USA? Criteria (2) and (3) are not satisfied. But even if for a moment we accept such a diagnosis, what does it mean for civil rights and the politics in general? The problem is not whether Donald Trump suffers from a personality disorder. The problem is whether such a diagnosis has political significance. Should mental patients be prohibited to run for offices? The discussion is again merging morality with mental illness, and probably psychiatrists are caught in a trap of lay discussion but with psychiatric terms.

Several presidents of the USA were suffering from a mental disorder. Abraham Lincoln (1809–1865) had depressive episodes, Lyndon Johnson (1908–1973) was probably suffering from bipolar disorder, and John F. Kennedy and Bill Clinton manifested unusual hedonic tendencies. Many authors argue that Richard Nixon (1913–1994) was a narcissistic sociopath, while Ronald Reagan (1911–2004) suffered from Alzheimer's disease through much of his term and Woodrow Wilson (1856–1924) from vascular dementia. Several other leaders of other countries faced similar health problems, including Mahatma Gandhi (1869–1948) who was suffering from episodes of depression and had attempted suicide at a very young age, Sir Winston Churchill (1874–1965) who was probably suffering from a kind of bipolar type II illness with a heavy family history (including the suicide of his daughter), the USSR Secretary General Leonid Ilyich Brezhnev (1906–1982) who was suffering from alcohol and tranquilizer pills abuse and was demented during the last few years of his administration, and the Prime Minister of Greece Andreas Papandreou (1919–1996) who was suffering from bipolar disorder and was treated with lithium for years.

To add a personal opinion on the matter, a diagnosis of personality disorder could not be technically valid. There is probably an issue of values, at least to the extent someone disapproves the lifestyle which Donald Trump has adopted since younger age. He was a popular flamboyant celebrity, not the Mother Teresa type. Much of this lifestyle is probably a well-wrapped commercial product, and much of his behavior is promoting and selling that product. This path has made him President of the USA. It is unlikely that his personality traits are not in accord with this lifestyle; such an accord guarantees success, but still much of his behavior is probably deliberate and calculated. On the other hand, the senility of Mr. Trump is the probable key element, and an underlying vascular pathology could give rise to unusual behaviors, maybe by magnifying or setting out of control pre-existing personality traits, probably extreme. This is why he failed to control the situation and crossed the lines in such a way that made him lose re-election. Such a diagnosis could fulfill all the requirements for mental disorder, is largely free of value and moral load, and would be similar to the case of Woodrow Wilson. Still, one should have in mind that many such leaders failed to control their behavior and crossed similar lines; no need for vascular pathology to do that. The difference with Trump is the seemingly impulsive way of much of his behavior.

The case of President Trump is an excellent example of the dilemma psychiatry faces when it comes to diagnosis and society faces when it comes to dealing with problematic behaviors and mental illness, in a value-loaded background. Morality and values are pressing psychiatry to use its tools to solve the riddle. History teaches that solutions were never easy or straightforward and they always came at a price. On the other hand and from the reverse point of view, in the case of Donald Trump, a diagnosis of mental illness was used as a political weapon to defame him.

Is a Universal Definition of Mental Disorder Possible?

As said above, the WHO focuses the discussion on mental health and disability with mental health defined as a state of complete physical, mental, and social well-being and not merely the absence of disease or infirmity, with the person able to develop his/her own potential, with adequate coping, work productivity, and social life [182, 183]. The EPA defines mental health in a similar way to the WHO [193]. On the other hand, the APA puts the emphasis on the presence of distress/impairment which constitute the cardinal characteristic of individuals seeking professional help distress/impairment, focuses on the distinction between health and disease, and suggests that mental disorder is distinct from normal distress and cultural/subcultural way of behaving and is not determined in principle by the presence of conflict with society [186–192].

It is important to note that according to all the above definitions, the mere presence of symptoms alone does not generally justify by itself for the diagnosis of a mental disorder, at least officially. It is solidly proven that those asking for professional help do this because of the presence of distress and personalizing appraisals as well as global impairment [197, 199–201]. This, on the one hand, is convenient because it can be used as criterion to define mental disorder but on the other hand is both defensive and misleading and could eventually mean that mental disorder is a matter of choice.

So is there a way, in the absence of biological markers, to define mental disorder in a universal way, in such a manner that all mental disorders would fulfill the criteria of the definition (high sensitivity), but no other condition would do (high specificity)? Currently there are three major pillars to base such a definition: (a) symptoms and signs, (b) dysphoria, and (c) impairment and disability:

(a) Symptoms and signs: The major problem here is that of a vicious cycle of reasoning. Symptoms (from the Greek word *συμπίπτω* meaning tendency to co-occur) and signs (from the Latin word *signum* meaning to mark) are defined as manifestation of a medical disorder or of a non-pathological alteration in the body (e.g., pregnancy). All symptoms and signs are frequently but not always related to an underlying disorder. For example, coughing is a symptom related to a number of diseases, but statistically it is a behavior frequently observed without any medical condition present. Thus, as it is the rule in all medicine, the mere manifestation of "symptoms" or "signs" is not a definite clue for the presence of a mental disorder. However, in lack of pathognomonic or supportive biological markers, the presence of symptoms or signs is a necessary but not sufficient condition for the diagnosis of any mental disorder.
(b) Dysphoria: Although it looks like a symptom itself, dysphoria (from Greek *δυσφορία* meaning I carry a bad feeling) is characterized by a negative condition which makes the person to suffer and eventually to seek help. Not all symptoms are always related to suffering, but dysphoria does. Thus, the presence of dysphoria also fulfills the criterion for the presence of symptoms and/or signs. It should be mentioned that not all pathological mental conditions are associated with dysphoria. Some are neutral, while others, like euphoric mania, are clearly characterized by a pleasurable feeling.
(c) Impairment and/or disability: While somatic disability is more or less obvious, impairment and dysfunction as a result of mental disorder are often difficult to define and to diagnose. When it concerns easily observed and more or less value-free areas, e.g., work, it might be easier to assess its magnitude. However,

when it comes to value-loaded areas, e.g., interpersonal relationships and behaviors, then things might become extremely difficult and complex. Such is the case of narcissistic personality disorder where the impairment is isolated in interpersonal relationships with observed output in other areas, e.g., work being unaffected or even higher than average. Also, very frequently when an impairment in these areas is observed, it is secondary to the collapse of interpersonal relationships. Another special case is that mild hypermania in the frame of bipolar-II disorder which might manifest increased mental efficacy and increased functioning at least for short periods of time. This is not to be taken as an example of generalized dispute of the basis fact that mental disorder leads to disability. Some authors take this argument too far [272], and this kind of over-generalization is not based on data but on lack of current knowledge on bipolar illness. At the end of the day, even mild hypomania leads to at least some kind of global impairment, in spite of a boost in certain isolated areas which nevertheless draw the attention.

On the basis of the above three criteria, one could easily argue that a mental disorder should fulfill two of them, and in this way, it could be sufficiently defined. At least one symptom will always be present (either a or b criteria). In those mental disorders with difficult definitions (as it is with narcissistic personality or bipolar-II disorder), a careful consideration of the situation would be important. Theoretically, and in a value-free approach, narcissistic personality cannot be considered on the basis of the distress it might cause to other people, as long as this has no impact on the person itself in terms of suffering or observed impairment in the long term. Long-term observation is an important element here. But in case criteria (b) and (c) are not fulfilled, then it is probably an issue of morality but hardly of medicine, and this also leads to the assumption that neither criterion (a) would apply. This does not mean that morality plays any role in the diagnosis or the treatment of mental disorders. On the contrary it means that this is an "either mental disorder or moral issue" situation.

An interesting question arises when only criterion (c) is fulfilled (not because of a somatic disorder) but neither (a) nor (b). In such cases, either the diagnosis of no somatic disorder is present should be challenged, or the presence subtle symptoms suggesting an abnormal psychological functioning (e.g., indifference, apathy, neglect, etc.) should be pursuit. However, one should have always in mind that technically, (a) criterion is either included or implied by (b) and (c). This might lead to the conclusion that the presence of either criterion (b) or (c) could be sufficient for the definition and diagnosis of mental disorder. This theoretically could be the case, but in practice criterion (a) and the demand for two not only one criterion to be fulfilled could serve as a safety mechanism to avoid over-diagnosis.

This approach is in accord with the assumption that severe mental disorder is always characterized by significant distress or impairment of personal functioning [3, 127, 204] accompanied by a pervasive inability of the person to react appropriately and to adjust to common social, occupational, and personal challenges. In a philosophical-cognitive view, there is some kind of severe disruption of "intentionality" as a core feature of expressed behavior [205].

On the contrary, the "health pole" is characterized by the opposite features, but the crucial element is not the presence of a state of unchallenged happiness. This is a misunderstanding, emerging from happiness as the core meaning and aim of life according to Aristotle and the philosophers of the Enlightenment [206]. According to them, "happiness" is quite different from "hedonism" (from the Greek word *ηδονή* meaning the lust or pleasure). In fact, happiness was considered to be the aim of life, mainly in order to downgrade the role of religious-based meanings in life. Happiness was a more pragmatic as well as humanistic proposal, but the meaning of happiness remained open to debate, leaving space for religious as well as secular interpretations. For many, happiness constituted a basic human right, and society started changing towards a structure that could allow pursuit of happiness. Change was radical; not only the seeking of happiness became perfectly legitimate, but on the other hand not being happy was now considered to be a problem in need for a solution. Further on this path, the concept of "happiness" would be later utilized in several efforts to define mental disorder.

It should be noted however that often the diagnosis of all three criteria (a, b, and c) is marginal, that is, you cannot be absolutely certain that each of them is fulfilled. So in such a situation, e.g., you have a person which is sad but not sure whether this is definitely disproportional after taking into consideration what happens in his/her life and although the person finds it difficult to work and feels miserable because of this, his objective productivity has not been affected. At this grey zone, it is impossible to distinguish between sub-threshold, attenuated, or prodromal forms of mental disorders and normal reactions to environmental situations. This leads to under- or over-diagnosis depending on the psychiatrist and the individual case in a rather unpredictable way. Interestingly, more concern exists concerning over-diagnosis and inappropriate medicalization than under-diagnosis [299–301].

After having completed the writing of this chapter, and while I was doing the final corrections, my eye was caught by an article by Alison Escalante in the *Forbes* [302]. It was uploaded in August 11, 2020, and within a week it had almost 400k views. It discusses an article recently published in the *American Journal of Physical Anthropology* [303] which questions whether mental disorders are medical disorders, by using "evolutionary psychiatry" as a vehicle. This approach is not new and had been already extensively discussed [110]. Neither the fact that all the arguments in that article are already answered in the literature, none of them is new and all are naïve, trivial (and at points the style is definitely "accusatory"), nor the obvious low competency of the authors in issues pertaining to mental health (and probably to health issues in general) did not preclude that article to hit the headlines in a prestigious magazine like the *Forbes*. An article with the opposite arguments and conclusions would never have done it, and we have seen this effect again and again [304]. This is strong proof of the desperate need our society has to find arguments and maybe any kind of argument, no matter its validity, that could help society deny the very existence of mental illness.

References

1. Kenny AJP. Mental health in Plato's Republic. Proc Br Acad. 1969;55:229–53.
2. Seeskin K. Plato and the origin of mental health. Int J Law Psychiatry. 2008;31(6):487–94. https://doi.org/10.1016/j.ijlp.2008.09.004.
3. Stein DJ, Phillips KA, Bolton D, Fulford KW, Sadler JZ, Kendler KS. What is a mental/psychiatric disorder? From DSM-IV to DSM-V. Psychol Med. 2010;40(11):1759–65. https://doi.org/10.1017/S0033291709992261.
4. MacIntyre A. A short history of ethics. South Bend, IN: University of Notre Dame Press; 1998.
5. Sidgwick H. Outlines of the history of ethics for English readers. London: MacMillan; 1888.
6. Newhauser RE. The seven deadly sins: from communities to individuals (Studies in medieval and reformation traditions, volume 123). Leiden: Brill; 2007.
7. Menninger K. Changing concepts of disease. Ann Intern Med. 1948;29(2):318–25. https://doi.org/10.7326/0003-4819-29-2-318.
8. Menninger K. The contribution of psychoanalysis to American psychiatry. Bull Menninger Clin. 1954;18(3):85–91.
9. Menninger K. The vital balance: the life process in mental health and illness. New York: Viking Penguin; 1963a.
10. Menninger KA. Freud and American psychiatry. J Am Psychoanal Assoc. 1956;4(4):614–25. https://doi.org/10.1177/000306515600400403.
11. Menninger KA. The vital balance: the life process in mental health and illness. New York: Penguin Books; 1963 (1977).
12. Berk H, Jaso H. Memorantum; 1976.
13. Lieberman JA. Shrinks: the untold story of psychiatry. New York: Back Bay Books; 2016.
14. Alarcon RD. Culture, cultural factors and psychiatric diagnosis: review and projections. World Psychiatry. 2009;8(3):131–9. https://doi.org/10.1002/j.2051-5545.2009.tb00233.x.
15. Kleinman A. Anthropology and psychiatry. The role of culture in cross-cultural research on illness. Br J Psychiatry. 1987;151:447–54. https://doi.org/10.1192/bjp.151.4.447.
16. Littlewood R. From categories to contexts: a decade of the 'new cross-cultural psychiatry'. Br J Psychiatry. 1990;156:308–27. https://doi.org/10.1192/bjp.156.3.308.
17. Maj M. Bereavement-related depression in the DSM-5 and ICD-11. World Psychiatry. 2012;11(1):1–2. https://doi.org/10.1016/j.wpsyc.2012.01.001.
18. Belsky J, Jonassaint C, Pluess M, Stanton M, Brummett B, Williams R. Vulnerability genes or plasticity genes? Mol Psychiatry. 2009;14(8):746–54. https://doi.org/10.1038/mp.2009.44.
19. Rutter M, Moffitt TE, Caspi A. Gene-environment interplay and psychopathology: multiple varieties but real effects. J Child Psychol Psychiatry. 2006;47(3–4):226–61. https://doi.org/10.1111/j.1469-7610.2005.01557.x.
20. Ruse M. Entry for "reductionism". In: Honderich T, editor. The Oxford companion to philosophy. 2nd ed. Oxford: Oxford University Press; 2005.
21. Carnap R. The logical structure of the world and pseudoproblems in philosophy. Berkeley, CA: University of California Press; 1928/1967.
22. Neurath O. Sociology and physicalism. In: Ayer A, editor. Logical positivism. New York: The Free Press; 1931/1959.
23. Neurath O. Philosophical papers 1913-1946. Dordrecht: Reidel; 1983.
24. Blackburn S. The Oxford dictionary of philosophy. Oxford: Oxford University Press; 2008. https://doi.org/10.1093/acref/9780199541430.001.0001.
25. Schaffner K. Reduction and reductionism in psychiatry. In: Fulford KWM, Davies M, Gipps GT, et al., editors. The Oxford handbook of philosophy and psychiatry. Oxford: Oxford University Press; 2013. https://doi.org/10.1093/oxfordhb/9780199579563.001.0001.
26. Nagel T. Mind and cosmos: why the materialist neo-Darwinian conception of nature is almost certainly false. Oxford: Oxford University Press; 2012.
27. Godfrey-Smith P. Philosophy of biology. Princeton: Princeton University Press; 2013.

28. Murphy N. Reductionism and emergence. A critical perspective. In: Murphy N, Knight C, editors. In human identity at the intersection of science, technology and religion. Burlington, VT: Ashgate; 2010.
29. Kauffman S. Beyond reductionism: reinventing the sacred. Edge; 2006.
30. Longo G, Montévil M, Kauffman S. No entailing laws, but enablement in the evolution of the biosphere. In: Proceedings of the 14th annual conference companion on genetic and evolutionary computation. GECCO'12, New York, NY, USA; 2012. ACM. p. 1379–92.
31. Ulanowicz R. Ecology: the ascendant perspective. New York: Columbia University Press; 1997.
32. Ulanowicz RE. Ecosystem development: symmetry arising. Symmetry Cult Sci. 1996;7(3):321–34.
33. Silberstein M, McGeever J. The search for ontological emergence. Philos Quart. 1999;49(195):182–200.
34. Ruse M. Do organisms exist? Am Zool. 1989;29(3):1061–6.
35. Scott A. Reductionism revisited. J Conscious Stud. 2004;11(2):51–68.
36. Clayton P, Davies P. The re-emergence of emergence: the emergentist hypothesis from science to religion. New York: Oxford University Press; 2006.
37. Jørgensen S. Integration of ecosystem theories: a pattern. 3rd ed. Dordrecht: Kluwer Academic Publishers; 2002.
38. Anderson PW. More is different. Science. 1972;177(4047):393–6. https://doi.org/10.1126/science.177.4047.393.
39. Dawkins R. The Blind Watchmaker: why the evidence of evolution reveals a universe without design. New York: W. W. Norton; 2015.
40. Dawkins R. The selfish gene, Oxford Landmark Science. 4th ed. Oxford University Press; 2016.
41. Kukla A. Antirealist explanations of the success of science. Philos Sci. 1996;63(3):S298–305. https://doi.org/10.1086/289964.
42. Pope ML. Personal construction of formal knowledge. Interchange. 1982;13(4):3–14. https://doi.org/10.1007/Bf01191417.
43. Andersen P, Emmeche C, Finnemann N, Christiansen P. Downward causation: minds, bodies and matter. Aarhus: Aarhus University Press; 2001.
44. Campbell DT. 'Downward causation' in hierarchically organized biological systems. In: Ayala F, Dobzhansky T, editors. Studies in the philosophy of biology. Stuttgart: Macmillan; 1974. p. 179–86.
45. Campbell DT. Levels of organization, downward causation, and the selection-theory approach to evolutionary epistemology. In: Tobach E, Greenberg G, editors. Scientific methodology in the study of mind: evolutionary epistemology. Hillsdale, NJ: Erlbaum; 1990. p. 1–17.
46. Carroll J. Anti-reductionism. In: Beebee H, Hitchcock C, Menzies P, editors. The Oxford handbook of causation. Oxford handbooks online. New York: University Press; 2009.
47. Dossey L. Reinventing medicine: beyond mind-body to a new era of healing. San Francisco: Harper San Francisco; 1999.
48. Hulswit M. How causal is downward causation? J Gen Philos Sci. 2005;36(2):261–87.
49. Schaffner KF. Theory structure in the biomedical sciences. J Med Philos. 1980;5(1):57–97. https://doi.org/10.1093/jmp/5.1.57.
50. Schaffner KF. Exemplar reasoning about biological models and diseases: a relation between the philosophy of medicine and philosophy of science. J Med Philos. 1986;11(1):63–80. https://doi.org/10.1093/jmp/11.1.63.
51. Schaffner KF. Discovery and explanation in biology and medicine, vol. 1. Chicago: University of Chicago Press; 1993.
52. Schaffner KF. Reduction: the Cheshire cat problem and a return to roots. Synthese. 2006;151(3):377–402. https://doi.org/10.1007/s11229-006-9031-2.
53. Schaffner KF. Ernest Nagel and reduction. J Philos. 2012;109(8–9):534–65. https://doi.org/10.5840/jphil20121098/926.

54. Cantor N, Smith EE, French RS, Mezzich J. Psychiatric diagnosis as prototype categorization. J Abnorm Psychol. 1980;89(2):181–93. https://doi.org/10.1037//0021-843x.89.2.181.
55. Frances A, Pincus HA, Widiger TA, Davis WW, First MB. DSM-IV: work in progress. Am J Psychiatry. 1990;147(11):1439–48. https://doi.org/10.1176/ajp.147.11.1439.
56. Lilienfeld SO, Marino L. Mental disorder as a Roschian concept: a critique of Wakefield's "harmful dysfunction" analysis. J Abnorm Psychol. 1995;104(3):411–20. https://doi.org/10.1037//0021-843x.104.3.411.
57. Lilienfeld SO, Marino L. Essentialism revisited: evolutionary theory and the concept of mental disorder. J Abnorm Psychol. 1999;108(3):400–11. https://doi.org/10.1037//0021-843x.108.3.400.
58. Kingma E. Naturalistic accounts of mentla disorder. In: Fulford KWM, Davies M, Gipps GT, et al., editors. The Oxford handbook of philosophy and psychiatry. Oxford: Oxford University Press; 2013. https://doi.org/10.1093/oxfordhb/9780199579563.001.0001.
59. Bolton D, Hill J. Mind, meaning, and mental disorder: the nature of causal explanation in psychology and psychiatry. New York, NY: Oxford University Press; 1996.
60. Fulford KWM, Thornton T, Graham G. Karl Jaspers and general psychopathology. In: Fulford KWM, Thornton T, Graham G, editors. The Oxford textbook of philosophy and psychiatry. Oxford: Oxford University Press; 2006b.
61. Kendler KS, Campbell J. Expanding the domain of the understandable in psychiatric illness: an updating of the Jasperian framework of explanation and understanding. Psychol Med. 2014;44(1):1–7. https://doi.org/10.1017/S0033291712003030.
62. Jacob F. Evolution and tinkering. Science. 1977;196(4295):1161–6. https://doi.org/10.1126/science.860134.
63. Kendell RE, Cooper JE, Gourlay AJ, Copeland JR, Sharpe L, Gurland BJ. Diagnostic criteria of American and British psychiatrists. Arch Gen Psychiatry. 1971;25(2):123–30. https://doi.org/10.1001/archpsyc.1971.01750140027006.
64. Rosenhan DL. On being sane in insane places. Science. 1973;179(4070):250–8. https://doi.org/10.1126/science.179.4070.250.
65. Slater L. Opening skinner's box. London: Bloomsbury; 2004.
66. Spitzer RL, Lilienfeld SO, Miller MB. Rosenhan revisited: the scientific credibility of Lauren Slater's pseudopatient diagnosis study. J Nerv Ment Dis. 2005;193(11):734–9. https://doi.org/10.1097/01.nmd.0000185992.16053.5c.
67. Zimmerman M. Pseudopatient or pseudoscience: a reviewer's perspective. J Nerv Ment Dis. 2005;193(11):740–2. https://doi.org/10.1097/01.nmd.0000186016.78212.d8.
68. Slater L. Reply to Spitzer and colleagues. J Nerv Ment Dis. 2005;193(11):743–4. https://doi.org/10.1097/01.nmd.0000185883.10503.98.
69. Zachar P. Alternative perspectives on psychiatric validation: DSM, ICD, RDoC, and beyond; 2015.
70. Popper K. The logic of scientific discovery. New York, NY: Basic Books; 1959.
71. Haslam N. Reliability, validity and the mixed blessings of operationalism. In: Fulford KWM, Davies M, Gipps GT, et al., editors. The Oxford handbook of philosophy and psychiatry. Oxford: Oxford University Press; 2013. https://doi.org/10.1093/oxfordhb/9780199579563.001.0001.
72. Fulford KWM, Thornton T, Graham G. Natural classification, realism and psychiatric science. In: Fulford KWM, Thornton T, Graham G, editors. The Oxford textbook of philosophy and psychiatry. Oxford: Oxford University Press; 2006c.
73. Cloninger CR. Establishment of diagnostic validity in psychiatric illness: Robins and Guze's method revisited. In: Robins L, Barrett J, editors. The validity of psychiatric diagnosis. New York, NY: Raven Press; 1989. p. 9–18.
74. Cronbach LJ, Meehl PE. Construct validity in psychological tests. Psychol Bull. 1955;52(4):281–302. https://doi.org/10.1037/h0040957.
75. Robins E, Guze SB. Establishment of diagnostic validity in psychiatric illness: its application to schizophrenia. Am J Psychiatry. 1970;126(7):983–7. https://doi.org/10.1176/ajp.126.7.983.

76. Kendell R. The role of diagnosis in psychiatry. Oxford: Blackwell Scientific Publications; 1975a.
77. Sadler JZ, Wiggins OP, Schwartz MA. Philosophical perspectives on psychiatric diagnostic classification. Baltimore, MD: Johns Hopkins University Press; 1994.
78. Hempel C. Aspects of scientific explanation and other essays in the philosophy of science, vol. 1. New York: The Free Press; 1965.
79. Fulford K, Sartorius N. A secret history of ICD and the hidden future of DSM. In: Broome M, Bortolotti L, editors. Psychiatry as cognitive neuroscience: philosophical perspectives. Oxford: Oxford University Press; 2009. p. 29–48.
80. Fulford KWM, Davies M, Gipps GT, Graham G, Sadler J, Stanghellini G, Thornton T. The next hundred years: watching our Ps and Q. In: Fulford KWM, Davies M, Gipps GT, et al., editors. The Oxford handbook of philosophy and psychiatry. Oxford: Oxford University Press; 2013. https://doi.org/10.1093/oxfordhb/9780199579563.001.0001.
81. Klerman GL. The evolution of a scientific nosology. In: Shershow J, editor. Schizophrenia: science and practice. Cambridge, MA: Harvard University Press; 1978. p. 99–121.
82. Aronowitz R. Framing disease: an underappreciated mechanism for the social patterning of health. Soc Sci Med. 2008;67(1):1–9. https://doi.org/10.1016/j.socscimed.2008.02.017.
83. Parnas J, Sass LA, Zahavi D. Recent developments in philosophy of psychopathology. Curr Opin Psychiatry. 2008;21(6):578–84. https://doi.org/10.1097/yco.0b013e32830e4610.
84. Potter NN. Recent developments in philosophy of mind and psychopathology. Curr Opin Psychiatry. 2010;23(6):542–5. https://doi.org/10.1097/YCO.0b013e32833d41ed.
85. Kirk S, Kutchins H. The selling of DSM: the rhetoric of science in psychiatry. New York: Aldine de Gruyter; 1992.
86. Mishler E. Research interviewing: context and narrative. Cambridge, MA: Harvard University Press; 1991.
87. Spitzer RL. Psychiatric diagnosis: are clinicians still necessary? Compr Psychiatry. 1983;24(5):399–411. https://doi.org/10.1016/0010-440x(83)90032-9.
88. Othmer E, Othmer S. The clinical interview using DSM-IV-TR: Vol 1: fundamentals. Arlington, VA: American Psychiatric Publishing; 2002.
89. Geerts E, Brüne M. On the role of ethology in clinical psychiatry: what do ontogenetic and causal factors tell us about ultimate explanations of depression? In: Adriaens P, De Block A, editors. Maladapting mindsphilosophy, psychiatry, and evolutionary theory. Oxford: Oxford University Press; 2013. https://doi.org/10.1093/med/9780199558667.001.0001.
90. Stanghellini G. Philosophical resources for the psychiatric intervies. In: Fulford KWM, Davies M, Gipps GT, et al., editors. The Oxford handbook of philosophy and psychiatry. Oxford: Oxford University Press; 2013. https://doi.org/10.1093/oxfordhb/9780199579563.001.0001.
91. Gadamer HG. Truth and method (Weinsheimer and D.G. Marshall, Trans.). 2nd ed. New York, NY: The Continuum Publishing Company; 2004.
92. Heidegger M. Being and time. New York: Harper Perennial; 2008.
93. Rorty R. Philosophy and the mirror of nature. Princeton, NJ: Princeton University Press; 1981.
94. Fulford KWM, Thornton T, Graham G. Philosophical methods in mental health practice and research. In: Fulford KWM, Thornton T, Graham G, editors. The Oxford textbook of philosophy and psychiatry. Oxford: Oxford University Press; 2006d.
95. Fulford KWM, Thornton T, Graham G. Values in psychiatric diagnosis. In: Fulford KWM, Thornton T, Graham G, editors. The Oxford textbook of philosophy and psychiatry. Oxford: Oxford University Press; 2006e.
96. Laudan L, editor. Mind and medicine: problems of explanation and evaluation in psychiatry and the biomedical sciences. Berkeley: University of California Press; 1983.
97. Sadler J. Values in psychiatric diagnosis and classification. In: Fulford KWM, Davies M, Gipps GT, et al., editors. The Oxford handbook of philosophy and psychiatry. Oxford: Oxford University Press; 2013a. https://doi.org/10.1093/oxfordhb/9780199579563.001.0001.
98. Stanghellini G. The grammar of the psychiatric interview. A plea for the second-person mode of understanding. Psychopathology. 2007;40(2):69–74. https://doi.org/10.1159/000098486.

99. Monti MR, Stanghellini G. Psychopathology: an edgeless razor? Compr Psychiatry. 1996;37(3):196–204. https://doi.org/10.1016/s0010-440x(96)90036-x.
100. Richardson J. Handbook of qualitative research methods for psychology and the social sciences. Chichester: Wiley-Blackwell; 1996.
101. McGuffin P, Farmer A. Polydiagnostic approaches to measuring and classifying psychopathology. Am J Med Genet. 2001;105(1):39–41.
102. Van Praag HM, Asnis GM, Kahn RS, Brown SL, Korn M, Friedman JM, Wetzler S. Nosological tunnel vision in biological psychiatry. A plea for a functional psychopathology. Ann N Y Acad Sci. 1990;600:501–10. https://doi.org/10.1111/j.1749-6632.1990.tb16905.x.
103. Smolik P. Validity of nosological classification. Dialogues Clin Neurosci. 1999;1(3):185–90.
104. Merriam-Webster Editorial Staff. Merriam-Webster's collegiate dictionary. 11th ed. Springfield, MA: Merriam-Webster; 2003.
105. Newman-Dorland W. Dorland's illustrated medical dictionary. 31st ed. Philadelphia, PA: Saunders; 2007.
106. Weiner E, Simpson J. The compact Oxford english dictionary. Oxford: Oxford University Press; 1991.
107. Kendell RE. The concept of disease and its implications for psychiatry. Br J Psychiatry. 1975b;127:305–15. https://doi.org/10.1192/bjp.127.4.305.
108. Shah P, Mountain D. The medical model is dead--long live the medical model. Br J Psychiatry. 2007;191:375–7. https://doi.org/10.1192/bjp.bp.107.037242.
109. De Block A, Adriaens P. Why philosophers of psychiatry should care about evolutionary theory. In: Adriaens P, De Block A, editors. Maladapting minds: philosophy, psychiatry, and evolutionary theory. Oxford: Oxford University Press; 2011.
110. Fountoulakis KN. Ethology, evolutionary psychology, sociobiology, and evolutionary psychiatry. In: Fountoulakis KN, Nimatoudis I, editors. Psychobiology of behaviour. Cham: Springer International Publishing; 2019. p. 345–95. https://doi.org/10.1007/978-3-030-18323-3_12.
111. Fulford KWM, Van Staden CW. Values-based practice: Topsy-Turvy take-home messages from ordinary language philosophy (and a few next steps). In: Fulford KWM, Davies M, Gipps GT, et al., editors. The Oxford handbook of philosophy and psychiatry. Oxford: Oxford University Press; 2013. https://doi.org/10.1093/oxfordhb/9780199579563.001.0001.
112. Austin JL. A plea for excuses: the presidential address. Proc Aristot Soc. 1956;57:1–30.
113. Ghaemi N. The concepts of psychiatry: a pluralistic approach to the mind and mental illness. Baltimore, MD: Johns Hopkins University Press; 2007.
114. Ghaemi N. On depression: drugs, diagnosis, and despair in the modern world. Baltimore, MD: Johns Hopkins University Press; 2013.
115. Ghaemi SN. Toward a hippocratic psychopharmacology. Can J Psychiatry. 2008;53(3):189–96. https://doi.org/10.1177/070674370805300309.
116. Graham G. The disordered mind. New York, NY: Routledge; 2010.
117. Kincaid H. Do we need theory to study disease? Lessons from cancer research and their implications for mental illness. Perspect Biol Med. 2008;51(3):367–78. https://doi.org/10.1353/pbm.0.0019.
118. Varga S. Naturalism, interpretation, and mental disorder. Oxford: Oxford University Press; 2015. https://doi.org/10.1093/med/9780198747253.001.0001.
119. Beebee H, Sabbarton-Leary N. Are psychiatric kinds real? Eur J Anal Philos. 2010;6(1):11–27.
120. Boyd R. What realism implies and what it does not. Dialectica. 1989;43(1–2):5–29. https://doi.org/10.1111/j.1746-8361.1989.tb00928.x.
121. Boyd R. Realism, anti-foundationalism and the enthusiasm for natural kinds. Philos Stud. 1991;61(1):127–48.
122. Kendler KS, Zachar P, Craver C. What kinds of things are psychiatric disorders? Psychol Med. 2011;41(6):1143–50. https://doi.org/10.1017/S0033291710001844.
123. Parnas J, Nordgaard J, Varga S. The concept of psychosis: a clinical and theoretical analysis. Clin Neuropsychiatry. 2010;7(2):32–7.
124. Samuels R. Delusion as a natural kind. In: Broome M, Bortolotti L, editors. Psychiatry as cognitive neuroscience: philosophical perspectives. Oxford: Oxford University Press; 2009. p. 49–80.

125. Rosch E. Principles of categorization. In: Rosch E, Lloyd B, editors. Cognition and categorization. Hillsdale, NJ: Lawrence Erlbaum; 1975. p. 27–48.
126. Wittgenstein L. Philosophical investigations. 4th ed. Hoboken: Wiley-Blackwell; 2009.
127. Stein DJ. What is a mental disorder? A perspective from cognitive-affective science. Can J Psychiatry. 2013;58(12):656–62. https://doi.org/10.1177/070674371305801202.
128. Kendler K, Parnas J, editors. Philosophical issues in psychiatry II: nosology. Oxford: Oxford University Press; 2013. https://doi.org/10.1093/med/9780199642205.001.0001.
129. Kendler K, Parnas J, editors. Philosophical issues in psychiatry III: the nature and sources of historical change. Oxford: Oxford University Press; 2015. https://doi.org/10.1093/med/9780199642205.001.0001.
130. Eysenck HJ. Classification and the problems of diagnosis. In: Eysenck HJ, editor. Handbook of abnormal psychology. London: Pitman Medical Publishing Company Ltd.; 1960.
131. Scheff TJ. The labelling theory of mental illness. Am Sociol Rev. 1974;39(3):444–52.
132. Sedgwick P. Illness--mental and otherwise. Stud Hastings Cent. 1973;1(3):19–40.
133. Foucault M. Histoire de la folie à l'âge classique. Paris: Plon; 1961.
134. Foucault M. Madness and civilization: a history of insanity in the age of reason. New York: Vintage Books; 1988.
135. Foucault M, Howard R, Cooper D. Madness and civilization. Milton Park: Taylor & Francis; 1961. https://doi.org/10.4324/9780203278796.
136. Campbell EJ, Scadding JG, Roberts RS. The concept of disease. Br Med J. 1979;2(6193):757–62. https://doi.org/10.1136/bmj.2.6193.757.
137. Laing R. The politics of experience. Harmondsworth: Penguin Books; 1967.
138. Szasz T. The myth of mental illness: foundations of a theory of personal conduct. New York: Hober-Harper; 1961.
139. Szasz TS. The myth of mental illness. New York: Harper and Row; 1974.
140. Keil G, Keuck L, Hauswald R, editors. Vagueness in psychiatry. Oxford: Oxford University Press; 2016. https://doi.org/10.1093/med/9780198722373.001.0001.
141. Boorse C. On the distinction between disease and illness. Philos Public Aff. 1975;5(1):49–68.
142. Boorse C. A second rebuttal on health. J Med Philos. 2014;39(6):683–724. https://doi.org/10.1093/jmp/jhu035.
143. Murphy D. The medical model and the philosophy of science. In: Fulford KWM, Davies M, Gipps GT, et al., editors. The Oxford handbook of philosophy and psychiatry. Oxford: Oxford University Press; 2013. https://doi.org/10.1093/oxfordhb/9780199579563.001.0001.
144. Roe K, Murphy D. Function, dysfunction, adaptation? In: Adriaens P, Block A, editors. Maladapted minds. Oxford: Oxford University Press; 2011. p. 216–37.
145. Bortolotti L. Delusions and other irrational beliefs. Oxford: Oxford University Press; 2009. https://doi.org/10.1093/med/9780199206162.001.1.
146. Wakefield JC. The concept of mental disorder. On the boundary between biological facts and social values. Am Psychol. 1992;47(3):373–88. https://doi.org/10.1037//0003-066x.47.3.373.
147. Wakefield JC. Limits of operationalization: a critique of Spitzer and Endicott's (1978) proposed operational criteria for mental disorder. J Abnorm Psychol. 1993;102(1):160–72. https://doi.org/10.1037//0021-843x.102.1.160.
148. Wakefield JC. The DSM's theory-neutral nosology is scientifically progressive: response to Follette and Houts (1996). J Consult Clin Psychol. 1998;66(5):846–52. https://doi.org/10.1037//0022-006x.66.5.846.
149. Wakefield JC, First MB. Diagnostic validity and the definition of mental disorder: a program for conceptually advancing psychiatry. Can J Psychiatry. 2013a;58(12):653–5. https://doi.org/10.1177/070674371305801201.
150. Wakefield JC, First MB. The importance and limits of harm in identifying mental disorder. Can J Psychiatry. 2013b;58(11):618–21. https://doi.org/10.1177/070674371305801107.
151. Wakefield JC, Schmitz MF, Baer JC. Did narrowing the major depression bereavement exclusion from DSM-III-R to DSM-IV increase validity?: evidence from the National Comorbidity Survey. J Nerv Ment Dis. 2011;199(2):66–73. https://doi.org/10.1097/NMD.0b013e31820840c5.

152. Wakefield JC, Schmitz MF, First MB, Horwitz AV. Extending the bereavement exclusion for major depression to other losses: evidence from the National Comorbidity Survey. Arch Gen Psychiatry. 2007;64(4):433–40. https://doi.org/10.1001/archpsyc.64.4.433.
153. Cosmides L, Tooby J. From evolution to behavior: evolutionary psychology as the missing link. In: Dupre J, editor. The latest on the best: essays on evolution and optimality. Cambridge, MA: MIT Press; 1987. p. 277–306.
154. Kitcher P. The lives to come: the genetic revolution and human possibilities. New York, NY: Touchstone; 1996.
155. Murphy D. Psychiatry in the scientific imagine. Cambridge, MA: MIT Press; 2006.
156. Murphy D, Stich S. Darwin in the madhouse: evolutionary psychology and the classification of mental disorders. In: Carruthers P, Chamberlain A, editors. Evolution and the human mind: modularity, language and meta-cognition. Cambridge: Cambridge University Press; 2000. p. 62–92.
157. Murphy D, Woolfolk RL. Conceptual analysis versus scientific understanding: an assessment of Wakefield's folk psychiatry. Philos Psychiatry Psychol. 2000;7(4):271–93.
158. Tooby J, Cosmides L. On the universality of human nature and the uniqueness of the individual: the role of genetics and adaptation. J Pers. 1990;58(1):17–67. https://doi.org/10.1111/j.1467-6494.1990.tb00907.x.
159. Kendler KS. The nature of psychiatric disorders. World Psychiatry. 2016;15(1):5–12. https://doi.org/10.1002/wps.20292.
160. Schaler JA. Szasz under fire: the psychiatric abolitionist faces his critics. Chicago, IL: Open Court Publishers; 2004.
161. Wakefield JC. Darwin, functional explanation, and the philosophy of psychiatry. In: Adriaens P, De Block A, editors. Maladapting mindsphilosophy, psychiatry, and evolutionary theory. Oxford: Oxford University Press; 2013. https://doi.org/10.1093/med/9780199558667.001.0001.
162. Wakefield JC. The biostatistical theory versus the harmful dysfunction analysis, part 1: is part-dysfunction a sufficient condition for medical disorder? J Med Philos. 2014;39(6):648–82. https://doi.org/10.1093/jmp/jhu038.
163. Wakefield JC, Conrad JA. Does the harm component of the harmful dysfunction analysis need rethinking?: Reply to Powell and Scarffe. J Med Ethics. 2019;45(9):594–6. https://doi.org/10.1136/medethics-2019-105578.
164. Roe K, Murphy D. Function, dysfunction, and adaptation? In: Adriaens P, De Block A, editors. Maladapting minds philosophy, psychiatry, and evolutionary theory. Oxford: Oxford University Press; 2013. https://doi.org/10.1093/med/9780199558667.001.0001.
165. Ruse M. Are homosexuals sick? In: Caplan A, Engelhardt H, McCartney J, editors. Concepts of health and disease: interdisciplinary perspectives. Reading, MA: Addison-Wesley; 1981. p. 693–723.
166. Mealey L. The sociobiology of sociopathy: an integrated evolutionary model. Behav Brain Sci. 2010;18(3):523–41. https://doi.org/10.1017/s0140525x00039595.
167. Fulford KWM. Moral theory and medical practice. Noûs. 1993;27(3):401–3.
168. Bowen R. The making of DSM-III: a diagnostic manual's conquest of American psychiatry By Hanna S. Decker Oxford, UK: Oxford University Press, 2013. £35.00. ISBN 9780195382235. Br J Psychol. 2013;104(4):588–90. https://doi.org/10.1111/bjop.12053.
169. Decker HS. The making of DSM-III: a diagnostic manual's conquest of American Psychiatry. New York: Oxford University Press; 2013.
170. Jutel A. Classification, disease, and diagnosis. Perspect Biol Med. 2011;54(2):189–205. https://doi.org/10.1353/pbm.2011.0015.
171. Mayes R, Horwitz AV. DSM-III and the revolution in the classification of mental illness. J Hist Behav Sci. 2005;41(3):249–67. https://doi.org/10.1002/jhbs.20103.
172. Fodor J. Psychological explanation: An introduction to the philosophy of psychology. New York, NY: Crown Publishing Group/Random House; 1968a.
173. Fodor JA. The appeal to tacit knowledge in psychological explanation. J Philos. 1968b;65(20):627. https://doi.org/10.2307/2024316.

174. Shepherd M. The extent of mental disorder. Beyond the layman's madness. Can Psychiatr Assoc J. 1976;21(6):401–9. https://doi.org/10.1177/070674377602100607.
175. Papineau D. Mental disorder, illness and biological dysfunction. In: Griffiths A, editor. Philosophy, psychology and psychiatry, Royal Institute of Philosophy Supplement 37. Cambridge: Cambridge University Press; 1994. p. 73–84.
176. Colombetti G. Psychopathology and the enactive mind. In: Fulford KWM, Davies M, Gipps GT, et al., editors. The Oxford handbook of philosophy and psychiatry. Oxford: Oxford University Press; 2013. https://doi.org/10.1093/oxfordhb/9780199579563.001.0001.
177. Boorse C. What a theory of mental-health should be. J Theory Soc Behav. 1976;6(1):61–84. https://doi.org/10.1111/j.1468-5914.1976.tb00359.x.
178. Graham G. Ordering disorder: mental disorder, brain disorder and therapeutic intervention. In: Fulford KWM, Davies M, Gipps GT, et al., editors. The Oxford handbook of philosophy and psychiatry. Oxford: Oxford University Press; 2013. https://doi.org/10.1093/oxfordhb/9780199579563.001.0001.
179. Hare RM. Descriptivism. Proc Br Acad. 1963;49:115–34.
180. Fulford KWM, Thornton T, Graham G. From bioethics to value-based parctice in psychiatric diagnosis. In: Fulford KWM, Thornton T, Graham G, editors. The Oxford textbook of philosophy and psychiatry. Oxford: Oxford University Press; 2006a.
181. U.S. Department of Health and Human Services. Mental health: a report of the Surgeon General. Rockville: Public Health Service; 1999.
182. World Health Organization. Promoting mental health: concepts, emerging evidence, practice. Geneva: World Health Organization; 2004.
183. World Health Organization. Constitution of the World Health Organization—basic documents. 46th ed. Geneva: World Health Organization; 2006.
184. Deci EL, Ryan RM. Hedonia, eudaimonia, and well-being: an introduction. J Happiness Stud. 2006;9(1):1–11. https://doi.org/10.1007/s10902-006-9018-1.
185. Dobson JF. Herophilus of Alexandria. Proc R Soc Med. 1925;18(Sect Hist Med):19–32.
186. Klein DF. A proposed definition of mental illness. New York, NY: Raven Press; 1978. p. 41–71.
187. Bolton D. What is mental illness? In: Fulford KWM, Davies M, Gipps GT, et al., editors. The Oxford handbook of philosophy and psychiatry. Oxford: Oxford University Press; 2013. https://doi.org/10.1093/oxfordhb/9780199579563.001.0001.
188. Spitzer RL, Endicott I. Medical and mental disorder: proposed definition and criteria. In: Spitzer RL, Klein D, editors. Critical issues in psychiatric diagnosis. New York, NY: Raven Press; 1978. p. 15–40.
189. Spitzer RL, Williams JB. The definition and diagnosis of mental disorder. In: Grove W, editor. Deviance and mental illness. Beverly Hills, CA: Sage; 1982. p. 15–31.
190. Spitzer RL, Williams JB. Basic principles in the development of DSM-III. In: Mezzich J, von Cranach M, editors. International classification in psychiatry: unity and diversity. Cambridge: Cambridge University Press; 1988. p. 81–6.
191. American Psychiatric Association. Diagnostic and statistical manual of mental disorders, fourth edition text revision (DSM-IV-TR). DSM library. Washington, DC: American Psychiatric Association; 2000. https://doi.org/10.1176/appi.books.9780890420249.dsm-iv-tr.
192. American Psychiatric Association. Diagnostic and statistical manual of mental disorders, fifth edition (DSM-5). DSM library. Washington, DC: American Psychiatric Association; 2013. https://doi.org/10.1176/appi.books.9780890420249.dsm-iv-tr.
193. Galderisi S, Heinz A, Kastrup M, Beezhold J, Sartorius N. Toward a new definition of mental health. World Psychiatry. 2015;14(2):231–3. https://doi.org/10.1002/wps.20231.
194. DeVylder JE, Oh HY, Corcoran CM, Lukens EP. Treatment seeking and unmet need for care among persons reporting psychosis-like experiences. Psychiatr Serv. 2014;65(6):774–80. https://doi.org/10.1176/appi.ps.201300254.
195. Johns LC, van Os J. The continuity of psychotic experiences in the general population. Clin Psychol Rev. 2001;21(8):1125–41. https://doi.org/10.1016/s0272-7358(01)00103-9.

196. van Os J, Linscott RJ, Myin-Germeys I, Delespaul P, Krabbendam L. A systematic review and meta-analysis of the psychosis continuum: evidence for a psychosis proneness-persistence-impairment model of psychotic disorder. Psychol Med. 2009;39(2):179–95. https://doi.org/10.1017/S0033291708003814.
197. Baumeister D, Sedgwick O, Howes O, Peters E. Auditory verbal hallucinations and continuum models of psychosis: a systematic review of the healthy voice-hearer literature. Clin Psychol Rev. 2017;51:125–41. https://doi.org/10.1016/j.cpr.2016.10.010.
198. Kaymaz N, Drukker M, Lieb R, Wittchen HU, Werbeloff N, Weiser M, Lataster T, van Os J. Do subthreshold psychotic experiences predict clinical outcomes in unselected non-help-seeking population-based samples? A systematic review and meta-analysis, enriched with new results. Psychol Med. 2012;42(11):2239–53. https://doi.org/10.1017/S0033291711002911.
199. Lovatt A, Mason O, Brett C, Peters E. Psychotic-like experiences, appraisals, and trauma. J Nerv Ment Dis. 2010;198(11):813–9. https://doi.org/10.1097/NMD.0b013e3181f97c3d.
200. Underwood R, Kumari V, Peters E. Appraisals of psychotic experiences: an experimental investigation of symptomatic, remitted and non-need-for-care individuals. Psychol Med. 2016a;46(6):1249–63. https://doi.org/10.1017/S0033291715002780.
201. Underwood R, Kumari V, Peters E. Cognitive and neural models of threat appraisal in psychosis: a theoretical integration. Psychiatry Res. 2016b;239:131–8. https://doi.org/10.1016/j.psychres.2016.03.016.
202. Peters E, Ward T, Jackson M, Morgan C, Charalambides M, McGuire P, Woodruff P, Jacobsen P, Chadwick P, Garety PA. Clinical, socio-demographic and psychological characteristics in individuals with persistent psychotic experiences with and without a "need for care". World Psychiatry. 2016;15(1):41–52. https://doi.org/10.1002/wps.20301.
203. Peters E, Ward T, Jackson M, Woodruff P, Morgan C, McGuire P, Garety PA. Clinical relevance of appraisals of persistent psychotic experiences in people with and without a need for care: an experimental study. Lancet Psychiatry. 2017;4(12):927–36. https://doi.org/10.1016/S2215-0366(17)30409-1.
204. Bolton D. What is mental disorder?: An essay in philosophy, science, and values. New York: Oxford University Press; 2008.
205. Bolton D, Hill J. Mind, meaning, and mental disorder: the nature of causal explanation in psychology and psychiatry. Oxford: Oxford University Press; 2003.
206. McMahon DM. The pursuit of happiness in history. In: Boniwell I, Davis S, editors. Oxford handbook of happiness. Oxford: Oxford University Press; 2012.
207. Zachar P, Krueger R. Personality disorder and validity: a history of controversy. In: Fulford KWM, Davies M, Gipps GT, et al., editors. The Oxford handbook of philosophy and psychiatry. Oxford: Oxford University Press; 2013. https://doi.org/10.1093/oxfordhb/9780199579563.001.0001.
208. Turner MA. Psychiatry and the human sciences. Br J Psychiatry. 2003;182:472–4. https://doi.org/10.1192/bjp.182.6.472.
209. Millon T. Authoritarianism, intolerance of ambiguity, and rigidity under ego- and task-involving conditions. J Abnorm Psychol. 1957;55(1):29–33. https://doi.org/10.1037/h0040352.
210. Millon T. On the renaissance of personality assessment and personality theory. J Pers Assess. 1984;48(5):450–66. https://doi.org/10.1207/s15327752jpa4805_1.
211. Millon T. Classification in psychopathology: rationale, alternatives, and standards. J Abnorm Psychol. 1991;100(3):245–61. https://doi.org/10.1037//0021-843x.100.3.245.
212. Church J. Boundary problems: negotiating the challenges of responsibility and loss. In: Fulford KWM, Davies M, Gipps GT, et al., editors. The Oxford handbook of philosophy and psychiatry. Oxford: Oxford University Press; 2013. https://doi.org/10.1093/oxfordhb/9780199579563.001.0001.
213. Lewis AJ. Health as a social concept. Br J Sociol. 1955;4:109–24.
214. John ZS. Vice and the diagnostic classification of mental disorders: a philosophical case conference. Philos Psychiatry Psychol. 2008;15(1):1–17. https://doi.org/10.1353/ppp.0.0152.

215. Sadler JZ. Values and psychiatric diagnosis. Oxford: Oxford University Press; 2005.
216. Sadler JZ. Vice and mental disorders. In: Fulford KWM, Davies M, Gipps GT, et al., editors. The Oxford handbook of philosophy and psychiatry. Oxford: Oxford University Press; 2013b. https://doi.org/10.1093/oxfordhb/9780199579563.001.0001.
217. Digman JM. Higher-order factors of the Big Five. J Pers Soc Psychol. 1997;73(6):1246–56. https://doi.org/10.1037//0022-3514.73.6.1246.
218. De Raad B, Barelds DP. A new taxonomy of Dutch personality traits based on a comprehensive and unrestricted list of descriptors. J Pers Soc Psychol. 2008;94(2):347–64. https://doi.org/10.1037/0022-3514.94.2.347.
219. Fountoulakis KN, Gonda X. Modeling human temperament and character on the basis of combined theoretical approaches. Ann Gen Psychiatry. 2019;18:21. https://doi.org/10.1186/s12991-019-0247-1.
220. DeYoung CG, Peterson JB, Higgins DM. Higher-order factors of the Big Five predict conformity: are there neuroses of health? Pers Individ Diff. 2002;33(4):533–52.
221. Akiskal H, Akiskal K. Epilogue. The Renaissance of the ancient concept of temperament (with a focus on affective temperaments). In: Mai M, Akiskal H, Mezzich J, Okasha A, editors. Personality disorders. Hoboken: Wiley; 2005. p. 479–500.
222. Cohen P. Personality development in childhood: old and new findings. In: Cloninger C, editor. Personality and psychopathology. Washington, DC: American Psychiatric Press; 1999. p. 101–27.
223. Krueger RF, Eaton NR, Clark LA, Watson D, Markon KE, Derringer J, Skodol A, Livesley WJ. Deriving an empirical structure of personality pathology for DSM-5. J Pers Disord. 2011;25(2):170–91. https://doi.org/10.1521/pedi.2011.25.2.170.
224. Krueger RF, South S, Johnson W, Iacono W. The heritability of personality is not always 50%: gene-environment interactions and correlations between personality and parenting. J Pers. 2008;76(6):1485–522. https://doi.org/10.1111/j.1467-6494.2008.00529.x.
225. Charland L. Character: moral treatment and the personality disorders. In: Radden J, editor. The philosophy of psychiatry: a companion. International perspectives in philosophy and psychiatry. New York, NY: Oxford University Press; 2004a. p. 64–77.
226. Charland LC. Moral nature of the DSM-IV Cluster B personality disorders. J Pers Disord. 2006;20(2):116–25. ; discussion 115–181. https://doi.org/10.1521/pedi.2006.20.2.116.
227. Peter Z, Nancy Nyquist P. Personality disorders: moral or medical kinds—or both? Philos Psychiatry Psychol. 2010;17(2):101–17. https://doi.org/10.1353/ppp.0.0290.
228. Zachar P. The clinical nature of personality disorders: answering the neo-Szazian critique. Philos Psychiatry Psychol. 2011;18:191–202.
229. Widiger TA, Mullins-Sweatt SN. Five-factor model of personality disorder: a proposal for DSM-V. Annu Rev Clin Psychol. 2009;5:197–220. https://doi.org/10.1146/annurev.clinpsy.032408.153542.
230. Clark LA. Temperament as a unifying basis for personality and psychopathology. J Abnorm Psychol. 2005;114(4):505–21. https://doi.org/10.1037/0021-843X.114.4.505.
231. Dolan-Sewell RT, Krueger RF, Shea MT. Co-occurrence with syndrome disorders. In: Livesley WJ, editor. Handbook of personality disorders: theory, research, and treatment. Guilford Press: New York, NY; 2001. p. 84–104.
232. Zachar P. Personality disorder: philosophical problems. In: Schramme T, Edwards S, editors. Handbook of the philosophy of medicine. Dordrecht: Springer Netherlands; 2017. p. 1005–24. https://doi.org/10.1007/978-94-017-8688-1_77.
233. McGlashan TH, Grilo CM, Skodol AE, Gunderson JG, Shea MT, Morey LC, Zanarini MC, Stout RL. The Collaborative Longitudinal Personality Disorders Study: baseline Axis I/II and II/II diagnostic co-occurrence. Acta Psychiatr Scand. 2000;102(4):256–64. https://doi.org/10.1034/j.1600-0447.2000.102004256.x.
234. Oldham JM, Skodol AE, Kellman HD, Hyler SE, Doidge N, Rosnick L, Gallaher PE. Comorbidity of axis I and axis II disorders. Am J Psychiatry. 1995;152(4):571–8. https://doi.org/10.1176/ajp.152.4.571.

235. Clark LA. Assessment and diagnosis of personality disorder: perennial issues and an emerging reconceptualization. Annu Rev Psychol. 2007;58:227–57. https://doi.org/10.1146/annurev.psych.57.102904.190200.
236. Lenzenweger MF. Schizotaxia, schizotypy, and schizophrenia: Paul E. Meehl's blueprint for the experimental psychopathology and genetics of schizophrenia. J Abnorm Psychol. 2006;115(2):195–200. https://doi.org/10.1037/0021-843X.115.2.195.
237. Meehl PE. Schizotaxia, schizotypy, schizophrenia. Am Psychol. 1962;17(12):827–38. https://doi.org/10.1037/h0041029.
238. Aniskiewicz AS. Autonomic components of vicarious conditioning and psychopathy. J Clin Psychol. 1979;35(1):60–7. https://doi.org/10.1002/1097-4679(197901)35:1<60::aid-jclp2270350106>3.0.co;2-r.
239. Blair RJ. Neurocognitive models of aggression, the antisocial personality disorders, and psychopathy. J Neurol Neurosurg Psychiatry. 2001;71(6):727–31. https://doi.org/10.1136/jnnp.71.6.727.
240. Blair RJ. The amygdala and ventromedial prefrontal cortex in morality and psychopathy. Trends Cogn Sci. 2007;11(9):387–92. https://doi.org/10.1016/j.tics.2007.07.003.
241. Blair RJ. Neuroimaging of psychopathy and antisocial behavior: a targeted review. Curr Psychiatry Rep. 2010;12(1):76–82. https://doi.org/10.1007/s11920-009-0086-x.
242. Blair RJ, Jones L, Clark F, Smith M. The psychopathic individual: a lack of responsiveness to distress cues? Psychophysiology. 1997;34(2):192–8. https://doi.org/10.1111/j.1469-8986.1997.tb02131.x.
243. Blair RJR. Responsiveness to distress cues in the child with psychopathic tendencies. Pers Individ Diff. 1999;27(1):135–45. https://doi.org/10.1016/S0191-8869(98)00231-1.
244. Fine C, Kennett J. Mental impairment, moral understanding and criminal responsibility: psychopathy and the purposes of punishment. Int J Law Psychiatry. 2004;27(5):425–43. https://doi.org/10.1016/j.ijlp.2004.06.005.
245. Sinnott-Armstrong W. Moral psychology. The evolution of morality: adaptations and innateness. Cambridge, MA: MIT Press; 2008a.
246. Sinnott-Armstrong W. Moral psychology. The neuroscience of morality: emotion, brain disorders, and development. Cambridge, MA: MIT Press; 2008b.
247. Blair RJ. Normative theory or theory of mind? A response to Nichols. In: Sinnott-Armstrong W, editor. Moral psychology, the cognitive science of morality: intuition and diversity. Cambridge, MA: MIT Press; 2008. p. 275–8.
248. Hare RD. Psychopaths and their nature: implications for the mental health and criminal justice systems. In: Millon T, Simonsen E, Birket-Smith M, Davis R, editors. Psychopathy: antisocial, criminal, and violent behavior. New York, NY: Guilford Press; 1998. p. 188–213.
249. Widiger TA. Psychopathy and DSM-IV psychopathology. In: Patrick J, editor. Handbook of psychopathy. New York, NY: Guilford; 2006. p. 156–71.
250. Fountoulakis KN, Leucht S, Kaprinis GS. Personality disorders and violence. Curr Opin Psychiatry. 2008;21(1):84–92. https://doi.org/10.1097/YCO.0b013e3282f31137.
251. Hinshaw SP, Cicchetti D. Stigma and mental disorder: conceptions of illness, public attitudes, personal disclosure, and social policy. Dev Psychopathol. 2000;12(4):555–98. https://doi.org/10.1017/s0954579400004028.
252. Thornicroft G. Shunned: discrimination against people with mental illness. Oxford: Oxford University Press; 2006.
253. Kleinman A. Rethinking psychiatry: from cultural category to personal experience. New York: Free Press; 1991.
254. Littlewood R. Psychiatry's culture. Int J Soc Psychiatry. 1996;42(4):245–68. https://doi.org/10.1177/002076409604200402.
255. Sartorius N, Jablensky A, Korten A, Ernberg G, Anker M, Cooper JE, Day R. Early manifestations and first-contact incidence of schizophrenia in different cultures. A preliminary report on the initial evaluation phase of the WHO Collaborative Study on determinants of outcome of severe mental disorders. Psychol Med. 1986;16(4):909–28. https://doi.org/10.1017/s0033291700011910.

256. Vaillant GE. Positive mental health: is there a cross-cultural definition? World Psychiatry. 2012;11(2):93–9. https://doi.org/10.1016/j.wpsyc.2012.05.006.
257. Harrison G, Hopper K, Craig T, Laska E, Siegel C, Wanderling J, Dube KC, Ganev K, Giel R, an der Heiden W, Holmberg SK, Janca A, Lee PW, Leon CA, Malhotra S, Marsella AJ, Nakane Y, Sartorius N, Shen Y, Skoda C, Thara R, Tsirkin SJ, Varma VK, Walsh D, Wiersma D. Recovery from psychotic illness: a 15- and 25-year international follow-up study. Br J Psychiatry. 2001;178(6):506–17. https://doi.org/10.1192/bjp.178.6.506.
258. Leff J, Sartorius N, Jablensky A, Korten A, Ernberg G. The International Pilot Study of Schizophrenia: five-year follow-up findings. Psychol Med. 1992;22(1):131–45. https://doi.org/10.1017/s0033291700032797.
259. World Health Organization. Schizophrenia: an international follow up study. New York: Wiley; 1979.
260. Cohen A, Patel V, Thara R, Gureje O. Questioning an axiom: better prognosis for schizophrenia in the developing world? Schizophr Bull. 2008;34(2):229–44. https://doi.org/10.1093/schbul/sbm105.
261. Karagianis J, Novick D, Pecenak J, Haro JM, Dossenbach M, Treuer T, Montgomery W, Walton R, Lowry AJ. Worldwide-schizophrenia outpatient health outcomes (W-SOHO): baseline characteristics of pan-regional observational data from more than 17,000 patients. Int J Clin Pract. 2009;63(11):1578–88. https://doi.org/10.1111/j.1742-1241.2009.02191.x.
262. Kulhara P, Shah R, Grover S. Is the course and outcome of schizophrenia better in the 'developing' world? Asian J Psychiatr. 2009;2(2):55–62. https://doi.org/10.1016/j.ajp.2009.04.003.
263. Padma TV. Developing countries: the outcomes paradox. Nature. 2014;508(7494):S14–5. https://doi.org/10.1038/508S14a.
264. Engel GL. The need for a new medical model: a challenge for biomedicine. Science. 1977;196(4286):129–36. https://doi.org/10.1126/science.847460.
265. Engel GL. The biopsychosocial model and the education of health professionals. Ann N Y Acad Sci. 1978;310:169–87. https://doi.org/10.1111/j.1749-6632.1978.tb22070.x.
266. Engel GL. The clinical application of the biopsychosocial model. Am J Psychiatry. 1980;137(5):535–44. https://doi.org/10.1176/ajp.137.5.535.
267. Engel GL. Sounding board. The biopsychosocial model and medical education. Who are to be the teachers? N Engl J Med. 1982;306(13):802–5. https://doi.org/10.1056/NEJM198204013061311.
268. Bolton D, Gillett G. The biopsychosocial model of health and disease new philosophical and scientific developments. London: Palgrave Macmillan; 2019.
269. Borrell-Carrio F, Suchman AL, Epstein RM. The biopsychosocial model 25 years later: principles, practice, and scientific inquiry. Ann Fam Med. 2004;2(6):576–82. https://doi.org/10.1370/afm.245.
270. Saad M, de Medeiros R, Mosini AC. Are we ready for a true biopsychosocial-spiritual model? The many meanings of "spiritual". Medicines (Basel). 2017;4(4):79. https://doi.org/10.3390/medicines4040079.
271. Sulmasy DP. A biopsychosocial-spiritual model for the care of patients at the end of life. Gerontologist. 2002;42 Spec No 3:24–33. https://doi.org/10.1093/geront/42.suppl_3.24.
272. Crisp R. Commentary: value-based practice by a different route. In: Fulford KWM, Davies M, Gipps GT, et al., editors. The Oxford handbook of philosophy and psychiatry. Oxford: Oxford University Press; 2013. https://doi.org/10.1093/oxfordhb/9780199579563.001.0001.
273. Coltheart M, Langdon R, McKay R. Delusional belief. Annu Rev Psychol. 2011;62:271–98. https://doi.org/10.1146/annurev.psych.121208.131622.
274. Andreasen NC, Nopoulos P, O'Leary DS, Miller DD, Wassink T, Flaum M. Defining the phenotype of schizophrenia: cognitive dysmetria and its neural mechanisms. Biol Psychiatry. 1999;46(7):908–20. https://doi.org/10.1016/s0006-3223(99)00152-3.
275. Andreasen NC, O'Leary DS, Cizadlo T, Arndt S, Rezai K, Ponto LL, Watkins GL, Hichwa RD. Schizophrenia and cognitive dysmetria: a positron-emission tomography study of dysfunctional prefrontal-thalamic-cerebellar circuitry. Proc Natl Acad Sci U S A. 1996;93(18):9985–90. https://doi.org/10.1073/pnas.93.18.9985.

276. Andreasen NC, Paradiso S, O'Leary DS. "Cognitive dysmetria" as an integrative theory of schizophrenia: a dysfunction in cortical-subcortical-cerebellar circuitry? Schizophr Bull. 1998;24(2):203–18. https://doi.org/10.1093/oxfordjournals.schbul.a033321.
277. Maher B. Anomalous experience and delusional thinking: the logic of explanations. In: Oltmanns T, Maher B, editors. Delusional beliefs. New York, NY: Plenum Press; 1988. p. 15–33.
278. Kaprinis GS, Fountoulakis KN, Kaprinis SG. Arguments against the cognitive dysmetria hypothesis of schizophrenia. Percept Mot Skills. 2002;94(3 Pt 1):975–84. https://doi.org/10.2466/pms.2002.94.3.975.
279. Fountoulakis KN, Dragioti E, Theofilidis AT, Wikilund T, Atmatzidis X, Nimatoudis I, Thys E, Wampers M, Hranov L, Hristova T, Aptalidis D, Milev R, Iftene F, Spaniel F, Knytl P, Furstova P, From T, Karlsson H, Walta M, Salokangas RKR, Azorin JM, Bouniard J, Montant J, Juckel G, Haussleiter IS, Douzenis A, Michopoulos I, Ferentinos P, Smyrnis N, Mantonakis L, Nemes Z, Gonda X, Vajda D, Juhasz A, Shrivastava A, Waddington J, Pompili M, Comparelli A, Corigliano V, Rancans E, Navickas A, Hilbig J, Bukelskis L, Injac Stevovic L, Vodopic S, Esan O, Oladele O, Osunbote C, Rybakowski J, Wojciak P, Domowicz K, Figueira ML, Linhares L, Crawford J, Panfil AL, Smirnova D, Izmailova O, Lecic-Tosevski D, Temmingh H, Howells F, Bobes J, Garcia-Portilla MP, Garcia-Alvarez L, Erzin G, Karadag H, De Sousa A, Bendre A, Hoschl C, Bredicean C, Papava I, Vukovic O, Pejuskovic B, Russell V, Athanasiadis L, Konsta A, Stein D, Berk M, Dean O, Tandon R, Kasper S, De Hert M. Staging of schizophrenia with the use of PANSS: an international multi-center study. Int J Neuropsychopharmacol. 2019;22(11):681–97. https://doi.org/10.1093/ijnp/pyz053.
280. Cartwright SA. Report on the diseases and physical peculiarities of the negro race. New Orleans Med Surg J. 1851;7:707–9.
281. Engelhardt HT. Ideology and etiology. J Med Philos. 1976;1(3):256–68. https://doi.org/10.1093/jmp/1.3.256.
282. Bayer R. Homosexuality and American psychiatry: the politics of diagnosis. New York: Basic Books; 1981.
283. Mechanic D. Definitions and perspectives. In: Horowitz A, Scheid T, editors. A handbook for the study of mental health. Cambridge: Cambridge University Press; 1999.
284. Bateson G, Jackson DD, Haley J, Weakland J. Toward a theory of schizophrenia. Behav Sci. 2007;1(4):251–64. https://doi.org/10.1002/bs.3830010402.
285. Fromm-Reichmann F. Notes on the development of treatment of schizophrenics by psychoanalytic psychotherapy. Psychiatry. 1948;11(3):263–73. https://doi.org/10.1080/00332747.1948.11022688.
286. Harrington A. The fall of the schizophrenogenic mother. Lancet. 2012;379(9823):1292–3. https://doi.org/10.1016/s0140-6736(12)60546-7.
287. Morris JA. The fall of the schizophrenogenic mother. Lancet. 2012;380(9837):110. https://doi.org/10.1016/S0140-6736(12)61180-5.
288. Neill J. Whatever became of the schizophrenogenic mother? Am J Psychother. 1990;44(4):499–505. https://doi.org/10.1176/appi.psychotherapy.1990.44.4.499.
289. Cohen CI, Timimi S, editors. Liberatory psychiatry: philosophy, politics and mental health. Cambridge: Cambridge University Press; 2008. https://doi.org/10.1017/CBO9780511543678.
290. Grob GN. Mental health policy in America: myths and realities. Health Aff (Millwood). 1992;11(3):7–22. https://doi.org/10.1377/hlthaff.11.3.7.
291. Porter R. A social history of madness: the world through the eyes of the insane. New York, NY: Weidenfeld & Nicolson; 1988.
292. Rafter NH. Creating born criminals. Urbana, IL: University of Illinois Press; 1997.
293. Rothman DJ. Conscience and convenience: the asylum and its alternatives in progressive America. Hawthorne, NY: Aldine de Gruyter; 2002a.
294. Rothman DJ. The discovery of the asylum: social order and disorder in the new republic. Hawthorne, NY: Aldine de Gruyter; 2002b.
295. Charland LC. As autonomy heads into harm's way. Philos Psychiatry Psychol. 2004b;11(4):361–3. https://doi.org/10.1353/ppp.2005.0007.

296. Charland LC. A madness for identity: psychiatric labels, consumer autonomy, and the perils of the internet. Philos Psychiatry Psychol. 2004c;11(4):335–49. https://doi.org/10.1353/ppp.2005.0006.
297. Kenny L, Hattersley C, Molins B, Buckley C, Povey C, Pellicano E. Which terms should be used to describe autism? Perspectives from the UK autism community. Autism. 2016;20(4):442–62. https://doi.org/10.1177/1362361315588200.
298. Lee B, editor. The dangerous case of Donald Trump: 27 Psychiatrists and mental health experts assess a President. New York: Thomas Dunne Books; 2017.
299. Frances A. The new crisis of confidence in psychiatric diagnosis. Ann Intern Med. 2013;159(10):720. https://doi.org/10.7326/0003-4819-159-10-201311190-00021.
300. Kogge W, Richter M. Synthetic biology and its alternatives. Descartes, Kant and the idea of engineering biological machines. Stud Hist Philos Biol Biomed Sci. 2013;44(2):181–9. https://doi.org/10.1016/j.shpsc.2013.03.013.
301. Moynihan R, Heath I, Henry D. Selling sickness: the pharmaceutical industry and disease mongering. BMJ. 2002;324(7342):886–91. https://doi.org/10.1136/bmj.324.7342.886.
302. Escalante A. Researchers doubt that certain mental disorders are disorders at all. Forbes; 2020.
303. Syme KL, Hagen EH. Mental health is biological health: why tackling "diseases of the mind" is an imperative for biological anthropology in the 21st century. Am J Phys Anthropol. 2020;171 Suppl 70(S70):87–117. https://doi.org/10.1002/ajpa.23965.
304. Fountoulakis KN, Hoschl C, Kasper S, Lopez-Ibor J, Moller HJ. The media and intellectuals' response to medical publications: the antidepressants' case. Ann Gen Psychiatry. 2013;12(1):11. https://doi.org/10.1186/1744-859X-12-11.

Chapter 14
Mental Health

"Mental health" is a concept which is often utilized to denote the field of medicine served by psychiatry, mostly because it is an elegant term with a positive aura, points to health rather than to disease, and is overinclusive of disciplines allied to psychiatry. On the other hand, conceptually, it indeed focuses on the achievement of optimum mental health rather than on the treatment of mental disorders.

As discussed in Chap. 13, the World Health Organization (WHO) creator of the classification system International Classification of Diseases (ICD)-10 defined mental health as not merely the absence of mental illness, and additionally it does not identify with "well-being" [25, 28]. According to the WHO, the basic concept is that mental health could be defined as "a state of complete physical, mental and social well-being and not merely the absence of disease or infirmity" [29]. The definition becomes more sophisticated by suggesting that "mental health is defined as a state of well-being in which every individual realizes his or her own potential, can cope with the normal stresses of life, can work productively and fruitfully, and is able to make a contribution to her or his community" [28]. In this frame, the true nature of "mental health" is the ability to respond appropriately and to adjust to common social, occupational, and personal challenges [25, 28] but also has political implications. The European Psychiatric Association (EPA) attempted to define mental health as "a dynamic state of internal equilibrium which enables individuals to use their abilities in harmony with the universal values of society. Basic cognitive and social skills; ability to recognize, express and modulate one's own emotions, as well as empathize with others; flexibility and ability to cope with adverse life events and function in social roles; and harmonious relationship between body and mind represent important components of mental health which contribute, to varying degrees, to the state of internal equilibrium" [9].

This conceptualization of mental health is often but not always or necessarily accompanied by positive feelings and well-being or professional success, especially during specific phases of life like adolescence or when facing difficult circumstances like migration or catastrophic situations [6]. A drawback in these definitions

K. N. Fountoulakis, *Psychiatry*, https://doi.org/10.1007/978-3-030-86541-2_14

is that they are loaded with moral as well as ethical (e.g., hedonism) [6] and cultural issues [7]. They could also be contaminated by political views, since they imply to some extent that the political system might have an impact on mental health. If such an assumption is accepted without caution, it could lead to all kinds of interpretations and abuse of psychiatry.

Apart from the theory that there is no mental disorder and subsequently there is no such thing as mental health, in epistemological and philosophical terms, there are different approaches to mental health and its components.

One approach proposed the presence of three domains:

- Self-realization (referring to the ability of individuals who are able to fully exploit their own potential).
- Sense of mastery over their environment.
- Sense of autonomy (referring to the ability to identify, confront, and solve problems) [14].

Another theory suggested the presence of three components of mental health:

- Emotional well-being (referring to happiness, interest in life, and satisfaction).
- Psychological well-being (which links most parts of one's own personality, being good at managing the responsibilities of daily life, having good relationships with others, and being satisfied with one's own life).
- Social well-being (positive functioning, social contribution, social integration, social actualization, and social coherence) [9, 17].

Often, the discussion on the definitions of mental health becomes impossible with the destructive involvement of theoretical existential elaborations on the nature of freedom and control over the environment, especially during catastrophic situations [22].

In the literature there is much consensus that basic cognitive and social skills constitute an important component of mental health, since they determine everyday life functioning, behavior, and the ability to cope with various and continuous challenges [2, 10, 21, 26]. Such cognitive-emotional skills are:

- The ability for empathy (the ability to experience and understand what others feel without confusion between oneself and others) [5].
- The ability to carefully analyze social patterns and events [12].
- Features of flexibility (the ability to revise a course of action or idea and adapt to changes) [18].
- The ability for emotional regulation is an additional important component since it determines social adjustment and interaction with others and acts as a mediator for the coping and the adjustment to stress [11, 20, 23]. Emotional dysregulation possibly constitutes a risk factor for the development of a mental disorder but also for a medical condition [1, 3, 4, 13, 15, 16].
- A harmonious relationship and interaction between body and mind and perception of self as a unity with unique features and characteristics is also important [8].

However, the above reflect only one aspect of mental health and concern the average person. The cognitive-emotional characteristics described above could well describe also the psychopath who uses his/her empathic skills to manipulate others or to adapt and avoid detection and persecution or criminal persons who have a sense of well-being and are in perfect harmony with self while killing others.

In line with this, often persons with good mental health could feel sad, unwell, angry or unhappy, or find it very difficult to accept changes and adapt to undesirable situations, and this should be considered as an important part of the inner experience and behavior of a human being.

After taking a note of these important exceptions, in general, the definition and conceptualization of mental health include a positive affective state with a significant sense of mastery over the environment [19, 27]. Such a general approach is in accord with the definition of recovery after the experience of mental disorder; recovery is conceptualized as the attainment of a fulfilled and valued life by building on those functions spared by the disorder, in spite of the fact that other functions are permanently impaired [24].

References

1. Araragi N, Lesch KP. Serotonin (5-HT) in the regulation of depression-related emotionality: insight from 5-HT transporter and tryptophan hydroxylase-2 knockout mouse models. Curr Drug Targets. 2013;14(5):549–70. https://doi.org/10.2174/1389450111314050005.
2. Artero S, Touchon J, Ritchie K. Disability and mild cognitive impairment: a longitudinal population-based study. Int J Geriatr Psychiatry. 2001;16(11):1092–7. https://doi.org/10.1002/gps.477.
3. Barlow DH, Allen LB, Choate ML. Toward a unified treatment for emotional disorders. Behav Ther. 2004;35(2):205–30. https://doi.org/10.1016/s0005-7894(04)80036-4.
4. Campbell-Sills L, Barlow DH, Brown TA, Hofmann SG. Acceptability and suppression of negative emotion in anxiety and mood disorders. Emotion. 2006;6(4):587–95. https://doi.org/10.1037/1528-3542.6.4.587.
5. Decety J, Smith KE, Norman GJ, Halpern J. A social neuroscience perspective on clinical empathy. World Psychiatry. 2014;13(3):233–7. https://doi.org/10.1002/wps.20146.
6. Deci EL, Ryan RM. Hedonia, eudaimonia, and Well-being: an introduction. J Happiness Stud. 2006;9(1):1–11. https://doi.org/10.1007/s10902-006-9018-1.
7. Dobson JF (1925) Herophilus of Alexandria. Proc R Soc med. 1925;18(sect Hist med):19–32.
8. Fuchs T, Schlimme JE. Embodiment and psychopathology: a phenomenological perspective. Curr Opin Psychiatry. 2009;22(6):570–5. https://doi.org/10.1097/YCO.0b013e3283318e5c.
9. Galderisi S, Heinz A, Kastrup M, Beezhold J, Sartorius N. Toward a new definition of mental health. World Psychiatry. 2015;14(2):231–3. https://doi.org/10.1002/wps.20231.
10. Gigi K, Werbeloff N, Goldberg S, Portuguese S, Reichenberg A, Fruchter E, Weiser M. Borderline intellectual functioning is associated with poor social functioning, increased rates of psychiatric diagnosis and drug use--a cross sectional population based study. Eur Neuropsychopharmacol. 2014;24(11):1793–7. https://doi.org/10.1016/j.euroneuro.2014.07.016.
11. Gross JJ, Muñoz RF. Emotion regulation and mental health. Clin Psychol Sci Pract. 1995;2(2):151–64. https://doi.org/10.1111/j.1468-2850.1995.tb00036.x.

12. Heinz A, Kluge U. Anthropological and evolutionary concepts of mental disorders. J Specul Philos. 2010;24(3):292–307. https://doi.org/10.1353/jsp.2010.0013.
13. Helmers KF, Mente A. Alexithymia and health behaviors in healthy male volunteers. J Psychosom Res. 1999;47(6):635–45. https://doi.org/10.1016/s0022-3999(99)00067-7.
14. Jahoda M. Current concepts of positive mental health. Basic Books. 1958; https://doi.org/10.1037/11258-000.
15. Kauhanen J, Kaplan GA, Cohen RD, Julkunen J, Salonen JT. Alexithymia and risk of death in middle-aged men. J Psychosom Res. 1996;41(6):541–9. https://doi.org/10.1016/s0022-3999(96)00226-7.
16. Kennedy SE, Koeppe RA, Young EA, Zubieta JK. Dysregulation of endogenous opioid emotion regulation circuitry in major depression in women. Arch Gen Psychiatry. 2006;63(11):1199–208. https://doi.org/10.1001/archpsyc.63.11.1199.
17. Keyes CLM. Mental health as a complete state: how the Salutogenic perspective completes the picture, Bridging Occupational, Organizational and Public Health. Netherlands: Springer; 2013. https://doi.org/10.1007/978-94-007-5640-3_11.
18. Klanker M, Feenstra M, Denys D. Dopaminergic control of cognitive flexibility in humans and animals. Front Neurosci. 2013;7:201. https://doi.org/10.3389/fnins.2013.00201.
19. Lamers SM, Westerhof GJ, Bohlmeijer ET, ten Klooster PM, Keyes CL. Evaluating the psychometric properties of the mental health continuum-short form (MHC-SF). J Clin Psychol. 2011;67(1):99–110. https://doi.org/10.1002/jclp.20741.
20. McCarthy CJ, Lambert RG, Moller NP. Preventive resources and emotion regulation expectancies as mediators between attachment and college students' stress outcomes. Int J Stress Manag. 2006;13(1):1–22. https://doi.org/10.1037/1072-5245.13.1.1.
21. Moritz DJ, Kasl SV, Berkman LF. Cognitive functioning and the incidence of limitations in activities of daily living in an elderly community sample. Am J Epidemiol. 1995;141(1):41–9. https://doi.org/10.1093/oxfordjournals.aje.a117344.
22. Murphy HB. The meaning of symptom-check-list scores in mental health surveys: a testing of multiple hypotheses. Soc Sci Med. 1978;12(2A):67–75. https://doi.org/10.1016/0271-7123(78)90031-7.
23. Schwartz D, Proctor LJ. Community violence exposure and children's social adjustment in the school peer group: the mediating roles of emotion regulation and social cognition. J Consult Clin Psychol. 2000;68(4):670–83. https://doi.org/10.1037//0022-006x.68.4.670.
24. Slade M, Amering M, Farkas M, Hamilton B, O'Hagan M, Panther G, Perkins R, Shepherd G, Tse S, Whitley R. Uses and abuses of recovery: implementing recovery-oriented practices in mental health systems. World Psychiatry. 2014;13(1):12–20. https://doi.org/10.1002/wps.20084.
25. U.S. Department of Health and Human Services. Mental health: a report of the surgeon general. New York: Public Health Service, Rockville; 1999.
26. Warren EJ, Grek A, Conn D, Herrmann N, Icyk E, Kohl J, Silberfeld M. A correlation between cognitive performance and daily functioning in elderly people. J Geriatr Psychiatry Neurol. 1989;2(2):96–100. https://doi.org/10.1177/089198878900200209.
27. Waterman AS. Two conceptions of happiness: contrasts of personal expressiveness (eudaimonia) and hedonic enjoyment. J Pers Soc Psychol. 1993;64(4):678–91. https://doi.org/10.1037/0022-3514.64.4.678.
28. World Health Organization. Promoting mental health: concepts, emerging evidence, practice. Geneva: World Health Organization; 2004.
29. World Health Organization. Constitution of the World Health Organization – basic documents. 46th ed. Geneva: World Health Organization; 2006.

Chapter 15
Disability in Mental Disorders

Disability is one of the core characteristics of mental disorders. In Chap. 13 it has been discussed in detail why it is one of the three criteria for the definition and the diagnosis of mental disorders. It is the reason we are discussing mental disorder within the discipline of medicine and not in the frame of literature or sociology.

The term "disability" is used to denote a large variety of impairments, limitations, and restrictions in activities and participation because of any cause, located primarily within the individual person.

It could include an:
- Intellectual
- Mental
- Physical
- Sensory impairment or
- Some combination of them

And it could be:
- Acquired
- Congenital or
- Developmental in origin.

Disability is a complex phenomenon which also involves features of the societal environment in interaction with the personal features causing disability. The presence of bodily impairment (including mental) concerns the medical model of disability, according to which disability is within the person; is directly caused by disease, trauma, or other medical conditions; and requires professional medical treatment.

On the other hand, the limitations imposed on people by the constraints of an ableist society concern the social model which sees disability as a matter of the full integration of individuals into society. According to the social adapted model, although a person's disability does exist within the person, often the surrounding

K. N. Fountoulakis, *Psychiatry*, https://doi.org/10.1007/978-3-030-86541-2_15

society and environment are more limiting than the disability itself. Stigma plays a key role in these cases. It is important to note that continued disability is linked to institutionalization, discrimination, and social exclusion [8, 10, 11].

Although the basic definition suggests that disability comes from the characteristics of the person and is modified by the environment, it is not infrequent to find disabilities completely determined by the societal environment, for which the personal features of the individual serve only as a trigger, but they do not play a major role by themselves in the actual development of the disability.

The basic definition is reflected in the example of people with paraparesis who find it difficult to circulate around modern cities, but this disability is significantly influenced, to the degree of complete compensation, by specific measures taken concerning the existence of special structures, mechanisms, and in general ways of easy access. Subsequently, in these people, while their impairment is present, the actual disability could be minimized with proper intervention. Interestingly, during the recent COVID-19 epidemic, there was a concern that the use of face masks could worsen the disability of deaf people because it would make lips reading impossible. This could serve as a wonderful example of how disability is frequently environmentally determined and how changes in the environment could induce disability in unexpected ways.

However, on the other hand, people with some kind of mild issue in appearance, e.g., big nose or prominent ears, do not carry any kind of impairment per se, since there is no primary bodily or mental dysfunction; however they might be the focus of bullying and harassment since childhood, and consequently they might develop some kind of disability which would not have been in place if social interaction was proper. In these people the opposite happens; while there is no actual impairment present, disability develops as a consequence of social attitude, stigma, and discrimination. One such prominent example in the area of mental health is homosexuality, which until a few decades ago could trigger fierce social rejection and discrimination, causing much trouble and impairment to individuals.

People with disabilities may experience a narrower margin of health, both because of poverty and social exclusion, also because they may be vulnerable to secondary medical conditions, and because they also face barriers in accessing the health and rehabilitation services they need.

The World Health Organization (WHO) established the International Classification of Functioning, Disability and Health (ICF) [15]. The joint use of the ICD-10 [14] and the ICF serves this purpose [7]. The WHO also developed a standard instrument, the World Health Organization Disability Assessment Schedule (WHODAS) [13].

The WHODAS captures an individual's level of functioning in six major life domains:

1. Cognition (understanding and communication).
2. Mobility (ability to move and get around).
3. Self-care (ability to attend to personal hygiene, dressing, and eating and to live alone).

4. Getting along (ability to interact with other people).
5. Life activities (ability to carry out responsibilities at home, work, and school).
6. Participation in society (ability to engage in community, civil, and recreational activities) [4].

In sharp contrast to the rest of medical conditions, mental disorders are sometimes related with higher functioning at least in the short term. Mild hypomania in the frame of benign bipolar-II disorder is sometimes accompanied by higher creativity, non-conformity, goal-striving, meticulousness, or empathy. Narcissistic personalities could present as extremely successful professionals with poor interpersonal relationships, but in many cases, it is difficult if not impossible to distinguish between the objective status and the subjective reports by peers which often are directed by envy. However these are rather isolated areas loaded with conceptual problems and do not reflect the wider picture, which is that world-wide there seems to be greater disability related to mental conditions than to physical conditions, particularly concerning the effect on social roles and personal relationships [12]. At the end of the day, all mental disorders are characterized either by dysphoria or disability no matter what was the situation during individual episodes (see also Chap. 13).

Disability-adjusted life year (DALY) is a measure of global disability and burden and corresponds to the number of years of life lost due to premature death or poor health and disability (Fig. 15.1). One DALY represents the loss of the equivalent of 1 year of full health. Using DALYs, the burden of diseases that cause early death but little disability can be compared to that of diseases that do not cause death but do cause disability. Globally, 60% of DALYs are due to premature mortality. This does not reflect how disabling a specific medical condition is, since the method heavily depends on the prevalence of the condition as well as the average age at onset. This is why schizophrenia is not the most disabling condition, in spite of its devastating impact on the lives of patients and the relatively young age at onset; what lowers its impact on the society is the fact that it is relatively rare. It is reported to cause 15.5 million DALYs world-wide, while bipolar disorder which probably has a two to

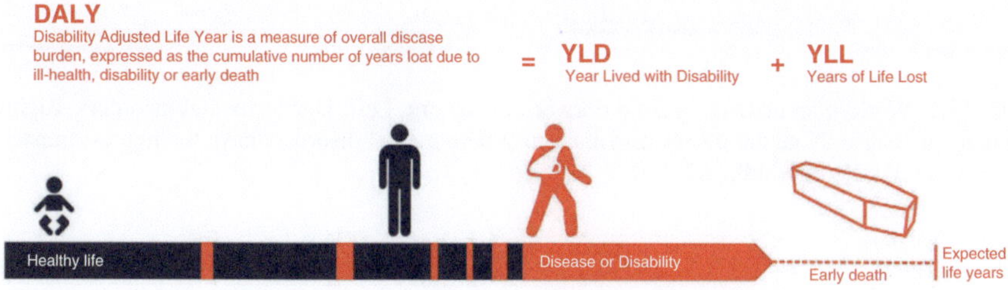

Fig. 15.1 Infographic of the concept of disability-adjusted life years (DALYs)

three times higher prevalence follows with nine million. However it is depression that is the leading cause with 54 million DALYs, while anxiety disorders account for approximately three to seven million DALYs each [3].

There is a large discrepancy in the absolute number but also in the proportion of DALYs caused by mental disorders in different regions of the world (Fig. 15.2). In high-income countries, the impact is by far greater [1, 5, 6]. World-wide mental disorders are ranked as the fourth cause of DALYs responsible for 6.35% of total DALYs. However, in Europe and the USA, they are the leading causes of disability accounting for 8.75% and 12.5%, respectively. The sum of DALYs of mental and neurological disorders (in which dementias are included) accounts for approximately 11% in Europe and 16% the USA. With adding also the DALYs from intentional injuries (suicide attempts and violence), the disability linked to any mental or neurological condition and related issues becomes one the three leading causes of DALYs everywhere in the world except Africa (Table 15.1). In terms of the contribution of specific disorders, depression seems to be the leading cause of DALYs, followed by anxiety, intentional self-harm, and dementia. It is interesting to note that if one adds dementias and self-inflicted injuries in the statistics of mental disorders alone, the percentage of DALYs increases by 2.5% world-wide and by more than 4.5% in Europe and 6% in the USA (Table 15.2). The impact of suicide and self-inflicted injuries is widely neglected, but statistics suggest that suicide attempts and self-harm are significant causes of disability, while suicide is the leading cause of death among teenagers and adults under 35 [9] with an estimate of two million deaths and 10–20 million non-fatal attempts every year world-wide [2].

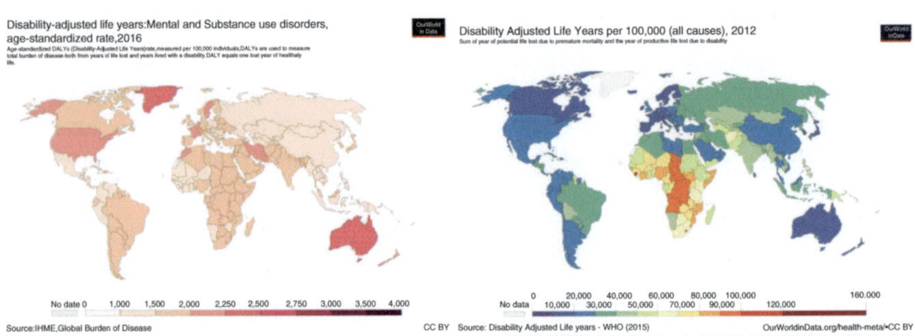

Fig. 15.2 World map of DALYs from ourworldindata.org. Left: Due to mental disorders. Right: Due to any cause. Note the mirror distribution: Where mental disorders have the highest impact, the impact of other morbidities is low

Table 15.1 Disease and injury regional estimates for percentage of total DALYs caused by each category of medical conditions for year 2015 from the World Health Organization (www.who.int). Archived from the original on May 6, 2018

		World	Europe	USA	Americas	Southeast Asia	Africa
1	Cardiovascular diseases	15.28	26.50	16.7	14.72	14.60	5.57
2	Infectious and parasitic diseases	13.47	2.67	1.4	3.56	13.28	32.17
6 + 11 + 13	Mental plus neurological disorders plus intentional injuries	12.49	16.69	23.50[a]	21.56[a]	11.01	6.47
6 + 13	Mental disorders plus intentional injuries	9.29	11.28	15.92[a]	16.23[a]	8.10	5.10
3	Malignant neoplasms	9.13	17.19	15.8	12.39	6.15	3.17
4	Neonatal conditions	8.36	2.05	1.5	3.56	10.25	13.10
5	Unintentional injuries	7.95	6.90	5.9	7.26	8.42	8.14
6	Mental and substance use disorders	6.35	8.75	12.5	10.49	5.61	3.09
7	Respiratory infectious	5.58	2.39	1.8	3.24	5.11	10.48
8	Respiratory diseases	4.14	3.63	6.2	4.73	5.53	1.60
9	Musculoskeletal diseases	4.04	5.77	7.3	5.86	4.01	1.24
10	Digestive diseases	3.27	2.85	3.9	4.08	4.30	2.47
11	Neurological conditions	3.20	5.41	7.6	5.33	2.91	1.37
12	Nutritional deficiencies	2.94	1.49	1.4	2.22	3.53	4.33
13	Intentional injuries	2.94	2.52	3.4	5.74	2.49	2.02
14	Diabetes mellitus	2.65	2.58	4.8	5.02	2.82	1.14
15	Congenital anomalies	2.43	1.55	1.6	2.45	2.36	3.15
16	Sense organ diseases	2.34	2.25	1.4	2.39	3.03	0.95
17	Genitourinary diseases	2.31	2.44	2.8	2.98	2.72	1.16
18	Endocrine, blood, immune disorders	1.11	0.71	1.4	1.13	0.77	1.89
19	Skin diseases	0.87	0.74	0.8	1.06	0.88	0.58
20	Maternal conditions	0.74	0.06	0.1	0.20	0.55	1.96
21	Oral conditions	0.64	1.17	1.1	1.08	0.52	0.23
22	Other neoplasms	0.21	0.30	0.3	0.39	0.11	0.11
23	Sudden infant death syndrome	0.07	0.08	0.2	0.11	0.05	0.07

[a]Mental disorders are leading causes of disability in the Americas and the USA and within the three leading causes in Europe

Table 15.2 Mental disorders, neurological conditions, and intentional injuries DALYs regional estimates for 2015 from the World Health Organization (www.who.int). Archived from the original on May 6, 2018

		World	Europe	USA	Americas	Southeast Asia	Africa
	Mental and substance use disorders						
1.	Depressive disorders	2.03	2.58	3.32	3.15	1.94	1.15
	(a). Major depressive disorder	1.66	2.10	2.76	2.63	1.60	0.99
	(b). Dysthymia	0.38	0.48	0.56	0.51	0.34	0.16
2.	Bipolar disorder	0.34	0.44	0.54	0.55	0.31	0.17
3.	Schizophrenia	0.59	0.61	1.12	0.81	0.48	0.18
4.	Alcohol use disorders	0.49	1.29	0.91	0.89	0.33	0.20
5.	Drug use disorders	0.66	1.29	3.20	1.47	0.38	0.23
	(a). Opioid use disorders	0.48	0.94	2.19	0.99	0.30	0.21
	(b). Cocaine use disorders	0.04	0.07	0.45	0.20	0.01	0.00
	(c). Amphetamine use disorders	0.05	0.14	0.16	0.09	0.02	0.00
	(d). Cannabis use disorders	0.02	0.03	0.07	0.04	0.02	0.01
	(e). Other drug use disorders	0.06	0.12	0.32	0.16	0.04	0.01
6.	Anxiety disorders	0.92	1.15	1.84	1.97	0.77	0.42
7.	Eating disorders	0.06	0.10	0.21	0.13	0.04	0.02
8.	Autism and Asperger syndrome	0.38	0.40	0.46	0.48	0.38	0.22
9.	Childhood behavioral disorders	0.24	0.21	0.28	0.33	0.25	0.19
	(a). Attention-deficit/ hyperactivity syndrome	0.02	0.02	0.03	0.03	0.02	0.01
	(b). Conduct disorder	0.22	0.18	0.25	0.30	0.23	0.17
10.	Idiopathic intellectual disability	0.29	0.24	0.18	0.21	0.38	0.17
11.	Other mental and behavioral disorders	0.36	0.46	0.49	0.48	0.34	0.15
	Neurological conditions						
1.	Alzheimer's disease and other dementias	1.09	2.41	3.99	2.08	0.64	0.18
2.	Parkinson's disease	0.10	0.28	0.48	0.25	0.04	0.01
3.	Epilepsy	0.55	0.45	0.30	0.51	0.60	0.63
4.	Multiple sclerosis	0.06	0.15	0.34	0.16	0.05	0.01
5.	Migraine	1.00	1.37	1.30	1.41	1.17	0.41
6.	Non-migraine headache	0.16	0.23	0.20	0.23	0.17	0.06
7.	Other neurological conditions	0.25	0.51	0.97	0.70	0.24	0.08
	Intentional injuries						
	Self-harm	1.41	1.86	2.23	1.66	1.91	0.74
	Interpersonal violence	1.04	0.52	1.12	4.02	0.56	1.03
	Collective violence and legal intervention	0.49	0.14	0.03	0.05	0.02	0.24

References

1. Benyamina A, Blecha L, Reynaud M. Global burden of disease in young people aged 10-24 years. Lancet. 2012;379(9810):29. https://doi.org/10.1016/S0140-6736(12)60021-X.
2. Bertolote JM, Fleischmann A. Suicide and psychiatric diagnosis: a worldwide perspective. World Psychiatry. 2002;1(3):181–5.
3. Collins PY, Patel V, Joestl SS, March D, Insel TR, Daar AS, Scientific Advisory B, The Executive Committee of the Grand Challenges on Global Mental Health, Anderson W, Dhansay MA, Phillips A, Shurin S, Walport M, Ewart W, Savill SJ, Bordin IA, Costello EJ, Durkin M, Fairburn C, Glass RI, Hall W, Huang Y, Hyman SE, Jamison K, Kaaya S, Kapur S, Kleinman A, Ogunniyi A, Otero-Ojeda A, Poo MM, Ravindranath V, Sahakian BJ, Saxena S, Singer PA, Stein DJ. Grand challenges in global mental health. Nature. 2011;475(7354):27–30. https://doi.org/10.1038/475027a.
4. Fountoulakis KN, Lekka E, Kouidi E, Chouvarda I, Deligiannis A, Maglaveras N. Development of the global disability scale (Glo.Di.S): preliminary results. Ann General Psychiatry. 2012;11(1):14. https://doi.org/10.1186/1744-859X-11-14.
5. Gore FM, Bloem PJ, Patton GC, Ferguson J, Joseph V, Coffey C, Sawyer SM, Mathers CD. Global burden of disease in young people aged 10-24 years: a systematic analysis. Lancet. 2011;377(9783):2093–102. https://doi.org/10.1016/S0140-6736(11)60512-6.
6. Harhay MO, King CH. Global burden of disease in young people aged 10-24 years. Lancet. 2012;379(9810):27–8.; author reply 28. https://doi.org/10.1016/S0140-6736(12)60019-1.
7. Kostanjsek N, Rubinelli S, Escorpizo R, Cieza A, Kennedy C, Selb M, Stucki G, Ustun TB. Assessing the impact of health conditions using the ICF. Disabil Rehabil. 2011;33(15–16):1475–82. https://doi.org/10.3109/09638288.2010.527032.
8. Linton S. Claiming disability: knowledge and identity. New York: New York University Press; 1998.
9. O'Connor R, Sheehy N (2000) Understanding suicidal behaviour. Wiley-Blackwell,
10. Oliver M. Social work with disabled people. London: Macmillan; 1983.
11. Oliver M. The social model of disability: thirty years on. Disability & Society. 2013;28(7):1024–6. https://doi.org/10.1080/09687599.2013.818773.
12. Ormel J, Petukhova M, Chatterji S, Aguilar-Gaxiola S, Alonso J, Angermeyer MC, Bromet EJ, Burger H, Demyttenaere K, de Girolamo G, Haro JM, Hwang I, Karam E, Kawakami N, Lepine JP, Medina-Mora ME, Posada-Villa J, Sampson N, Scott K, Ustun TB, Von Korff M, Williams DR, Zhang M, Kessler RC. Disability and treatment of specific mental and physical disorders across the world. Br J Psychiatry. 2008;192(5):368–75. https://doi.org/10.1192/bjp.bp.107.039107.
13. World Health Organization. WHO psychiatric disability assessment schedule. Geneva: WHO; 1988.
14. World Health Organization. The ICD-10 classification of mental and behavioural disorders: clinical descriptions and diagnostic guidelines. Geneva; 1992.
15. World Health Organization. International classification of functioning, disability and health. Geneva: WHO; 2001.

Chapter 16
The Mind-Brain Problem and the Problem of Free Will

The mind-brain problem and free will, its meaning, its historical origins, and its significance to philosophy, religion, and society are all important for psychiatry. It is important for psychiatrists as well as mental health workers to understand mental events as well as the limits of our understanding and the consequences of these limits in the way we perceive mental disorders and manage patients suffering from them as well as their families.

The efforts to understand the nature of human kind, and especially the nature of the human mind and behavior, can be traced several thousand years ago. They were characterized by complex turns and unexpected alliances and, especially in the beginning, by the lack of the scientific eye and method. Free will is at the center of human existence and human experience; as such lay persons but especially patients are interested to know about it. Their concern is how free will could be affected by mental illness but also from medication. Activists consider free will to be the ultimate human characteristic, indestructible, and almost sacred.

The law often demands the assessment of capacities related to it and often in a bimodal (yes/no) way, and because of this, some authors even consider the compatibility question, that is, whether everything is pre-determined, to be particularly relevant to forensic psychiatry [204]. Free will is located at the core of morality. Morality prerequisites the existence of free will, that is, persons are responsible for their actions and therefore their actions characterize them and are also the source of important positive or negative consequences from the side of the others as individuals but also from the society as a whole [148, 167, 211]. It is impossible to create a system of morality without free will [175]. To be more pragmatic, the whole societal structure of moral responsibility, law and order, reward, and punishment, even the very societal structure itself, relies on the assumption that there is accountability because of the existence of free will during decision and choice-making.

The basic philosophical question (the "basic argument" according to Galen Strawson; born 1952) is whether something can have no cause, in other words whether it could be the cause of itself (causa sui). However this is exactly what

K. N. Fountoulakis, *Psychiatry*, https://doi.org/10.1007/978-3-030-86541-2_16

moral responsibility demands, and Strawson responds negative to this question [205]. This is what Robert Kane (born 1938) defines as "ultimate responsibility" [75, 122–129], and it is absolutely necessary as a prerequisite to justify the "ultimate punishment," which is the basis for many religious beliefs. The "original sin" in Christian theology is considered to reflect the prototypical "ultimate free will" and "ultimate punishment." Whether "ultimate responsibility" is a possibility even in philosophical terms is a matter of hot debate. Many authors propose less strong versions of free will and accountability in order to be able to justify responsibility to a reasonable extent [217]. This is again in accord with the Christian theology which according to both St Augustine and Martin Luther suggests that humans no longer have ultimate free will after the fall, and it is only because of God's grace that only a few could be eventually saved.

Before continuing we need to make a general comment. The topic of "free will" which partially overlaps with the "mind-brain" problem is the focus of diverse disciplines of human thought which try not only to provide with answers but even to understand each other and communicate. This stretches the limits of each discipline and to the eye of the neuroscientist or the clinical psychiatrist, and many proposals could not only balance at the borders of schizotypal thinking, but they could indeed be real schizotypal thinking themselves. This only reflects the agonizing efforts of human beings to understand their nature and destiny and often to tailor science to their psychological needs.

An important clarification would be that in the text that follows, the word "metaphysical" is used to denote what is not considered to follow the laws of nature or reason and cannot be the focus of scientific study and research. It will include religious, spiritual, paranormal, spacy, and all those approaches that defy or go against the basic principles of science. Maybe the word "supernatural" could be more suitable, but one problem here is that the word is rejected by those who really advocate for a metaphysical and essentially supernatural approach [80].

16.1 The Mind-Brain Problem

In this frame, the early attempts are sacred rather than scientific or philosophical and could be classified into [180]:

- Animistic.
- Mythological.
- Demonological (religiousness of late antiquity with strong ethical elements).

As described in detail in Chap. 3, originally in ancient Egypt and Mesopotamia, the observation of the seasonal changes in the waters of the rivers, the floods, and the subsequent changes in the soil led to the conceptualization of the human body as an analogue to the river valley and pointed to the importance of moisture but also to the balance of elements and mechanisms, leading to the first idea of the "equilibrium." Also, the very early moral ideas emerged in Egyptian mythology, suggesting that after death, the heart, which during life records all good and evil deeds, will be weighed against a feather to assess sin and guilt (Fig. 3.6). Later, it was Pythagoras (580–496 BC) who suggested for the first time that the brain is the organ where mental phenomena take place and proposed that the mathematical concepts of balance and ratio should be used to explain how and why physical characteristics of the body (e.g., moisture vs. dryness, cold vs. hot) lead to mental disorders. He was the first to conceive the concept of "bipol," and accordingly he arranged the characteristics and the elements. He went further and proposed a four-component structure (tetractys) of humans; one component was the body and the three other (three was another sacred number) corresponded to "νους" (mind, corresponding to reason and intelligence), "ψυχή" (soul, corresponding to affect), and "πνεύμα" (spirit, corresponding to the immortal part). But the real structure of the tetractys reflects two bipols: mortality (body) vs. immortality (spirit) and reason (mind) vs. passions (soul). Pythagorean metaphysical and religious beliefs also included immortality and the transmigration of souls, and these beliefs influenced his scientific approaches especially concerning mental phenomena and human behavior (Fig. 16.1).

The traditional dualistic approach (also known as "Cartesian dualism") was shaped by Rene Descartes (1596–1650; Figs. 6.1 and 6.2) and proposes that the mind is non-physical and its states are not identical with the physical states of the brain. Descartes also suggested that mind and brain have some kind of interaction through the pineal gland. The biggest problem with dualism was the inability to answer the simple question how it happens, and biological conditions (e.g., alcohol intoxication) can have such a clear and profoundly devastating effect on human decision-making. Benedict Spinoza (1632–1677; Fig. 6.5) rejected Cartesian dualism in favor of what was called "neutral monism." Further, he suggested that psychological function is a natural phenomenon and subsequently the human mind is subject to the laws of nature. In an ethical extension of these views, the concepts of "good" and "evil" are merely the products of the emotions of pleasure and pain. John Locke (1632–1704; Fig. 6.6) also rejected the presence of Descartes' "innate ideas" and suggested that only experience shapes the human mind. Similar concepts were expressed by Thomas Hobbes (1588–1679; Fig. 6.7). Locke and Hobbes

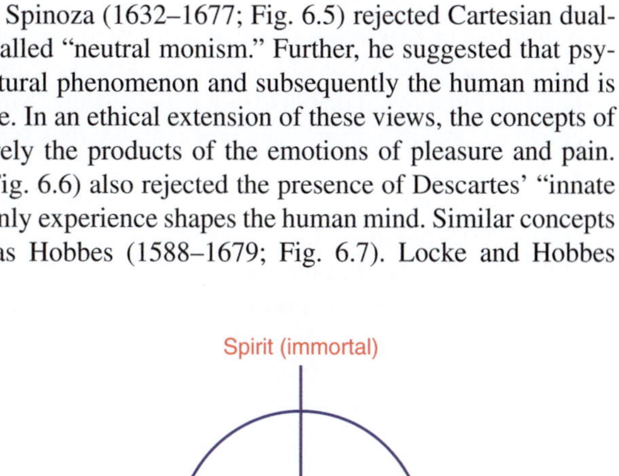

Fig. 16.1 The Pythagorean tetractys (tetrad)

seemed embarrassed when discussing the possible total lack of inherent ideas. Both philosophers utilized the “tabula rasa” theory to promote the idea of universal human rights.

Other compromising solutions appeared in the form of property dualism, which proposes that although mind emerges as a result of the proper and ultra-complex organization of the brain, mind, and brain/body have distinct properties, and consciousness cannot be reduced to neurobiology. The properties of the mind are “novel,” that is, they do not exist in the participating material compounds. However, this seems only superficial. Property dualism avoids direct answers. Almost all modern proposals [64, 65] boil down to this kind of dualism even when they try not to admit it. There are exceptions however since several modern authors suggest that quantum physics point to observable effects which cannot be explained by any narrative in space-time as we conceive it [89]. These “exotic” theoretical physics which are often hardly really scientific offer room for the presence of a “quasi-scientific” but essentially metaphysical immaterial (supernatural) agency and permit the survival of ideas like free will, responsibility, and religious faith [80, 207].

Overall, since the Renaissance, a number of basic frameworks were developed in order to understand the nature as well as the mind-brain relationship. The old distinction was that of dualistic vs. monistic approaches; however eventually there was a development of many mixed and, according to the author of the current book, pseudo-monistic approaches. This, currently, suggests that the only real distinction is not dualism vs. monism but rather scientific vs. unscientific theories.

The most important of these theoretical considerations are as follows:

- **Substance or Cartesian dualism** (after Rene Descartes; Figs. 6.1 and 6.2) proposes two distinct kinds of substances and that the mind exists outside the body/brain and in a realm distinct from that of nature. Mind and brain interact in the epiphysis, and their events and processes are thus synchronized. It is compatible with the religious belief on the immortality of the soul [51, 106]. More recent advocates of this approach such as the Nobel laureate Sir John Carew Eccles (1903–1997) proposed that brain and mind interact in the left prefrontal cortex or at the quantum level of the sub-cellular organelles [58–61]. This theory implies that at least partially the mind obeys to the laws of nature since it is influenced by the states of the body after interacting in the epiphysis. As a consequence, one could argue that the Cartesian idea does neither suggest a dual nature nor a clear distinction between mind and brain but a continuum from natural to metaphysical and essentially to the supernatural. It does not involve any divine power, so it is somehow flexible in its interaction with scientific theories. It is also closer to “property dualism” rather than Leibniz’s “psychophysical parallelism.”
- **Psychophysical parallelism** is a metaphysical theory mainly advocated by Gottfried Wilhelm von Leibniz (1646–1716; Fig. 6.8) who rejected Descartes’ proposal concerning the interaction of mind and brain in the epiphysis and suggested that God has created a pre-established synchronization between the two and, after that time point, mind and brain continue to function independently but in harmony because of this initial “calibration” [179]. Leibniz ideas are abso-

lutely dualistic as they suggest a radically different nature for brain and mind which are considered to be absolutely independent. It is also clearly unscientific since it utilizes a metaphysical explanation (divine or supernatural intervention) in order to fill in the gaps and shortcomings of this theory. Kant's ideas have elements from both Descartes and Leibniz, but Kant is much closer to Leibniz [217].

- **Occasionalism** is an idea which was first developed by Muslim philosophers and especially by Al-Ghazali (1058–1111) in Bagdad, and in Europe it was advocated among others by Louis de La Forge (1632–1666) and Father Nicolas Malebranche (1638–1715), who also rejected Descartes' ideas and suggested that God intervenes at each occasion an event occurs and synchronizes the functioning of the brain and mind [94, 152]. Malebranche went further and abused a pregnant dog in public specifically in order to show that it is a soulless automaton, a mere machine, not deserving sentiments or treating it as a living being. In essence it is a variety of those ideas which directly involve divinity into the everyday mental function of humans, was in complete accord with the religious ideas of that time, and is, of course, radically incompatible with science.
- **Property dualism** suggests that, although mind emerges as a result of the proper organization of the brain, mind and brain/body, have distinct properties, and consciousness cannot be reduced to neurobiology. The exact meaning and the consequences of the phrase "cannot be reduced to neurobiology" are important to discuss and understand. This theory is based on the assumption that only one kind of substance exists, but at the same time, there are two kinds of properties. A simple metaphor for this theory is water; it is composed by atoms of hydrogen and oxygen, and the creation of the water molecule can be causally traced back to them. However the properties of water cannot. These properties are "novel," that is, they do not exist in either of the participating compounds. However, this seems only superficial. As it is also evident in the metaphor with water, eventually all "novel" properties are based on already existing properties in the composing parts; it is the human inability to monitor complex interactions in the microcosm that gives birth to the illusion of property dualism, and at the end of the day, the core question is whether the laws of nature are followed. This is the core question, and property dualism avoids the direct answer to this issue; essentially all modern proposals (even when they try not to admit it) boil down to this kind of dualism, which, in a similar way to Cartesian dualism, argues for a continuum between natural and metaphysical/supernatural. Some authors try to utilize a compromise between property dualism and science by suggesting modern physics which seem somewhat "exotic" could substitute metaphysics. They argue that this new science of physics holds the key to this "quality shift" in properties and argue that when a theory applicable to the world below the Plank length (10^{-33} cm) exists, then our questions concerning consciousness will be answered.
  - *Emergent materialism (emergentism)* argues that increasingly complex organization of matter gives rise to novel and higher-level properties which are not reducible to their basic components. Accordingly, the complex organization of the brain gives rise to the mind, but the mind is not reducible to the neuro-

biology of the brain [64, 65]. The extreme version of emergentism proposes that mental states have novel and absolutely distinct properties and thus get too close to dualism.

- *Non-reductive physicalism* suggests that mental states are based and caused by neurobiological phenomena, but they are not ontologically reducible to neurobiology. What exactly this means and how it differs from other property dualism theories is obscure.

  Anomalous monism refers to the idea that although mental events are physical events, they are "anomalous" (from Greek *ανώμαλος* meaning not normal), that is, they do not strictly follow the laws of nature [45, 46]. It is a version of the identity theory. This approach denies the possibility of reducing psychological laws to physical laws. It is rather hard to think how this theory could be conceived. Davidson uses the phrase "there are no (strict) psychophysical laws," but it is unclear whether he believes that "soft" laws exist. On the other hand, the concept that there are no laws is also unclear. Technically this might mean that mental phenomena are governed by some kind of a random choice generator.

  Predicate dualism argues that there is only the physical ontological category of substances (only matter) and properties, but it is impossible to describe with human languages the mental events in terms of physical mechanisms. This is an easily observed fact complicated by the gaps in our knowledge. It inevitably gives rise to "folk psychology" a term used to describe human capacity to explain and predict the behavior and mental state of others [46, 76]. Especially Jerry Alan Fodor (1935–2017) argues that it is impossible to reduce theories on the basis of behavior of functionally defined kinds, which are most often multiply realized. This is in accord with the approach of functionalism, and an example he uses is that of "money" which are a functionally defined kind and can be multiply realized in the form of precious metals, exchangeable goods, or services. However, as in the example of water in property dualism, also here Fodor misses the core need for a "bridge" concept which is essentially a reduction of these material representations of "money" and reflects the basic property of "economic value," shared by all forms money can realize in. Fodor argues it is impossible to reduce a law about money to a law of physics [78], but this is probably wrong. On the other hand, there seems to be a gap between the materialistic neurobiological domain and the socio-political and economic domain, a gap the discipline of sociobiology has developed in order to fill [79]. The theory of predicate dualism has some similarities to the "Roschian concept" [141], which is in accord with the Wittgensteinian family resemblance concept, and has been proposed as a model for the definition of mental disorders (see Chap. 13). This concept suggests that a network of similarities rather than precise definitions is more adequate to describe a group of similar objects [181, 226]. The case of the mind-brain problem in fact suggests the presence of a continuum from the understandable (natural, brain) to the impossible to conceive (metaphysical, mind). It also has many agnostic elements since it describes and delineates the field, but it does not try to give real answers.

Biological naturalism suggests that mental states are causally but not ontologically reducible to neurobiology [42].

- *Epiphenomenalism* suggests that all mental states are causally related to physical states, but they do not exert any influence on them, and they are mere reflections (mental states are completely dependent on neurobiology, but they are causal dead-ends) [38, 86]. Some authors suggest that since a bulk of data suggests that conscious experience is created by non-conscious processes, there is a strong scientific support for epiphenomenalism, at least for some part of human internal experience.
- *Interactionism or interactionist dualism* proposes that body and mind are distinct and independent, but they interact and exert causal effects on one another. It is very much in accord with common sense and close to Cartesian dualism, but it is hardly scientific as a theory. It implies that the laws of nature work for the metaphysical and vice-versa the metaphysical exerts a direct effect on nature [179]. This is a convenient way to avoid the concept of "supernatural" but simultaneously keeping its benefits in place.
- *Panpsychist property dualism or panpsychism* is one of the oldest mind-body theories and in many aspects is very close to animism. Many philosophers including Plato, St Augustine, Spinoza, and William James have discussed topics pertaining to this area. According to it, all matter has a mental aspect, and mind or a mind-like aspect is a fundamental and ubiquitous feature of reality [133, 193, 228].

Ludwig Wittgenstein (1989–1951) criticized the Cartesian dualism and went on criticizing all kinds of dualism [15, 225], but he could not give any real solution himself which is not strange given the knowledge of his time and the fact that he was not a neuroscientist himself.

- **Materialism** is the philosophical monism, which proposes that matter is the single fundamental substance in nature. Accordingly, life, mind, and mental states but also human society are products of material interactions and strictly follow the laws of nature [119].

  Materialism includes the following basic currents:

- *Naive materialism* corresponds to an antiquated approach, which argued that the universe is composed of a limited number of elements (e.g., the four-element theory of Empedocles).
- *Metaphysical materialism* suggests that there are some kind of metaphysical properties in the mind that make it possible to understand itself as exclusively a product of matter and the laws of nature [23, 52]. Of course this is not real monistic materialism; it resembles at least some type of property dualism.
- *Dialectical materialism* considers all phenomena, either mental or sociopolitical, as results of the relationships and interactions among material parts of nature. It was mainly developed by Karl Marx (1818–1883; Fig. 8.20) and Friedrich Engels (1820–1895) on the ideas of Georg Wilhelm Friedrich Hegel

(1770–1831; Fig. 8.18). It gave birth to historical materialism. It rejects any metaphysical/supernatural component, and in this sense, it is in line with scientific principles.

- *Reductive materialism* considers all mental phenomena to be able to be reduced to simpler natural-material events in the brain [65, 66]. In its extreme form it gives birth to:
- *Eliminativism or eliminative materialism* has been proposed by Paul Churchland (born 1942) and eliminates all other concepts except matter and nature [37, 39, 142]. In principle, eliminative materialists argue that mental states do not exist per se, and therefore it has elements of epiphenomenalism. It is rather unpopular among philosophers since it leaves no space for complex elaborations. Churchland rejects the "folk" and "common-sense" approach to psychology, which says it has not progressed at all since the ancient Greek philosophers, insists that we should follow a hard scientific method in our efforts to solve the mind-brain problem, and rejects all common-sense and lay people's opinion and approaches. If such a scientific method is followed, then our endless list of entities will be shortened because most entities will either be reduced or eliminated [80].
- *Physicalism* is closely related to materialism, and the terms are often used interchangeably, but it has evolved on the basis of the development in physical sciences to incorporate more recent advances in our understanding of the nature (force and energy, time and space, etc.) [28, 230]. The basic principle is that no matter the nature of things (matter, energy, or else) the laws of nature should be taken into consideration and any metaphysical elements should be eliminated.

Neuroscience has progressed much during the last half century and has elucidated several aspects of the mind-brain problem at the level of atomic, molecular, and cellular interactions in the brain [131, 156]. This has made materialism the main source of scientific skepticism about free will, *because it precludes any metaphysical and supernatural interpretation of mental phenomena and claims that the laws of nature govern also human thinking and behavior* [75, 187]. Subsequently the subjective human experience is just psychological epiphenomena or illusions, simple by-products of brain functioning. The term "epiphenomenalism" was coined by Thomas Henry Huxley (1825–1895) in 1874 [117].

Materialism is closely related to science, and it is considered to be its result since it is supposed, that is, basic assumption is respecting the laws of nature and trying to interpret the universe as well as the human experience on the basis of these laws. Reductionism is at the core of scientific thinking and research and suggests that phenomena can be described in terms of other simpler or more fundamental phenomena. Therefore it implies that all complex systems are the sum of their parts and can be explained in terms of scientific laws and theories [110, 153]. The more events we can explain with simple fundamental ways, the more complex explanations are obsolete, a phenomenon often called "Occam's razor" [3, 184].

Ontological reductionism suggests that higher-level entities are nothing but the sum of their parts. This means that only the entities at the lowest level are real, while higher-level entities are composite derivatives [22]. Some authors argue that what is critical is the organization of higher levels [199, 209], but they neglect that this organization emerges from lower levels unless there is external intervention of some kind.

This brings discussion to the question whether in living systems and particularly in the human brain causation is downward or emerges from lower levels. Observation shows that living systems develop themselves and self-organize which easily leads to a non-reducibility argument concerning these systems [70, 121].

The next important question at the core of materialism concerns an important characteristic of living systems which are autopoietic, which means they develop themselves by creating the controls of the matter-energy flow and eventually make the self-organization possible. This could be easily said that it leads to a non-reducibility concerning these systems, a question which is central not only to the mind-brain problem but to the mystery of life itself [70, 121]. However, the rules of arguing are not easy. Is it possible that matter has self-organizing processes that seem to go against the law of entropy to take place? The theory of synergetics (cooperation) tries to explain how well-ordered structures seem to spontaneously arise in unstable systems through self-organization [98, 99]. However this theory is hypothetical and makes unproven assumptions on the role of certain variables in causation, while it introduces terms like "enslavement" in order to favor slowly changing variables (ordering parameters) against quicker variables [202].

The greatest problem with systems theory is that indeed there are systems including neural networks which are more than the sum of their parts and they seem to possess properties that emerge de novo. Subsequently their functioning cannot be reduced. However there is a fundamental flaw in this line of reasoning: all these systems are artificially made by humans; in other words there is a creator, an intelligent plan which creates these systems, and in this way, properties are inserted in a top-down way. Such systems do not exist in nature. The only exception is life, which in turn leads to the fundamental question: is life the result of bottom-up natural events governed entirely by natural laws or is there a creator and an intelligent plan (top down)? What systems theorists neglect is that you cannot have them both, that is, a bottom-up evolution and de novo emergence of properties.

The mind has of course additional characteristics, including the selection of focus of attention [84] and consciousness which differs from "free will" or "decision-making" [100, 160] leaving space to suggest that a possibility is that our awareness of intention may be constructed after the event, and its timing may be manipulated by contextual factors. In simple words, this means that the experience of "decision" is a psychological epiphenomenon (an illusion) which occurs after the actual decision has been made [136, 222]. An alternative interpretation could be that "free will" is the selective focus and choice of action in the form of a top-down control and independently of all impinging forces [84], while a more minimal approach suggests that we only have moral judgment on our already taken decisions (meta-responsibility) given that we know we cannot always take control of immediate responses [198] or that that free will concerns the modification of future decisions though synaptic change through memory [208]. However, these are simply convenient bypasses of the real problem.

- **Idealism** on the contrary considers mind and mental phenomena to be the first-order realities. Matter is subject to mind or even constitutes a simple epiphenomenon of complex mental processes. This is in accord with a number of theological doctrines especially East Asian. In this frame, reality is mentally constructed and immaterial [149, 194]. It is the "mirror image" of eliminative materialism, and it is true monistic since it rejects the existence of anything other than the mind. As a consequence, it rejects nature and its laws and embraces the metaphysical in an unconditional way:
- *Subjective idealism* questions the human ability to perceive the environment, and since this occurs through mental processes with a strong subjective element, subsequently matter cannot be given causal priority over mind.
- *Objective idealism* argues on the existence of an objective consciousness which exists before and, in some sense, independently of human ones. According to this, all entities are composed of mind or spirit [2, 28, 53, 176].
- *Transcendental idealism*, founded by Immanuel Kant (1724–1804; Fig. 7.5), maintains that by using a priori principles, the mind creates the world we perceive by connecting intuitions with concepts and putting them on a temporal relationship; this creates the perception of the world in the form of space and time [155]. According to Kant, freedom of will and choice identify with morality; the person should be radical and should not hesitate to act against existing inclinations, desires, communities, or traditions when reason dictates so. Kant combines this idea with a theory which is basically dualistic in nature [217].
- *Absolute idealism* was found by Georg Wilhelm Friedrich Hegel (1770–1831; Fig. 8.18) and constitutes an extreme form of "idealism." While idealism refers to the concept that the experience of reality by humans is a subjective mental construct, "absolute idealism" breaks the boundaries between opposing concepts with "sublation" (in German: Aufhebung, integration without elimination or reduction), and thus Hegel is considered as the originator of the thesis, antithesis, and synthesis triad. By following this line of thinking, from Plato to Leibniz and Kant, and by accepting a higher order of existence for the human soul in comparison to the material world, Hegel subsequently put an absolute emphasis in human self-determination and free will. However, what is unique with Hegel is that instead of rejecting the Cartesian and Kantian dualism, he introduced a synthesizing approach of mind and brain (monism). He started with the definition of "being" and "nothing," and while Kant argued that humans can have knowledge only of a "finite world of appearances" (subjective), Hegel developed the idea that the "absolute" (monistic approach to the mind/brain problem) excludes multiplicity (dualism); therefore subjective and objective are one but in the sense that the mind determines matter. In simple words this means that the mind creates reality, and subsequently reality is not perceived by the brain, but it is an absolute and complete product of the mind. Thus, "reality" was defined as "what determines itself." On the contrary "finite" things (according to Descartes) define each other by their borders and are essentially independent from the mind of the observer. Behind, this is the idea that the "finite" (object) has to become "infinite" (subjective creation) in order to reach reality (subjectively created)

[146]. In this frame the brain and the mind are one with properties which are not under the laws of nature (a synthesis leading to monism). Hegel synthesized mind and body dualism in "true infinity," thus suggesting that it is possible to conceive it rather than accepting it without understanding. According to Hegel the mind and the brain are neither incompatible nor compatible, but the mind self-transcends the brain.

In all its jargon, idealism is certainly correct in the way it describes how the human brain perceives the environment, and to the extent, it accepts that there is an objective external reality which contributes to the subjective experience. We know that the brain "creates" an internal "proxy" of the environment on the basis of input from the sensory organs. This representation is in fact a distorted image with much subjective input and interpretation often relying too much on assumptions. However, reliability and validity of the result are so big that they permit the person to survive through complex challenges in a generally hostile environment. This was much more obvious some millennia ago, but it is still evident. In evolutionary terms, there is no need the internal representation of the environment to be an absolutely exact replica of the actual external environment. This would consume too much of resources, and evolution is an economic and saving process and invests as much as needed for adequate function and survival of the species, but no more than that. No evolutionary pressure exists beyond the point of adequate adjustment and survival. Thus, the way humans (and animals in general) perceive the environment is an admixture of reliable and valid (objective) charting in combination with input coming from internal sources like pre-existing inclinations, memories, etc. (subjective). Evolutionary pressure improves both components to the extent the cost-to-benefit ratio is favorable. It is interesting that the common denominator of this human experience is big enough across individuals, so that different persons manage to communicate and even more coordinate their actions. As a result of the above, the biggest problem with idealism is that it cannot explain how differently and independently constructed realities, which correspond minimally or not at all to an external object, seem to agree to a significant degree among a large number of people. If there is no external object and everything is a kind of fantasy (matter is the product of the mind), then we all live in a dream (similar to the *Matrix* movie), and all discussions have no meaning. If fact even these discussions are the product of our minds; they are not "real." It is a dead-end argument, and it is the same also in the case of the existence of an external object with minimal effect on the construction of reality. If on the other hand there is at least partial objectivity in the way the mind perceives the environment, and of course external objects do exist in reality, then this objectivity is shared by all humans, or at least there is a minimum of it which is shared. This is not rejected by most idealists, but the extent to which much of the processing is also shared is rejected. But this is exactly what we scientifically know today, but we did not during the era Kant and others lived. And if we add an evolutionary perspective into our thinking (which was not available in Kant's time), then the conclusion is that the mind adjusts and responses to the challenges of the objective environment ("reality") and not vice versa. Mind cannot be primary and matter secondary. Today we also know that the environment pre-existed the mind and that the mind is probably an evolutionary product of the struggle of the living organisms to survive, multiply, and expand. Thus, idealism is incompatible with Darwinian evolution, which gives primacy to matter, and demands the acceptance of creationist approaches. This is a detail not discussed at all, since most idealist philosophers are also atheists.

- **Neutral monism** was proposed by Ernst Mach (1838–1916), William James (1842–1910), and Bertrand Russell (1872–1970) and argues that that there is only one kind of elements, and they are neither mental nor physical. Matter and mind are two of the many facets of things, which exist in the form of neutral elements. Their ideas are based on perception as the basis of human understanding of the world [5]. While it looks like a monistic theory, the fact that it does not exclude metaphysics puts it close to a dualistic approach without a sharp distinction of substances.
- **Functionalism** suggests that the state of the mind could be independent from the physical state of the brain and thus persons with quite different brain state could have the same mental state. This concept is compatible both with materialism and with dualism, and essentially it reflects a compromise. While its basic elements go back as far as Aristotle and Hobbes, in full it has been conceived during the end of the twentieth century [13, 111, 138, 169, 177, 189, 191]. However, this approach implies that two mental states in different persons could be absolutely identical, which is rather impossible, taken into consideration the unconceivable number of possible states and certainly it cannot be proven. While technically as a theory meets Popper's falsification criterion, practically it is impossible to test. Then the next theoretical question is that if indeed mental states do not correspond to brain states, then what is their origin and which laws govern them? This leads to some type of dualism, and it is a limitation to the concept that mental states should be characterized in terms of functioning alone. As a school, functionalism originated in the USA in the late nineteenth century by William James, James Rowland Angell, George H. Mead, Archibald L. Moore, and John Dewey (see Chap. 9 of the current book) as a reaction to the German school of structuralism.

For someone self-identified as a scientist, like the author of the current book, it is impressive that today there are still reputable philosophers of mind who are proud to be considered dualists, that is, they accept the metaphysical (and essentially supernatural) in their way of thinking [80]. One such example is David Chalmers (born 1966) [29, 30, 32, 101, 188], who characterizes his view as "naturalistic dualism," and essentially it seems to be a variant of anomalous monism. Interesting are his ideas concerning that zombies (mindless human beings with a fully functioning brain) are logically possible and subsequently qualia and sentience are not fully explained by brain function alone. This is a top-down theory: since he accepts dualism, separation of the two components (mind and brain) is possible, while both retain their function. Well, nobody has ever observed this. The concept of "zombies" was conceived for the first time by Robert Kirk (born 1933) [130]. In this frame, Chalmers argues that consciousness is a fundamental property ontologically autonomous of any known (or even possible) physical properties [31]. This is of course a leap of faith since all the neuroscientific data are against the possibility for the existence of "zombies" and there is not the slightest even remote clue that they could constitute a possibility. Nobody has ever seen a zombie, and we can say with confidence that nobody will never see one. In a panpsychic extension of his theories, he even suggests that all information-bearing systems may be conscious and claims that his views are the end result of a line of thinking that goes back to Leibniz's "mill" argument [137].

There are two basic and widely followed beliefs which make dualism appealing and very important both for lay people and for intellectuals. These issues also explain why philosophers of the "age of reason" were very reluctant to accept the materialism and the scientific method when concerning mind and behavior:

- *Immortality of the soul and life after death.* Dualism is absolutely necessary for such supernatural beliefs to survive. Although there are some quasi-materialistic and pseudo-scientific arguments that the mind could be like a kind of software that could survive the death of the body and even be transmitted and stored elsewhere [44], there is absolutely no hinds how this might happen, and on the contrary, the way brain works plus the laws of nature, as we currently understand them, rather preclude such things from happening. Such ideas are excellent as the basis for sci-fi adventures, e.g., in the *Star Trek* universe (which by the way is one of the favorite shows of the author of the current book), but they can serve recreation alone; they have absolutely no contact at all with scientific conceptualizations. So, this is the main reason why dualism survives no matter the accumulating evidence against. The claim that recent neuroscientific data do not reject dualism [44] is essentially mistaken. Recent advances do not wipe out the metaphysical possibility of dualism as they cannot do so with any metaphysical or even supernatural explanation to the extent human beings choose to believe it. What they do however is to explain in terms of natural laws the functioning of the brain step by step, thus pushing the borderline of metaphysical explanation further away.
- *Free will.* In a deterministic universe (see below), dualism is the only solution for the presence of a "true" free will, the way humanistic and existential philosophers think of it and conceive it. A metaphysical/supernatural mind, which makes decisions independently of natural law and even against the needs of the body, is at the core of many philosophical and religious doctrines. Even in clinical psychiatry, many mental health professionals find it more convenient to utilize a mind-brain dichotomy when elaborating on diagnosis. Thus, while personality and values are characterized by the presence of "free will," on the contrary, mental illness is characterized by the enslavement by neurobiology. There are empirical data in support of such a way of think-

ing in professionals at clinical settings, with dualism serving not only as a compromise between their scientific training and personal philosophical or religious beliefs but also as a tool to diagnose individual patients; those considered to be responsible for their actions are "neurotic" and those not responsible are "psychotic." The first category might not be ill at all, and psychotherapy is the most appropriate way of treatment, while for the later, pharmacotherapy should be the most efficacious intervention [147]. Although such research provides with some valuable insight on the way mental health professionals understand their discipline and arrive at diagnostic and therapeutic decisions and simultaneously maintain their own values system, the real meaning of such findings and their interpretations is complex and of questionable reliability and validity.

16.2 The Origins of the Concept of "Free Will"

The concept of "free will" concerns an old debate on the very nature of human existence. While a basic definition could be that it refers to the ability of humans to choose between different possible actions unimpeded [161], the details and the essence of this human characteristic remain obscure. Several philosophers including Arthur Schopenhauer (1788–1860; Fig. 16.2) base their concept of free will on the ability of the mind to make decisions based on reason, thus implying that passions are those that limit free will [185]. Free will is still the center of discussion and debate between philosophy, religion, and science, with profound political implications. Therefore, in order to elaborate on it, one needs first to explore the historical roots of the concept.

The roots of discussions concerning free will started already since antiquity although with a different basis, and it took centuries for the concept and the field to mature [82]. The elementary existential questions whether humans do have real control over their actions and to what extent emerged already since the antiquity and occupied the thoughts of Plato, Aristotle, Epictetus, and especially the Stoic philosophers with conflicting opinions [14, 154, 186]. Both Plato and Aristotle probably shared the practical plain man's belief in free will, and in accord with the bipartite or tripartite nature of the soul, they recognized reason as the source of free will but also desire which was independent of reason [14, 182]. This was in contrast to the Socratic teachings that reason can and should prevail over passions. Aristotle pointed to the distinction of things we do εκόντες (hekontes meaning willingly) and "άκοντες" (akontes meaning unwillingly). While the traditional translation (probably because of Cicero) suggested that Aristotle was meaning voluntary vs. involuntary, Aristotle probably was meaning reason vs. passions. Overall, Socrates, Plato, and Aristotle as well as the Stoic philosophers suggested that a wise and virtuous person is obliged to make the right thing for

Fig. 16.2 Arthur Schopenhauer (1788–1860)

the right reasons, leaving little space for true free will. According to this approach, the Stoics suggested that it is only "εξουσία αυτογραφίας" (exousia autopragias, the ability to act on one's own) which reflected freedom and independence [14, 55]. These concepts identified bad decisions and wrongdoing with foolishness and the inability to distinguish between good and evil, right and wrong, and important and unimportant, and this could maybe constitute an early cognitive theory of personality. The moral-political consequence of these theories was that they considered the foolish person as being enslaved to evil instead of following the divine law. This is a conceptualization of behavior and mental health which is also discussed in Chap. 14. On the other hand, Gnostic philosophers built their arguments against the existence of a good God by exposing the inevitable failure of decision-making and the absence of true free will in humans plus the resulting suffering.

Origen of Alexandria (c. 184–c. 253), also known as Origen Adamantius, was probably the first Christian author ever to elaborate on free will [18]. However, the very term "free will" (liberum arbitrium) was introduced by Christian philosophers in the fourth century AD and was formally described by Thomas Aquinas (1225–1274; Fig. 4.7). According to his views, the essence of human existence was the immortal soul which consisted of intellect and will. This was the early mentioning of the importance of "free will" and as such is somewhat a precursor to modern existential theories (see also Chap. 4) [54].

Therefore, the concept of free will has emerged more as a social and theological/religious need to define the responsibilities of the individual towards the society and God, rather than as the product of scientific curiosity. By late antiquity it was reduced to the freedom of the mind rather than freedom of action, and this was reflected also in Christian theology. It is also determined that behavior can be corrected with the application of the right measures from the side of the society or of the authority in general. This approach was more evident in western societies, and subsequently, the concept as it is discussed today emerged in Europe. For society, it serves as the aim to justify the power of law and the obligation of all the members of the society to follow rules and laws and behave according to moral standards; otherwise they will be held responsible of their inappropriate behavior, with possible punishment resulting as a consequence. For religion, the need was similar, that is, to constitute the person absolutely responsible for the following or not of the theological teachings and commands [216]. It is interesting to note that for those the concept served, the approach was practical and rather straightforward. In spite of theological discussions, the church utilized an approach implying the full existence of human free will, often beyond true human capacity; on the other hand, already since antiquity mental illness and conditions like rage were considered by the legal system as restricting the ability to free will and therefore unlawful deeds during such mental states should be punished in a light way or not at all. This is an indirect proof that the theoretical concept of free will emerged out of a practical need, and this practical need had priority with the use of common reason, over the theoretical elaborations.

In the teachings and philosophy of Christianity, Judaism, and Islam, free will has a prominent position because it implies that humans have the ability to choose good from evil and are responsible for their choices in front of God. The question was whether and how "human will" could go against the "will of God" or the divine plan for the world. Even the disobedience of the original sin essentially refers to human free will and the ability to have an informed choice between good and evil.

For Christianity, leading figures were St Augustine, Thomas Aquinas, Luis de Molina, Jacobus Arminius, Martin Luther, and John Calvin (Fig. 16.3). St Augustine was a supporter of divine predestination and influenced all the others, especially Protestants [105]. According to him, the original sin was committed by a rational being who was wise and virtuous, and this is why the act resulted in the absolute punishment. But after the fall, humans were no longer free to liberate themselves. Thus, St Augustine had a strong commitment in the ultimate need for God's grace, and he fought with the sect of Pelagians who were advocating for human free will [27]. Centuries later, the counter-reformation 19th ecumenical council of the

Fig. 16.3 From left to right: Saint Augustine of Hippo (354–430). Thomas Aquinas (1225–1274). Martin Luther (1483–1546). John Calvin (1509–1564)

Catholic Church at Trent, Italy (1545–1563) declared the strong presence of human free will, in clear contrast to St Augustine teachings.

Maimonides was the leading figure for Judaism and was strongly in favor of free will, while in Islam Abu al-Ash'ari (9th–tenth century) argued towards a higher balance between human free will and God's will (jabr) which however was challenged by Shia theologists [201, 220, 227]. In sharp contrast, the Hindu philosophy essentially rejects the "free will" concept [132], while Buddhism suggests free will is inherently conditioned and therefore not fundamentally free [157].

Lay person's opinion is clearly supportive of the presence of free will since this is what is in accord with the everyday human internal experience [26, 134]. This is probably the result of the mere consciousness of behavior although the process of decision is often more or less "automatic" and subconscious [200]. Essentially it reflects a false sensation or seeming experience [40, 115]. Some research data has shown that people with strong beliefs in the existence of free will seem also to have more pro-social personality traits [72].

The above clearly suggest that the concept of "free will" is socially derived and emerged in order to serve social needs and more specifically to define the relationship of the individual person with other members of the society, the rights and obligations of the person, as well as obedience to the rules of the community and subsequent punishment when these rules are violated. It is essentially a criterion of whether corrective measures should be applied or other action from the side of the society is more appropriate.

Such a social construct is unlikely to fit within the rules of scientific discussion, and subsequently it might not be possible to give a scientific definition or explanation concerning what "free will" stands for. Science is not per se obliged to respond to non-scientific questions and topics. However, since the issue of "free will" is so important for human thought and society and since it reflects on fundamental aspects of human experience and almost comes to define what is human and what is not, scientific thinking and method is obliged to elaborate on the issue and provide with both answers as well as questions and help in the further understanding of human nature. Beyond this, psychiatry seems obliged to reply to this question since mental disorders are almost by definition "disorders of free will," not only in the way lay persons perceive it but also according to legislation. Also, those philosophers and philosophies with high intellectual investment in free will also embrace anti-psychiatry, or their line of thoughts leads to anti-psychiatric approaches (see also Chap. 23) and reject the existence of mental disorders per se.

16.3 Basic Driving Ideas and Concepts

Determinism Vs. Probabilism and Randomness

At the roots of the mind-brain problem is the question whether nature and its laws are deterministic or not. This has been put through as the core issue by William James [214]. However, what exactly could the alternative to determinism be and what consequences this could have for the topic under discussion is elusive. An additional problem is that there are some examples which can be interpreted according to a deterministic or an indeterminate theory, depending on the assumptions used to interpret the conditions [21].

Determinism

The basis assumption of determinism is that for each event there is some theory or system of laws such that the occurrence of that event is derivable from those laws together with initial conditions [10]. It suggests that only one course of events is possible. It includes:

- "Causal determinism" which suggests that every event is the product of previous events as mediated by the laws of nature. Causal determinists suggest that there is nothing uncaused or self-caused. The universe is a complex domino with only the initial cause of the big-bang in question.
  - "Nomological" or "scientific" determinism is the most important type of causal determinism.
- "Necessitarianism" suggests there is only one way of things to be and everything happens for a reason and by necessity.
- "Pre-determinism" argues that all events are determined in advance.
- "Fatalism" is a form of teleological determinism which includes "hard theological" determinism and the idea of "predestination."

The view scientists had on the characteristics of natural laws was until recently based on the advances of the seventeenth, eighteenth, and nineteenth centuries and especially on the work of Sir Isaac Newton (1643–1727; Fig. 16.4) and the elaboration with the development of specific mathematical functions to calculate the variables related to physical events by Pierre-Simon, marquis de Laplace (1749–1827;

Fig. 16.4 Left: Sir Isaac Newton (1643–1727). Right: Pierre-Simon, marquis de Laplace (1749–1827)

Fig. 16.4). According to this view, the nature is deterministic with strict laws which mediate the cause-result effect [135]. The same cause would always produce the same result. In spite of its complexity, the world was viewed as a deterministic system. An allegory of this concept could be a chess match; no matter how complex could be and no matter how enormous the variants and options to play, at the end of the day, there is a finite number of combinations, and with players with a sufficient process power (like modern mainframe computers), the chess game collapses into the question which player will make the first move; this player will eventually win. Although for the average human, chess seems to be a very creative game to play; in reality it is absolutely deterministic. Throwing dice on the other hand produces "random" results, impossible to predict and determine, but this is largely because you need sophisticated equipment to monitor the variables determining the action of throwing the dice and calculating speed and trajectory. At the end of the day, playing is impossible since everything could be pre-calculated, predicted, and "determined."

A radical shift in the understanding of nature came with quantum mechanics which suggested an "exotic" view of nature, not compatible with the everyday human experience, but quantum non-deterministic randomness is just as bad as determinism for free will [207]. However there exist two distinct conceptualizations:

- A deterministic approach championed by Albert Einstein (1879–1955; Fig. 16.5) and also in accord with the ideas of Charles Darwin and Stephen Hawking [6, 68, 107]. The famous phrase by Einstein that "God does not throw dice" expresses the idea that hidden variables and laws explain any situation in a deterministic way.

Fig. 16.5 Niels Bohr with Albert Einstein at Paul Ehrenfest's home in Leiden (December 1925)

- The Copenhagen interpretation proposed by Niels Bohr (1885–1962; Fig. 16.5) and Werner Heisenberg (1901–1976) which suggests that physical systems have only probabilities not specific properties.

Theological Determinism

It includes two main types: the hard idea that God has a divine plan which is in effect and pre-determines everything and the softer idea that the wisdom of God is so great that He knows everything that will eventually happen.

The whole religious approach is based on moral responsibility, reward, and punishment, as a result of the existence of free will during decision and choice- making. It is in accord with the concept of "ultimate responsibility" by Robert Kane [75,

122–129] which justifies the "ultimate punishment," which is the basis for many religious beliefs. As already discussed, the "original sin" in Christian theology is considered to reflect the prototypical "ultimate free will" and "ultimate punishment." Whether "ultimate responsibility" is a possibility even in philosophical terms is a matter of hot debate. Many authors propose less strong versions of free will and accountability in order to be able to justify responsibility to a reasonable extent [217]. This is again in accord with the Christian theology which according to both St Augustine and Martin Luther suggests that humans no longer have ultimate free will after the fall and it is only because of God's grace that only a few could be eventually saved.

Finally it should be pointed that "determinism" does not coincide with "predictability." When the events are complex enough, they could well exceed human capacity for calculations and possibly only nature in the sense of a cosmic super-computer can calculate the outcome. Given the complexity, the human approach could be only probabilistic, but this is only a practical restriction in the understanding of an otherwise deterministic phenomenon.

Probabilism

Probabilism is based on an ancient Greek doctrine (*πιθανισμός*). It provides solution to problems where real knowledge is not certain or reliable or for whatever reason cannot lead to a reliable specific prediction [47]. It is unclear whether it is as strong as determinism in terms of nature (i.e., certain natural laws are probabilistic per se) or whether it reflects limits in the human ability to understand nature and its laws in the frame of their overwhelming complexity. This means that when an infinite number of events co-occur and interact and each one has an infinite complexity, then it is impossible for us to calculate the outcome the way Laplace did for simpler events. The only solution to this problem is calculation of probabilities for the outcome. It was abused as a term by the Catholic Church, and by the mid-seventeenth century, it was discredited as scandalous "laxism."

As said above, two distinct conceptualizations for nature exist:

- A deterministic approach championed by Albert Einstein (1879–1955; Fig. 16.5).
- The Copenhagen interpretation proposed by Niels Bohr (1885–1962; Fig. 16.5) and Werner Heisenberg (1901–1976) which suggests that physical systems have only probabilities not specific properties. According to this interpretation because these probabilities are based on random quantum events which when averaged at a large scale they eliminate each other, determinism seems dominant at the macroscopic level. This is called "adequate determinism" (the laws of quantum mechanics asymptotically approach the laws of classical mechanics), and it suggests, for example, that the sun is predicted that it will most likely rise tomorrow in spite of the unpredictability of the position of its particles. However, at the same time, there is a probabilistic rather than deterministic environment at the sub-atomic level.

However, probabilism is quite different from "randomness"; it simply means that it is impossible for us to calculate the functions of an event, and for practical reasons, we are able to predict the outcome with a large uncertainty. If an event is random, then chances are equally distributed. If an astronomical number of events co-occur, their combined output will certainly look random to us although it probably is not. The word "unpredictable" is likely more suitable instead of "random."

Randomness

It corresponds to the lack of predictability in events. According to Ramsey theory, ideal randomness is impossible especially for large structures (complete disorder is impossible) [171]. While man-made random generation machines produce only "pseudo-random" series of numbers, there are a number of physical phenomena which are expected to be truly random, although this is a matter of debate. They include phenomena like atmospheric and thermal noise and external electromagnetic and quantum phenomena, e.g., cosmic background radiation or radioactive decay as measured over short timescales [95].

However, the "hidden-variable theories" reject the view that nature contains irreducible randomness. That is, what we observe and perceive as random is simply the result of our gap in knowledge and our inability to precisely calculate the outcome because of high complexity.

At the end of the day, the basic question whether there is a fundamental randomness in the universe, which is restricted by natural laws or on the contrary the essence of the universe is deterministic and randomness is an epiphenomenon of our inability to calculate extremely complex events, still remains unanswered.

16.4 Determinism and Free Will

While the original conception of "free will" conceptualized the ability of humans to make decisions in the frame or even against divine will and God's plan, since the Enlightenment, philosophical theories replaced God with nature, but nevertheless they were confronted with the idea that there is some outside limiting force, corresponding to the deterministic character of the laws of nature. In other words, the question which was unresolved for centuries was whether it was possible to conceive mind in terms of brain alone and simultaneously accept that the brain function, which is subject to the laws of nature, could be free to choose its products (i.e., thoughts and actions). This seemed an oxymoron since for centuries the scientific world was certain that the nature and universe and subsequently their governing laws have a deterministic character and previous events strictly predict future events. For centuries the brain could be conceptualized exclusively as an "automaton," what in today's terms one could call a processor with properties of a "Turing machine."

No matter scientific or theological, determinism suggests that only one course of events is possible, which is inconsistent with the existence of free will [9]. This problem has been identified in ancient Greek philosophy [14] and remains a major focus of philosophical debate [126, 128]. The view that free will is in principle incompatible with determinism is called "incompatibilism" and includes many sub-theories and variations. It reflects the conflict between the subjective feeling of freedom and self-determination every human being has, with the observation of a nature perfectly regulated by deterministic laws. It has become known as the "problem of free will" or "the dilemma of determinism" and has moral, legal, and socio-political implications [74, 104].

The most important problem with determinism is that it challenges the notion of moral as well as legal responsibility of the person for his own actions [113, 114, 221]. The foundations of law and morality, the very core of our societal structure, seem incompatible with the combination of materialism with determinism [4]. Also, the libertarian approaches require indeterminism of some kind and to a minimum degree in the processes of decision-making [41].

On the other hand, compatibilists argue that determinism does not preclude free will and permits the existence of an unbiased and morally responsible agent, a point of view specifically rejected by Sean Spence who concludes that everything represents natural events without any real moral value [197].

The frustration of philosophers and thinkers was even greater because a close analysis of the philosophical theories of the Enlightenment reveals that the main target was freedom, not truth. The main conditions limiting freedom at that time were the personalized God (metaphysics and religion) and the authoritarian regimes (politics). While fighting kings was intellectually relatively easy, a total abolishment of the supernatural had as a consequence the acceptance of death as the final end, in the frame of pure materialism. Most philosophers chose to abolish religion alone and keep metaphysics in place. In this frame, happiness was introduced as the meaning and aim of life. This did not imply any hedonistic feature; instead it was utilized in order to constitute a replacement for the abolished Divine will and plan, as well as for the will of the authoritarian regimes. Philosophy during the whole duration of the twentieth century tried to reconcile and synthesize the impossible (Fig. 16.6), and the main debating schools were psychoanalysis vs. existential theorists with the first arguing for the presence of subconscious which limits or even destroys free will, while the later were arguing that only conscious processes exist in the mind [108]. Non-metaphysical approaches tend to consider free will as being simply an illusion [103, 192]. Several solutions to the problem have been proposed, and one of them concerns the separation of "freedom of will" from "freedom of action" [212], but this approach, although it serves theology, simply avoids confronting the fundamental problem. Behaviorist approaches have also been proposed [178] but contributed little or not at all to the issue. The most popular solutions to the determinism problem were built upon the traditional dualistic approach and contain metaphysical elements by suggesting the existence of a "mind" defiant of natural laws [190].

Fig. 16.6 In the early twentieth-century philosophy tried to reconcile and synthesize the impossible

But then, if one accepts dualism, the next question is how the mind decides? Are there any rules governing it? If the answer is yes, then the conclusion is that there is some kind of non-natural metaphysical rules which guide the mind on how to function. A decision concerns on choosing among options to respond to a stimuli. In order to break these rules, another set of rules should be present in order to support such a decision. But then, "will" is not free, because it is restricted by rules it cannot break completely. At the end of the day, it seems like a babushka doll.

On the other hand, while determinism has been put through as the core issue by William James [214], it is unlikely that today we can consider it as such. Determinism as well as probabilism and randomness all refer to an external locus of control, a set of rules (or non-rules) which are not under the individual's control and will never be. Determinism suggests that the outcome is absolutely pre-determined while probabilism that while the individual event is bind to probabilities the big outcome is deterministically defined. Even the individual event is determined but in such a complex way that it is impossible for humans to calculate and predict precisely. If on the other hand no such rules exist, then decision is made on the basis of random choice. Again, this is nothing like "free will." Throwing dice is nothing close to free will; it reflects indifference concerning the outcome. So, neither metaphysical assumptions nor the rejection of determinism has any real benefic on our quest for an answer to the riddle of "free will." Neither of them leaves space for "free will."

16.5 Mind, Brain, and Free Will in the Frame of Modern Neuroscientific Knowledge

Our contemporary knowledge can provide some insight on how decisions are made within the human brain, but essentially we have progressed little since the Renaissance idea that consciousness and free will are the ability of humans to control their instincts and achieve morality [199]. According to this idea, "free will" was based on reason and is not completely free to break from it. Many authors feel uncomfortable with this lack of progress and seek ways to bypass it, by adopting theories and proposals of weak scientific value, and subsequently they label as "nihilistic" or "pessimistic" the criticism they receive.

There is a debate whether the human brain can completely understand itself and subsequently whether consciousness could be scientifically studied. A negative answer could be derived by the application of the concept of "radical uncertainty" which extends "ad infinitum," as proposed by Kurt Gödel (1906–1978). Gödel argued on the need for a mathematically structured philosophical methodology which eventually would lead to the need for higher levels in order to prove lower ones, and this would be an eternal (ad infinitum) process with final uncertainty or inability to explain and understand [90–92]. It is exactly what Bertrand Russell suggested as the aim of philosophy, that is, to reconciliate us with the need to tolerate uncertainty in our explanation and understanding of the world [183].

Indeed, some philosophers, e.g., Colin McGinn (born 1950) believe that there are limits to human cognitive capacities which preclude the comprehension of what consciousness really is, while others like Searle and Nagel disagree. The third approach, suggested by the Nobel laurate Gerald Maurice Edelman (1929–2014), is that consciousness is simply the outcome of the computational workings of the association areas of the brain, and in this frame it could be the focus of scientific research [62–64, 66]. A special case is the so-called "Fodorian pessimism" after Jerry Alan Fodor (1935–2017) whose work on the modularity of mind [76, 77] exerted a great influence on the philosophy of the mind and has implications also concerning the definition of mental disorder (see Chap. 13). Fodor proposes that peripheral systems for motor control and low-level perception are modular in structure, highly specialized, and they only utilize those information relevant to their function. On the contrary higher neurocognitive functions are not modular. His theory on higher functions has many pro and against arguments and findings, but what is important is that Fodor argues that the human mind can understand only its modular part. As a consequence, understanding of higher functions including self, consciousness, and free will is beyond human mental abilities by definition [77].

In the same line of thinking, the old, early nineteenth-century questions on the way and the mechanisms through the human mind perceives and interprets other humans could be answered. While folk psychology appeals to general and common-sense theory of behavior and motivation [87], it is evident that this is achieved through the running of a simulation routine in the mind of the observer. The level of mental functioning of this routine is a matter of debate with some authors suggesting it involves a high-level cognitive process [25, 85, 93, 116], while others argue in favor of a lower-level sensory-motor process as being responsible [88]. At any case, it seems that social cognition is probably somehow biologically embedded in the human brain [93].

Before free will is consciousness, which denotes that the person is aware of itself and of the environment. The word "consciousness" comes from the Latin conscious, meaning "knowing, being aware" which in turn comes from conscire (con meaning "together" + scire meaning "know"). Probably this is a loan translation of the Greek word *συνείδηση* (sinidisi) which comes from the verb *σύνοιδα* (meaning "I know well"). From the Latin scire also come the words "sense" and "science." Probably it originated from the Proto-Indo-European root "skei-" (meaning to cut, split), and from this also the Greek word *σχίζω* (schizo meaning I tear apart; also part of the word "schizophrenia") comes from. So far there was no success in the efforts to identify neural mechanisms that give rise to consciousness [43, 48].

A basic approach to consciousness is its conceptualization as a state of awareness. Modern philosophers of mind such as John Searle (born 1932) and Thomas Nagel (born 1937) ascribe three main features to awareness: subjectivity, unity, and intentionality:

- Subjectivity poses the greatest scientific challenge since it could be in direct conflict with the principles of conducting research and conceptualizing in a scientific way.
- Unity refers to the fact that experiences surface into consciousness as a unified whole and all the various senses are merged into a single conscious experience of reality.
- Intentionality refers to the attribution of meaning beyond the moment of experience.

A very short description of the "mind apparatus" is that there is a complex network which receives sensory information, and amygdala adds an emotional component which is essentially a primitive kind of assessment and decision (e.g., fear in the sight of a snake), while the hippocampus provides with feedback from previous experiences and memories on the basis of current sensory input and emotional state [210]. While until this level the module theory of Fodor is probably right, from this level up, the functioning becomes far more complex. The representation of the environment undergoes several loops of processing before reaching the frontal lobe where it is considered against affect (current emotional state) and previous memories and other mental components (e.g., response options), which are retrieved and are held in consciousness. Affects provide the person with fast decisions which serve survival. Some of these decisions are easy to understand (e.g., fear of animals), but others are incomprehensible probably because they serve hidden evolutionary needs (e.g., aesthetics and attraction to the opposite sex). These emotional processes are evolutionary older and are characterized by speed and dominance. They utilize pre-determined strong assumptions which are biased towards the triggering of those behaviors that serve survival. Logical thinking aims towards an "objective" assessment of the situation. The two processes, although independent in principle, interact and influence each other. The emotional status causes bias in logical thinking, and logical analysis triggers emotions.

One astonishing characteristic of the human brain is that it is capable of performing complex process of information in extremely short time but is incapable of performing much less demanding process of information. It is as if it is calibrated to perform on high demands. This is because the essential architecture and the basic principles on which the brain function is based on are completely different from that of electronic computers and specifically of what is called "Turing machines." The principles of the so-called "non-Turing" machines have only recently begun to emerge, and their main characteristic is that their functions are not deterministic. The function of the brain as a non-Turing computational machine is expected by some authors to solve the incompatibility problem, since it allows space for some kind of "free will" in the frame of a non-deterministic environment. What kind of these non-Turing machines will be and what their structure and functional principles could be remains to be determined.

One perfect example of this radical shift of how the mind and the brain exist in one and develop in a non-deterministic way are Noam Chomsky's (born 1928) theories of language and its development, including the universal grammar theory, the generative grammar theory, the Chomsky hierarchy, and the minimalist program [118, 162, 168, 218]. According to him, there is an intrinsic element concerning the grammatical structure, and this is biologically embedded in humans. Therefore, children need only minimal input from the environment so that to fully develop complex language skills ("poverty of the stimulus" argument) [11, 12, 33–36, 71, 83, 229].

Since determinism was and still is a hard and unsolved problem for philosophers elaborating on "free will," a compromise came with the idea of a two-stage process, originally proposed by William James. These two stages include one undetermined and unpredictable which creates the wide variability of options from which the person can choose and a second stage which is more rational or governed by rules and thus more or less predictable [49, 57, 122–129, 143]. In accord with this, a number of authors proposed that exotic quantum physics could serve to understand the way human mind works. They suggest that genuine indeterminacy at the quantum level in the right places could play a crucial role [1, 19, 89, 131, 206], and in the words of Martin Heisenberg, they will constitute "an interplay between chance and lawfulness in the brain" [109]. One of the most important such theories is that of Sir Roger Penrose (born 1931) and Stuart Hameroff (born 1947) who independently developed theories to link consciousness to quantum events. They suggest that quantum fluctuations in the brain could be amplified by deterministic chaos and lead to behavioral choices which could be not predictable [102, 163–166]. Their ORCH-OR (orchestrated objective reduction) theory locates these quantum events in the microtubules of the neurons [166]. However this model is based on hypothetical physical theories that do not yet exist and assumes local quantum effects, we don't know they exist, and even if they do exist, their relationship to major mental events is unknown and there is no clue on how to connect them [96].

Taken all the above together, the author's opinion is that "non-Turing machines" could probably serve as an operational model of brain function, but it is highly unlikely they will serve the solving of the "free will" riddle, simply because their functioning is also guided by rules and limitations (see above the determinism vs. probabilism vs. randomness discussion). Additionally, the supporting models based on advanced and hypothetical physics do not seem to add anything strong or important to the debate.

In experimental psychology terms, the most discussed experiment on free will was conducted by Benjamin Libet (1916–2007) and his colleagues in the 1980s [139, 140]. These researchers reported that the readiness potential preceded the report of decision to move by 0.5 seconds, suggesting that decision had been completed at a subconscious level before reaching consciousness. Libet went further and interpreted his work as leaving no space for the initial decision but argued that still there is the ability of the mind to "veto" this early formulation of decision. According to Libet this "veto" is the expression of free will. However this is evidently another impressive but methodologically flawed and poorly interpreted study, and the results are most likely a methodological artifact [97, 144]. There is no need for exotic interpretations or new flamboyant theories [163, 164, 170]. The biggest problem in the interpretation of these results is the complex way the events and their products after initial processing enter the brain cortex. The usual way to interpret is linear, but this is naïve and reveals lack of knowledge on how the brain works. Today we know that there are many parallel gates to the cortex each permitting different input related to the event. When synchronization of these gates is problematic, peculiar experiences enter the consciousness, with "déjà vu" being the most impressive. "Déjà vu" is the experience of having experienced the same event before, and it is probably the result of desynchronization of input to the cortex; while the perception of the new event enters first and products of the mnemonic process enter afterwards, in this case entering is reverse, giving the false experience that memories of the event already existed [67]. Other similar studies do exist and suggest that there could be detectable changes in brain activity occurring before the experiencing of decision and are predictive of decision itself [195]; however in the opinion of the author of the current book, all these experimental conditions do not really reflect "will" or "choice" the way we discuss it in socio-legal terms.

The psychological aspect of "free will" is also of prime importance. Research on the opinion of lay people suggests that overall, it seems the prevailing belief among lay people is that of a "naive compatibilistic free will" [7] and that people are more eager to attribute "free will" to themselves in comparison to others, whose behavior considers to be predictable [172, 173].

There are some psychological research data suggesting that the concept of free will truly serves its social mission. Belief in human free will and moral responsibility are related to extrovert personality traits [73] and predict better job performance [203]. Interestingly, there are some data suggesting that loss of belief in free will results in higher chances of cheating [215], increased aggression, and less helpful behavior [8]. However, the Vohs and Schooler experiment was carried out in only 30 subjects concerning the first experiment and 122 in five groups concerning the second. With the p-level of <0.01 chosen as the cut-off, and by taking into consideration the number of statistical tests carried out, probably none of the outcomes would be significant after correcting for multiple comparisons, not to mention the fundamental question whether such laboratory experiments reflect any real-life behaviors. What is also interesting in that particular study was that the participants were tested concerning their beliefs on the existence of free will only after and not before the intervention, raising doubts concerning the composition of the study sample and the possible selectivity in reporting of results. The results of the study by Baumeister et al. were based on 70 subjects for the first, 52 for the second, and 56 for the third experiment, and they were definitely negative after controlling for multiple testing, and they also lack controlling of the study sample. Both studies manifest ethical problems concerning the way subjects were informed on the aims and the process, making the validity of their consent to participate questionable.

16.6 Towards a Modern Understanding of "Free Will"

As already said, much of proposals and theories are peculiar extensions of scientific knowledge or intellectual exercises at the borderlines of reason, to such an extent that they often resemble schizotypal thinking. The combination of metaphysical ideals and beliefs with science or quasi-science and balancing between them to make theories work seems to be the rule rather than the exception [158].

But this has not stopped the concept of "free will" from evolving in the last decades especially in the light of the developments in neurosciences, and currently there are various competing conceptions of it [126, 128, 159, 223, 224]. According to Henrik Walter (born 1962), there are three main elements of what we conceive as "free will": (a) the ability to act otherwise and have alternative options to choose, (b) the dependence of choice on an intelligible reason, and (c) the person is the originator (causal source) of the action [124, 217]. All three are a matter of hard debate and also the exact meaning of the words, and the concepts included are also debatable and opinions differ significantly [124, 150, 224].

The ability to act otherwise is purely hypothetical; what has been done cannot be undone, and no situation is exactly similar to a situation of the past. The fact that alternatives might exist does not mean that the person could had acted otherwise.

Reason is a different story. In practical terms, today the concept of "will" should be the focus rather than the concept of "free." In neuroscientific terms it would be more suitable to talk about "decisions" rather than "will" which today refers also to instinctual impulses. This is probably the deep meaning of the philosophical debate, whether the individual can make rational decisions or is swept by instinctual wills. The obvious answer is "both" but also that rational decisions clearly do exist. This was the core idea during the Renaissance and early modern times that human decisions are free because they are based on reason and therefore no supervision or guidance by either supernatural (God) or earthy (kings) powers was necessary or acceptable.

However, the meaning of "rational" is not identical to the meaning of "free." Rational is strongly bound to some kind of calculations based on assumptions and rules and emerges as response to events (internal or external). However, it also points to the possibility that the use of the term "will" is erroneous; "choice" or "reason" might be a better word, and "choice based on reason" might substitute "free will." This is because "will" stands for "decisions" made also from impulses, wishes, and affect (pre-determined values), which are by definition not free according not only to the neurobiological model but also to philosophy and religion which consider them to be "passions" and literally to be enemies of free will. On the contrary, the suggestion by several philosophers that free will is based on reason [50, 124, 217] could be defined and conceptualized on this base and is fully supported by current neurobiological understanding of the brain. However, an important problem is that decisions based on reason constitute a small minority of everyday decisions and of course are not "completely" free. Analytical reasoning is by far a rare mechanism for the human brain to arrive at decisions. In this general frame, Gilbert Ryle (1900–1976) argues there is no such thing as a "will," and instead the discussion should focus on voluntary versus involuntary actions. He also discusses the role of consciousness in the formulation of will [151]. Discussion on how rationally unexplained decisions of self-sacrifice or self-harm are made for the sake of collective fit or the survival of the species or family in expense of the survival of the individual can be found elsewhere [79], and they are informative of the complex non-linear and unexpected way human psychology works.

Then the third element is also crucial. Could the person be the source of the action? Again, reasoning is bound by rules and exists only in response to events (internal or external). Does this mean that in order to be "free," and the person to be the source, a decision should be independent of events (and the needs and values of the individual person)? In that case the probable answer is that this kind of decision is random and random decision is not really free, since it is again determined by chance but more important is not under the control of the individual. On the other hand, unless someone accepts a Cartesian dualistic approach, it is not possible to "create something out of nothing," and it is not possible to apply an external "force" (or agent) which is not based on some kind of pre-existing values. Even in the case a dualistic approach is accepted, no answer is obtained; this simply pushes the boundaries from the natural to the metaphysical/supernatural realm where logic is vicious and answers exist before questions. In other words, conscious free will, that is, decision-making after conscious logical process of the circumstances and the available alternatives, is not compatible with our current neuroscientific knowledge [80, 196].

As a consequence, it seems reasonable to conclude that, strictly speaking, and with the utilization of "ultimate" definitions, the concept of free will cannot be satisfied, and this does not depend on the nature of reality. It is important to realize that "free will" is a concept which serves social needs, not a scientific question. This is why it is so problematic to phrase scientific questions concerning free will and also to incorporate science-derived answers into our humanistic or human-centered universe [56].

As our world is most likely deterministic in deep principle, the experience of "free will" or more appropriately of "choice based on reason" is probably the result of the complexity of the function of the human mind. Each decision is the end-product of an astronomical number of material interactions of unimaginable complexity. A measure of this complexity can be shown with the following example. Let's take not the whole human brain but only one of the two optic nerves. This part of the nervous system contains between 770,000 and 1.7 million nerve fibers [120], and to make calculations easy, let's assume it is one million. Now if each of these neurons can take only two distinct states (e.g., yes/no, fire/silent, etc.) which is a very simplistic approach if one takes into consideration the neuronal function, then the total number of states the optic nerve can take is $2^{1,000,000}$. This number is larger than $10^{250,000}$ which should be compared with 10^{80}–10^{97} which correspond to estimations of the number of all elementary particles of the visible universe. The human optic nerve could carry signals reflecting the perception of almost ten million different colors. Interestingly the wealth in color perception increases only 23 times ($2^{23} = 8,388,608$) the complexity of the system in comparison to black and white vision. These mathematics are more complex variants of "chessboard mathematical problems" and suggest that the complexity of the network rather than the variability of its units contributes more to the resulting overall complexity of the system, meaning that the interplay among neurons rather than the processes inside the neurons is what determines the magnitude of complexity. This observation somewhat justifies the ideas of property dualism and functionalism, but not to their full extent.

Then, having put all the evidence and knowledge and thoughts and ideas on the table, it is time to revisit the basis assumptions and elements of free will. As said before, according to Henrik Walter (born 1962), there are three main elements of what we conceive as "free will" [124, 217]:

(a) *The ability to act otherwise and have alternative options to choose.* Is this statement valid? What has been done cannot be changed, and therefore this statement either expresses my arbitrary certainty that I could have done otherwise although I did not, or it implies that in an identical condition in the future I am certain I can do otherwise. This in turn raises the question whether there are two moments in time which can be considered to be "identical." The clear answer is that there are no such moments because of the complexity of the universe but also because all humans accumulate experience and therefore the later moment includes the experience of the earlier and utilizes the knowledge of the consequences of the first decision. Thus earlier experiences bind later decisions, and circumstances are never either identical or comparable. This first element has no hard validity for the free will debate.
(b) *The dependence of choice on an intelligible reason.* As clearly shown above, the vast majority of everyday decisions are made not based on reason but based on a semi-automatic sub-conscious level and are based on affects and basic needs as well as previous memories and experiences. The important question is whether the conscious effortful mental process could override the resolutions dictated by the emotions and the impulses. The answer to this question is difficult, and this struggle between the spirit and passions has been the field of fruitful philosophical discussions since antiquity. The pragmatic answer is that analytical mental processes (based on reason) might be able to overcome these more basic mental processes but not all and not always, and this is true for all human beings.
(c) *The person is the originator (causal source) of the action.* This statement bears many different explanations and meanings. In materialistic terms, the person is

always the originator of his actions, and by definition these actions constitute responses to some kind of external events, even in those cases the triggering event seems to be internal.

The scale of complexity described above, considered together with the limited number of experimental trials we can conduct, and the limited number of subjects and observations we can gather, makes impossible the study of these phenomena with the standard reductionist approach we currently can afford and which science normally follows when studying events. Currently we lack the tools both theoretical and operational to dig deeper into these mechanisms and to understand them: far less to extrapolate subjective properties (consciousness) from the properties of objects (interconnected nerve cells).

But coming back to the human subjective experience, no interaction and no product of interactions in the brain are identical to a previous or to a future one. Each moment in time is unique and resembles no other before or after. Each combination of the values of an infinite number of variables is unique and will never exist again. In this sense, every experience of decision is new and clearly is different from all the previous decisions. This is not only in the cosmic sense but also at the practical-psychological level. Each time I make a decision concerning the "same" situation (which is never the same as explained before but more or less is experienced as the same), I always consider past experiences and choices. The different constellation of situations this experience of decision-making is based on makes it unique and also makes it informed in a much different way from the past ones. This uniqueness in experience creates the subjective experience of "choice" and gives birth to the idea of "free will."

16.7 Free Will and Mental Illness

Philosophers discussing free will are primarily interested in discussing and elaborating on the issue in healthy adult subjects with full mental capacity and even with higher than average intelligence, morality, or social status. They are much less interested in discussing free will in deprived individuals, except for those politically suppressed. This is a more or less romantic and idealistic approach which implies that humans are inherently able to achieve anything and only circumstances restrict them. Mental illness was an unsolved and potentially lethal problem for most philosophical theories of existence, human freedom, and free will.

The reality is clear: people with mental illnesses are characterized by decreased volition, difficulty in filtering and altering their behavior, and difficulty in adjustment and functioning. Both reasoning and emotional processing and the functioning of the instincts seem to be problematic and even more problematic seems to be the interplay and interdependence among them.

Mental disorders might bind the person's ability to decide and act in a specific way or preclude specific options of action or they might affect the ability to acting for intelligible (understandable) reasons and eventually deprive a person from being the causal originator of its actions [24, 213, 217]. A core characteristic of mental patients is lack of adaptability and reduced copying abilities, and they tend to manifest repeatedly the same behaviors and to utilize the same malfunctioning mechanisms to arrive at a decision. The core feature of "intentionality" in behavior seems to be greatly disrupted [16]. Additionally, in mental disorders, it seems that mental functions have qualitative differences from their normal counterparts, and a cardinal feature is the decrease or even total lack of flexibility to incorporate new input and exert conscious control, as in the case of delusions, in sharp contrast to beliefs which are their normal counterparts [17]. Especially psychotic disorders are considered to have a potentially destructive effect on human volition [69, 112] and the debate which happens with other less severe mental disorders is still ongoing, not precluding also insanity defense for non-psychotic illnesses [69]. An example is that depressive patients tend to utilize depressive biases and schemata, to systematically retrieve negative memories, and as a result they often do not act or act in an inappropriate and self-destructive way. A similar pattern is observed with all mental disorders, including addictions of all kinds [81]. Thus, one of the most typical and universal characteristic of mental disorders is dysfunctional decision-making and subsequently loss of at least part of free will [126].

The debate is even more intense in forensic grounds since it is sometimes said that the mental disorder and not the person is what had caused the crime, meaning that the person was not the agent of the decisions or the genuine source of the offense [20].

In contrast, existential philosophers argue that mental disorder is a matter of personal choice, and in fact it could be exactly the ultimate act of freedom this exercise of free choice against an undesirable or hostile world. According to this approach, patients free themselves from the suppression of the environment through mental illness [174]. This is of course in accord with the anti-psychiatric approach, but clinical as well as neurobiological and pharmacological data do not support it. The opinion of the author of the current book is that the presence of mental disorder as well as the restrictions in mental ability and subsequently in free will in children and the elderly pose destructive limitations in existential theories, and this is probably why these philosophers went to extremes to eliminate these problems since they were unable to solve. Side effects of this "elimination" effort is the celebration of mental disorders as the ultimate expression of freedom and human free will or the embrace of ideas concerning sexual rights for underaged children (see also Chap. 23).

The most reliable and valid interpretation of the relationship between free will (or choice based on reason) and mental illness is that the former is restricted and malfunctions in the case of the later, this is true for all decision-making processes. Again, this is not an "all-or-nothing" phenomenon but depending on the severity of mental disorder, that is, the "degrees of freedom" are restricted to a variable degree, and the availability of response options narrows. In clinical terms, will/choice is as free as large is the variety of options under consideration for a response to a specific situation through not only reasoning but through any decision-making process the human brain utilizes (Fig. 16.7). Effortful processes are more often necessary to serve decisions which normally do not require or require minimum effort (Fig. 16.8). Often, the choice not to undergo treatment is seen not as the product of free will, but this is a "hot" topic of debate during the last decades. It is interesting to note that ideas reflecting a reduced "free will" in mental patients are also embedded in legislation and social rules already since antiquity, and as a general principle, persons are considered to be free and responsible unless they suffer from a mental disorder [145, 219].

Fig. 16.7 Will is as free as large is the variety of options under consideration for a response to a specific situation. Note: The figure cannot capture the fact that the user (person in the picture) and the options (clothes) are a single entity, not a separate user and separate options

Fig. 16.8 Free will is restricted and malfunctions in the case of mental disorder. The availability of response options narrows. Effortful processes are more often necessary. Note: The figure cannot capture the fact that the user (person in the picture) and the options (clothes) are a single entity, not a separate user and separate options

References

1. Acin A, Masanes L. Certified randomness in quantum physics. Nature. 2016;540(7632): 213–9. https://doi.org/10.1038/nature20119.
2. Altmann M. The Palgrave handbook of German idealism. Basingstoke: Palgrave Macmillan; 2014.
3. Ariew R. Ockham's razor: a historical and philosophical analysis of Ockham's principle of parsimony. University of Illnois at Urbana-Champaign; 1976.
4. Baggini J. Freedom regained:the possibility o free will. Granta Books; 2015.
5. Banks EC. The realistic empiricism of Mach, James, and Russell: neutral monism reconceived. Cambridge University Press; 2014.
6. Barlow N. The autobiography of Charles Darwin 1809–1882. With the original omissions restored. Collins, London: Edited and with appendix and notes by his grand-daughter Nora Barlow; 1958.
7. Baumeister RF, Crescioni AW, Alquist JL. Free will as advanced action control for human social life and culture. Neuroethics-Neth. 2010;4(1):1–11. https://doi.org/10.1007/s12152-010-9058-4.
8. Baumeister RF, Masicampo EJ, Dewall CN. Prosocial benefits of feeling free: disbelief in free will increases aggression and reduces helpfulness. Personal Soc Psychol Bull. 2009;35(2):260–8. https://doi.org/10.1177/0146167208327217.
9. Baumeister RF, Monroe AE. Recent research on free will. In: Olson JM, Zanna MP, editors. Advances in experimental social psychology, advances in experimental social psychology. New York: Academic Press; 2014. p. 1–52. https://doi.org/10.1016/b978-0-12-800284-1.00001-1.
10. Berofsky B. Determinism. In: Audi R, editor. The Cambridge dictionary of philosophy. Cambridge: Cambridge University Press; 1995. p. 199–200.
11. Berwick RC, Friederici AD, Chomsky N, Bolhuis JJ. Evolution, brain, and the nature of language. Trends Cogn Sci. 2013;17(2):89–98. https://doi.org/10.1016/j.tics.2012.12.002.
12. Berwick RC, Pietroski P, Yankama B, Chomsky N. Poverty of the stimulus revisited. Cogn Sci. 2011;35(7):1207–42. https://doi.org/10.1111/j.1551-6709.2011.01189.x.
13. Block N. Readings in the philosophy of psychology, vol. 1 and 2. Cambridge, MA: Harvard University Press; 1980.
14. Bobzien S. Determinism and freedom in stoic philosophy. Oxford University Press; 1998.
15. Bolton D. An approach to Wittgenstein's philosophy. London: The Macmillan Press LtD; 1979.
16. Bolton D, Hill J. Mind, meaning, and mental disorder: the nature of causal explanation in psychology and psychiatry. Mind, meaning, and mental disorder: the nature of causal explanation in psychology and psychiatry. New York: Oxford University Press; 1996.
17. Bortolotti L. Delusions and other irrational beliefs. Oxford University Press. 2009; https://doi.org/10.1093/med/9780199206162.001.1.
18. Boys-Stones G. Human autonomy and divine revelation in Origen. In: Swain S, Harrison S, Elsner J, editors. Severan Culture. Cambridge: Cambridge University Press; 2007. p. 488–99.
19. Brassard G, Raymond-Robichaud P. Can free will emerge from determinism in quantum theory? In: Suarez A, Adams P, editors. Is science compatible with free will? New York: Springer; 2013.
20. Buchanan A, Zonana H. Mental disorder as the cause of a crime. Int J Law Psychiatry. 2009;32(3):142–6. https://doi.org/10.1016/j.ijlp.2009.02.007.
21. Butterfield J. Determinism. In: Craig E, editor. Routledge encyclopedia of philosophy, vol. 3. London: Routledge; 1998. p. 33–9.
22. Campbell DT. 'Downward causation' in hierarchically organised biological systems. In: Ayala FJ, Dobzhansky T, editors. Studies in the philosophy of biology. Berkeley: University of California Press; 1974. p. 179–86.

23. Carr HW. Metaphysics and materialism. Nature. 1921;108(2712):247–8. https://doi.org/10.1038/108247a0.
24. Carroll BT, Kirkhart R, Ahuja N, Soovere I, Lauterbach EC, Dhossche D, Talbert R. Katatonia: a new conceptual understanding of catatonia and a new rating scale. Psychiatry (Edgmont (Pa: Township)). 2008;5(12):42–50.
25. Carruthers P. How we know our own minds: the relationship between mindreading and metacognition. Behav Brain Sci. 2009;32(2):121–38.; discussion 138–182. https://doi.org/10.1017/S0140525X09000545.
26. Caruso G. Free will and consciousness: a determinist account of the illusion of free will. Lexington Books; 2012.
27. Chadwick H. John Ferguson: Pelagius: a historical and theological study. (Kaye prize essay, 1952.) pp. x + 206. Cambridge: Heffer, 1956. Cloth, 15s. Net. Class Rev. 2009;8(1):87. https://doi.org/10.1017/s0009840x00164066.
28. Chalmers D. The conscious mind. New York: Oxford University Press; 1996a.
29. Chalmers D. The character of consciousness. Oxford University Press; 2010.
30. Chalmers D. Constructing the world. Oxford University Press; 2014.
31. Chalmers DJ. Facing up to the problem of consciousness. J Conscious Stud. 1995;2(3):200–19.
32. Chalmers DJ. The conscious mind: in search of a fundamental theory. Oxford University Press; 1996b.
33. Chomsky N. The development of grammar in child language: formal discussion. Monogr Soc Res Child Dev. 1964;29:35–9.
34. Chomsky N. The United States and the universality of human rights. Int J Health Serv. 1999;29(3):623–6. https://doi.org/10.2190/V3QN-ELK5-LF26-46KF.
35. Chomsky N. Universals of human nature. Psychother Psychosom. 2005;74(5):263–8. https://doi.org/10.1159/000086316.
36. Chomsky N. Language architecture and its import for evolution. Neurosci Biobehav Rev. 2017;81:295–300. https://doi.org/10.1016/j.neubiorev.2017.01.053.
37. Churchland PM. Eliminative materialism and the propositional attitudes. J Philos. 1981;78(2):67–90. https://doi.org/10.2307/2025900.
38. Churchland PM. Matter and consciousness. Cambridge, MA: MIT Press; 1984.
39. Churchland PM. The engine of reason, the seat of the soul: a philosophical journey into the brain. Cambridge, MA: MIT Press; 1995.
40. Clark TW. Fear of mechanism. A compatibilist critique of the volitional brain. J Conscious Stud. 1999;6(8–9):279–93.
41. Clarke R. Libertarian accounts of free will, vol. 4. Oxford University Press USA; 2003.
42. Collins C. Searle on consciousness and dualism. Int J Philos Stud. 1997;5(1):15–33. https://doi.org/10.1080/09672559708570843.
43. Colombetti G. Psychopathology and teh enactive mind. In: Fulford KWM, Davies M, Gipps R, editors. The Oxford handbook of philosophy and psychiatry. Oxford University Press; 2013.
44. Cooper R. Psychiatry and philosophy of science. Acumen Publishing. 2012; https://doi.org/10.1017/upo9781844653980.
45. Davidson D. Mental Events. In: Foster L, Swanson JW, editors. Essays on actions and events. Clarendon Press; 1970. p. 207–24.
46. Davidson D. Essays on actions and events. Oxford University Press; 1980.
47. de Finetti B. Probabilism: a critical essay on the theory of probability and on the value of science. Erkenntnis (1975-). 1989;31(2/3):169–223.
48. De Sousa A. Towards an integrative theory of consciousness: part 1 (neurobiological and cognitive models). Mens sana monographs. 2013;11(1):100–50. https://doi.org/10.4103/0973-1229.109335.
49. Dennett DC. Brainstorms: philosophical essays on mind and psychology, vol. 1. Bradford Books; 1978.
50. Dennett DC. Elbow room: the varieties of free will worth wanting. Cambridge, MA: MIT Press; 1984.

51. Descartes R, Cottingham J, Stoothoff R, Murdochtrans D. The philosophical writings of René Descartes (1641). Cambridge: Cambridge University Press; 1984.
52. Dewey J. The metaphysical assumptions of materialism. J Specul Philos. 1882;16(2): 208–13.
53. Dicker G. Berkeley's idealism: a critical examination. Oxford: Oxford University Press; 2011.
54. Dihle A. The theory of will in classical antiquity / Albrecht Dihle. Sather classical lectures; v. 48. Berkeley: University of California Press; 1982.
55. Dobbin R. Proharesis in Epictetus. Anc Philos. 1991;11:111–35.
56. Double R. The non-reality of free will. Oxford: Oxford University Press; 1990.
57. Doyle RO. The two-stage model to the problem of free will. In: Suarez A, Adams P, editors. Is science compatible with free will? New York: Springer; 2013.
58. Eccles J. Brain, speech, and consciousness. The understanding of the brain. New York: McGraw-Hill; 1973.
59. Eccles J. A unitary hypothesis of mind-brain interaction in the cerebral cortex. Proc R Soc Lond B Biol Sci. 1990;240(1299):433–51. https://doi.org/10.1098/rspb.1990.0047.
60. Eccles J, Popper K. The self and its brain: an argument for interactionism Routledge; 1984.
61. Eccles JC. Do mental events cause neural events analogously to the probability fields of quantum mechanics? Proc R Soc Lond B Biol Sci. 1986;227(1249):411–28. https://doi.org/10.1098/rspb.1986.0031.
62. Edelman G. The mindful brain: cortical organization and the group-selective theory of higher brain function. Oxford: MIT Press; 1978.
63. Edelman G. The remembered present: a biological theory of consciousness. New York: Basic Books; 1989.
64. Edelman G. Bright air, brilliant fire: on the matter of the mind. New York: Basic books; 1992.
65. Edelman G. A universe of consciousness: how matter becomes imagination. New York: Basic books; 2001.
66. Edelman G, Changeux J. The brain. London: Transaction Publishers; 2001.
67. Efron R. Temporal perception, aphasia and D'ej'a vu. Brain. 1963;86:403–24. https://doi.org/10.1093/brain/86.3.403.
68. Einstein A. The world as I see it. London: Filiquarian Publishing; 2006.
69. Elliott C. The rules of insanity: moral responsibility and the mentally ill offender. Albany: State university of New York Press; 1996.
70. Ellis GFR. Top-down causation and the human brain. In: Murphy N, Ellis GFR, O'Connor T, editors. Downward causation and the neurobiology of free will, Understanding complex systems. Berlin: Springer; 2009. p. 63–81.
71. Everaert MBH, Huybregts MAC, Berwick RC, Chomsky N, Tattersall I, Moro A, Bolhuis JJ. What is language and how could it have evolved? Trends Cogn Sci. 2017;21(8):569–71. https://doi.org/10.1016/j.tics.2017.05.007.
72. Feldman G, Chandrashekar SP. Laypersons' beliefs and intuitions about free will and determinism: new insights linking the social psychology and experimental philosophy paradigms. Soc Psychol Personal Sci. 2018;9(5):539–49. https://doi.org/10.1177/1948550617713254.
73. Feltz A, Cokely ET. Do judgments about freedom and responsibility depend on who you are? Personality differences in intuitions about compatibilism and incompatibilism. Conscious Cogn. 2009;18(1):342–50. https://doi.org/10.1016/j.concog.2008.08.001.
74. Fischer J. God, foreknowledge and freedom. Stanford, CA: Stanford University Press; 1989.
75. Fischer JM, Kane R, Pereboom D, Vargas M. Four views on free will, vol. 117. London: Wiley-Blackwell; 2007.
76. Fodor J. Psychological explanation: an introduction to the philosophy of psychology. New York: Crown Publishing Group/Random House; 1968.
77. Fodor J. The modularity of mind: an essay on faculty psychology. Cambridge, MA: MIT Press; 1983.
78. Fodor JA. Special sciences (or: the disunity of science as a working hypothesis). Synthese. 1974;28(2):97–115.

79. Fountoulakis KN. Ethology, evolutionary psychology, Sociobiology, and evolutionary psychiatry. In: Fountoulakis KN, Nimatoudis I, editors. Psychobiology of behaviour. Cham: Springer; 2019. p. 345–95. https://doi.org/10.1007/978-3-030-18323-3_12.
80. Fountoulakis KN. Being strictly scientific is the only way forward for psychiatry. Acta Psychiatr Scand. 2021;143(3):187–8. https://doi.org/10.1111/acps.13284.
81. Frankfurt HG. Freedom of the will and the concept of a person. 68%J journal of. Philosophy. 1971;1:5–20.
82. Frede M. A free will: origins of the notion in ancient thought. 1st ed. Berkeley: University of California Press; 2011.
83. Friederici AD, Chomsky N, Berwick RC, Moro A, Bolhuis JJ. Language, mind and brain. Nat Hum Behav. 2017;1(10):713–22. https://doi.org/10.1038/s41562-017-0184-4.
84. Frith CD. Free will top-down control in the brain. In: Murphy N, GFR E, O'Connor T, editors. Downward causation and the neurobiology of free will. Understanding complex systems. Berlin: Springer; 2009. p. 199–209.
85. Frith CD, Frith U. Interacting minds--a biological basis. Science. 1999;286(5445):1692–5. https://doi.org/10.1126/science.286.5445.1692.
86. Gallagher S. Where's the action?: epiphenomenalism and the problem of free will. In: Banks W, Pockett S, Gallaghe S, editors. Does consciousness cause behavior? An investigation of the nature of intuition. Cambridge, MA: MIT Press; 2006. p. 109–24.
87. Gallagher S. Intersubjectivity and psychopathology. In: Fulford KWM, Davies M, G.T. Gipps R et al. (eds) The Oxford handbook of philosophy and psychiatry. Oxford University Press: Oxford; 2013. https://doi.org/10.1093/oxfordhb/9780199579563.001.0001.
88. Gallese V. The roots of empathy: the shared manifold hypothesis and the neural basis of intersubjectivity. Psychopathology. 2003;36(4):171–80. https://doi.org/10.1159/000072786.
89. Gisin N. Are there quantum effects coming from outside space–time? Nonlocality, free will and "no many-worlds". In: Suarez A, Adams P (eds) Is science compatible with free will? Springer, New York, NY; 2013.
90. Godel K. The consistency of the axiom of choice and of the generalized continuum-hypothesis. Proc Natl Acad Sci U S A. 1938;24(12):556–7. https://doi.org/10.1073/pnas.24.12.556.
91. Gödel K. Gödel collected works. Oxford: Oxford University Press; 1953.
92. Gödel K. On formally undecidable propositions of principia Mathematica and related systems. New York: Dover Publications; 1962.
93. Goldman A, de Vignemont F. Is social cognition embodied? Trends Cogn Sci. 2009;13(4):154–9. https://doi.org/10.1016/j.tics.2009.01.007.
94. Griffel F. Al-Ghazali's philosophical theology. Oxford: Oxford University Press; 2010.
95. Groblacher S, Paterek T, Kaltenbaek R, Brukner C, Zukowski M, Aspelmeyer M, Zeilinger A. An experimental test of non-local realism. Nature. 2007;446(7138):871–5. https://doi.org/10.1038/nature05677.
96. Grush R, Churchland PS. Gaps in Penrose's toiling. J Conscious Stud. 1995;2(1):10–29.
97. Guggisberg AG, Mottaz A. Timing and awareness of movement decisions: does consciousness really come too late? Front Hum Neurosci. 2013;7:385. https://doi.org/10.3389/fnhum.2013.00385.
98. Haken H. Principles of brain functioning. Berlin: Springer; 1995.
99. Haken H, Stadler M, editors. Synergetics of cognition. Berlin: Springer; 1990.
100. Hallett M. Physiology of volition. In: Murphy N, Ellis GFR, O'Connor T, editors. Downward causation and the neurobiology of free will. Understanding complex systems. Berlin, Heidelberg: Springer; 2009. p. 127–43.
101. Hameroff S, Kaszniak A, Chalmers D, editors. Toward a science of consciousness III: the third Tucson discussions and debates. Cambridge, MA: The MIT Press; 1999.
102. Hameroff SR, Penrose R. Conscious events as orchestrated space-time selections. J Conscious Stud. 1996;3(1):36–53.
103. Harris S. The moral landscape: how science can determine human values. Free Press; 2011.
104. Harris S. Free will. Free Press; 2012.

105. Harrison S. Augustine's way into the will: the theological and philosophical significance of de libero Arbitrio. Oxford University Press UK; 2006.
106. Hart WD. Dualism. In: Guttenplan S, editor. A companion to the philosophy of mind. Oxford: Blackwell; 1996.
107. Hawking S (1994) Black holes and baby universes and other essays. Bantam,
108. Heidegger M. Being and time. New York: Harper Perennial; 2008.
109. Heisenberg M (2013) The origin of freedom in animal behaviour. In: Suarez A, Adams P (eds) Is science compatible with free will? Springer, New York, NY. ,
110. Hempel C. Aspects of scientific explanation. New York: Free Press; 1965.
111. Hill CS. Qualitative characteristics, type materialism and the circularity of analytic functionalism. Behav Brain Sci. 2010;16(1):50–1. https://doi.org/10.1017/s0140525x00028867.
112. Hoglund P, Levander S, Anckarsater H, Radovic S. Accountability and psychiatric disorders: how do forensic psychiatric professionals think? Int J Law Psychiatry. 2009;32(6):355–61. https://doi.org/10.1016/j.ijlp.2009.09.004.
113. Honderich T. The consequences of determinism: a theory of determinism, vol. 2. Oxford: Oxford University Press; 1990.
114. Honderich T. How free are you? The determinism problem. Oxford: Oxford University Press; 1993.
115. Hume D. Enquiry concerning. In: Beauchamp TL, editor. Human Understanding. Oxford: Oxford University Press; 1999.
116. Hutto D. Folk psychological narratives. Cambridge, MA: The MIT Press; 2007. https://doi.org/10.7551/mitpress/7525.001.0001.
117. Huxley T. On the hypothesis that animals are automata, and its history. The Fortnightly Review. 1874;95:555–80.
118. Jager G, Rogers J. Formal language theory: refining the Chomsky hierarchy. Philos Trans R Soc Lond Ser B Biol Sci. 2012;367(1598):1956–70. https://doi.org/10.1098/rstb.2012.0077.
119. Jaworski W. Why materialism is false, and why it has nothing to do with the mind. Philosophy. 2016;91(2):183–213. https://doi.org/10.1017/S0031819116000036.
120. Jonas JB, Schmidt AM, Muller-Bergh JA, Schlotzer-Schrehardt UM, Naumann GO. Human optic nerve fiber count and optic disc size. Invest Ophthalmol Vis Sci. 1992;33(6):2012–8.
121. Juarrero A. Top-down causation and autonomy in complex systems. In: Murphy N, Ellis GFR, O'Connor T, editors. Downward causation and the neurobiology of free will. Understanding complex systems. Berlin: Springer; 2009. p. 83–102.
122. Kane R. Free will and values, vol. 4. New York: State University of New York Press; 1985.
123. Kane R. Two kinds of incompatibilism. Philos Phenomenol Res. 1989;50(2):219–54.
124. Kane R. The significance of free will, vol. 2. New York: Oxford University Press; 1996.
125. Kane R. Responsibility, luck, and chance: reflections on free will and indeterminism. J Philos. 1999;96(5):217–40. https://doi.org/10.2307/2564666.
126. Kane R. A contemporary introduction to free will. Oxford: Oxford University Press; 2005.
127. Kane R. Free will and the dialectic of selfhood: can one make sense of a traditional free will requiring ultimate responsibility? J Ideas Y Valores. 2009;58(141):25–43.
128. Kane R. The Oxford handbook of free will. 2nd ed. Oxford: Oxford University Press; 2012. https://doi.org/10.1093/oxfordhb/9780195399691.001.0001.
129. Kane R. Can a traditional libertarian or Incompatibilist free will be reconciled with modern science? Steps toward a positive answer. In: Suarez A, Adams P, editors. Is science compatible with free will? New York: Springer; 2013.
130. Kirk R, Squires JER. Zombies v. materialists. Aristotelian Society Supplementary. 1974;48(1):135–64.
131. Koch C. Free will, physics, biology, and the brain. In: Murphy N, Ellis G, O'Connor T, editors. Downward causation and the neurobiology of free will. Cham: Springer; 2009. p. 31–52.
132. Koller J. Asian philosophies. 5th ed. Prentice Hall; 2007.
133. Lamme VAF. Challenges for theories of consciousness: seeing or knowing, the missing ingredient and how to deal with panpsychism. Philos Trans R Soc Lond Ser B Biol Sci. 2018;373(1755). https://doi.org/10.1098/rstb.2017.0344.

134. Lamont C. Freedom of choice affirmed. Boston: Beacon Press; 1969.
135. Laplace P. Pierre-Simon Laplace philosophical essay on probabilities (translated from the fifth French edition of 1825 with notes by the translator). New York: Springer; 1995.
136. Lau HC. Volition and the function of consciousness. In: Murphy N, Ellis G, O'Connor T, editors. Downward causation and the neurobiology of free will. Understanding complex systems. Berlin: Springer; 2009. p. 153–69.
137. Leibniz GW. La Monadologie, edition établie par E. Paris: Boutroux; 1991.
138. Lewis DK. An argument for the identity theory. J Philos. 1966;63(1):17–25. https://doi.org/10.2307/2024524.
139. Libet B. Unconscious cerebral initiative and the role of conscious will in voluntary action. Behav Brain Sci. 2010;8(4):529–39. https://doi.org/10.1017/s0140525x00044903.
140. Libet B, Gleason CA, Wright EW, Pearl DK. Time of conscious intention to act in relation to onset of cerebral activity (readiness-potential). The unconscious initiation of a freely voluntary act. Brain. 1983;106(Pt 3):623–42. https://doi.org/10.1093/brain/106.3.623.
141. Lilienfeld SO, Marino L. Mental disorder as a Roschian concept: a critique of Wakefield's "harmful dysfunction" analysis. J Abnorm Psychol. 1995;104(3):411–20. https://doi.org/10.1037//0021-843x.104.3.411.
142. Matthews E. Mental disorder: can Merleau-Pondy take us beyond the 'mind-brain' problem? In: Fulford KWM, Davies M, Gipps R, editors. The Oxford handbook of philosophy and psychiatry. Oxford: Oxford University Press; 2013. https://doi.org/10.1093/oxfordhb/9780199579563.001.0001.
143. Mele AR. Autonomous agents: from self control to autonomy, vol. 96. Oxford University Press; 1995.
144. Mele AR. Free will and neuroscience: revisiting Libet's studies. In: Suarez A, Adams P, editors. Is science compatible with free will? New York: Springer; 2013.
145. Meynen G. Free will and mental disorder: exploring the relationship. Theor Med Bioeth. 2010;31(6):429–43. https://doi.org/10.1007/s11017-010-9158-5.
146. Miller A. Hegels science of logic. New York: Humanities Press International Inc.; 1993.
147. Miresco MJ, Kirmayer LJ. The persistence of mind-brain dualism in psychiatric reasoning about clinical scenarios. Am J Psychiatry. 2006;163(5):913–8. https://doi.org/10.1176/ajp.2006.163.5.913.
148. Morse SJ. The non-problem of free will in forensic psychiatry and psychology. Behav Sci Law. 2007;25(2):203–20. https://doi.org/10.1002/bsl.744.
149. Muehlmann R. Berkeley's ontology and the epistemology of idealism. Can J Philos. 1978;8(1):89–111.
150. Muller S, Walter H. Reviewing autonomy: implications of the neurosciences and the free will debate for the principle of respect for the patient's autonomy. Camb Q Healthc Ethics. 2010;19(2):205–17. https://doi.org/10.1017/S0963180109990478.
151. Murphy N, Ellis GFR, O'Connor T, editors. Downward causation and the neurobiology of free will. Berlin: Springer; 2009.
152. Nadler S. The Cambridge companion to Malebranche. Cambridge companions to philosophy. Cambridge: Cambridge University Press; 2006. https://doi.org/10.1017/ccol0521622123.
153. Nagel E. The structure of science. New York: Harcourt, Brace, and World; 1961.
154. Nagel T. Freedom, The view from nowhere. Oxford: Oxford University Press; 1989.
155. Neujahr P. Kant's idealism. Macon: Mercer University Press; 1995.
156. Newsome WT. Human freedom "emergence". In: Murphy N, Ellis GFR, O'Connor T, editors. Downward causation and the neurobiology of free will. Understanding complex systems. Berlin: Springer; 2009.
157. Nicholas G, Kjellberg P. Buddhism and the freedom of the will: Pali and Mahayanist responses. In: Campbell J, O'Rourke M, Shier D, editors. Freedom and determinism. Cambridge, MA: MIT Press; 2004.
158. O'Connor T. Persons and causes: the metaphysics of free will, vol. 438. Oxford University Press USA; 2000.

159. O'Connor T, Franklin C, editors. Free will. Stanford: The Stanford Encyclopedia of Philosophy; 2020. https://plato.stanford.edu/archives/spr2020/entries/freewill/.
160. O'Connor T. Conscious willing and the emerging sciences of brain and behavior. In: Murphy N, Ellis GFR, O'Connor T, editors. Downward causation and the neurobiology of free will. Understanding complex systems. Berlin: Springer; 2009. p. 173–86.
161. Omoregie J. Freewill: the degree of freedom within. London: AuthorHouseUK; 2015.
162. Ottl B, Jager G, Kaup B. Does formal complexity reflect cognitive complexity? Investigating aspects of the Chomsky hierarchy in an artificial language learning study. PLoS One. 2015;10(4):e0123059. https://doi.org/10.1371/journal.pone.0123059.
163. Penrose R. The emperor's new mind: concerning computers, minds, and the laws of physics. New York: Oxford University Press; 1989.
164. Penrose R. Mechanisms, microtubules, and the mind. J Conscious Stud. 1994;1:241–9.
165. Penrose R. Shadows of the mind: a search for the missing science of consciousness. Oxford University Press; 1996.
166. Penrose R, Hameroff S. What 'Gaps'? Reply to Grush and Churchland. J Conscious Stud. 1995;2:99–112.
167. Pereboom D. Living without free will. Cambridge studies in philosophy. Cambridge: Cambridge University Press; 2009. https://doi.org/10.1017/cbo9780511498824.
168. Piattelli-Palmarini M. Ever since language and learning: afterthoughts on the Piaget-Chomsky debate. Cognition. 1994;50(1–3):315–46. https://doi.org/10.1016/0010-0277(94)90034-5.
169. Polger TW. Are sensations still brain processes? Philos Psychol. 2011;24(1):1–21. https://doi.org/10.1080/09515089.2010.533263.
170. Popper KR, Eccles JC. The self and its brain. London: Springer; 1977.
171. Promel HJ. Complete disorder is impossible: the mathematical work of Walter Deuber. Combinatorics Probability & Computing. 2005;14(1–2):3–16. https://doi.org/10.1017/S0963548304006674.
172. Pronin E. The introspection illusion. In: Advances in experimental social psychology, vol 41. Advances in experimental social psychology. Academic Press, New York; 2009. pp. 1–67. https://doi.org/10.1016/s0065-2601(08)00401-2.
173. Pronin E, Kugler MB. People believe they have more free will than others. Proc Natl Acad Sci U S A. 2010;107(52):22469–74. https://doi.org/10.1073/pnas.1012046108.
174. Ratcliffe M. Depression and the phenomenology of free will. In: Fulford KWM, Davies M, Gipps R et al. (eds) The Oxford handbook of philosophy and psychiatry. Oxford University Press: Oxford; 2013. https://doi.org/10.1093/oxfordhb/9780199579563.001.0001.
175. Ravven H. The self beyond itself: an alternative history of ethics, the new brain sciences, and the myth of free will. New York: The New Press; 2013.
176. Redding P. Continental Idealism: Leibniz to Nietzsche. London: Routledge; 2009.
177. Rey G. Contemporary philosophy of mind: a contentiously classical approach, vol. 425. London: Wiley-Blackwell; 1997.
178. Ringen JD. Adaptation, teleology, and selection by consequences. J Exp Anal Behav. 1993;60(1):3–15. https://doi.org/10.1901/jeab.1993.60-3.
179. Robinson H. Dualism. In: Stich S, Warfield T, editors. The Blackwell guide to philosophy of mind. Oxford: Blackwell; 2003. p. 85–101.
180. Roccatagliata G. Storia de la psichiatria antica. Milano, Italy: Hoepli; 1973.
181. Rosch E. Principles of categorization. In: Rosch E, Lloyd B, editors. Cognition and categorization. Hillsdale, NJ: Lawrence Erlbaum; 1975. p. 27–48.
182. Ross WD. Aristotle, vol. 24. London: Routledge; 1923.
183. Russell B. A history of western philosophy and its connection with political and social circumstances from the earliest times to the present day. New York, NY: Simon and Schuster; 1945.
184. Schaffer J. What not to multiply without necessity. Australas J Philos. 2015;93(4):644–64.
185. Schopenhauer A. Essay on the freedom of the will. Mineola: Dover Publications; 1960.
186. Searle J. The problem of free will. Freedom and neurobiology: reflections on free will, language, and political power. Toronto: Columbia University Press; 2013.

187. Searle JR. Minds, brains and science, vol. 1. London: Harvard University Press; 1984.
188. Shear J, editor. Explaining consciousness: the hard problem. Cambridge, MA: MIT Press; 1996.
189. Shoemaker S. Identity, cause, and mind. Cambridge: Cambridge University Press; 1984.
190. Sluyter A. Neo-environmental determinism, intellectual damage control, and nature/society science. Antipode. 2003;35(4):813–7. https://doi.org/10.1046/j.1467-8330.2003.00354.x.
191. Smart JJC. Sensations and brain processes. Philos Rev. 1959;68(2):141–56. https://doi.org/10.2307/2182164.
192. Smart JJC. Free-will, praise and blame. Mind. 1961;70(279):291–306.
193. Smith CU. Charles Darwin, the origin of consciousness, and panpsychism. J Hist Biol. 1978;11(2):245–67. https://doi.org/10.1007/BF00389301.
194. Solomon R. In the Spirit of Hegel. Oxford: Oxford University Press; 1985.
195. Soon CS, Brass M, Heinze HJ, Haynes JD. Unconscious determinants of free decisions in the human brain. Nat Neurosci. 2008;11(5):543–5. https://doi.org/10.1038/nn.2112.
196. Spence SA. Free will in the light of neuropsychiatry. Philos Psychiatry Psychol. 1996;3(2):75–90.
197. Spence SA. The actor's brain. Oxford: Oxford University Press; 2009a. https://doi.org/10.1093/med/9780198526667.001.0001.
198. Spence SA. Thinking beyond the Bereitschaftspotential: consciousness of self and others as a necessary condition for change. In: Murphy N, GFR E, O'Connor T, editors. Downward causation and the neurobiology of free will. Understanding complex systems. Berlin: Springer; 2009b. p. 211–23.
199. Sperry RW. A modified concept of consciousness. Psychol Rev. 1969;76(6):532–6. https://doi.org/10.1037/h0028156.
200. Spinoza B. Ethics. Sunnyvale: Loki's Publishing; 2017.
201. Staley K. Aquinas: compatibilist or libertarian. The Saint Anselm Journal. 2005;2(2):74.
202. Stephan A. Ordner, Sklaven and andere Systemgeister [orderlies, slaves and other ghosts in the system]. Ethik and Sozialwissenschaften. 1996;7:642–3.
203. Stillman TF, Baumeister RF, Vohs KD, Lambert NM, Fincham FD, Brewer LE. Personal philosophy and personnel achievement: belief in free will predicts better job performance. Soc Psychol Personal Sci. 2010;1(1):43–50. https://doi.org/10.1177/1948550609351600.
204. Stone AA. The ethical boundaries of forensic psychiatry: a view from the ivory tower. Bull am Acad psychiatry law 12:209–19, 1984. J Am Acad Psychiatry Law. 2008;36(2):167–74.
205. Strawson G. The impossibility of ultimate moral responsibility. In: Pereboom D, editor. Free will Hackett; 2009. p. 289–307.
206. Suarez A. Free will and nonlocality at detection as basic principles of quantum physics. In: Suarez A, Adams P, editors. Is science compatible with free will? New York: Springer; 2013.
207. Suarez A, Adams P, editors. Is science compatible with free will? New York: Springer; 2013.
208. Tse PU. The neural basis of free will: criterial causation. Cambridge, MA: The MIT Press; 2015.
209. Van Gulick R. Who's in charge here? And who's doing all the work? In: Heil J, Mele A, editors. Mental causation. Oxford: Clarendon Press; 1995. p. 233–56.
210. Van Hoesen GW. The modern concept of association cortex. Curr Opin Neurobiol. 1993;3(2):150–4. https://doi.org/10.1016/0959-4388(93)90202-a.
211. Van Inwagen P. An essay on free will, vol. 34. Oxford: Oxford University Press; 1986.
212. VanArragon R. Key terms in philosophy of religion. New York: Continuum International Publishing Group; 2010.
213. Verdellen CW, Hoogduin CA, Kato BS, Keijsers GP, Cath DC, Hoijtink HB. Habituation of premonitory sensations during exposure and response prevention treatment in Tourette's syndrome. Behav Modif. 2008;32(2):215–27. https://doi.org/10.1177/0145445507309020.
214. Viney DW. James,William, on free will and determinism. Journal of Mind and Behavior. 1986;7(4):555–65.
215. Vohs KD, Schooler JW. The value of believing in free will: encouraging a belief in determinism increases cheating. Psychol Sci. 2008;19(1):49–54. https://doi.org/10.1111/j.1467-9280.2008.02045.x.

216. Walker D. The decline of hell: seventeenth-century discussions of eternal tornment Routledge · and Kegan. London; 1964.
217. Walter H. Neurophilosophy of free will: from libertarian illusions to a concept of natural autonomy. Cambridge, MA: MIT Press; 2009.
218. Wang L, Bastiaansen M, Yang Y, Hagoort P. Information structure influences depth of syntactic processing: event-related potential evidence for the Chomsky illusion. PLoS One. 2012;7(10):e47917. https://doi.org/10.1371/journal.pone.0047917.
219. Watson G. Free will, vol. 47. vol 3. Oxford: Oxford University Press; 2003.
220. Watt M. Free-will and predestination in early Islam. London: Luzac & Co; 1948.
221. Weatherford R. The implications of determinism. London: Routledge; 1991.
222. Wegner DM. The illusion of conscious will. Cambridge, MA: MIT Press; 2002.
223. Widerker D. Moral responsibility and alternative possibilities. London: Routledge; 2017. https://doi.org/10.4324/9781315248660.
224. Widerker D, McKenna M, editors. Moral responsibility and alternative possibilities: essays on the importance of alternative possibilities. London: Ashgate; 2003.
225. Wittgenstein L. Philosophical investigations. Oxford: Blackwell; 1958.
226. Wittgenstein L. Philosophical investigations. 4th ed. London: Wiley-Blackwell; 2009.
227. Wolfson H. The philosophy of kalam. Cambridge: Harvard University Press; 1976.
228. Woodward WR. Fechner's panpsychism: a scientific solution to the mind-body problem. J Hist Behav Sci. 1972;8:367–86.
229. Yang C, Crain S, Berwick RC, Chomsky N, Bolhuis JJ. The growth of language: universal grammar, experience, and principles of computation. Neurosci Biobehav Rev. 2017;81(Pt B):103–19. https://doi.org/10.1016/j.neubiorev.2016.12.023.
230. Zuboff A. One self, the logic of experience. Inquiry-an interdisciplinary. J Philos. 1990;33(1):39–68. https://doi.org/10.1080/00201749008602210.

Part III
The Identity of Modern Psychiatry

Chapter 17
Introduction to the Identity of Modern Psychiatry

The vast majority of psychiatrists, at least during the 1970s, originally wanted to become psychoanalysts. This was what attracted them to the field, and most of them obtained some kind of formal or informal training in it. After all, Freudian theories marked the transformation of psychiatry from an asylum-oriented and isolated discipline, neglected and despised by others, to a prominent socially appealing and politically strong field. The problem was that psychoanalysis went too far; it rejected the scientific method and went on to become a sociopolitical reform force rather than a medical or psychological discipline. Of course, today it is impossible to even think of how our world was before Freud; the impact of his theory on our view of the world as well as of ourselves was enormous, with both positive as well as dark aspects.

The darkest side of the psychoanalytic wave was the rejection of the scientific method and the establishment of a theory in which logic worked in a cyclical closed circuit. It was always self-confirming, no matter the results and no matter the data. It was more of a theology or a political ideology, with faithful followers and sinful villains. How sad the villains were often the parents of patients most usually the mothers who should not only endure the tragedy of their child's mental illness but also the indignity of being blamed for it. The concept of "schizophrenogenic mother" was such an example, while also homosexuals as well as other groups fell victim to similar doctrines. But as psychoanalysis is not a scientific field, it soon found its way out and reconciled with its victims, only to loosen its doctrines and allow the ideological enemy, that is, existentialism, to fuse with classical psychoanalysis into new theoretical adventures, often with the additional adoption of Marxist elements.

With the field in full disorder and without any rules of the game on the table, there was an unprecedented explosion in the number of charlatans treating patients. Charlatans existed since antiquity, but now they had a glamorous and polished image. Shining complex and jargon, but greatly appealing sociopolitical theories, are widely accepted by lay people, the media, and intellectuals. Psychiatry finds it

K. N. Fountoulakis, *Psychiatry*, https://doi.org/10.1007/978-3-030-86541-2_17

extremely difficult to isolate itself from these movements and to make this clear to the public. This is not only because the public is attracted to these theories but also because many psychiatrists are dedicated to them and also because appealing theories and practices bring money. One such example is Wilhelm Reich's orgone theory and treatment. It is astonishing and impossible to understand how it happened, and thousands of educated and intelligent people from all professions and fields of science could have ever believed that an invisible network of orgasmic energy was the key to mental health, not to mention climate control.

The bitter truth is that unfortunately, methods whose ineffectiveness and lack of scientific nature were ridiculously obvious were often incorporated within mainstream psychiatry or were comfortably tolerated. Wilhelm Reich is not the only such example, and today with the Internet and the social media, the situation is far worse. Also mainstream television as well as newspapers was always the perfect tool for fuzzy theories to spread.

The worst is that in spite of recent developments, psychiatrists could not agree concerning basic things: what psychiatry is, what mental illness is, and which treatment is best for the patient. And the responsibility for this chaotic environment relies solely on us.

The next few chapters will deal with exactly this issue. The first of them will attempt a very short review of the current status of psychiatric theory and practice, based on the review of the literature as well as the clinical experience of the author. This review was not intended to be very hard and strict, but rather reasonable, and every effort was made so that it would reflect the actual status of our discipline. The other chapters will investigate and describe how contemporary psychiatry is related to the other specialties of medicine, social and political forces, including the anti-psychiatric and other social and political movements.

One of the chapters deals with the status of psychiatry as a profession. Again this is something unique among medical specialties. That specialty defines also a profession is something taken as granted for them, and usually no special elaboration on the issue is needed. But for psychiatry things are different since the setting often determines the orientation of the professional as well as the subgroup of patients he/she will treat and how.

Chapter 18
Psychiatry: The State of the Art

18.1 Psychiatry: Art or Science?

As both a researcher and a clinician, the author is always struggling to carry any new research findings and developments in his everyday clinical practice. The most difficult in this effort is to select and apply only those developments with proven value and clinical utility, probably not in a perfect way but at least to a reasonable extend. Otherwise there is a risk to fill your practice with useless, time-consuming, and possibly harmful elements for the patient, which might also carry an economic burden [41].

This chapter reflects the personal combined academic opinion and clinical experience of the author and is the result of decades of academic and clinical work until today.

Whether medicine is an art or a science is a long debate. The oldest mentioning can be traced in the Hippocratic texts as follows:

Original Greek	Latin	English (translation by the author)
Ὁ βίος βραχύς, ἡ δὲ τέχνη μακρή, ὁ δὲ καιρὸς ὀξύς, ἡ δὲ πεῖρα σφαλερή, ἡ δὲ κρίσις χαλεπή	*Vita brevis,* *Ars longa,* *Occasio praeceps,* *Experimentum periculosum,* *Iudicium difficile*	Life is short, But the craft takes long, The opportunity has a narrow window, Experience is misleading, And judgment is difficult

The most frequent translation of this text is in favor of fine arts; reversing the text, it will be read as: *ars longa, vita brevis*. In reality, however, the ancient Greek word "*τέχνη*" does not mean only "fine arts," and especially at that time period, it meant "technique" or "craft." Of course, again at that time, craft was something similar to what we consider today as highly skilled work probably with university degree, not craft according to modern meaning. Today a number of activities, e.g., software development, could be included under this umbrella in its ancient

K. N. Fountoulakis, *Psychiatry*, https://doi.org/10.1007/978-3-030-86541-2_18

meaning. During the Renaissance in North Italy, the disciplines were divided into two broad categories: those under a "Maestro" and those under a "Dottore."

The term "Maestro" means the Master both in knowledge but also the supervisor or the head of the work. It comes from the Latin word *magisterium* (meaning the chief, head, director, teacher), which comes from the Proto-Indo-European word *mag-yos* and the root word "meg." From the same root, via ancient Greek and Persian come also the words magic, mega-, etc. This group of words denotes the head of a group of skilled craftsmen and extends to fine arts.

"Dottore" or Doctor in English comes from the Medieval Latin word *doctor* (meaning the religious teacher, adviser, or scholar). The verb *docere*, which means "to teach" or "to make acceptable," is related to the ancient Greek word *δοκείν* (dokin, meaning to accept and believe) and comes from the Indo-European root word *dek*, which means "to take, accept." This group of words denotes teaching of advanced and abstracted concepts and includes "dogma," "descent," "décor," "discipline," "dignity," "docile," "orthodox," "paradox", etc.

From all specialties of medicine, psychiatry is the one which retains many features of skilled craft and is in desperate seeking for those elements that will elevate it to the status of precise science.

Many authors idealize "the art of psychiatry" [18]. But then, the question that emerges is, do we really have several hundred thousand fine artists who create masterpieces of art corresponding each to an individual patient? Unfortunately, if we follow this line of thinking, then we must accept that our work is more similar to that of a blue collar worker rather than to a scientist or an artist. It is not rational to suggest that several million pieces of fine art are created everyday around the world in psychiatric settings. Like it or not, pieces of fine art are unique; medical practice is massive.

Then, where are we standing now? Psychiatry has a long history from the asylum to psychoanalysis, to pharmacotherapy and biological psychiatry, and to social and community interventions. What would the result be if one attempts to describe where we are, what we should follow in our everyday clinical practice, what seems to survive and what does not, and what we should teach our students and residents?

To answer this question, we need to clarify how much of our current "knowledge" is truly based on scientific data and which part of our knowledge these data concern. Probably our evidence will be far from excellent to support or reject a theory or a treatment method, but we need to chart which of them have at least some reasonable support, or they should definitely be rejected because under the modern scientific view, they make no sense. We will probably find themselves in the position to admit that in reality, we know little concerning many of the things we take as standard knowledge in our field.

However, as Thomas Samuel Kuhn (1922–1996) argued, scientists normally work within paradigms, but communication across these paradigms is usually problematic [80]. This is especially true for psychiatry, and according to Kuhn, this is a sign of immaturity of a field. According to this line of thinking, only when anomalies accumulate beyond a certain point does a state of crisis appear for the specific scientific field and does the paradigm shift with new theories emerging and dominate. Such a crisis occurred in psychiatry in the 1960s and led to the development of modern classification systems, while paradigm shifts were also psychoanalysis and psychopharmacology. An interesting aspect of Kuhn's ideas is that paradigms are "incommensurable," which means that there are no external validating criteria to compare; this is because perception, meaning, and the interpretation are paradigm-dependent.

Thus, although there is much hope that the fragmented parts of our discipline will eventually reconcile in a "theory of everything" for psychiatry [70, 71, 73], it is doubtful that such a theory, when it arrives, will include all our current ideas, inclinations, and orientations. Also, we need to see how reasonable and testable is the rest of our "knowledge," which is based on authority, tradition, or loans from other disciplines, mainly the social sciences and psychology. To do that, we must investigate as separately as possible different areas of the theory and practice of psychiatry. This is what follows in this particular chapter.

18.2 Clinical Assessment and Diagnosis

The problems with the very idea and definition of mental disorder have been discussed in previous chapters. Since the basic concepts are problematic to some extent, the tools and procedures that are based on them cannot be free of problems either. The biggest single problem concerning both the conceptualization and the process of diagnosis is the lack of any specific biological marker, which would be both pathognomonic and independent from the clinical picture. In all the rest of medicine, such markers are the rule rather than the exception. Unfortunately, in psychiatry, there are no markers like thyroid hormone levels (TSH) for thyroid disease or blood sugar for diabetes mellitus. These laboratory tests indicate the diagnosis, regardless of whether clinical symptoms are present. In contrast, in psychiatry, there are no pathognomonic laboratory exams even in the most severe and debilitating cases of mental illness. This is sad since the sister specialty of neurology obtained its first independent biomarker, EEG, in epilepsy almost a century ago [21]. Psychometric and neuropsychological testing of any kind does not have the power of real laboratory testing. Psychometric evaluation is almost useless in everyday clinical practice for the experienced clinician in the assessment of each patient. Neuropsychological testing, on the other hand, could provide some valuable information that is difficult or impossible to obtain during clinical interview; however the results of this kind of assessment need careful consideration against the clinical picture by trained and experienced users, and their added value is relatively weak. Projective testing like the Rorschach plates are often used as if they were independent laboratory testing, but in fact their value is so much under dispute that they are considered to be pseudo-science by many authors.

Since there are no laboratory means available and the diagnosis is made exclusively on clinical terms, the next question is how this is best achieved today. First it is important to clearly state that currently, the state of the art and the accumulated knowledge from hard scientific data strongly support two things:

- First, mental disorder exists and is responsible for a disproportionally high percentage of global disability in comparison with other medical conditions.
- Second, mental disorder definitely includes a number of different diagnostic entities. The true number is a matter of debate, and comorbidity is high, but still, mental disorder is not a single nosological entity.

Subsequently, and since modern ways of treating the patient tend to be somewhat specific for nosological categories, putting the correct diagnosis is essential. Probably it might make little real difference whether the diagnosis is very precise and utilizes fine sub-categories or is limited to broad categories. For example, non-affective psychoses are generally treated the same way as schizophrenia, depressive disorders the same way as major depression, and bipolar spectrum disorders the same way as bipolar disorder. However, differentiating between major broad diagnostic categories is essential.

The diagnosis is based exclusively on information gathered by the psychiatrist on the basis of clinical interview and observation of the patient's behavior, as well as from information provided by the family, friends, and peers. Family history data, although frequently much appreciated by the clinician, have no effect on the diagnosis and contribute little for the prognosis. For example, in a patient with a clinical picture of depression, a family history of schizophrenia in first-degree relatives will change neither the diagnosis nor the treatment. It might alert the clinician that in the future, psychotic symptoms might emerge, but this is hypothetical and statistical, not clinically significant.

The terms used to describe psychiatric symptoms as well as diagnostic entities used to be vaguely defined and were utilized in an idiosyncratic way. This is not the case anymore. It is now widely accepted that mental disorders manifest characteristic sets of symptoms and behaviors that are in principle accessible to observation and question-and-answer techniques [99]. Both classification systems utilize specific terminology, and several lists of symptoms and terms with specific and analytic descriptions have been published [122]. In spite of their differences, DSM and ICD classification systems both constitute an excellent basis to guide the diagnostic process [102]. Of course, every scientific tool manifests shortcomings, problems, and gaps; however most of the critique I have heard in my practice was coming from colleagues with minimal knowledge of these systems. As a result, most of that critique was rather poetic or impressionistic rather than pragmatic and scientific. It is true that modern diagnostic approaches are based on "de-narratization," that is, by the lack of attention to the content and the meaning of thoughts, feelings, and preferences of the patient as well as to the relationship with the therapist. Although many authors suggest that this is a problem [104, 108], the evidence does not support these concerns.

An important detail is that although the DSM is by far more popular in academic cycles and it is the one mostly used in teaching and research, knowledge of the ICD is absolutely necessary for the clinician since all countries (including the USA) are obliged to provide statistical data to the WHO on the basis of ICD classification, and all state documents should include ICD codes. On the other hand, in the literature, treatment indications are developed on the basis of DSM classification.

The next step would be to gather the relevant data by examining the patient and interviewing family members in a reliable way, avoiding biases, omissions, and especially avoiding the "halo effect"[96, 117]. This term refers to the tendency of the examiner to rate later items on the basis of previous ones or on the basis of his/her general idea concerning the diagnosis. Because of the "halo effect," the examiner neglects to investigate in depth the possible answers to every item. Avoiding these biases and reducing variability of diagnosis are best achieved with the use of one of the available semi-structured interviews [76, 91, 110]; the most widely known and used among them are SCAN v. 2.1 [122] and the SCID [98]. The utilization of neuropsychological testing might give important information but of limited validity, and its usefulness concerns mainly the diagnosis of neurocognitive disorders.

The most advanced method to achieve the diagnosis is not to focus on those clinical elements that confirm the diagnosis in mind. The best strategy is to identify a central most possible diagnosis and two alternatives and to try to gather data that confirm each of them independently. In this way, bias and "halo effect" are minimized.

Conclusively, today the state of the art concerning the diagnostic approach includes the following:

- Diagnosing the patient according to both the ICD and DSM classification systems.
- Use of a semi-structured clinical interview.
- Utilization of terms and concepts already in place by classification systems and their accompanying nomenclature, textbooks, and interviews and avoiding the use of idiosyncratic glossary.
- The technique should include every effort to avoid biases and especially the "halo effect."
- Utilization of laboratory examinations exclusively in order to exclude a somatic disorder as the cause of symptoms.

18.3 Biological Markers

Since the first reports on the possible usefulness of the dexamethasone suppression test [22, 23] and the disappointment that followed the initial enthusiasm, the validation of biological markers became more careful and precise [66, 95]. Several other tests of various modalities, including genetic, neuroendocrine, psychophysiological, and imaging, have been proposed, but so far, none of them survived [2, 5, 9, 10, 16, 17, 33, 64, 77, 88, 106, 107, 116].

It is, however, encouraging that research continues to produce new findings and new proposals, often with astonishing results [19, 125]. The hope is, it is not far the time when the first clinically useful marker will be found.

Currently, the only reliable biomarker is response or non-response to medication treatment. However, because of the complexity of the clinical picture and course and the high placebo response, in combination with high partial adherence rates, it is often difficult to differentiate between true responders and non-responders.

This is disappointing of course, especially when considering that the sister medical specialty of neurology obtained its first laboratory diagnostic method almost a century ago when Hans Berger (1873–1941), who by irony was a German psychiatrist, recorded the first human EEGs in 1924, and a decade later, its definite usefulness in the diagnosis of epilepsy was proven [21].

18.4 Treatment of Mental Disorders

The backbone of the methodology to validate the efficacy of a treatment method today is the randomized placebo-controlled trial under double-blind conditions. There are several major problems with this method: The first is the high cost of these trials. The second is that as a result, most of these trials concerning pharmaceutical agents are designed and sponsored by the pharmaceutical industry, on the basis of strict directives by regulatory bodies. Although it is difficult or almost impossible to cheat, cases of data fabrication have been detected, and in any case, this kind of research is not independent; instead, it tends to focus on areas of commercial interest and to avoid areas of possible financial harm for the sponsor [8]. Recently, the wide availability of molecular scanners questions the future of this kind of trials [40]. The third very important issue is that it is almost impossible even at the theoretical level to create a placebo condition to test the efficacy of psychotherapies, while psychotherapy trials suffer from significant bias due to allegiance issues [28, 35, 36, 93]. Unfortunately, most trials show moderate or small treatment effects for treatment options, and the generalizability of the results is questionable. In the literature, there are many opinions against the usefulness of this kind of trials; some authors argued that because they are constructed on the basis of a null hypothesis (that active agent and placebo have equal efficacy), the rejection of this hypothesis does not necessarily mean that the treatment "works" [25, 65]. This is in sharp contrast to the theory behind the null hypothesis testing [38, 94], which clearly suggests that when it is rejected, then the alternative hypothesis is considered to be true [92, 120]. Of course, a prerequisite is that both hypotheses should be rigidly described and structured, but this is almost always the case in trials of psychiatric treatments, especially pharmaceutical ones.

Another interesting suggestion is that since the placebo arm usually shows a significant improvement from baseline, an appropriate option should be to prescribe placebo and enhance the placebo response, instead of prescribing psychopharmacological medication [25]. This is a clear declaration of lack of knowledge on the nature of the placebo effect, which is mainly based on the blindness of the trial (although some placebo effect has been shown in open trials as well) and reflects the effects of a number of variables, including the natural course of the illness and psychological factors (triggered by blindness and the possibility to receive an active agent). The problem is that you cannot prescribe "placebo" to a patient without informing him/her that it is in fact placebo, that is, telling the patient clearly that it does not work and that we do know that.

Psychotherapy

It is impossible to test the efficacy of psychoanalysis in a scientific way; this makes several authors question its scientific nature. The opinion that it could become scientific in nature [81] is probably wishful thinking. In a rare occasion, Aaron Beck tried to test the psychoanalytic theory that depression was "anger or hostility turned inwards" and that depressed patients wanted to punish themselves. However, his research rejected it [12–15]. As a result, since the 1980s, a number of pragmatic and empirically derived psychotherapeutic methods have adjusted to the DSM and ICD classification systems and somehow fulfilled modern standards in order to be considered as treatment options. So far, there is a bulk of data supporting the efficacy of a number of psychotherapies, including cognitive behavioral therapy, cognitive analytic therapy, interpersonal therapy, behavioral activation therapy, and mindfulness therapy [27, 28, 31, 57, 100].

However, meta-analyses revealed that the data are frequently of low quality [109]. Furthermore, there are two main specific problems with these data: First, all trials are open or single blind at best. Second, in most trials, the study samples are probably contaminated with non-clinical cases because often the inclusion criteria demand score above a certain cut-off in psychometric scales rather than clinical diagnosis. Control conditions are also problematic; for example, the waiting list is probably not placebo but is instead nocebo (worsens the patient) [54]. In many trials, psychotherapy is not given as monotherapy; all too often, it is given on top of treatment as usual with medication, and this is not clearly stated in the published papers. Even so, there is strong evidence that CBT is not better than psychological placebo [54]. An additional problem seems to be that all types of psychotherapies seem to perform equally, and they do not differ from supportive psychotherapy [100, 109]. There are some data supporting the traditional recommendation for a combined application of both psychotherapy and pharmacotherapy [56].

In conclusion, there are some data in favor of the efficacy of the more recently developed methods of psychotherapy, but their efficacy has not yet been proven beyond doubt. Although they are frequently recommended as monotherapy, the best practice would be to use them in combination with pharmacotherapy.

A clinician should be trained and competent to deliver at least one of these psychotherapies.

Pharmacotherapy

A number of psychopharmacological agents are available today, but their proven indications are limited. This is a result of the mechanism and the prerequisites an agent should fulfill in order to receive a label to justify its use in the treatment of a specific mental disorder. In spite of what is often discussed in the media, that the proliferation of diagnostic categories with subsequent DSM and ICD editions is because of the influence of the pharmaceutical industry, in reality this proliferation restricts the target population for medication treatment. The cost for testing forces the industry to focus on a limited number of specific disorders and mainly on schizophrenia, bipolar disorder, unipolar depression, and dementia. The data for the rest of mental disorders are rare and inadequate to rely on for a clinical decision. Essentially most mental disorders are orphan in terms of pharmacotherapy. The relatively moderate effect size and the high rates of treatment resistance suggest that most psychopharmaceutical agents probably do not target the core biochemical features of mental disorders but rather offer improvement of symptoms through indirect pathways and effects. This is not unusual in medicine; many treatment options do not target the core physiology of a disease but are able to exert a strong treatment effect in an indirect way. Such an example is the efficacy of diuretics to alleviate the burden associated with congestive heart failure, without exerting any direct effect on the heart.

On the other hand, the data are strong, and the efficacy of antidepressants for the treatment of unipolar depression is solidly proven [29, 45, 46, 48, 55, 97] as well as of antipsychotics for schizophrenia [62, 82–87]. Additionally, a heterogeneous group of agents have solid data for the treatment of bipolar disorder [43, 44, 49–51].

Unfortunately, the clinician is more frequently obliged to utilize his/her experience and prescribe treatment in an idiosyncratic way than to rely on data and guidelines. This is because although there is an in-principle proven efficacy of psychopharmacological agents, generalizability to wider populations is problematic.

Overall, despite the magnitude of the efficacy medium for all classes of agents, and although they make a significant difference in the lives of patients, we are in a desperate need for better compounds, with greater efficacy in wider domains.

A reverse of "Holmes' Rule" (after Oliver Wendell Holmes 1841–1935) states that "medications are guilty until proven innocent." This is already in place as both an attitude and as legislation and practice, mainly as a result of the thalidomide case. An agent should be proven safe before going into the market, and especially for psychopharmaceutical agents, the mechanisms are heavily focusing on safety rather than efficacy [59–61].

Given they constitute the only therapeutic option proven beyond doubt, it is not acceptable for a clinician not to utilize pharmacotherapy. On the contrary, competence in pharmacotherapy is a core skill for the clinical psychiatrist.

Rehabilitation

With the arrival of pharmacotherapy and the release of a great number of previously hospitalized patients in the community, the need for their better integration was pressing. Psychiatric rehabilitation is exactly the process of restoring social functioning and promoting well-being in mental patients.

Since the beginning, the theoretical base for the rehabilitation of mental patients has been the community support theory, which is in accordance with the community integration theories, psychosocial theories, and rehabilitation and educational paradigms. It includes a variety of rehabilitation services like community residential services, workplace accommodations, supported employment or education, social firms, assertive community treatment (or outreach) teams, and others.

In spite of the huge theoretical and ethical support and the large economical investment from most developed countries on this issue, little systematic research has been carried out, and the hard data available are limited. Overall there is some support to the idea that rehabilitation helps stabilize the illness and somewhat improves functioning [11, 26], although it is questionable how this makes a real difference in the lives of patients [20, 75]. The assessment of the outcome and the validation of the procedures and interventions are pressing pending issues.

Prevention and Persons at Risk

Prevention is already one of the cornerstones of medicine since the era of Hippocrates. The basic problem is that prevention needs the identification of predictive markers of any kind, which are strongly related to the development of an illness, but can be traced before the onset of symptoms and signs. The relationship should be so strong that the presence of the marker is highly predictive of the emergence of the medical condition in the future. Such a marker is the glucose tolerance test for diabetes mellitus, while in turn high blood sugar is a marker that predicts the later development of cardiovascular problems. Unfortunately, such markers do not exist for mental disorders.

There are several reports suggesting that mental illness could be related to a number of biological or environmental risk factors. It is sad that in spite of the often exploding enthusiasm, none of these theories proved to be strong enough to serve as a paradigm shift in our prevention practice. As an example, most recently it has been reported that vitamin D supplementation during gestation and early years reduces the risk for schizophrenia [39, 58, 90]. Given the large number of positive reports that eventually were not confirmed, one should be cautious. Often studies present negative trials as “promising” [78].

Even worse, the outcomes are often fuzzy, and many authors and studies deal not with mental health prevention but with the promotion of psychological well-being.

This has its origins in the Menninger psychoanalytical approach after WWII, and of course, it is misleading.

In this frame, the data are currently not sufficient to arrive at a definite conclusion concerning the true efficacy of prevention programs. There are some data to consider [4, 119, 123], but overall the programs are empirically derived and designed on the basis of informed knowledge rather than hard data. The possibility that the universal, the selective, and the indicated preventive interventions are more or less equally efficacious [112, 121] constitutes a reason for caution. The basic assumption is that a generally healthy lifestyle is beneficial for mental health [113]. However, this has not been proven, and in fact, it is often an arbitrary wishful over-interpretation of study results. Childhood trauma or loss of a parent at a very young age are considered by many authors to be of prime importance [37, 72], and it is probably the only cue with some kind of data support.

The most widely used interventions at school aim to promote psychological well-being and increase the ability to achieve developmental milestones, develop positive copying skills, and foster a healthy lifestyle. The universal primary preventive interventions at the level of the population promote a generic healthy lifestyle and also target at reasonable risk factors, including violence, trauma, smoking and substance abuse, and regular exercise. For individuals at a high risk of developing mental disorders, the interventions include more targeted actions on specific domains, including early pharmacotherapy or psychotherapy, dietary supplements, and social support and guidance. School interventions overall produced unimpressive results. One exception is tobacco, alcohol, and substance abuse, which clearly constitutes independent risk factors for the development of mental disorders, but success in the campaigns against them is questionable. Cardiovascular risk factors constitute risk factors for psychogeriatric disorders. Subsequently, intervention in these areas is of prime importance and with benefits proven beyond doubt, but interestingly, although their importance has been documented beyond doubt, cardiovascular risk factors are rarely included in mental health prevention programs.

The population clearly at a higher risk is the offspring of patients with mental disorders, especially of those with psychotic disorders. However, interventions aimed at preventing the development of any mental disorder yielded inconsistent and rather disappointing results [111]. This differs from early intervention, which refers to intervention as early as possible but only after at least some clinical symptoms emerge. Early intervention seems to be strongly beneficial and improves the long-term outcome of the patient.

A separate domain of prevention programs concerns not only the prevention of deaths from suicide but also the reduction of harm and disability that come with self-destructive behaviors and suicidal attempts. Overall, most suicide prevention programs failed to reduce deaths by suicide, although there might be a positive effect on ideation and attempts. It is impressive that there seems to be a clear and strong anti-suicidal protective effect in real life, not virtual community networking in elderly people and war veterans [42, 124].

18.5 Digital Psychiatry

Digital technology created much hope that psychiatry would at last have the adequate tools to bridge the gap between psychophysiology and clinical symptomatology, and with the accumulation of giga-datasets, a quantum leap is eminent for mental health [7]. Digital technology combined with clinical observation and subjective self-report in parallel with advanced computational and analytical methods would be a paradigm shift for the field [67].

Even the delivery of psychotherapy seems to have potential in digital technology; the elimination of distance could radically improve the mental health care of isolated populations and make access easier for many, especially in countries with shortage of psychiatrists [69]. The importance and usefulness of this method has been clearly shown also during the lockdown period during the COVID-19 pandemic in 2020. This is called "telepsychiatry," and along with social media, mobile applications, artificial intelligence, and machine learning are the main areas where an unforeseen change is expected to occur [63]. The first digitally functioning precise treatment algorithms have been recently developed by the CINP (Fig. 18.1) [43, 49–51].

Mobile apps are of particular interest, and they are spreading rapidly [3]. A few years ago, it was estimated that there were already several thousands of mental health care-related apps [118]. The combination of smartphones and apps could be of particular importance in the fight against alcohol and substance abuse and the prevention of suicide. They could also improve medication adherence and provide forms of psychoeducation in a user-friendly fashion especially to adolescents and young adults [34].

There is a pressing need for the development of guidelines of how to use this new technology; these guidelines should focus not only on guiding psychiatrists but also on helping patients and their families act in this chaotic and unregulated environment [89]. While patients are eager to make digital technologies part of their

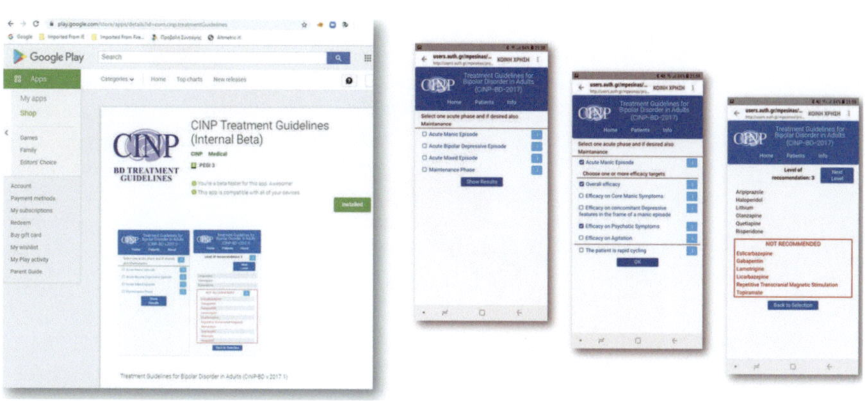

Fig. 18.1 The CINP digital application for mobile phones for the implementation of treatment guidelines for bipolar disorder

treatment, the attitudes among psychiatrists vary from dread to enthusiastic adoption [105]. Clearly the potential is there, but we have little idea concerning the precise areas of benefit and the pathways we should follow with this digital revolution. Focused research is absolutely necessary; a problem could be that digital technology develops and also spreads faster than our ability to research its consequences [115].

18.6 Precision and Personalized Psychiatry (3P)

Personalized medicine is the result of a long way from prescientific through evidence-based to precision medicine, because absolute precision leads to personalization and because each person significantly differs from all the others (Fig. 18.2). In this frame the emphasis is put on the achievement of a high level of exactness in measurement, which eventually will lead to a fully personalized diagnosis and treatment (Fig. 18.3) [6, 24, 30, 47, 101].

When such procedures are available, a radical change in psychiatric training will be necessary, since psychiatrists, as all doctors, will increasingly find themselves needing to interpret and take into consideration a vast variety of results from tests and methods with whom they had little contact during their residency years or their formal training (Fig. 18.4). Thus, such an evolution will not just be quantitatively "more precise" than contemporary psychiatry but instead will be radically qualitatively different, in accordance with the development of a "paradigm shift" as coined by Thomas Kuhn (1922–1996) in order to refer to a fundamental change in the basic

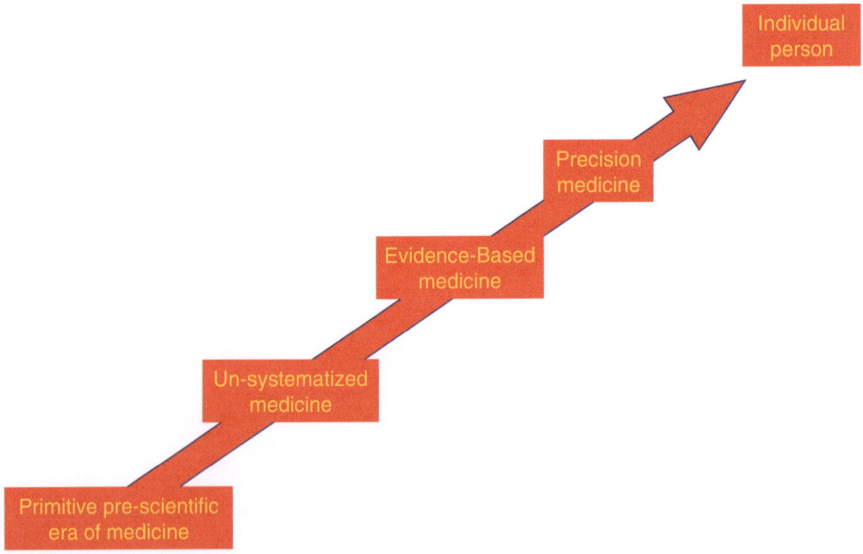

Fig. 18.2 The evolution of precision and personalized medicine

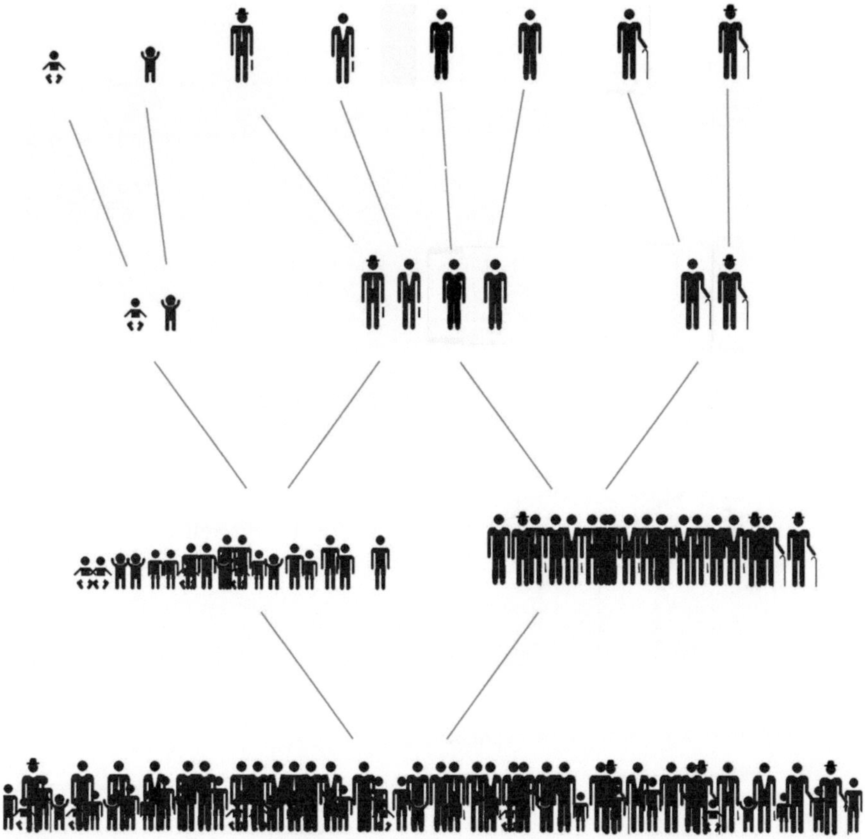

Fig. 18.3 From large to more defined groups and eventually to individual person-oriented diagnosis and treatment

concepts and experimental practices of a scientific discipline [79]. Such a paradigm shift has already occurred in oncology (Fig. 18.5).

But while precision and personalized medicine have at least a robust conceptual context, the issue is far more complicated with psychiatry, where most fundamental questions are still unanswered [53]. This seems an oxymoron taking into consideration the long history of personalized approach in psychiatry, especially in the frame of psychoanalysis. There is an important difference here: It is radically different to talk about personalized psychiatry on the basis of observable clinical symptoms and the neurobiological characteristics of mental disorder in an individual patient, and different on the basis of his/her values, dreams, wishes, preferences, past trauma, and in general his internal subjective life experience. However, in modern scientific terms, psychiatry has not yet benefited from the advancements technology brought in diagnosis and treatment, as it is the case with other clinical specialties. For this reason, in the event psychiatry manages to embrace the precision approach, this will be even more transformative than it was in other fields of medicine since it will lead to informed decisions at a qualitatively different level [52].

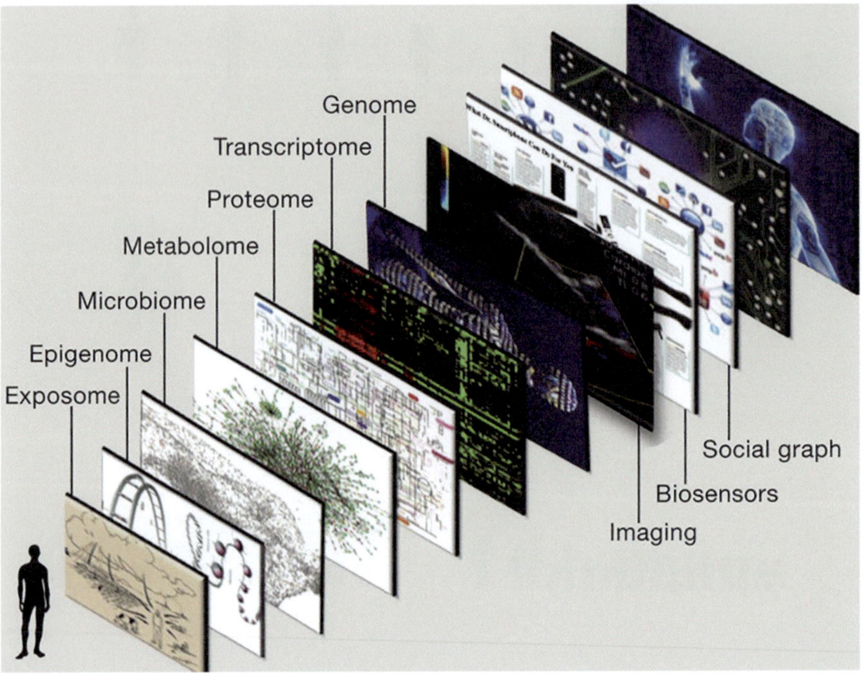

Fig. 18.4 Geographic Information System of a Human Being (from Topol EJ. Cell 2014;157(1):241–53. Reproduced after permission)

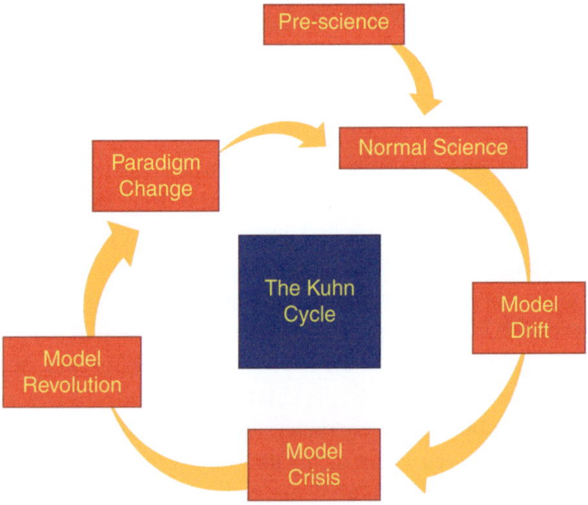

Fig. 18.5 The Kuhn cycle of the paradigm shift in science

Currently the biggest problems psychiatry faces are as follows:

- Definition of mental disorder is problematic, with a wide gray zone between illness and health [74].
- Most psychiatric disorders probably represent a mix of natural diagnostic categories.
- Psychiatry relies on a detailed clinical assessment with careful consideration of the clinical picture, long-term course, previous treatment history, and family/genetic history, but these data are often incomplete, conflicting, vague, and unreliable.
- Clinical predictors of treatment response or side effects do not exist.
- Mental disorders have not always been considered to be diseases of the brain, and many scholars still refuse to consider them as such.
- Their neurobiological substrate has been questioned in favor for a psychosocial etiopathogenesis.
- Psychopharmacological treatments have been discovered in a nonsystematic, serendipitous way, and the understanding of their mode of action is still unsatisfactory.
- Psychotherapies have weak data support.
- Available treatments seem to concern broad categories rather than specific disorders, and research does not cover all areas and topics of interest [44].
- Biological markers are not clinically useful [2, 5, 10, 16, 17, 33, 68, 77].
- The research data are insufficient [44].
- The research data are often over-interpreted.
- In many instances, the data suggest a less-than-desirable response rate.

Currently, 3P is a dream rather than a reality of the near future [1]. Like all precision medicine, it will be a laborious and costly task; thus, the partnership of scientists with industry and the commercialization of new methods and technologies will be an important element for success. Additionally, the governments should succeed in their obligation to provide with the appropriate legal framework, which will both support development and progress but will also protect the rights and the privacy of patients and their families [47]. Eventually, while wishes and perspectives are appealing, it is only after materialization and application in the everyday clinical practice that precision and personalized psychiatry will fulfill its promises and role. The collaboration among the academia, the industry, and the governments is absolutely necessary since each one of them has unique and complementary roles, abilities, and interests [6, 32, 101, 103, 114].

References

1. Alda M. Personalized psychiatry: many questions, fewer answers. J Psychiatry Neurosci. 2013;38(6):363–5. https://doi.org/10.1503/jpn.130221.
2. Allen NC, Bagade S, McQueen MB, Ioannidis JP, Kavvoura FK, Khoury MJ, Tanzi RE, Bertram L. Systematic meta-analyses and field synopsis of genetic association studies

in schizophrenia: the SzGene database. Nat Genet. 2008;40(7):827–34. https://doi.org/10.1038/ng.171.

3. Anguera JA, Jordan JT, Castaneda D, Gazzaley A, Arean PA. Conducting a fully mobile and randomised clinical trial for depression: access, engagement and expense. BMJ Innov. 2016;2(1):14–21. https://doi.org/10.1136/bmjinnov-2015-000098.
4. Arango C, Diaz-Caneja CM, McGorry PD, Rapoport J, Sommer IE, Vorstman JA, McDaid D, Marin O, Serrano-Drozdowskyj E, Freedman R, Carpenter W. Preventive strategies for mental health. Lancet Psychiatry. 2018;5(7):591–604. https://doi.org/10.1016/S2215-0366(18)30057-9.
5. Arranz MJ, de Leon J. Pharmacogenetics and pharmacogenomics of schizophrenia: a review of last decade of research. Mol Psychiatry. 2007;12(8):707–47. https://doi.org/10.1038/sj.mp.4002009.
6. Ashley EA. The precision medicine initiative: a new national effort. JAMA. 2015;313(21):2119–20. https://doi.org/10.1001/jama.2015.3595.
7. Baker JT, Germine LT, Ressler KJ, Rauch SL, Carlezon WA Jr. Digital devices and continuous telemetry: opportunities for aligning psychiatry and neuroscience. Neuropsychopharmacology. 2018;43(13):2499–503. https://doi.org/10.1038/s41386-018-0172-z.
8. Ban TA. Prolegomenon to the clinical prerequisite: psychopharmacology and the classification of mental disorders. Prog Neuro-Psychopharmacol Biol Psychiatry. 1987;11(5):527–80. https://doi.org/10.1016/0278-5846(87)90019-4.
9. Bandelow B, Baldwin D, Abelli M, Bolea-Alamanac B, Bourin M, Chamberlain SR, Cinosi E, Davies S, Domschke K, Fineberg N, Grunblatt E, Jarema M, Kim YK, Maron E, Masdrakis V, Mikova O, Nutt D, Pallanti S, Pini S, Strohle A, Thibaut F, Vaghi MM, Won E, Wedekind D, Wichniak A, Woolley J, Zwanzger P, Riederer P. Biological markers for anxiety disorders, OCD and PTSD: a consensus statement. Part II: neurochemistry, neurophysiology and neurocognition. World J Biol Psychiatry. 2017;18(3):162–214. https://doi.org/10.1080/15622975.2016.1190867.
10. Barnett JH, Smoller JW. The genetics of bipolar disorder. Neuroscience. 2009;164(1):331–43. https://doi.org/10.1016/j.neuroscience.2009.03.080.
11. Baronet AM, Gerber GJ. Psychiatric rehabilitation: efficacy of four models. Clin Psychol Rev. 1998;18(2):189–228. https://doi.org/10.1016/s0272-7358(97)00106-2.
12. Beck AT. Thinking and depression. I. Idiosyncratic content and cognitive distortions. Arch Gen Psychiatry. 1963;9:324–33. https://doi.org/10.1001/archpsyc.1963.01720160014002.
13. Beck AT. Thinking and depression. Ii. Theory and therapy. Arch Gen Psychiatry. 1964;10:561–71. https://doi.org/10.1001/archpsyc.1964.01720240015003.
14. Beck AT. Depression: causes and treatment. Philadelphia, PA: University of Pennsylvania Press; 1972.
15. Beck AT, Haigh EA. Advances in cognitive theory and therapy: the generic cognitive model. Annu Rev Clin Psychol. 2014;10(1):1–24. https://doi.org/10.1146/annurev-clinpsy-032813-153734.
16. Biomarkers Definitions Working G. Biomarkers and surrogate endpoints: preferred definitions and conceptual framework. Clin Pharmacol Ther. 2001;69(3):89–95. https://doi.org/10.1067/mcp.2001.113989.
17. Bird A. Perceptions of epigenetics. Nature. 2007;447(7143):396–8. https://doi.org/10.1038/nature05913.
18. Bloch S. The art of psychiatry. World Psychiatry. 2005;4(3):130–4.
19. Bodkin JA, Coleman MJ, Godfrey LJ, Carvalho CMB, Morgan CJ, Suckow RF, Anderson T, Ongur D, Kaufman MJ, Lewandowski KE, Siegel AJ, Waldstreicher E, Grochowski CM, Javitt DC, Rujescu D, Hebbring S, Weinshilboum R, Rodriguez SB, Kirchhoff C, Visscher T, Vuckovic A, Fialkowski A, McCarthy S, Malhotra D, Sebat J, Goff DC, Hudson JI, Lupski JR, Coyle JT, Rudolph U, Levy DL. Targeted treatment of individuals with psychosis carrying a copy number variant containing a genomic triplication of the glycine decarboxylase gene. Biol Psychiatry. 2019;86(7):523–35. https://doi.org/10.1016/j.biopsych.2019.04.031.

20. Bond GR, Drake RE, Becker DR, Mueser KT. Effectiveness of psychiatric rehabilitation approaches for employment of people with severe mental illness. Journal of Disability Policy Studies. 2016;10(1):18–52. https://doi.org/10.1177/104420739901000104.
21. Britton J, Frey L, Hopp J, et al. Appendix 6. A brief history of EEG. In: Louis E, Frey L, editors. Electroencephalography (EEG): an introductory text and atlas of Normal and abnormal findings in adults, children, and infants [internet]. Chicago: American Epilepsy Society; 2016.
22. Carroll BJ. Dexamethasone suppression test in depression. Lancet. 1980;2(8206):1249. https://doi.org/10.1016/s0140-6736(80)92507-6.
23. Carroll BJ, Martin FI, Davies B. Resistance to suppression by dexamethasone of plasma 11-O.H.C.S. levels in severe depressive illness. Br Med J. 1968;3(5613):285–7. https://doi.org/10.1136/bmj.3.5613.285.
24. Chambers DA, Feero WG, Khoury MJ. Convergence of implementation science, precision medicine, and the learning health care system: a new model for biomedical research. JAMA. 2016;315(18):1941–2. https://doi.org/10.1001/jama.2016.3867.
25. Charland L. Why psychiatry should fear medicalization. In: Fulford KWM, Davies M, G.T. Gipps R et al. (eds) The Oxford handbook of philosophy and psychiatry. Oxford University Press: Oxford. https://doi.org/10.1093/oxfordhb/9780199579563.001.0001.
26. Cheng JF, Chen CY, Lin MC, Huang XY. To explore the efficacy of community rehabilitation for facilitating daily function among patients with mental illness. Perspect Psychiatr Care. 2018;54(4):580–5. https://doi.org/10.1111/ppc.12283.
27. Churchill R, Moore TH, Caldwell D, Davies P, Jones H, Furukawa TA, Lewis G, Hunot V. Cognitive behavioural therapies versus other psychological therapies for depression. Cochrane Database Syst Rev. 2010;9. https://doi.org/10.1002/14651858.CD008698.
28. Churchill R, Moore TH, Furukawa TA, Caldwell DM, Davies P, Jones H, Shinohara K, Imai H, Lewis G, Hunot V. 'Third wave' cognitive and behavioural therapies versus treatment as usual for depression. Cochrane Database Syst Rev. 2013;10:CD008705. https://doi.org/10.1002/14651858.CD008705.pub2.
29. Cipriani A, Salanti G, Furukawa TA, Turner E, Ioannidis JPA, Geddes JR. Network meta-analysis of antidepressants - Authors' reply. Lancet. 2018;392(10152):1012–3. https://doi.org/10.1016/S0140-6736(18)31780-X.
30. Collins FS, Varmus H. A new initiative on precision medicine. N Engl J Med. 2015;372(9):793–5. https://doi.org/10.1056/NEJMp1500523.
31. Cuijpers P, Noma H, Karyotaki E, Cipriani A, Furukawa TA. Effectiveness and acceptability of cognitive behavior therapy delivery formats in adults with depression: a network meta-analysis. JAMA Psychiat. 2019;76(7):700–7. https://doi.org/10.1001/jamapsychiatry.2019.0268.
32. Cunningham P. The health 202: NIH wants 1 million Americans to contribute to new pool of gene data. Washington Post. 2018;(2018-01-16).
33. de Leon J. Pharmacogenomics: the promise of personalized medicine for CNS disorders. Neuropsychopharmacology. 2009;34(1):159–72. https://doi.org/10.1038/npp.2008.147.
34. Depp CA, Ceglowski J, Wang VC, Yaghouti F, Mausbach BT, Thompson WK, Granholm EL. Augmenting psychoeducation with a mobile intervention for bipolar disorder: a randomized controlled trial. J Affect Disord. 2015;174:23–30. https://doi.org/10.1016/j.jad.2014.10.053.
35. Dragioti E, Dimoliatis I, Evangelou E. Disclosure of researcher allegiance in meta-analyses and randomised controlled trials of psychotherapy: a systematic appraisal. BMJ Open. 2015a;5(6):e007206. https://doi.org/10.1136/bmjopen-2014-007206.
36. Dragioti E, Dimoliatis I, Fountoulakis KN, Evangelou E. A systematic appraisal of allegiance effect in randomized controlled trials of psychotherapy. Ann General Psychiatry. 2015b;14:25. https://doi.org/10.1186/s12991-015-0063-1.
37. Dube SR, Anda RF, Felitti VJ, Chapman DP, Williamson DF, Giles WH. Childhood abuse, household dysfunction, and the risk of attempted suicide throughout the life span: findings from the adverse childhood experiences study. JAMA. 2001;286(24):3089–96. https://doi.org/10.1001/jama.286.24.3089.

38. Everitt B. The Cambridge dictionary of statistics. Cambridge: Cambridge University Press; 1998.
39. Eyles DW, Trzaskowski M, Vinkhuyzen AAE, Mattheisen M, Meier S, Gooch H, Anggono V, Cui X, Tan MC, Burne THJ, Jang SE, Kvaskoff D, Hougaard DM, Norgaard-Pedersen B, Cohen A, Agerbo E, Pedersen CB, Borglum AD, Mors O, Sah P, Wray NR, Mortensen PB, McGrath JJ. The association between neonatal vitamin D status and risk of schizophrenia. Sci Rep. 2018;8(1):17692. https://doi.org/10.1038/s41598-018-35418-z.
40. Fountoulakis KN. Are molecular scanners going to make double-blind placebo-controlled trials impossible? Evid Based Ment Health. 2016;19(4):128. https://doi.org/10.1136/eb-2016-102486.
41. Fountoulakis KN. Being strictly scientific is the only way forward for psychiatry. Acta Psychiatr Scand. 2021;143(3):187–8. https://doi.org/10.1111/acps.13284.
42. Fountoulakis KN, Gonda X, Rihmer Z. Suicide prevention programs through community intervention. J Affect Disord. 2011;130(1–2):10–6. https://doi.org/10.1016/j.jad.2010.06.009.
43. Fountoulakis KN, Grunze H, Vieta E, Young A, Yatham L, Blier P, Kasper S, Moeller HJ. The International College of Neuro-Psychopharmacology (CINP) treatment guidelines for bipolar disorder in adults (CINP-BD-2017), part 3: the clinical guidelines. Int J Neuropsychopharm. 2017a;20(2):180–95. https://doi.org/10.1093/ijnp/pyw109.
44. Fountoulakis KN, Moeller HJ, Kasper S. Personalised and precision psychiatry: what do the CINP bipolar guidelines suggest? Int J Psychiatry Clin Pract. 2019;23(1):80–1. https://doi.org/10.1080/13651501.2018.1470246.
45. Fountoulakis KN, Moller HJ. Efficacy of antidepressants: a re-analysis and re-interpretation of the kirsch data. Int J Neuropsychopharm. 2011;14(3):405–12. https://doi.org/10.1017/S1461145710000957.
46. Fountoulakis KN, Moller HJ. Are antidepressants clinically useful? Conclusion of a decade of debate. World Psychiatry. 2014;13(2):201–2. https://doi.org/10.1002/wps.20112.
47. Fountoulakis KN, Stahl SM. Precision and personalized assessment, diagnosis and treatment in psychiatry. CNS Spectr:1–7; 2020. https://doi.org/10.1017/S1092852920000139.
48. Fountoulakis KN, Veroniki AA, Siamouli M, Moller HJ. No role for initial severity on the efficacy of antidepressants: results of a multi-meta-analysis. Ann General Psychiatry. 2013;12(1):26. https://doi.org/10.1186/1744-859X-12-26.
49. Fountoulakis KN, Vieta E, Young A, Yatham L, Grunze H, Blier P, Moeller HJ, Kasper S. The International College of Neuropsychopharmacology (CINP) treatment guidelines for bipolar disorder in adults (CINP-BD-2017), part 4: unmet needs in the treatment of bipolar disorder and recommendations for future research. Int J Neuropsychopharm. 2017b;20(2):196–205. https://doi.org/10.1093/ijnp/pyw072.
50. Fountoulakis KN, Yatham L, Grunze H, Vieta E, Young A, Blier P, Kasper S, Moeller HJ. The International College of Neuro-Psychopharmacology (CINP) treatment guidelines for bipolar disorder in adults (CINP-BD-2017), part 2: review, grading of the evidence, and a precise algorithm. Int J Neuropsychopharm. 2017c;20(2):121–79. https://doi.org/10.1093/ijnp/pyw100.
51. Fountoulakis KN, Young A, Yatham L, Grunze H, Vieta E, Blier P, Moeller HJ, Kasper S. The International College of Neuropsychopharmacology (CINP) treatment guidelines for bipolar disorder in adults (CINP-BD-2017), part 1: background and methods of the development of guidelines. Int J Neuropsychopharm. 2017d;20(2):98–120. https://doi.org/10.1093/ijnp/pyw091.
52. Friston KJ. Precision psychiatry. Biol Psychiatry Cogn Neurosci Neuroimaging. 2017;2(8):640–3. https://doi.org/10.1016/j.bpsc.2017.08.007.
53. Fulford KWM, Thornton T, Graham G. Knowledge, research and evidence-based medicine. In: Fulford KWM, Thornton T, Graham G, editors. The Oxford textbook of philosophy and psychiatry. Oxford: Oxford University Press; 2006.
54. Furukawa TA, Noma H, Caldwell DM, Honyashiki M, Shinohara K, Imai H, Chen P, Hunot V, Churchill R. Waiting list may be a nocebo condition in psychotherapy trials: a contri-

bution from network meta-analysis. Acta Psychiatr Scand. 2014;130(3):181–92. https://doi.org/10.1111/acps.12275.
55. Furukawa TA, Salanti G, Atkinson LZ, Leucht S, Ruhe HG, Turner EH, Chaimani A, Ogawa Y, Takeshima N, Hayasaka Y, Imai H, Shinohara K, Suganuma A, Watanabe N, Stockton S, Geddes JR, Cipriani A. Comparative efficacy and acceptability of first-generation and second-generation antidepressants in the acute treatment of major depression: protocol for a network meta-analysis. BMJ Open. 2016;6(7):e010919. https://doi.org/10.1136/bmjopen-2015-010919.
56. Furukawa TA, Watanabe N, Churchill R. Combined psychotherapy plus antidepressants for panic disorder with or without agoraphobia. Cochrane Database Syst Rev. 2007;1:CD004364. https://doi.org/10.1002/14651858.CD004364.pub2.
57. Furukawa TA, Weitz ES, Tanaka S, Hollon SD, Hofmann SG, Andersson G, Twisk J, DeRubeis RJ, Dimidjian S, Hegerl U, Mergl R, Jarrett RB, Vittengl JR, Watanabe N, Cuijpers P. Initial severity of depression and efficacy of cognitive-behavioural therapy: individual-participant data meta-analysis of pill-placebo-controlled trials. Br J Psychiatry J Ment Sci. 2017;210(3):190–6. https://doi.org/10.1192/bjp.bp.116.187773.
58. Ghaderi A, Banafshe HR, Mirhosseini N, Moradi M, Karimi MA, Mehrzad F, Bahmani F, Asemi Z. Clinical and metabolic response to vitamin D plus probiotic in schizophrenia patients. BMC Psychiatry. 2019;19(1):77. https://doi.org/10.1186/s12888-019-2059-x.
59. Ghaemi N. The concepts of psychiatry: a pluralistic approach to the mind and mental illness. Baltimore, MD: Johns Hopkins University Press; 2007.
60. Ghaemi N. On depression: drugs, diagnosis, and despair in the modern world. On depression: drugs, diagnosis, and despair in the modern world. Baltimore, MD: Johns Hopkins University Press; 2013.
61. Ghaemi SN. Toward a Hippocratic psychopharmacology. Can J Psychiatr. 2008;53(3):189–96. https://doi.org/10.1177/070674370805300309.
62. Haddad PM, Correll CU. The acute efficacy of antipsychotics in schizophrenia: a review of recent meta-analyses. Ther Adv Psychopharmacol. 2018;8(11):303–18. https://doi.org/10.1177/2045125318781475.
63. Hariman K, Ventriglio A, Bhugra D. The future of digital psychiatry. Curr Psychiatry Rep. 2019;21(9):88. https://doi.org/10.1007/s11920-019-1074-4.
64. Hashimoto E, Riederer PF, Hesselbrock VM, Hesselbrock MN, Mann K, Ukai W, Sohma H, Thibaut F, Schuckit MA, Saito T. Consensus paper of the WFSBP task force on biological markers: biological markers for alcoholism. World J Biol Psychiatry. 2013;14(8):549–64. https://doi.org/10.3109/15622975.2013.838302.
65. Healy D. Mania: a short history of bipolar disorder. Baltimore, MD: Johns Hopkins University Press; 2008.
66. Hirschfeld RM, Koslow SH, Kupfer DJ. The clinical utility of the dexamethasone suppression test in psychiatry. Summary of a National Institute of Mental Health workshop. JAMA. 1983;250(16):2172–4.
67. Hsin H, Fromer M, Peterson B, Walter C, Fleck M, Campbell A, Varghese P, Califf R. Transforming psychiatry into data-driven medicine with digital measurement tools. NPJ Digit Med. 2018;1:37. https://doi.org/10.1038/s41746-018-0046-0.
68. Insel TR. The NIMH research domain criteria (RDoC) project: precision medicine for psychiatry. Am J Psychiatry. 2014;171(4):395–7. https://doi.org/10.1176/appi.ajp.2014.14020138.
69. Jain N, Singh H, Koolwal GD, Kumar S, Gupta A. Opportunities and barriers in service delivery through mobile phones (mHealth) for severe mental illnesses in Rajasthan, India: a multi-site study. Asian J Psychiatr. 2015;14:31–5. https://doi.org/10.1016/j.ajp.2015.01.008.
70. Kandel E. Psychiatry, psychoanalysis, and the new biology of mind. Washington, DC: American Psychiatric Publishing; 2005.
71. Kandel ER. Psychotherapy and the single synapse. The impact of psychiatric thought on neurobiologic research. N Engl J Med. 1979;301(19):1028–37. https://doi.org/10.1056/NEJM197911083011904.

72. Kelleher I, Keeley H, Corcoran P, Ramsay H, Wasserman C, Carli V, Sarchiapone M, Hoven C, Wasserman D, Cannon M. Childhood trauma and psychosis in a prospective cohort study: cause, effect, and directionality. Am J Psychiatry. 2013;170(7):734–41. https://doi.org/10.1176/appi.ajp.2012.12091169.
73. Kendler KS. The structure of psychiatric science. Am J Psychiatry. 2014;171(9):931–8. https://doi.org/10.1176/appi.ajp.2014.13111539.
74. Kendler KS. The nature of psychiatric disorders. World psychiatry: official journal of the World Psychiatric Association (WPA). 2016;15(1):5–12. https://doi.org/10.1002/wps.20292.
75. Killaspy H, Marston L, Green N, Harrison I, Lean M, Cook S, Mundy T, Craig T, Holloway F, Leavey G, Koeser L, McCrone P, Arbuthnott M, Omar RZ, King M. Clinical effectiveness of a staff training intervention in mental health inpatient rehabilitation units designed to increase patients' engagement in activities (the Rehabilitation Effectiveness for Activities for Life [REAL] study): single-blind, cluster-randomised controlled trial. Lancet Psychiatry. 2015;2(1):38–48. https://doi.org/10.1016/S2215-0366(14)00050-9.
76. Kirk S, Kutchins H. The selling of DSM: the rhetoric of science in psychiatry. New York: Aldine de Gruyter; 1992.
77. Kraemer HC, Schultz SK, Arndt S. Biomarkers in psychiatry: methodological issues. Am J Geriatr Psychiatry. 2002;10(6):653–9. https://doi.org/10.1097/00019442-200211000-00004.
78. Krivoy A, Onn R, Vilner Y, Hochman E, Weizman S, Paz A, Hess S, Sagy R, Kimhi-Nesher S, Kalter E, Friedman T, Friedman Z, Bormant G, Trommer S, Valevski A, Weizman A. Vitamin D supplementation in chronic schizophrenia patients treated with clozapine: a randomized, double-blind, placebo-controlled clinical trial. EBioMedicine. 2017;26:138–45. https://doi.org/10.1016/j.ebiom.2017.11.027.
79. Kuhn T. The structure of scientific revolutions. 4th ed. Chicago: The University of Chicago Press; 1962.
80. Kuhn T. The structure of scientific revolutions. 50th anniversary. 4th ed. Chicago: University of Chicago Press; 2012.
81. Lacewing M. Could psychoanalysis be a science? In: Fulford KWM, Davies M, G.T. Gipps R et al. (eds) The Oxford handbook of philosophy and psychiatry. Oxford University Press; 2013. https://doi.org/10.1093/oxfordhb/9780199579563.001.0001.
82. Leucht S, Arbter D, Engel RR, Kissling W, Davis JM. How effective are second-generation antipsychotic drugs? A meta-analysis of placebo-controlled trials. Mol Psychiatry. 2009a;14(4):429–47. https://doi.org/10.1038/sj.mp.4002136.
83. Leucht S, Cipriani A, Spineli L, Mavridis D, Orey D, Richter F, Samara M, Barbui C, Engel RR, Geddes JR, Kissling W, Stapf MP, Lassig B, Salanti G, Davis JM. Comparative efficacy and tolerability of 15 antipsychotic drugs in schizophrenia: a multiple-treatments meta-analysis. Lancet. 2013;382(9896):951–62. https://doi.org/10.1016/S0140-6736(13)60733-3.
84. Leucht S, Kissling W, Davis JM. Second-generation antipsychotics for schizophrenia: can we resolve the conflict? Psychol Med. 2009b;39(10):1591–602. https://doi.org/10.1017/S0033291709005455.
85. Leucht S, Pitschel-Walz G, Abraham D, Kissling W. Efficacy and extrapyramidal side-effects of the new antipsychotics olanzapine, quetiapine, risperidone, and sertindole compared to conventional antipsychotics and placebo. A meta-analysis of randomized controlled trials. Schizophr Res. 1999;35(1):51–68. https://doi.org/10.1016/s0920-9964(98)00105-4.
86. Leucht S, Tardy M, Komossa K, Heres S, Kissling W, Davis JM. Maintenance treatment with antipsychotic drugs for schizophrenia. Cochrane Database Syst Rev. 2012a;5:CD008016. https://doi.org/10.1002/14651858.CD008016.pub2.
87. Leucht S, Tardy M, Komossa K, Heres S, Kissling W, Salanti G, Davis JM. Antipsychotic drugs versus placebo for relapse prevention in schizophrenia: a systematic review and meta-analysis. Lancet. 2012b;379(9831):2063–71. https://doi.org/10.1016/S0140-6736(12)60239-6.
88. Lewczuk P, Riederer P, O'Bryant SE, Verbeek MM, Dubois B, Visser PJ, Jellinger KA, Engelborghs S, Ramirez A, Parnetti L, Jack CR Jr, Teunissen CE, Hampel H, Lleo A, Jessen F, Glodzik L, de Leon MJ, Fagan AM, Molinuevo JL, Jansen WJ, Winblad B, Shaw LM,

Andreasson U, Otto M, Mollenhauer B, Wiltfang J, Turner MR, Zerr I, Handels R, Thompson AG, Johansson G, Ermann N, Trojanowski JQ, Karaca I, Wagner H, Oeckl P, van Waalwijk van Doorn L, Bjerke M, Kapogiannis D, Kuiperij HB, Farotti L, Li Y, Gordon BA, Epelbaum S, Vos SJB, Klijn CJM, Van Nostrand WE, Minguillon C, Schmitz M, Gallo C, Lopez Mato A, Thibaut F, Lista S, Alcolea D, Zetterberg H, Blennow K, Kornhuber J, Members of the Wfsbp Task Force Working on this Topic: Peter Riederer CGDKALMFT. Cerebrospinal fluid and blood biomarkers for neurodegenerative dementias: an update of the consensus of the task force on biological markers in psychiatry of the world Federation of Societies of biological psychiatry. World J Biol Psychiatry. 2018;19(4):244–328. https://doi.org/10.1080/15622975.2017.1375556.

89. Marzano L, Bardill A, Fields B, Herd K, Veale D, Grey N, Moran P. The application of mHealth to mental health: opportunities and challenges. Lancet Psychiatry. 2015;2(10):942–8. https://doi.org/10.1016/S2215-0366(15)00268-0.
90. McGrath J, Saari K, Hakko H, Jokelainen J, Jones P, Jarvelin MR, Chant D, Isohanni M. Vitamin D supplementation during the first year of life and risk of schizophrenia: a Finnish birth cohort study. Schizophr Res. 2004;67(2–3):237–45. https://doi.org/10.1016/j.schres.2003.08.005.
91. Mishler E. Research interviewing: context and narrative. Research interviewing: context and narrative. Cambridge, MA: Harvard University press; 1991.
92. Moore D, McCabe G. Introduction to the practice of statistics. 4th ed. New York: W.H. Freeman and Co; 2003.
93. Munder T, Gerger H, Trelle S, Barth J. Testing the allegiance bias hypothesis: a meta-analysis. Psychother Res. 2011;21(6):670–84. https://doi.org/10.1080/10503307.2011.602752.
94. Neyman J, Pearson ES. On the problem of the most efficient tests of statistical hypotheses. Philos Trans Royal Soc B: Biol Sci A. 1933;231(694–706):289–337.
95. Nierenberg AA. How to evaluate a diagnostic marker test. JAMA. 1988;259(11):1699–702. https://doi.org/10.1001/jama.1988.03720110061036.
96. Nisbett RE, Wilson TD. The halo effect: evidence for unconscious alteration of judgments. J Pers Soc Psychol. 1977;35(4):250–6. https://doi.org/10.1037/0022-3514.35.4.250.
97. Ogawa Y, Furukawa TA, Takeshima N, Hayasaka Y, Atkinson LZ, Tanaka S, Cipriani A, Salanti G. Efficacy of antidepressants over placebo is similar in two-armed versus three-armed or more-armed randomized placebo-controlled trials. Int Clin Psychopharmacol. 2018;33(2):66–72. https://doi.org/10.1097/YIC.0000000000000201.
98. Osorio FL, Loureiro SR, Hallak JEC, Machado-de-Sousa JP, Ushirohira JM, Baes CVW, Apolinario TD, Donadon MF, Bolsoni LM, Guimaraes T, Fracon VS, Silva-Rodrigues APC, Pizeta FA, Souza RM, Sanches RF, Dos Santos RG, Martin-Santos R, Crippa JAS. Clinical validity and intrarater and test-retest reliability of the structured clinical interview for DSM-5 - clinician version (SCID-5-CV). Psychiatry Clin Neurosci. 2019;73(12):754–60. https://doi.org/10.1111/pcn.12931.
99. Othmer E, Othmer S. The clinical interview using DSM-IV-TR: Vol 1: fundamentals. Arlington, VA: American Psychiatric Publishing; 2002.
100. Pompoli A, Furukawa TA, Imai H, Tajika A, Efthimiou O, Salanti G. Psychological therapies for panic disorder with or without agoraphobia in adults: a network meta-analysis. Cochrane Database Syst Rev. 2016;4:CD011004. https://doi.org/10.1002/14651858.CD011004.pub2.
101. The Precision Medicine Initiative. 2015. https://obamawhitehouse.archives.gov/node/333101. Accessed June 10, 2018.
102. Regier DA. State-of-the-art psychiatric diagnosis. World Psychiatry. 2004;3(1):25–6.
103. Roesler J, Tyler L, Canady K, Zhong L (2013) Intellectual property issues impacting the future of personalized medicine. AIPLA's MidWinter Institute for CLE Credits.
104. Rosch E. Cognitive reference points. Cogn Psychol. 1975;7(4):532–47.
105. Sabin JE, Harland JC. Professional ethics for digital age psychiatry: boundaries, privacy, and communication. Curr Psychiatry Rep. 2017;19(9):55. https://doi.org/10.1007/s11920-017-0815-5.

106. Schmitt A, Martins-de-Souza D, Akbarian S, Cassoli JS, Ehrenreich H, Fischer A, Fonteh A, Gattaz WF, Gawlik M, Gerlach M, Grunblatt E, Halene T, Hasan A, Hashimoto K, Kim YK, Kirchner SK, Kornhuber J, Kraus TFJ, Malchow B, Nascimento JM, Rossner M, Schwarz M, Steiner J, Talib L, Thibaut F, Riederer P, Falkai P, Members of the WTFoBM. Consensus paper of the WFSBP task force on biological markers: criteria for biomarkers and endophenotypes of schizophrenia, part III: molecular mechanisms. World J Biol Psychiatry. 2017;18(5):330–56. https://doi.org/10.1080/15622975.2016.1224929.
107. Schmitt A, Rujescu D, Gawlik M, Hasan A, Hashimoto K, Iceta S, Jarema M, Kambeitz J, Kasper S, Keeser D, Kornhuber J, Koutsouleris N, Lanzenberger R, Malchow B, Saoud M, Spies M, Stober G, Thibaut F, Riederer P, Falkai P, Markers WTFoB. Consensus paper of the WFSBP task force on biological markers: criteria for biomarkers and endophenotypes of schizophrenia part II: cognition, neuroimaging and genetics. World J Biol Psychiatry. 2016;17(6):406–28. https://doi.org/10.1080/15622975.2016.1183043.
108. Schwartz MA, Wiggins OP. Typifications. The first step for clinical diagnosis in psychiatry. J Nerv Ment Dis. 1987;175(2):65–77. https://doi.org/10.1097/00005053-198702000-00001.
109. Shinohara K, Honyashiki M, Imai H, Hunot V, Caldwell DM, Davies P, Moore TH, Furukawa TA, Churchill R. Behavioural therapies versus other psychological therapies for depression. Cochrane Database Syst Rev. 2013;10:CD008696. https://doi.org/10.1002/14651858.CD008696.pub2.
110. Spitzer RL. Psychiatric diagnosis: are clinicians still necessary? Compr Psychiatry. 1983;24(5):399–411. https://doi.org/10.1016/0010-440x(83)90032-9.
111. Stafford MR, Jackson H, Mayo-Wilson E, Morrison AP, Kendall T. Early interventions to prevent psychosis: systematic review and meta-analysis. BMJ. 2013;346:f185. https://doi.org/10.1136/bmj.f185.
112. Stockings EA, Degenhardt L, Dobbins T, Lee YY, Erskine HE, Whiteford HA, Patton G. Preventing depression and anxiety in young people: a review of the joint efficacy of universal, selective and indicated prevention. Psychol Med. 2016;46(1):11–26. https://doi.org/10.1017/S0033291715001725.
113. Strong WB, Malina RM, Blimkie CJ, Daniels SR, Dishman RK, Gutin B, Hergenroeder AC, Must A, Nixon PA, Pivarnik JM, Rowland T, Trost S, Trudeau F. Evidence based physical activity for school-age youth. J Pediatr. 2005;146(6):732–7. https://doi.org/10.1016/j.jpeds.2005.01.055.
114. Terry SF. Obama's precision medicine initiative. Genet Test Mol Biomarkers. 2015;19(3):113–4. https://doi.org/10.1089/gtmb.2015.1563.
115. Thibaut F. Digital applications: the future in psychiatry? Dialogues Clin Neurosci. 2016;18(2):123.
116. Thibaut F, Boutros NN, Jarema M, Oranje B, Hasan A, Daskalakis ZJ, Wichniak A, Schmitt A, Riederer P, Falkai P, Markers WTFoB. Consensus paper of the WFSBP task force on biological markers: criteria for biomarkers and endophenotypes of schizophrenia part I: neurophysiology. World J Biol Psychiatry. 2015;16(5):280–290. https://doi.org/10.3109/15622975.2015.1050061.
117. Thorndike EL. A constant error in psychological ratings. J Appl Psychol. 1920;4(1):25–9. https://doi.org/10.1037/h0071663.
118. Torous JB, Chan SR, Yellowlees PM, Boland R. To use or not? Evaluating ASPECTS of smartphone apps and Mobile Technology for Clinical Care in psychiatry. J Clin Psychiatry. 2016;77(6):e734–8. https://doi.org/10.4088/JCP.15com10619.
119. van der Gaag M, Smit F, Bechdolf A, French P, Linszen DH, Yung AR, McGorry P, Cuijpers P. Preventing a first episode of psychosis: meta-analysis of randomized controlled prevention trials of 12 month and longer-term follow-ups. Schizophr Res. 2013;149(1–3):56–62. https://doi.org/10.1016/j.schres.2013.07.004.
120. Weiss N (2015) Introductory statistics. 10th.
121. Werner-Seidler A, Perry Y, Calear AL, Newby JM, Christensen H. School-based depression and anxiety prevention programs for young people: a systematic review and meta-analysis. Clin Psychol Rev. 2017;51:30–47. https://doi.org/10.1016/j.cpr.2016.10.005.

122. WHO. Schedules for Clinical Assessment in Neuropsychiatry (SCAN) v.2.1. Geneva: World Health Organization; 1994.
123. World Health Organization. Prevention of mental disorders: effective interventions and policy options: summary report. 2004. https://www.who.int/mental_health/evidence/en/prevention_of_mental_disorders_sr.pdf.
124. Zalsman G, Hawton K, Wasserman D, van Heeringen K, Arensman E, Sarchiapone M, Carli V, Hoschl C, Barzilay R, Balazs J, Purebl G, Kahn JP, Saiz PA, Lipsicas CB, Bobes J, Cozman D, Hegerl U, Zohar J. Suicide prevention strategies revisited: 10-year systematic review. Lancet Psychiatry. 2016;3(7):646–59. https://doi.org/10.1016/S2215-0366(16)30030-X.
125. Zetterberg H. Psychosis and biological markers. Lancet Psychiatry. 2017;4(1):3–5. https://doi.org/10.1016/S2215-0366(16)30407-2.

Chapter 19
Psychiatry among the Other Medical Specialties

19.1 The Rise of Psychiatry as a Medical Discipline

Since antiquity, psychiatry was considered to be part of medicine. Mania and melancholia are two of the most ancient terms in medical nomenclature, recognized also by Hippocrates (460–370 BC; Fig. 3.16). The diagnosis and treatment of mental disorders and the attitude of the society toward mental patients usually followed the general rules of medicine and health. Religious and spiritual approaches were applied to other medical conditions also. However, mental patients were more often victims of stigmatization, marginalization, and prerecession, but this was something that has also been happening to other groups of patients, including those with leprosy, psoriasis, epilepsy, and cancer, among others.

The men who determined the new medical specialty were Philippe Pinel (1745–1826; Figs. 7.18 and 7.19) and especially Jean-Philippe Esquirol (1772–1840; Fig. 8.4). They soon became the prototype of the psychiatric specialist. The school Esquirol founded determined the basic characteristics of the new medical specialty. He was the first to adopt a clinical descriptive approach and remained extremely cautious of any psychological interpretation, to the extent that he was criticized for creating a "psychiatry without psychology" [30]. Of course the "godfather" of the specialty was Johann Reil (1759–1813; Fig. 8.1), who coined the term "psychiatry" and replaced "alienists" with "psychiatrists." In England and France, the separation of psychiatry occurred sometime in the mid-nineteenth century [9, 15]. However, as Kraepelin wrote, during most of the nineteenth century, mental health was mainly in the hands of the supervisors, attendants, and administrators of the asylums, and the books on mental disorders were written either by doctors who were not psychiatrists or by philosophers and theologians who had never seen a single patient [14].

The man who triggered a dramatic change in the profession was Sigmund Freud (1856–1939; Fig. 9.38), not because of the scientific value of his theories but because of the effect they had both on the clinical practice and the environment of

K. N. Fountoulakis, *Psychiatry*, https://doi.org/10.1007/978-3-030-86541-2_19

psychiatrists as well as the change in lay people's perception of psychiatry. Freudian approaches permitted the majority of psychiatrists to leave asylums and obtain a clinical practice similar to that of doctors of other specialties; at the same time, psychoanalytical ideas had a pervasive influence on lay culture, thus making the concept of mental illness and the therapist less propulsive and more acceptable. Some aspects of psychiatry were even glorified because of psychoanalysis. But the most important and enduring contribution of psychoanalysis was the rejection of metaphysical explanations for human behavior and morality.

Unfortunately, this also had a dark side. In spite of the claims for a scientific psychology [5], Freudian theories and psychoanalysis were deeply hostile to empirical verification, and in many ways, they demanded "faith" instead of data. This, in conjunction with the vast popular acceptance and a supportive socio-ideological current, caused psychiatry to derail from its medical essence and standards. For several decades, other physicians considered psychiatrists to be those medical students who wanted to become "fugitives from medicine" or those with personal mental problems who sought shelter in psychiatry. The psychoanalytical movement was too much involved in socio-political affairs and was self-absorbed in the experience of omnipotence, which came with wider public acceptance and admiration.

As discussed in previous chapters, this situation created a crisis in the early 1970s and a pressing need for reform [16]. The man who shaped psychiatry into its current modern image was Robert Spitzer (1932–2015; Fig. 11.42). He supervised the creation of DSM-III and the introduction of the operationalized diagnostic criteria. On the other hand, none of the previously mentioned prominent figures managed to make any direct improvement for the lives of mental patients. Such a dramatic change came with psychopharmacology and the availability of treatment with antipsychotics, antidepressants, lithium, and other psychoactive substances. Here, it is possible to mention only a few names that can be credited with this advancement. The process took several decades; it is a child of the dye biochemistry and industry, and it demanded huge investments mainly from the side of private chemical and pharmaceutical companies. It is disappointing that the two Nobel awards given to Julius Wagner-Jauregg (1857–1940; Fig. 10.24) for pyrotherapy and to Egas Moniz (1874–1955; Fig. 10.40) for psychosurgery concern failed treatment options. The real heroes of psychiatric treatment were not honored in such a prominent way with the exception of Arvid Carlsson (1923–2018; Fig. 11.22). These real heroes include (but the list is not exhaustive) Ugo Cerletti (1877–1963; Fig. 10.44) and Lucio Bini (1908–1964; Fig. 10.44) for the development of ECT; Henri Laborit (1914–1995; Fig. 11.6), Pierre Deniker (1917–1998; Fig. 11.6), and Jean Delay (1907–1987; Fig. 11.6) for the introduction of antipsychotics; the Lange brothers Carl (1834–1900; Fig. 9.14) and Frederik (1842–1907), John Cade (1912–1980; Figs. 10.59 and 11.25), and Mogens Schou (1918–2005; Fig. 11.25) for the introduction of lithium; and Roland Kuhn (1912–2005; Fig. 11.19) for the introduction of antidepressants.

So, after these four major developments (recognition of mental illness, opening and wider acceptance of the psychiatric treatment, improvement of diagnostic procedures, and development of effective treatments), contemporary psychiatry today is a specialty of medicine aiming to study, prevent, and treat mental disorders in humans [8, 10, 28]. It utilizes research in the field of medicine, neuroscience, biology, and pharmacology as well as psychology and sociology [21], and additionally, it has been considered to constitute the middle ground between neurology and psychology [26]. It is interesting that historically, psychology is the child of psychiatry rather than of philosophy; the most prominent founding fathers of psychology were doctors, including Williams James, Wilhelm Wundt, Gustav Fechner, and Ernst Weber, not to mention the psychoanalysts Sigmund Freud, Carl Jung, and Alfred Adler and the cognitive behaviorist Aaron Beck.

19.2 The Voices against

It is important to discuss the opposite opinion, which rejects the biomedical model and also the scientific method altogether, although it rarely declares this openly and clearly. Essentially, the consideration of psychiatry as part of medicine, and also as a branch of natural sciences, has many opponents [13, 17, 23, 27]. Although some authors argue that the attacks on psychiatry were because of the dominant medical model for mental disorder [7], this is essentially wrong. During the 1960s and the 1970s, the dominant model for psychiatry was psychoanalysis, which, somehow was connected with medical doctors/psychoanalysts. The medical model emerged after psychoanalysis was discredited because it had trapped psychiatry in the most severe crisis in its history (see Chap. 11). The medical model was the response to antipsychiatric attacks, not their original cause. Before of the clear adoption of the biomedical model, during the late nineteenth century and the first half of the twentieth century, there was a great schism in psychiatry with two opposing approaches. The first was heralded by Emil Kraepelin and concerned an emphasis on the systematic approach to psychiatric diagnosis and classification and the second by Sigmund Freud (1856–1939; Fig. 9.38), who developed psychoanalysis on the basis of interpretation rather than observation and rejected the concept of psychiatric diagnosis. However, this was only a break that lasted a few decades, and soon afterwards, psychiatry returned to the nosological approach, with Robert Spitzer (1932–2015; Fig. 11.42) being the champion of this return to the roots. The introduction of classification systems based on empirical data made psychiatric diagnoses more comparable to the diagnoses in the other medical specialties and in close link with the scientific way and the rules of science [12, 25]. Eventually, psychiatry should be viewed as a kind of clinical cognitive neuroscience [18]. Following the biomedical model means that the assumption of an underlying neurobiological dysfunction is responsible for the emergence of mental disorders [18], and on the basis of this dysfunction, a differentiation between dysfunctional brains and unpopular behavior can be made, and an appropriate treatment can be applied where and when necessary, although practically this is still very difficult and maybe impossible because our knowledge is still limited [24],and data seem too often to depend on theoretical approaches [6].

Nevertheless, a crucial question would be, is it possible to reject the medical model for mental illness? Where would this lead to? The real problem is that many scholars feel that the medical model restricts human freedom and implies a sense of weakness, pessimism, or lack of control. Especially the "locus of control" for the persons' being itself seems to exist outside the will of the person, and his/her fate is controlled by others who are "professionals in the field." This latter "professional relationship" between the therapist and the patient triggers more feelings of lost independence and restricted freedom. This is mainly why the anti-medical approach focuses not on empirical data but on subjective experience and interpretation and rejects abstraction in favor of individualized understanding. It favors generic non-scientific interventions close to lay persons' beliefs, including diet, exercise, and societal change. In this way the "control" and the "power" is moved away from the "elite of specialists" and is given back to the person.

Such approaches exist also in other fields of medicine and recently are spreading at a worrying rate. They are very difficult to tackle because they a priori reject the scientific way; they do not engage in discussions that follow the rules of science or even at least those of analytical philosophy. Arguments are of poor quality if they exist at all. They often utilize pseudo-philosophical or conspiracy theories and retreat only when the everyday evidence are observable with the naked eye and are impossible to neglect; even then, they seem extraordinarily resilient, and this was proven in a horrifying way during the recent COVID-19 crisis. Even then, as our recent experience has proven, they find a way to combine extremely opposing theories and all kinds of irrational thoughts under the same roof as well as accuse people of science.

But until today, the attacks on psychiatry had little impact, and some authors suggest that the reason is that these attacks came mainly from outside psychiatry itself [2]. These authors forget that Thomas Szasz, David Cooper, Ronald Laing, Peter Breggin, Marius Romme, and many others were psychiatrists formally trained in psychiatry and that their attacks on mainstream psychiatry preceded those by most philosophers, and at any case, they started more than half a century ago. On the other hand, there is much truth that the battle frontlines are more than one, and besides psychiatry vs. anti-psychiatry, today a front where old-fashion psychiatry (all-inclusive and strongly psycho-socially oriented) clashes with scientifically rigorous Psychiatry (evidence-based and subsequently bio-medically oriented), while very close to us, scientific psychology clashes with lay mental counseling. These battle fronts will probably be the intellectual battleground of the future.

A new issue is proposed, that of an ill over-medicalization [2], but this is erroneous; the real question is whether psychiatry will keep pace with the rest of medicine and its evolution towards a strict science by following the "evidence-based approach." In contrast, today the banal old approach, that the clash is medical vs. psychological vs. social, seems to be the shameful result of trade unionism.

Today, it is the rule rather than the exception that antipsychiatric and antiscientific approaches are shaped into a "call to arms to protect the scientific and ethical integrity of psychiatry from a new enemy: profligate, industry-driven, commercial medicalization, and its more nefarious social consequences" [2]. But corruption and breaking of the law are not an issue to debate philosophically. It is a practical issue of law and regulation enforcement. The argument that there can be "no psychiatry that can aspire to pure science without any abuses of power or commercial interests" [2] is irrational. It is in line with the core of anti-psychiatric logic, which by definition identifies the scientific knowledge with power imbalance and abuse and calls for its destruction [4].

19.3 The Delineation of the Field of Psychiatry

Like most medical specialties, psychiatry includes a number of subspecialties and fields (Table 19.1). Psychiatry differs from neurology not only in the way mental disorders differ from neurological disorders but also in the way residents in psychiatry receive their training with emphasis on the doctor–patient relationship and with training in psychotherapy [26]. On the other hand, psychiatry differs from psychology, which in most regions of the world does not belong to health sciences academically, in spite of the fact that many psychologists deal with patients. Departments of psychology usually belong to schools of philosophy or social sciences, but in many

Table 19.1 Subspecialties and fields of psychiatry

• Addiction psychiatry
• Biological psychiatry
• Child and adolescent psychiatry
• Consultation–liaison psychiatry
• Emergency psychiatry
• Forensic psychiatry
• Geriatric psychiatry (Psychogeriatrics or Geropsychiatry)
• Intellectual disability
• Military psychiatry
• Neuropsychiatry
• Neuropsychology
• Perinatal and infant psychiatry
• Psychodynamic psychiatry/psychotherapy
• Psychological medicine
• Psychological psychiatry/psychotherapy
• Psychometrics
• Psychopharmacology
• Psychosomatic medicine
• Rehabilitation and social psychiatry
• Social and community psychiatry
• Transcultural psychiatry
• Trauma and disaster psychiatry

universities, they are independent. Psychology covers a variety of topics concerning human behavior, but it is the field of clinical psychology that is closer to psychiatry. It would not be irrational to suggest that the faculties of clinical psychology should be part of schools of health sciences the way pharmacology is. The main difference between psychiatrists and psychologists is that a psychiatrist is a physician specialized in the treatment of mental disorders, trained, and able to apply all medical procedures both for the diagnosis and the treatment of mental disorders, including laboratory testing and medication treatment as well as psychotherapy. Although psychology is increasingly re-orienting toward neurobiology, the work of a psychologist is limited to non-invasive techniques and psychotherapy, in spite of the fact that there was a trend toward licensing psychologists even to prescribe antidepressants. This is currently approved in five US states (Iowa, Idaho, Illinois, New Mexico, and Louisiana) as well as in the Public Health Service, the Indian Health Service, the US Military, and Guam, often after license from a medical board, but has not been adopted by any agency outside the USA [19]. This is in the frame of a controversial issue concerning whether prescription is an exclusive right of medical doctors. So far, this right has been challenged only in the USA, UK, and Canada [20]. For many authors this reflects a wider stigmatization attitude within the administration, which views mental patients as second-class patients, since for no other patient group with severe disability, there is approval of medication treatment by non-physicians in a completely independent way. There is a tendency to give prescription privileges to other health professionals (e.g., nurses) but not for severe medical conditions.

19.4 Doctors' Necessary Professional Skills in Mental Health

In terms of skills, certified psychiatrists should be able to diagnose mental disorders not only in general terms but also according to the classification systems (especially ICD, since it is used internationally by mental health systems) and to apply treatments, mainly pharmacological but also some kind of psychotherapy. They should be able to understand the subjective experience of patients and to put it in perspective together with a treatment plan. Although psychiatry remains significantly different from the other specialties of medicine, during the last few decades, it became more biological and subsequently less conceptually isolated from the others. The World Psychiatric Association (WPA) and the World Federation for Medical Education have produced a core curriculum for undergraduate training in psychiatry [29], while more recent documents have been produced by the WPA concerning under- and postgraduate psychiatric training. An important change is that the demand for personal psychotherapy is not as strong as it used to be 20 or 30 years ago [3].

On the other hand, since mental health issues are the most common principle diagnoses for Medicaid superutilizers in the USA [11], there is a need for all medical doctors to have basic knowledge of psychiatry and how to handle mental patients. In this frame, new medical curriculums introduce students to mental patients earlier in their studies and also in a wider range of settings. Such an earlier and more intense exposure improves their attitude and skills, diminishes stigma toward mental illness within the medical profession, and also improves the status of psychiatry as a medical specialty among the other specialties. In the past this interaction was mediated mainly by Consultation–Liaison services at the time when psychiatry was pushing to find its place among the other specialties of general hospitals. Consultation–Liaison flourished in the 1980s and 1990s, but it is no longer considered to be the pride of general hospital psychiatry.

In spite of the narrowing of the gap, psychiatry will always manifest radical differences from the rest of medicine. Mental disorder is loaded with a complex value system, while somatic disorders might have no value load at all. However, there is a tendency even for somatic disorders to obtain values as scientific advances develop [7]. A practical difference that characterizes psychiatry is that too much happens privately between the patient and the doctor, and this results, among other things, in the emergence of very special ethical issues [1, 22]. Another difference is that too many skills important for psychiatrists are rather very difficult or impossible to teach, and training has little effect. For other medical specialties, these are considered "soft skills," but for a psychiatrist, interpersonal skills and resilience to existential stress are of prime importance. Both are difficult to acquire and are mostly trait characteristic of a person; this is one of the reasons psychiatry will probably continue to qualitatively differ from the rest of medicine.

References

1. Adams G, Cook M. Beginning psychiatry. Bulletin of the Royal College of Psychiatrists. 1984;8:53–4.
2. Charland L. Why psychiatry should fear medicalization. In: Fulford KWM, Davies M, Gipps R et al. (eds) The Oxford handbook of philosophy and psychiatry. Oxford University Press. 2013. https://doi.org/10.1093/oxfordhb/9780199579563.001.0001.
3. Cox J, Marks M, Marteau L, Steiner J. Personal psychotherapy in the training of a psychiatrist. Bulletin of the Royal College of Psychiatrists. 1982;6:38–42.
4. Foucault M, Howard R, Cooper D. Madness and civilization. Taylor & Francis. 1961; https://doi.org/10.4324/9780203278796.
5. Freud S, Bonaparte M, Freud A, Kris E, Mosbacher E, Strachey J. Project for a scientific psychology. In: Bonaparte M, Freud A, Kris E, Mosbacher E, Strachey J, editors. The origins of psycho-analysis: letters to Wilhelm Fliess, drafts and notes: 1887–1902. New York: Basic Books; 1954. p. 347–445. https://doi.org/10.1037/11538-013.
6. Fulford KWM, Thornton T, Graham G. Psychopathology and the theory dependence of data. In: Fulford KWM, Thornton T, Graham G, editors. The Oxford textbook of philosophy and psychiatry. Oxford: Oxford University Press; 2006.
7. Fulford KWM, Van Staden CW. Values-based practice: Topsy-Turvy take-home messages from ordinary language philosophy (and a few next steps). In: Fulford KWM, Davies M, Gipps R et al. (eds) The Oxford handbook of philosophy and psychiatry. Oxford University Press: Oxford. 2013. https://doi.org/10.1093/oxfordhb/9780199579563.001.0001.
8. Gask L. Listening to patients. Br J Psychiatry. 1997;171:301–2. https://doi.org/10.1192/bjp.171.4.301.

9. Goldstein J. Console and classify: the French psychiatric profession in the nineteenth century. Cambridge: Cambridge University Press; 1987.
10. Guze SB. Psychobiology of personality. Am J Psychiatr. 1992;149(8):1107–8. https://doi.org/10.1176/ajp.149.8.1107.
11. Jiang H, Barrett M, Sheng M. Characteristics of hospital stays for nonelderly Medicaid super-utilizers. HCUP Statistical Brief #184. 2012; http://www.hcup-us.ahrq.gov/reports/statbriefs/sb184-Hospital-Stays-Medicaid-Super-Utilizers-2012.pdf.
12. Kendell RE. The concept of disease and its implications for psychiatry. Br J Psychiatry. 1975;127:305–15.
13. Kirk S, Kutchins H. The selling of DSM: the rhetoric of science in psychiatry. New York: Aldine de Gruyter; 1992.
14. Kraepelin E. Hundert Jahre Psychiatrie. Springer; 1918.
15. Leigh D. The historical development of British psychiatry vol I: eighteenth and nineteenth centuries. Oxford: Pergamon Press; 1961.
16. Lieberman JA. Shrinks: the untold story of psychiatry. London: Back Bay Books; 2016.
17. Monti MR, Stanghellini G. Psychopathology: an edgeless razor? Compr Psychiatry. 1996;37(3):196–204. https://doi.org/10.1016/s0010-440x(96)90036-x.
18. Murphy D. The medical model and the philosophy of science. In: Fulford KWM, Davies M, Gipps R et al. (eds) The Oxford handbook of philosophy and psychiatry. Oxford University Press: Oxford; 2013. https://doi.org/10.1093/oxfordhb/9780199579563.001.0001.
19. NIMH (2007) Information about mental illness and the brain. The teacher's guide. http://science-education.nih.gov/supplements/nih5/Mental/guide/info-mental-c.htm.
20. O'Connor SA. Should prescribing authority be shared with nonphysicians?: YES. Can Fam Physician. 2009;55(12):1176–8.
21. Pietrini P. Toward a biochemistry of mind? Am J Psychiatry. 2003;160(11):1907–8. https://doi.org/10.1176/appi.ajp.160.11.1907.
22. Radden J, Sadler JZ. The virtuous psychiatrist: character ethics in psychiatric practice. New York: Oxford university press; 2010.
23. Richardson J. Handbook of qualitative research methods for psychology and the social sciences. Wiley-Blackwell. London; 1996.
24. Roe K, Murphy D. Function, dysfunction, adaptation? In: Adriaens P, Block A, editors. Maladapted minds. Oxford: Oxford University Press; 2011. p. 216–37.
25. Shepherd M. The extent of mental disorder. Beyond the layman's madness. Can Psychiatr Assoc J. 1976;21(6):401–9. https://doi.org/10.1177/070674377602100607.
26. Shorter E. A history of psychiatry: from the era of the asylum to the age of Prozac. New York: Wiley; 1997.
27. Stanghellini G. The grammar of the psychiatric interview. A plea for the second-person mode of understanding. Psychopathology. 2007;40(2):69–74. https://doi.org/10.1159/000098486.
28. Storrow H. Outline of clinical psychiatry. New York: Appleton-Century-Crofts; 1969.
29. WPA, WFME. WPA Recommendations: Principles and Priorities for a Framework for Training Psychiatrists; 1998.
30. Zilboorg G. A history of medical psychology. New York: Norton; 1941.

Chapter 20
Psychiatry as a Professional Activity

As all medical specialties, psychiatry functions also as a profession under the legislation and the ethics that determine the medical profession. For any medical specialty, the issues that are very important are the number of professionals, their professional obligations and privileges, and their position and status in the healthcare system, among others.

Well within the frame of wider sociopolitical changes, and especially with the dawn of democracy and the concept of the responsible citizen, the Hippocratic physicians recognized that, on the one hand, their patients had the right to choose any type of treatment and any therapist they wished, and, on the other hand, the physicians themselves rather than some kind of divinity should take responsibility for the treatment outcome. This was the earliest conceptualization of medicine as a responsible humanistic profession, and it is reflected in the Hippocratic oath.

The enormous variation across the world in the numbers of psychiatrists and other mental health professionals is shown in Table 20.1 and Fig. 20.1 for the years 2011–2014. According to these data, in 2014 almost half of the world population lived in a region where there was less than 1 psychiatrist per 100,000 people.

Today, in many countries, the majority of people are treated by their general practitioners or by other medical doctors who are not psychiatrists. Referral to a specialist is controlled by these gatekeepers, but it is evident that the views of the patient, the gatekeepers, and the health systems often differ significantly. This happens in rich countries in order to reduce the cost and in poor countries because there are not enough specialists available [1].

In rich countries the large effort to reduce the costs has led to a growth of managed care systems and to the provision of mental health care and treatment by nonmedical therapists (nurses, psychologists, and social workers) with the psychiatrist playing the role of a supervising consultant. This is of course in sharp contrast to the psychoanalytical tradition of psychiatry as well as to the basic conceptualization of psychiatry as based on the direct patient–doctor relationship. Restrictions already in place concerning the use of resources are problematic when considered from the

K. N. Fountoulakis, *Psychiatry*, https://doi.org/10.1007/978-3-030-86541-2_20

Table 20.1 WHO data on the rates of mental health professionals per country http://www.who.int/gho/mental_health/human_resources/psychiatrists_nurses/en/s

Country	Psychiatrists working in the mental health sector (per 100,000)	Nurses working in the mental health sector (per 100,000)	Social workers working in the mental health sector (per 100,000)	Psychologists working in the mental health sector (per 100,000)	year
Switzerland	41.42			40.78	2011
Monaco	40.98	66.99	3.94	23.78	2014
Norway	29.69	123.08	26.45	54.28	2014
Luxembourg	21.15				2011
Belgium	20.32				2014
Netherlands	20.10	2.87		90.76	2014
Iceland	19.74	37.96	10.63	31.89	2011
Austria	19.71			79.88	2011
Finland	18.37			56.95	2014
Sweden	18.31	52.90			2014
Lithuania	17.76				2011
Uruguay	16.91				2014
San Marino	15.80	12.64	9.48	9.48	2014
Germany	15.23	56.06			2011
United Kingdom	14.63	67.35	1.99	12.83	2014
France	14.12	90.86	0.00	10.77	2014
Greece	14.09	50.57	6.92	12.13	2014
Estonia	13.87	21.42	1.71	7.79	2014
Canada	13.42	53.31	117.79	47.42	2014
United States of America	12.40	4.25	59.83	29.62	2014
Czech Republic	11.85	28.24	0.90	2.03	2011
Israel	11.57	11.63			2014
Russian Federation	11.06	33.24	1.76	2.61	2014
Latvia	10.85	30.76			2011
Italy	10.85	27.57	2.43	3.52	2014
Cuba	10.26	29.45	8.63		2014
Croatia	10.25	35.56	1.02	2.90	2011
Slovenia	10.21	89.57	1.49	3.61	2014
Ukraine	10.11	29.73	0.69	1.15	2011
Japan	10.10	102.55	6.06	3.99	2011
FYROM	9.98	26.92	0.83	1.47	2011
New Zealand	9.76	89.32		13.78	2011
Denmark	9.57	60.02	5.66	12.78	2014
Australia	9.16	70.91	8.56	16.68	2014
Montenegro	8.69	22.36	1.61	2.57	2014

Table 20.1 (continued)

Country	Psychiatrists working in the mental health sector (per 100,000)	Nurses working in the mental health sector (per 100,000)	Social workers working in the mental health sector (per 100,000)	Psychologists working in the mental health sector (per 100,000)	year
Spain	8.08	9.69	2.60	5.71	2014
Belarus	7.66	24.13	0.65	3.30	2014
Serbia	7.35	12.22	0.90	1.30	2014
Bulgaria	6.75	431.01	0.36	0.91	2011
Hungary	6.52	21.93	2.98	2.47	2011
Kazakhstan	6.26			0.29	2011
Ireland	6.06	112.83	3.84	3.53	2011
Andorra	6.00				2014
Romania	5.98	16.87	0.54	1.41	2014
Republic of Moldova	5.92	22.07	0.23	0.55	2014
Armenia	5.90				2014
Georgia	5.85	5.60	0.30	0.65	2014
Republic of Korea	5.12	21.61	6.01	1.12	2011
Poland	5.07	18.66	0.73	5.34	2014
Cook Islands	5.02	10.03	0.00	0.00	2011
Palau	4.87	68.19	68.19	0.00	2011
Bahrain	4.84	22.69	0.74	0.74	2014
Chile	4.66	1.59	2.18	5.19	2014
Portugal	4.49	12.46	0.70	2.24	2014
Barbados	4.29	41.71	1.95	9.35	2011
Bosnia and Herzegovina	4.00	12.94	1.39	3.16	2014
Panama	3.80	5.94	3.01	3.39	2014
Azerbaijan	3.72	6.40	0.07	0.38	2014
Maldives	3.70	0.00	0.00	1.14	2014
Kyrgyzstan	3.51	6.67	0.04	0.34	2011
Brazil	3.49	3.05	1.72	3.22	2014
Singapore	3.48	17.73	1.04	1.23	2014
Kuwait	3.33	19.23	0.35	2.10	2014
Brunei Darussalam	3.31	14.89	0.24	0.95	2014
Malta	3.17	66.83	5.12	4.39	2011
Trinidad and Tobago	3.12	56.61	3.05	0.00	2014
Qatar	2.95	8.82	0.13	1.28	2014
Dominica	2.77	11.06	1.38	1.38	2014
Cyprus	2.69	20.55	0.17	2.69	2014

(continued)

Table 20.1 (continued)

Country	Psychiatrists working in the mental health sector (per 100,000)	Nurses working in the mental health sector (per 100,000)	Social workers working in the mental health sector (per 100,000)	Psychologists working in the mental health sector (per 100,000)	year
Uzbekistan	2.68	8.96		0.02	2014
Tunisia	2.58		0.35	1.06	2014
Colombia	2.53			10.74	2014
Costa Rica	2.33	3.92	1.64	1.79	2011
Oman	2.32	12.71	0.26	0.38	2014
Algeria	2.29		0.08	3.43	2014
Tajikistan	2.16	7.14	0.00	0.01	2014
Seychelles	2.14	12.86	0.00	4.29	2014
Saudi Arabia	2.14	10.49	2.47	1.38	2014
Paraguay	2.00	4.00	0.41	3.41	2014
Saint Kitts and Nevis	1.91	13.37	0.00	5.73	2011
Grenada	1.88	63.97	2.82	0.00	2014
Iran (Islamic Republic of)	1.76	8.84	0.72	5.07	2014
Mauritius	1.62	9.33	0.23	0.08	2011
China	1.53	2.65		0.18	2011
Turkey	1.51		0.77	1.43	2014
Suriname	1.47	42.10	0.00	0.55	2014
Cabo Verde	1.37	2.93	0.20	3.51	2011
Albania	1.32	6.66	0.79	1.01	2014
Antigua and Barbuda	1.13	13.55	1.13	1.13	2011
Jamaica	1.11	8.90	0.57	0.43	2014
Ecuador	1.09	0.96	0.54	2.53	2014
Saint Lucia	1.09	26.14	1.09		2014
Dominican Republic	1.08	1.29	0.05	3.99	2014
Bolivia (Plurinational State of)	1.06	0.34	0.25	0.46	2011
Kiribati	1.00	6.03	0.00	0.00	2011
Tonga	0.95	10.40	0.00	0.95	2014
Saint Vincent and the Grenadines	0.91	17.37	1.83	3.66	2014
Viet Nam	0.91	2.92	0.00	0.09	2014
Micronesia (Federated States of)	0.90	6.30	0.00	0.00	2011

Table 20.1 (continued)

Country	Psychiatrists working in the mental health sector (per 100,000)	Nurses working in the mental health sector (per 100,000)	Social workers working in the mental health sector (per 100,000)	Psychologists working in the mental health sector (per 100,000)	year
Thailand	0.87	4.46	0.34	0.72	2014
Lebanon	0.87	0.77	2.32	1.65	2014
Peru	0.76	6.09	0.22	5.06	2014
Malaysia	0.76		0.84	0.89	2014
Egypt	0.69	3.10	0.29	0.12	2014
Mexico	0.67	2.81	0.52	2.11	2014
Belize	0.59	5.30	0.59	0.29	2014
Mongolia	0.52	7.70	0.15	0.19	2011
Samoa	0.52	2.09	0.52	0.00	2014
Sao Tome and Principe	0.51		0.51	1.52	2014
Jordan	0.51	4.72	0.16	0.27	2014
Guyana	0.50		0.37	0.00	2014
Philippines	0.46	0.49	0.07	0.07	2014
Morocco	0.45	4.10	0.03	0.05	2014
South Africa	0.39				2014
Vanuatu	0.39	2.32			2014
Honduras	0.38	2.14	1.17	0.69	2014
Iraq	0.37	1.46	0.22	0.09	2014
Sri Lanka	0.36	2.74	0.17	0.12	2014
Namibia	0.34	5.32	0.17	1.79	2014
Pakistan	0.31	15.43	2.32	1.09	2014
United Arab Emirates	0.30	2.12	0.25	0.51	2011
India	0.30	0.12	0.07	0.07	2014
Botswana	0.29	12.17	0.39	0.39	2014
Guatemala	0.29	1.31	0.07	0.53	2014
Indonesia	0.29	2.57	0.05	0.18	2014
Myanmar	0.29	0.25	0.01	0.00	2014
Bhutan	0.28	78.48	0.00	0.00	2011
Gabon	0.27	1.00	0.07	0.13	2011
Timor-Leste	0.26	1.65	0.09	0.09	2014
Cambodia	0.23	0.80			2011
Fiji	0.23	4.10	0.00	0.00	2011
Nepal	0.22				2014
Yemen	0.21	0.13	0.09	0.17	2011
Kenya	0.19		0.01		2011
Solomon Islands	0.18	8.38	0.00	0.00	2014

(continued)

Table 20.1 (continued)

Country	Psychiatrists working in the mental health sector (per 100,000)	Nurses working in the mental health sector (per 100,000)	Social workers working in the mental health sector (per 100,000)	Psychologists working in the mental health sector (per 100,000)	year
Swaziland	0.17	4.66	0.17	0.25	2011
El Salvador	0.16	0.48	0.05	0.08	2011
Comoros	0.14	0.43	0.00	0.00	2011
Senegal	0.14	0.34	0.06	0.02	2014
Bangladesh	0.13	0.27	0.00	0.03	2014
Benin	0.11	0.29	0.03	0.04	2014
Gambia	0.11	0.58	0.00	0.05	2014
Côte d'Ivoire	0.10	0.02	0.05		2014
Democratic Republic of the Congo	0.10	0.25	0.03	0.02	2014
Lesotho	0.10	4.67	0.10	0.29	2014
Nigeria	0.10	0.70	0.04	0.02	2014
Mauritania	0.09	1.34	0.30	0.15	2011
Uganda	0.09	0.76		0.02	2011
Papua New Guinea	0.09				2014
Zimbabwe	0.08	2.80	0.10	0.04	2014
Haiti	0.07	0.29	0.06	0.07	2014
Sudan	0.06	0.10	0.60	0.48	2011
Ghana	0.06	7.34	0.05	0.03	2014
Madagascar	0.06	0.35	0.02	0.05	2014
Rwanda	0.05	1.30	0.12	0.07	2011
Burkina Faso	0.05	0.56	0.01	0.03	2014
Mozambique	0.05	0.72	0.01	0.34	2014
Zambia	0.05	1.07	0.01	0.03	2014
Ethiopia	0.04	0.59	0.00	0.02	2011
Somalia	0.04	0.45	0.96	0.00	2011
Mali	0.04	0.08	0.01	0.01	2014
Togo	0.04	0.01	0.07	0.89	2014
Cameroon	0.03	0.15			2011
Congo	0.03	0.45	0.03	0.11	2011
Guinea	0.03	0.07	0.00	0.00	2011
Sierra Leone	0.03	1.31			2014
South Sudan	0.03	0.09		0.02	2014
Lao People's Democratic Republic	0.03	0.26	0.00	0.00	2014
Angola	0.02	0.61	0.01	0.03	2011

Table 20.1 (continued)

Country	Psychiatrists working in the mental health sector (per 100,000)	Nurses working in the mental health sector (per 100,000)	Social workers working in the mental health sector (per 100,000)	Psychologists working in the mental health sector (per 100,000)	year
Central African Republic	0.02	0.07	0.04	0.04	2011
Liberia	0.02	0.30	0.05		2014
Niger	0.02	0.10	0.02	0.03	2014
Chad	0.01	0.03	0.01	0.01	2011
Malawi	0.01	0.22	0.01	0.02	2011
Burundi	0.01	0.29	0.10	0.08	2014
United Republic of Tanzania	0.01	0.23	0.00	0.00	2014
Afghanistan	0.01	0.11	0.01	0.04	2014
Guinea-Bissau	0.00	0.61	0.00	0.06	2011
Nauru	0.00	9.75	0.00	0.00	2011
Niue	0.00	904.03	0.00	0.00	2011
Eritrea	0.00	0.99	0.02	0.05	2014
Marshall Islands	0.00	3.79	1.90	0.00	2014
Libya		5.18	0.58	1.04	2014
Syrian Arab Republic		0.83	0.05	0.12	2014

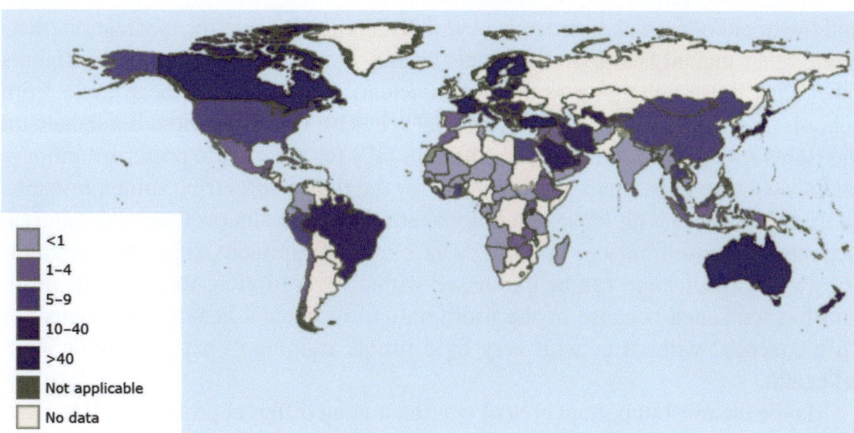

Fig. 20.1 Density of psychiatrist figures around the globe (WHO data)

traditional view. For example, it is considered impossible to have effective mental health care in spending only 15 min every 3 months per patient [2].

In Western countries a significant number of psychiatrists work in multidisciplinary teams alongside other mental health professionals, but the vast majority of psychiatrists worldwide work alone either in the public or in the private sector. While the psychiatrist used to be the leader of these multidisciplinary teams, recently this has been challenged by the status of any medical doctor in these teams. The management of these teams is a riddle to be solved [3]. Increasingly, the management of mental patients becomes fragmented with the psychiatrist making diagnostic and medication treatment decisions often without having ever seen the patient but exclusively on the basis of data provided by other mental health professionals, who additionally apply any psychotherapeutic interventions needed.

Psychotherapy is a form of treatment that can be applied by a psychiatrist or any mental health professional, provided he/she is specifically trained in this specific form of psychotherapy. In many countries this is not easy to control, especially with the recent proliferation of degrees by all sorts of new private universities and colleges around the world who provide flexible degrees with reference to the field of mental health. The most significant issue with these teams is medical responsibility of the patient. Essentially, the consulting psychiatrist is loaded with this responsibility in spite of the fact that he/she has never seen most of the patients and relies on the diagnostic skills of the other members of the team. This is particularly a problem since the mental health policy moves toward the direction of assuming responsibility also for "dangerous" patients and assessing the risk toward the society and others. This has recently led many psychiatrists to resist taking part in consultation activities for fear of litigation.

While psychiatrists are medical doctors, with a long training in the diagnosis and treatment of mental disorders and with the ability to prescribe medication treatment, other mental health professionals usually come from academic departments other than those belonging to the health sciences, and most of them come from schools of philosophy, education, or social sciences. Currently there is a debate on the rights for prescription, and this concerns GPs for the right to prescribe antipsychotics but also nurses and psychologists for the right to prescribe antidepressants. In many countries with loose regulations, any mental health professional can treat patients by himself/herself, and this is an important problem, especially in countries where medication can be purchased without prescription. Recently the problem has worsened because of the proliferation of "mental health consultants" or "life coaches" without or with very little formal training in psychology or mental health.

Maybe the most important area of conflict among different professionals participating in the same mental health team is the different opinions and considerations they have on the role of diagnosis. Although it is not exclusively a matter of difference in discipline, it is evident that most psychiatrists put high value on the usefulness of diagnosis, while many psychologists and social workers consider diagnosis to be a problem rather than a useful tool [4]. This often triggers endless arguments inside the team from persons with insufficient theoretical background and sometimes leads to bitter situations especially when personality clashes arise, masquerading as interprofessional conflict [5].

There seems little doubt that psychiatric treatment involves more use of drug treatments than in the past, but in private practice, it is not known what exactly happens since psychotherapy is more attractive to patients and more profitable for the psychiatrist. One unique characteristic of psychiatry is the powerful relationship between the doctor and the patient and the role this relationship plays for the treatment. It is an inherent element of the psychotherapeutic process, but it plays a strong role also during pharmacotherapy [5]. The problem arises when too powerful emotions appear from either side, no matter positive or negative (a phenomenon known as transference and countertransference). On the one hand, the patient might develop an undue dependence on the therapist, while on the other hand, there is a risk for malpractice because of diagnosis and treatment based on countertransference or even real abuse of the patient by the therapist. Since most of these problems arise in an undeliberate way, all psychiatrists should be trained and be able to recognize and manage them in order to avoid getting involved in professional and emotional difficulties [6].

Apart from treating mental disorders, there is a strong tendency worldwide that psychiatrists deal also with life issues outside the field of mental illness. It started with the banalization of psychoanalysis after WWII and continues today with a different form, but it is always preached by the media and has a place in lay culture. Probably it relates to the disappearance of the traditional family and the ongoing weakening of social networks with the transition from small villages/towns to mega-cities. In this frame many people consult with psychiatrists, psychologists, and other professionals (e.g., life coaches) concerning any kind of everyday psychological issues and problems [7, 8].

There are reports suggesting that psychiatrists experience higher levels of stress and burnout than other medical doctors [9], and probably the same factors leading to the choice of psychiatry as a career also make the person more vulnerable to stress and to the development of burnout [10]. This is in accordance with the "wounded healer" concept [11] and suggests that unfortunately, if we try to screen out vulnerable persons from the specialty, we risk losing key qualities like empathy and sensitivity, which are valuable for a psychiatrist [3, 12]. Maybe the most exceptionally stressful situation a psychiatrist could face is the suicide of one of his/her patients; more than half of psychiatrists reported such an experience, while one-third reported that they had been adversely affected emotionally by it [13]. Another one-third of psychiatrists reported that they have been threatened to be harmed or killed, or their family members [14]. It is interesting that younger, inexperienced, and male psychiatrists are at a higher risk of receiving such threats. Psychiatry may be the most interpersonally demanding occupation [15]. However, the problem becomes very complex when the psychiatrist himself/herself develops some kind of mental disorder, even a mild one [16].

Finally, in everyday psychiatric practice, the full array of medical ethics applies, but additionally, one should put in mind that mental patients, no matter their diagnosis, constitute a special vulnerable population that is also at a higher risk for abuse, not only by the society or their environments but also by their therapists. Not only are they having difficulty in claiming their legal rights but they are also often susceptible to weak defenses and poor judgment; they are easy victims in need for special protection and care. Both diagnosis and treatment and care should be value-driven and considered in a moral and ethical frame, which should be wider and stricter in comparison with the rest of medicine [12, 17].

References

1. Gater R, de Almeida e Sousa B, Barrientos G, Caraveo J, Chandrashekar CR, Dhadphale M, Goldberg D, al Kathiri AH, Mubbashar M, Silhan K, et al. The pathways to psychiatric care: a cross-cultural study. Psychol Med. 1991;21(3):761–74. https://doi.org/10.1017/s003329170002239x.
2. Hobson JA, Leonard J. Out of its mind: psychiatry in crisis – a call for reform. Cambridge, MA: Perseus Publishing; 2001.
3. Gask L, Sibbald B, Creed F. Evaluating models of working at the interface between mental health services and primary care. Br J Psychiatry. 1997;170:6–11. https://doi.org/10.1192/bjp.170.1.6.
4. Boyle M. Diagnosis. In: Newnes C, Dunn C, Holmes G, editors. This is madness: a critical look at psychiatry and the future of mental health services. Herefordshire: PCCS Books; 1999. p. 75–90.
5. Kleinman A. Rethinking psychiatry: from cultural category to personal experience. Free Press; 1991.
6. Rippere V, Williams R. Wounded healers: mental health workers' experiences of depression. Chichester: Wiley; 1985.
7. Cloud D. Control and consolation in american culture and politics: rhetoric of therapy. SAGE; 1998.
8. Moskowitz ES. In therapy we trust: America's obsession with self-fulfillment. In therapy we trust: America's obsession with self-fulfillment. Baltimore, MD: Johns Hopkins University Press; 2001.
9. Guthrie E, Black D. Psychiatric disorder, stress and burn-out. Adv Psychiatr Treat. 2018;3(5):275–81. https://doi.org/10.1192/apt.3.5.275.
10. Firth-Cozens J, Lema VC, Firth RA. Specialty choice, stress and personality: their relationships over time. Hosp Med (Lond, Engl 1998). 1999;60(10):751–5. https://doi.org/10.12968/hosp.1999.60.10.1222.
11. Bennet G. Patients and their doctors: the journey through medical care. London: Ballière Tindall; 1979.
12. Radden J, Sadler JZ. The virtuous psychiatrist: character ethics in psychiatric practice. New York, NY: Oxford University Press; 2010.
13. Dewar I, Eagles J, Klein S, Gray N, Alexander D. Psychiatric trainees' experiences of, and reactions to, patient suicide. Psychiatr Bull. 2018;24(1):20–3. https://doi.org/10.1192/pb.24.1.20.
14. Owen J. Death threats to psychiatrists. Psychiatr Bull. 1992;16:142–4.
15. Jones KS. The other end of the couch. Psychiatr Bull. 2018;22(8):515–6. https://doi.org/10.1192/pb.22.8.515.
16. Crane H. Depression. Interview by Helen Crane. BMJ. 2003;326(7402):1324. https://doi.org/10.1136/bmj.326.7402.1324.
17. Fulford KWM, Thornton T, Graham G. From bioethics to values-based practice. In: Fulford KWM, Thornton T, Graham G, editors. The oxford textbook of philosophy and psychiatry. Oxford University Press; 2006.

Chapter 21
Psychiatry among Human, Life and Social Sciences, Philosophy, and Religion

21.1 The Nature of the Social Status of Psychiatry

Why mental patients should not be treated by traditional healers, priests, or exotic holistic therapists? All of them are very experienced; they see maybe more mental patients than psychiatrists do, and they have a long tradition that goes back to antiquity, far older than the tradition of psychiatry. Like psychiatrists, they claim that they successfully treat patients, and many of their patients agree with that. They also educate their students, and they have some kind of discipline. What is more important is that they seem to be more accepted by both patients and lay people than psychiatrists. Their views and beliefs are appealing and widely accepted both by the public and intellectuals. So, what is the problem there? The answer is simple and straightforward but also unappealing and difficult for lay people to understand: When a debate is in place, the answer is always given by science. If science cannot give an answer, then anybody can say whatever he/she likes.

But answers based on science cannot be confronted by simple beliefs, thoughts, or preferences. Psychiatry, as part of medicine, is based on science, while the above-mentioned "healers" not only are not based on science but, even worse, scientifically gathered data though a long process and, after fighting with hard societal reactions, proved them to be inefficient as therapeutic approaches or simply wrong, or even harmful at cases. The problem is that not only lay persons but also many intellectuals find it difficult to understand the difference between science with its strict methodology and simple everyday unsystematic experience in combination with "loose wisdom" based on charming eloquency. Additionally, many who do understand the difference do not appreciate the value of logic and reasoning on the basis of data; they prefer judging on the basis of subjective feelings.

There is a big debate, and the conflict is deeper than it seems. It concerns the very nature of psychiatry and its relationship with non-medical intellectual areas. These areas perceive the advancement of biological theories and the increasing

K. N. Fountoulakis, *Psychiatry*, https://doi.org/10.1007/978-3-030-86541-2_21

medicalization of psychiatry as a threat. In this frame Leon Eisenberg (1922–2009) warned about the emergence of "mindlessness" in psychiatry, that is, the focus on brain by simultaneously neglecting the person and its experiences. Eisenberg adopted a middle position by also warning about the danger of "brainlessness" [25], but in spite of the popularity of his views, they are nothing more than a political compromise similar to the "bio-psychosocial" model [26–29]. In this debate, the question of whether the criteria for scientific reasoning are fulfilled is disregarded as misleading and irrelevant, and as an excuse, the complexity of human nature and experience are put forward.

But within our culture, the sciences are granted special status, they are taught at school, they define professional fields, and they constitute the basis for policy-making. Science has a higher status in lay people's opinion in comparison with philosophy and humanities and is considered highly reputable in contrast to pseudosciences like astrology. This sets another frame for discussion: classifying psychiatry within sciences and the medical discipline results in higher reputation and authority and better funding for mental health [21]. It is not by chance that the labeling of mental disorders as social problems kicks patients out of the expensive health system and puts them under the care of the much cheaper social services.

21.2 The Long and Difficult Relationship between Psychiatry and Philosophy

A detailed discussion on the philosophical questions pertaining to mental illness and its definition can be found in Chap. 13 of the current book and concering free will in Chap. 16. These two are key topics of intellectual debate since antiquity. While William James defined philosophy with the phrase "metaphysics means nothing but an unusually obstinate effort to think clearly"[55], we must put in mind that philosophy has often resulted in a nonsense effort to feel wise if not to show off about issues of which you do not have information or knowledge. This is not simply a bitter comment; it reflects the way things turn out. When enough information is gathered on a certain topic, such topic is absorbed by some specific discipline of science, and philosophy abandons the area. This was a repeated story in astronomy, physics, chemistry, and, of course, medicine. The only exception to this is bioethics, which often seem to grow in parallel with science, although the author of this book is reserved on whether this growth reflects a real flourishing of philosophical thinking or instead it reflects a "polished" way of regulating things in a practical way. This is in accordance with Ludwig Wittgenstein's approach that philosophical problems are merely "grammatical illusions" resulting from a "bewitchment of our intelligence by means of language" [122, 123].

Central to them is the philosophical question of whether reducibility is applicable in mental phenomena. Reductionism is a philosophical approach suggesting that observed phenomena can be described in terms of much simpler underlying processes [98]. It is based on three models: translation, derivation, and explanation [15, 87, 88]. There are several main facets of this theory [10], including ontological (sweeping) reductionism, suggesting that the entire reality is reducible to a finite number of simple elements [85, 102]; methodological reductionism, which refers to the scientific attempts to utilize reductionism in order to explain nature; theory reductionism, which demands that a newer theory brings progress by reducing the observed phenomena into more fundamental elements in comparison with older theories; epistemological reductionism, which refers to the use of levels of reduction in an effort to explain phenomena with subsequent levels leading to deeper understanding [2, 22, 23]; and creeping reductionism, which is dominant currently in mental health and psychiatry and constitutes a compromise, according to which there is partially applied reductionism to explain partial aspects but not the entire phenomenon, while the possibility of emergentism as the ultimate combining mechanism remains a valid option [102] (see also Chap. 13 of this book).

More specifically for the realm of brain and mind, there is a long tradition of scholars who consider the phenomena pertaining to this realm to be distinct from those in the natural sciences, and as a consequence, mental phenomena could not be subjective to reduction, especially reduction to biology. For psychiatry, the prospect of reductionism identifies with the scientific method and subsequently with the place of psychiatry among the other specialties of medicine and the disciplines of neurobiology and neurosciences [102–105].

A sweeping approach to mental illness would demand that every disorder should be reduced to mater and to processes at the molecular level or even deeper [105–107]. Although sophisticated models of overlapping prototypes and complex interactions could be proposed [14, 42, 71, 72], at the end of the day, the basic demands and consequences of the reductionistic method cannot be avoided. At least a creeping reductionist is already in place today concerning our dealing with the problem of mental phenomena and mental disorders, and the question is how far could this method go? In practice, all models manifest high variability and uncertainty; at least partially this is because of the way life has developed in the frame of evolution, and this is what the biologist and Nobel Laureate Francois Jacob (1920–2013) called "tinkering" [54].

But with a radically different way of thinking, Michel Foucault (1926–1984; Fig. 11.31) argued that the concept of "mental illness" has a particular history, and this history is the real determining factor that led to our current approach and understanding [34]. For Foucault this is a general idea, that is, all concepts have histories that determined them [33]. By this, he means that there are no scientific or objective reasons why a concept has evolved in a particular way; a different history, meaning that the balance of social forces is different, might mean a different outcome. According to Foucault, before the enlightenment, mental patients were viewed as "odd" rather than "ill"; they were tolerated and often considered as gifted or close to divinity. Of course nothing of these is even close to being a historical fact; Foucault adjusts history to his personal ideas and theories (see Chapters 3–6 of this book). He went on and argued that the worship of "reason" with the enlightenment led to the consideration of mental patients as being "unreasonable" and subsequently were unaccepted and stigmatized. Thus, they became the target for control of many professions and fields, including religion and medicine. According to Foucault, in historical terms, medicine has won the struggle for the controlling of this group; if religion had won, then our current approach and conceptualization would have been quite different. The core concept is that we are not progressing scientifically, but we are rather historically trapped in the current way of thinking [21]. Needless to say that many scholars criticized Foucault on the grounds that his history is slapdash [95], not to mention the double meaning of some of his ideas. For example, he suggests that concepts are the result of social processes, which implies that it could have been different, but at the same time, he also says that this was the only way things could evolve, thus canceling the essence of the previous argument. Sadly, in many rankings, including those based on bibliometrics, he gets the top position in the ranking of influential scholars. But after all, the current book also cites him abundantly.

A particular problem was always that mental disorders, unlike many medical conditions, are never asymptomatic. This is of course a consequence of the fact that there are no biological laboratory findings with a diagnostic or predictive value, and independent verification of the diagnosis is not possible. This, in turn, has led to the adoption of the phenomenological approach and ideas [94]. However, this embrace with philosophical movements brought psychopathology close to existential theories and eventually has opened the door to the anti-psychiatric approach. This occurs at least in part because there is no medical symptom or sign that has no "healthy equivalent." Coughing, sneezing, feeling tired, and various internal pains or sensations are all frequent normal experiences that mean nothing, unless they are found in a particular frame or out of proportion. This is also the case with psychiatric symptoms. Even hallucinations resemble the subjective experience of sensory perceptions with the difference that they experienced out of frame. But exactly this frame is, according to many philosophers, a matter of debate. This is in sharp contrast to the medical approach. While existential and anti-psychiatric ideas consider mental disorder to be a matter of choice and maybe the ultimate act of freedom against an oppressing reality (as Sartre suggested), the medical model considers mental patients to be the blameless victims of misfortune, similar to other patients suffering from somatic disorders and disabilities, and in need for a medically defined way of treatment.

Thus, several authors completely reject the biomedical model together with the scientific method, although they rarely declare this openly and clearly. Especially the variability of diagnoses was a central issue from outside criticism but also caused self-doubt in many psychiatrists and collectively in the whole field of mental health [62, 83, 96, 113]. However, until today, the attacks on psychiatry had little impact, and some authors suggest that the reason is that these attacks come mainly from scholars active in the field of social sciences and philosophy and outside psychiatry itself [16]. These authors forget that Thomas Szasz, David Cooper, Ronald Laing, Peter Breggin, Marius Romme, and many others were psychiatrists formally trained in psychiatry and that their attacks on mainstream psychiatry preceded those by most philosophers, and at any case, they started more than half a century ago. On the other hand, there is much truth in the argument that the lines of clashing are more than one, and besides psychiatry vs. anti-psychiatry, today a front where old-fashioned mental health (all-inclusive and strongly psycho-socially oriented) clashes with scientifically rigorous mental health (evidence-based and subsequently bio-medically oriented) exists. This front will probably be the intellectual battleground of the future. A new issue is proposed, that of an ill over-medicalization, with the "human face" of psychiatry and psychology being in retreat [16]. However this is erroneous; the real question is whether psychiatry will keep pace with the rest of medicine and its evolution toward a strict science by following the "evidence-based approach," and subsequently it will be able to provide reliable and valid treatment options in pace with the increasing demands and the hardening of the validation criteria. In contrast, the banal old approach today, that the clash is medical vs. psychological vs. social, seems to be the shameful result of trade unionism.

Today, it is the rule rather than the exception that anti-psychiatric and antiscientific approaches are shaped into a "call to arms to protect the scientific and ethical integrity of psychiatry from a new enemy: profligate, industry-driven, commercial medicalization, and its more nefarious social consequences" [16]. But corruption and breaking of the law are not an issue to debate philosophically. It is a practical issue of law and enforcement of regulations.

The argument that there can be "no psychiatry that can aspire to pure science without any abuses of power or commercial interests" [16] is irrational. It is in line with the core of anti-psychiatric logic that identifies by definition the scientific knowledge with power imbalance and abuse and calls for its destruction [34]. According to this line of thought, the complexity of human nature makes each individual person unique and precludes its classification into groups on the basis either of behavior or of inner experience. Thus, both reduction and generalizability are irrelevant, and the only way to approach psychological problems and mental disorders is through insightful understanding. This is unique in human scientific and philosophical fields, since almost all other disciplines face no opposition when they discuss human groups, e.g., social classes.

There is an ontological issue (about the nature) concerning the categories of mental disorders. Even today there are three major theoretical answers to the problem, stemming directly from Plato's teachings, suggesting that diagnostic categories could be:

(a) Platonic ideal archetypes, which do exist eternally and ideally in perfect form but in an incomplete form in everyday clinical practice.
(b) Mere names with some practical everyday utility but without any correspondence to reality (nominalism).
(c) They are real but general entities stemming out of real characteristics of real-world patients (neo-Aristotelianism) [59, 100].

An important development in the field was the emergence of analytic philosophy during the first two decades and throughout the first half of the twentieth century by Bertrand Russell (1872–1970), George Edward Moore (1873–1958), and Ludwig Wittgenstein (1989–1951). They all (see also Chap. 10 and Fig. 10.26) came to see philosophy not as the product of a foggy idealism but rather of a painstaking analytical thinking leading to the description of the logical structure of reality [5, 45, 50, 99, 123]. This was later called "analytic philosophy" and has logical rigor in writing as its cornerstone is not widely embraced by lay people and the general public; it is rather dry and technical, without any lyrical profundity.

Of course this was in sharp contrast to the way of thinking of the vast majority of scholars in the field of philosophy of mind, mental health, psychology, and sociology. Analytic philosophy is in contrast with the so-called continental philosophy, which includes all philosophical movements developed in continental Europe that do not utilize strict logical rules in arguing and in the use of language.

In clinical psychiatry, persistent distress and disability signify the presence of a mental disorder and the need for specialized treatment. The presence of either distress or disability seems to be the cornerstone in the definition of mental disorders, and it seems that in the long term, there is no mental patient without the experience of either of them or any mental disorder without either of them as its core characteristic. In the short term, however, this might not be the case. But, in a different context, this same suffering is considered to be an indicator of emotional struggle and of the need to address social, spiritual, or other types of personal problems, including values and moral issues [58, 116]. Several authors argue that these psychological processes are characteristic of the person and its interaction with society, and therefore, it is inappropriate to disengage the two in an effort to alleviate suffering. This is an idea with deep roots in religious concepts, puritan view of the world, and, one could say, reflects a masochistic anti-rationalistic attitude since it considers out-of-proportion and out-of-frame suffering as the necessary condition for the personal improvement.

This approach eventually lead many to an unscientific post-modern approach to mental disorders [13, 18, 20, 67, 69]. The first post-modern psychiatrist was probably Freud in the sense that he gave priority to interpretation instead of observation and to content instead of mechanism [76]. However, his original intention was to bring science to the field of psychology and the study of human behavior. At that time, simply kicking the metaphysical out was a move toward a more science-dominated field, but as the time passed, the developments in the allied fields and in mental health, in general, left psychoanalysis far behind. Unfortunately, today we know that psychoanalysis resisted research and aggressively attacked anyone who questioned its authority, an attitude that was evident especially immediately after WWII with the unfair attack on electroconvulsive treatment [109] but also later with the resistance to the development of DSM-III [70]. Psychoanalysis was never a true science, and in spite of wishful thoughts [68], it cannot become one. It was exactly as a reaction to the dominance of psychoanalysis that since the 1960s, a total rejection of psychiatry and the concept of mental illness and treatment occurred in the form of anti-psychiatry and various similar movements.

A particular issue for debate was the psychiatric interview, its nature, and technique. Traditionally, and since medico-biological markers are lacking, psychiatry is based on clinical interviews to assess the mental status of the patient and to establish the appropriate diagnosis. The question is how "objective" this could be and whether such an interview simply explores personal problems and afterward erroneously considers their characters to constitute symptoms in the medical sense [114]. Some scholars, like Jerry Fodor (1935–2017), argued that "explanation" is necessary in order to define the abnormal but not the normal [31, 32]. The meaning of the word "objective" was the focus of endless theoretical discussions. In scientific terms, however, it is reduced to the words "reliability" and "validity." The storm that hit psychiatry in the 1960s and 1970s led to the development of structured interview methods to assess operationalized diagnostic criteria. This increased the reliability of diagnosis, but the lack of biological markers made validity unknown [62, 80, 112]. However, many authors either do not know or neglect the fact that the increase in reliability also leads to an increase in validity independently of any progress in neurobiology. The essence of structured interviews concerns the assessment and scoring of symptoms on the basis of description, while the input of insight is kept at a minimum and is theoretically eliminated [89].

Many authors argued that structured interviews create a sort of "tunnel vision" or some kind of "procrustean errors." These lead to cherry picking from the variety in behavioral manifestations, and then the information is "stretched and trimmed" in order to fit operationalized criteria and diagnosis [75, 119]. While "tunnel vision" and "procrustean errors" are considered by many to reduce validity [111], this is wishful thinking rather than scientific fact. Psychiatric diagnosis has no proven validity, e.g., we do not know what exactly schizophrenia is in terms of brain or psychological function. Therefore, any claim that any method is able to reveal such mechanisms is erroneous. On the contrary, what we can surely do is to increase the reliability, that is, the agreement between clinicians and the stability of the diagnosis over time. This also leads to a higher validity by definition; if a lot of people agree that a certain quality is observable in a group of objects, it is highly likely that these objects share a common substrate or process. A side effect of these is a trade between reliability and wealth of information; only those information with specific characteristics are taken into consideration. But is it real "wealth" when it does not reliably contribute? This, in turn, creates the concern of "tunnel vision," but in reality, it is a "tunnel of clear vision" within thick mist. The dispute of the scientific method in psychiatry is based on wishful conclusions and exposes lack of knowledge of research and scientific methods and the problematic role of authority as the source of knowledge.

The major philosophical concern is that mental states are subjective, and their essence is impossible to be captured by simple observation. The meaning of these states to the experience of the individual person is best understood in the case of anxiety, which is on the one hand the commonest psychiatric problem, it is well understood physiologically especially in the frame of the fight-or-flight response, and on the other hand it has been the focus of philosophical thinking and it is considered to constitute the bridge between medicine and the humanities [43]. Especially existentialists focused on anxiety and considered it to be the result of alienation from others and the self [52, 60, 61, 117], and some of them tried to carry these concepts to psychiatry and psychopathology with varying success [8, 9, 11, 12, 44, 49, 56]. Other schools of thought argue that it is necessary not only to describe but also to understand an individual's experiences in the context of particular life history and circumstances and to utilize "idiographic" diagnostic schemas [20, 53, 78, 79].

Even language is considered to be a determining factor because of different meanings in verbal communication as expressed and perceived by different people [4, 46, 52, 97]. In structural interviews, shared meanings between the interviewer and interviewees are assumed, and some authors criticize that differences in linguistic meanings and other features are not investigated in the process of the interview. Critics insist on the need for laborious interpretation rather than simple description of responses. They argue that the technocratic approach to psychiatry has led to the "de-narratization" and the loss of the contact at the personal level of analysis of feelings, meanings, and values [20, 113]. These are fundamentally erroneous and against the scientific method. Interpretation at the level of the individual person has been proven to be unreliable, while structured interviews are tested for their reliability before given for wide everyday clinical use.

Overall, there are a number of claims that are clearly against all available hard data but are frequently used as arguments. The central claim is that subjective experiences are more specific and useful for establishing diagnosis. In this frame, it is theoriticized that only the laborious analysis of subjective experience can lead to the understanding of the patient's behavior and subsequently to diagnosis and therapeutic decision-making [114]. It is impressive how the conflicting and almost-hostile-to-each-other ideologies of existentialism and psychoanalysis join forces to attach scientific reductionism in psychiatry. Also, it is disappointing that 50–60 years after the major crisis in psychiatry, because of these unscientific and highly idiosyncratic approaches (see Chap. 11), highly regarded scholars neglect the evidence [114].

In addition to the above, contemporary classification systems have been criticized as reflecting a fundamentally Euro-American approach to mental illness, and critics argue that even when the diagnostic criteria are applicable across different cultures, this does not mean that there is a structural validity of the concept of specific mental illnesses across cultures [121].

Although these arguments are in contrast with the empirical data [64, 74, 101, 118], the pressure from advocacy groups for a more culturally sensitive approach is strong. Many authors argue that there is a large number of well-documented non-Western mental disorders that are still neglected and that even those included are often misinterpreted or misrepresented [7]. It is interesting that even the introduction of cultural factors in DSM-IV was perceived as being "culture-bound" since the basic features of major psychiatric diagnostic categories remained intact, and as the years passed, the cultural sensitivity of psychiatry as a discipline has declined, especially with the rise of biological psychiatry [64, 65].

On the other hand, the argument that contemporary concepts of mental illness are based on research in white middle-class persons from North America and Western Europe [17, 63, 64, 73] is no longer valid since a bulk of clinical data are available from around the world covering divergent races, cultures, and socioeconomic environments, and these data support the validity of major diagnostic categories and do not support the usefulness of any cultural features [1, 121].

The only mental issue that shows a clear geographical differentiation is suicidality, but the reasons behind this are unknown [35–41]. In spite of these, there is a prevailing belief, supported by some data, that in many countries, in everyday clinical practice, there is a prejudice against minority groups, and this is often semi-official with an institutional racism present within psychiatric services [77, 84, 108, 120].

21.3 Psychiatry and Religion

Psychiatry was always in a dynamic and peculiar relationship with religion since they share a number of areas of interest, like the assessment of behavior, inner preferences, values, and thoughts as well as alleviation of fear [6, 86]. The problem is that many religious, spiritual, or transpersonal experiences and beliefs formally meet many criteria of delusional or psychotic disorders [57, 91], but the metaphysical beliefs in the frame of any religion are not considered as reflecting a mental disorder. A problem arises when these beliefs have an out-of-frame and out-of-proportion impact on the person's everyday life and are in clear contrast with the expected beliefs and behaviors of his/her educational, cultural, and general religious/spiritual background. Religious ideas and unusual intense religiosity are often part of the delusion in patients with schizophrenia [81, 82, 110, 115]. Another problem is that often established common religious beliefs exert a distressing effect on the individual [19].

21.4 Psychiatry as a Science

As Thomas Samuel Kuhn (1922–s1996) argued, scientists normally work within paradigms, but communication across these paradigms is usually problematic [66]. This is particularly true concerning psychiatry, which is a field where multiple paradigms co-exist, and it is not only that communication among them is problematic but also that most of them are probably incompatible with each other. According to this line of thinking, only when anomalies accumulate beyond a certain point does a state of crisis appear for the specific scientific field and the paradigm shifts, with new theories emerging and dominating. Such a crisis occurred in psychiatry in the 1960s and led to the development of modern classification systems. Before this, other paradigm shifts were also psychoanalysis and psychopharmacology. An interesting aspect of Kuhn's ideas is that paradigms are "incommensurable," which means that there are no external validating criteria to compare; this is because the perception, meaning, and interpretation are paradigm-dependent.

In this almost chaotic environment, the problem of distinguishing science from pseudo-science (demarcation problem) is of utmost importance. Karl Popper defined science on the basis of the falsification criterion, that is, any scientific theory should allow for the design of a real or imaginary experimental condition that can lead to its rejection [51, 92]. This is the strongest definition to date and is not popular among philosophers because it is precise and reliable and leaves no space to jargon arguments. According to Popper himself, his approach leads to the classification of psychoanalysis as a pseudo-science [51, 92]. Critics of Popper's theory utilize the Quine–Duhem thesis or problem, named after Pierre Duhem (1861–1916) and Willard Van Orman Quine (1908–2000). According to them only whole theories but not isolated statements can be tested [3, 47, 48, 93]. On the other hand, Kuhn proposed that a field becomes scientific when its first paradigm emerges and "scientists" start solving puzzles on the basis of this paradigm [66]. However, this means that puzzle-solving defines scientific thought according to Kuhn, but this is not a strong proposal since, as Paul Feyerabend (1924–1994) showed, there are many non-scientific fields that utilize intellectual puzzle-solving within a taken-for-granted framework [30]. Theology is such a field as it is philosophy itself.

An approach of particular interest for psychiatry comes from the works of Wittgenstein on "family resemblance" [123]. Family resemblance suggests that there is no demarcation criterion or a strict definition of science, but there is a family of fields that share common characteristics, but members of the family might differ profoundly from each other, the way psychiatry differs from applied physics, but the family includes many other members like neurology, which bridge the differences. The same idea could be applied for the definition of mental disorders [21, 24, 90]. What this approach neglects is that, yes, a concept might not seem homogeneous like an apple or an orange, but it could look like more than a bunch of grapes. However, in all these families of concept, a key common feature always exists and bridges all concepts and brings them together, the same way the stem brings grapes together to form a bunch.

References

1. Alarcon RD. Culture, cultural factors and psychiatric diagnosis: review and projections. World Psychiatry. 2009;8(3):131–9. https://doi.org/10.1002/j.2051-5545.2009.tb00233.x.
2. Anderson PW. More is different. Science. 1972;177(4047):393–6. https://doi.org/10.1126/science.177.4047.393.
3. Ariew R. The Duhem thesis. Br J Philos Sci. 1984;35(4):313–25.
4. Askay R, Farquhar-Lacewing J. Being unconscious: Heidegger and Freud. In: Fulford KWM, Davies M, Gipps R et al. (eds) The Oxford handbook of philosophy and psychiatry. Oxford University Press: Oxford; 2013. https://doi.org/10.1093/oxfordhb/9780199579563.001.0001.
5. Baillie J. Introduction to Bertrand Russell. In: Baillie J, editor. Contemporary analytic philosophy. 2nd ed. Upper Saddle River: Prentice Hall; 1997.
6. Bhugra D. Religion and mental health. In: Bhugra D, editor. Psychiatry and religion: context, consensus and controversies. London: Routledge; 1996.
7. Bhugra D, Munro A. Troublesome disguises: underdiagnosed psychiatric syndromes. 1st ed. London: Wiley-Blackwell; 1997.
8. Binswanger L. Grundformen und Erkenntnis menschlichen Daseins. Zürich: Niehans; 1942.

9. Binswanger L. Being-in-the-world: selected papers of Ludwig Binswanger. New York: Basic books; 1963.
10. Blackburn S. The Oxford dictionary of philosophy. Oxford: Oxford University Press; 2008. https://doi.org/10.1093/acref/9780199541430.001.0001.
11. Boss M. Psychoanalysis and Daseinsanalysis. Oxford: Basic Books; 1963.
12. Boss M. Existential foundations of medicine and psychology. Lanham: Jason Aronson; 1977.
13. Bracken P, Thomas P. Postpsychiatry: a new direction for mental health. BMJ. 2001;322(7288):724–7. https://doi.org/10.1136/bmj.322.7288.724.
14. Cantor N, Smith EE, French RS, Mezzich J. Psychiatric diagnosis as prototype categorization. J Abnorm Psychol. 1980;89(2):181–93. https://doi.org/10.1037//0021-843x.89.2.181.
15. Carnap R. The logical structure of the world and Pseudoproblems in philosophy. Berkeley, California: University of California Press; 1928/1967.
16. Charland L. Why psychiatry should fear medicalization. In: Fulford KWM, Davies M, Gipps R et al. (eds) The Oxford handbook of philosophy and psychiatry. Oxford University Press: Oxford; 2013. https://doi.org/10.1093/oxfordhb/9780199579563.001.0001.
17. Clare A, editor. Psychiatry in dissent: controversial issues in thought and practice. 2nd ed. London: Routledge; 2001.
18. Clark LA. The role of moral judgment in personality disorder diagnosis. J Personal Disord. 2006;20(2):184–5. https://doi.org/10.1521/pedi.2006.20.2.184.
19. Clarke I. Psychosis and spirituality: consolidating the new paradigm. New York: John Wiley & Sons; 2010.
20. Cohen CI, Timimi S, editors. Liberatory psychiatry: philosophy, politics and mental health. Cambridge: Cambridge University Press; 2008. https://doi.org/10.1017/CBO9780511543678.
21. Cooper R. Psychiatry and philosophy of science. Durham: Acumen Publishing; 2012. https://doi.org/10.1017/upo9781844653980.
22. Dawkins R. The blind watchmaker: why the evidence of evolution reveals a universe without design. Norton & Company: W. W; 2015.
23. Dawkins R. The selfish gene. 4th ed. Oxford: Oxford Landmark Science; 2016.
24. Dupré J. The disorder of things: metaphysical foundations of the disunity of science. Harvard: Harvard University Press; 1993.
25. Eisenberg L. Mindlessness and brainlessness in psychiatry. Br J Psychiatry. 1986;148:497–508. https://doi.org/10.1192/bjp.148.5.497.
26. Engel GL. The need for a new medical model: a challenge for biomedicine. Science. 1977;196(4286):129–36. https://doi.org/10.1126/science.847460.
27. Engel GL. The biopsychosocial model and the education of health professionals. Ann N Y Acad Sci. 1978;310:169–87. https://doi.org/10.1111/j.1749-6632.1978.tb22070.x.
28. Engel GL. The clinical application of the biopsychosocial model. Am J Psychiatry. 1980;137(5):535–44. https://doi.org/10.1176/ajp.137.5.535.
29. Engel GL. Sounding board. The biopsychosocial model and medical education. Who are to be the teachers? N Engl J Med. 1982;306(13):802–5. https://doi.org/10.1056/NEJM198204013061311.
30. Feyerabend P. Against method: outline of an anarchistic theory of knowledge. London: New Left Books; 1975.
31. Fodor J. Psychological explanation: an introduction to the philosophy of psychology. New York: Crown publishing group/Random House; 1968a.
32. Fodor JA. The appeal to tacit knowledge in psychological explanation. J Philos. 1968b;65(20):627. https://doi.org/10.2307/2024316.
33. Foucault M. The order of things. London: Tavistock; 1970.
34. Foucault M, Howard R, Cooper D. Madness and civilization. Taylor & Francis. 1961; https://doi.org/10.4324/9780203278796.
35. Fountoulakis KN. Suicide and the economic situation in Europe: are we experiencing the development of a 'reverse stigma'? Br J Psychiatry. 2016;209(4):273–4. https://doi.org/10.1192/bjp.bp.115.171785.

36. Fountoulakis KN, Chatzikosta I, Pastiadis K, Zanis P, Kawohl W, Kerkhof AJ, Navickas A, Hoschl C, Lecic-Tosevski D, Sorel E, Rancans E, Palova E, Juckel G, Isacsson G, Jagodic HK, Botezat-Antonescu I, Rybakowski J, Azorin JM, Cookson J, Waddington J, Pregelj P, Demyttenaere K, Hranov LG, Stevovic LI, Pezawas L, Adida M, Figuera ML, Jakovljevic M, Vichi M, Perugi G, Andreassen OA, Vukovic O, Mavrogiorgou P, Varnik P, Dome P, Winkler P, Salokangas RK, From T, Danileviciute V, Gonda X, Rihmer Z, Forsman J, Grady A, Hyphantis T, Dieset I, Soendergaard S, Pompili M, Bech P. Relationship of suicide rates with climate and economic variables in Europe during 2000-2012. Ann General Psychiatry. 2016a;15:19. https://doi.org/10.1186/s12991-016-0106-2.
37. Fountoulakis KN, Gonda X. Differential correlation of suicide and homicide rates according to geographical areas: a study with population-level data. Psychiatry Res. 2017;249:167–71. https://doi.org/10.1016/j.psychres.2016.12.059.
38. Fountoulakis KN, Gonda X. Ancestry and different rates of suicide and homicide in European countries: a study with population-level data. J Affect Disord. 2018;232:152–62. https://doi.org/10.1016/j.jad.2018.02.030.
39. Fountoulakis KN, Kawohl W, Theodorakis PN, Kerkhof AJ, Navickas A, Hoschl C, Lecic-Tosevski D, Sorel E, Rancans E, Palova E, Juckel G, Isacsson G, Jagodic HK, Botezat-Antonescu I, Warnke I, Rybakowski J, Azorin JM, Cookson J, Waddington J, Pregelj P, Demyttenaere K, Hranov LG, Stevovic LI, Pezawas L, Adida M, Figuera ML, Pompili M, Jakovljevic M, Vichi M, Perugi G, Andreassen O, Vukovic O, Mavrogiorgou P, Varnik P, Bech P, Dome P, Winkler P, Salokangas RK, From T, Danileviciute V, Gonda X, Rihmer Z, Benhalima JF, Grady A, Leadholm AK, Soendergaard S, Nordt C, Lopez-Ibor J. Relationship of suicide rates to economic variables in Europe: 2000-2011. Br J Psychiatry. 2014;205(6):486–96. https://doi.org/10.1192/bjp.bp.114.147454.
40. Fountoulakis KN, Savopoulos C, Apostolopoulou M, Dampali R, Zaggelidou E, Karlafti E, Fountoukidis I, Kountis P, Limenopoulos V, Plomaritis E, Theodorakis P, Hatzitolios AI. Rate of suicide and suicide attempts and their relationship to unemployment in Thessaloniki Greece (2000-2012). J Affect Disord. 2015;174:131–6. https://doi.org/10.1016/j.jad.2014.11.047.
41. Fountoulakis KN, Savopoulos C, Zannis P, Apostolopoulou M, Fountoukidis I, Kakaletsis N, Kanellos I, Dimellis D, Hyphantis T, Tsikerdekis A, Pompili M, Hatzitolios AI. Climate change but not unemployment explains the changing suicidality in Thessaloniki Greece (2000-2012). J Affect Disord. 2016b;193:331–8. https://doi.org/10.1016/j.jad.2016.01.008.
42. Frances A, Pincus HA, Widiger TA, Davis WW, First MB. DSM-IV: work in progress. Am J Psychiatry. 1990;147(11):1439–48. https://doi.org/10.1176/ajp.147.11.1439.
43. Fulford KWM, Thornton T, Graham G. Experiences good and bad: an introduction to psychopathology, classification and diagnosis for philosophers. In: Fulford KWM, Thornton T, Graham G, editors. The Oxford textbook of philosophy and psychiatry. Oxford: Oxford University Press; 2006a.
44. Fulford KWM, Thornton T, Graham G. Karl Jaspers and general psychopathology. In: Fulford KWM, Thornton T, Graham G, editors. The Oxford textbook of philosophy and psychiatry. Oxford: Oxford University Press; 2006b.
45. Fulford KWM, Thornton T, Graham G. Philosophical outputs in mental health practice and research. In: Fulford KWM, Thornton T, Graham G, editors. The Oxford textbook of philosophy and psychiatry. Oxford: Oxford University Press; 2006c.
46. Gadamer HG. Truth and method (Weinsheimer and D.G. Marshall, trans.). 2nd edn. The Continuum Publishing company, New York; 2004.
47. Gillies D. Philosophy of science in the twentieth century: four central themes. London: Blackwell; 1993.
48. Gillies D. The Duhem thesis and the Quine thesis. In: Curd M, Cover J, editors. Philosophy of science: the central issues. New York: Norton; 1998. p. 302–19.
49. Glas G. Anxiety—animal reactions and the embodiment of meaning. In: Fulford KWM, Morris KJ, Sadler JZ, Stanghellini G, editors. Nature and narrative: an introduction to the new philosophy of psychiatry. Oxford: Oxford University Press; 2003. p. 231–49.

50. Glock H. What is analytic philosophy? Cambridge: Cambridge University Press; 2008.
51. Grant DC, Harari E. Psychoanalysis, science and the seductive theory of Karl Popper. Aust N Z J Psychiatry. 2005;39(6):446–52. https://doi.org/10.1080/j.1440-1614.2005.01602.x.
52. Heidegger M. Being and time. New York: Harper Perennial; 2008.
53. Igda Workgroup WPA. IGDA. 8: idiographic (personalised) diagnostic formulation. Br J Psychiatry Suppl. 2003;45:s55–7.
54. Jacob F. Evolution and tinkering. Science. 1977;196(4295):1161–6. https://doi.org/10.1126/science.860134.
55. James W. The principles of psychology. New York, NY: Henry Holt and Co; 1890.
56. Jaspers K. General psychopathology (trans: Hoenig J, Hamilton M). The Johns Hopkins University press (1997), Chicago; 1913.
57. Johnson CV, Friedman HL. Enlightened or delusional? J Humanist Psychol. 2008;48(4):505–27. https://doi.org/10.1177/0022167808314174.
58. Karasz A. Cultural differences in conceptual models of depression. Soc Sci Med. 2005;60(7):1625–35. https://doi.org/10.1016/j.socscimed.2004.08.011.
59. Kendell RE. The role of diagnosis in psychiatry. The role of diagnosis in psychiatry. Oxford, England: Blackwell Scientific Publications; 1975.
60. Kierkegaard S, Lowrie W. Fear and trembling and the sickness unto death. Princeton: Princeton University Press; 1941.
61. Kierkegaard S, Thomte R. Kierkegaard's Writings, VIII, Volume 8. Princeton: Princeton University Press; 2013. https://doi.org/10.2307/j.ctt24hrfg.
62. Kirk S, Kutchins H. The selling of DSM: the rhetoric of science in psychiatry. New York: Aldine de Gruyter; 1992.
63. Kleinman A. Anthropology and psychiatry. The role of culture in cross-cultural research on illness. Br J Psychiatry. 1987;151:447–54. https://doi.org/10.1192/bjp.151.4.447.
64. Kleinman A. Rethinking psychiatry: from cultural category to personal experience free press; 1991.
65. Kleinman A. Triumph or pyrrhic victory? The inclusion of culture in DSM-IV. Harv Rev Psychiatry. 1997;4(6):343–4. https://doi.org/10.3109/10673229709030563.
66. Kuhn T. The structure of scientific revolutions. 50th anniversary. 4th ed. Chicago: University of Chicago Press; 2012.
67. Kwate NO. The heresy of African-centered psychology. J Med Humanit. 2005;26(4):215–35. https://doi.org/10.1007/s10912-005-7698-x.
68. Lacewing M. Could psychoanalysis be a science? In: Fulford KWM, Davies M, Gipps R et al. (eds) The Oxford handbook of philosophy and psychiatry. Oxford University Press: Oxford; 2013. https://doi.org/10.1093/oxfordhb/9780199579563.001.0001.
69. Lewis B. Psychiatry and postmodern theory. J Med Human. 2000;21(2):71–84. https://doi.org/10.1023/a:1009018429802.
70. Lieberman JA. Shrinks: the untold story of psychiatry. Boston: Back Bay Books; 2016.
71. Lilienfeld SO, Marino L. Mental disorder as a Roschian concept: a critique of Wakefield's "harmful dysfunction" analysis. J Abnorm Psychol. 1995;104(3):411–20. https://doi.org/10.1037//0021-843x.104.3.411.
72. Lilienfeld SO, Marino L. Essentialism revisited: evolutionary theory and the concept of mental disorder. J Abnorm Psychol. 1999;108(3):400–11. https://doi.org/10.1037//0021-843x.108.3.400.
73. Littlewood R. From categories to contexts: a decade of the 'new cross-cultural psychiatry'. Br J Psychiatry. 1990;156:308–27. https://doi.org/10.1192/bjp.156.3.308.
74. Littlewood R. Psychiatry's culture. Int J Soc Psychiatry. 1996;42(4):245–68. https://doi.org/10.1177/002076409604200402.
75. McGuffin P, Farmer A. Polydiagnostic approaches to measuring and classifying psychopathology. Am J Med Genet. 2001;105(1):39–41.
76. McHugh P, Slavney P. The perspectives of psychiatry. 2nd ed. Baltimore: Johns Hopkins University Press; 1998.

77. McKenzie K, Bhui K. Institutional racism in mental health care. BMJ. 2007;334(7595):649–50. https://doi.org/10.1136/bmj.39163.395972.80.
78. Mezzich JE. Comprehensive diagnosis: a conceptual basis for future diagnostic systems. Psychopathology. 2002;35(2–3):162–5. https://doi.org/10.1159/000065138.
79. Mezzich JE, Kleinman A, Fabrega H, Parron DL. Culture and psychiatric diagnosis: a DSM-IV perspective. Washington: American Psychiatric Press Inc.; 1996.
80. Mishler E. Research interviewing: context and narrative. Cambridge: Harvard University press; 1991.
81. Mohr S, Borras L, Betrisey C, Pierre-Yves B, Gillieron C, Huguelet P. Delusions with religious content in patients with psychosis: how they interact with spiritual coping. Psychiatry. 2010a;73(2):158–72. https://doi.org/10.1521/psyc.2010.73.2.158.
82. Mohr S, Borras L, Rieben I, Betrisey C, Gillieron C, Brandt PY, Perroud N, Huguelet P. Evolution of spirituality and religiousness in chronic schizophrenia or schizo-affective disorders: a 3-years follow-up study. Soc Psychiatry Psychiatr Epidemiol. 2010b;45(11):1095–103. https://doi.org/10.1007/s00127-009-0151-0.
83. Monti MR, Stanghellini G. Psychopathology: an edgeless razor? Compr Psychiatry. 1996;37(3):196–204. https://doi.org/10.1016/s0010-440x(96)90036-x.
84. Morales A. Institutional racism in mental health and criminal justice. Soc Casework. 1978;59(7):387–95.
85. Nagel T. Mind and cosmos: why the materialist neo-Darwinian conception of nature is almost certainly false. Oxford: Oxford University Press; 2012.
86. Nelson SH, Torrey EF. The religious functions of psychiatry. Am J Orthopsychiatry. 1973;43(3):362–7. https://doi.org/10.1111/j.1939-0025.1973.tb00806.x.
87. Neurath O. Sociology and physicalism. In: Ayer A, editor. Logical Positivism. New York: The Free Press; 1931/1959.
88. Neurath O. Philosophical papers 1913–1946. Dordrecht: Reidel; 1983.
89. Othmer E, Othmer S. The clinical interview using DSM-IV-TR: Vol 1: fundamentals. Arlington, VA: American Psychiatric Publishing, Inc.; 2002.
90. Pickstone G. Ways of knowing: a new history of science, technology, and medicine. 2nd ed. Chicago: University of Chicago Press; 2001.
91. Pierre JM. Faith or delusion? At the crossroads of religion and psychosis. J Psychiatr Pract. 2001;7(3):163–72. https://doi.org/10.1097/00131746-200105000-00004.
92. Popper K. Conjectures and refutations. London: Routledge & Kegan Paul; 1963.
93. Quine WV. Main trends in recent philosophy: two dogmas of empiricism. The Philosophical Review. 1951;60(1):20–43. https://doi.org/10.2307/2181906.
94. Rashed M. Madness and the demand for recognition a philosophical inquiry into identity and mental health activism. Oxford: Oxford University Press; 2019. https://doi.org/10.1093/med/9780198786863.001.0001.
95. Reznek L. The philosophical Defence of psychiatry. London: Routledge; 1991.
96. Richardson J. Handbook of qualitative research methods for psychology and the social sciences. London: Wiley-Blackwell; 1996.
97. Rorty R. Philosophy and the Mirror of nature. Princeton, NJ: Princeton University Press; 1981.
98. Ruse M. Entry for "reductionism". In: Honderich T, editor. He Oxford companion to philosophy. 2nd ed. Oxford: Oxford University Press; 2005.
99. Russell B. A history of western philosophy and its connection with political and social circumstances from the earliest times to the present day. New York, NY: Simon and Schuster; 1945.
100. Sadler J, Wiggins O, Schwartz M. Philosophical perspectives on psychiatric diagnostic classification. Philosophical perspectives on psychiatric diagnostic classification. Baltimore, MD: Johns Hopkins University Press; 1994.
101. Sartorius N, Jablensky A, Korten A, Ernberg G, Anker M, Cooper JE, Day R. Early manifestations and first-contact incidence of schizophrenia in different cultures. A preliminary report on the initial evaluation phase of the WHO collaborative study on determinants of outcome of severe mental disorders. Psychol Med. 1986;16(4):909–28. https://doi.org/10.1017/s0033291700011910.

102. Schaffner K. Reduction and reductionism in psychiatry. In: Fulford KWM, Davies M, Gipps R et al. (eds) The Oxford handbook of philosophy and psychiatry. Oxford University Press: Oxford; 2013. https://doi.org/10.1093/oxfordhb/9780199579563.001.0001.
103. Schaffner KF. Theory structure in the biomedical sciences. J Med Philos. 1980;5(1):57–97. https://doi.org/10.1093/jmp/5.1.57.
104. Schaffner KF. Exemplar reasoning about biological models and diseases: a relation between the philosophy of medicine and philosophy of science. J Med Philos. 1986;11(1):63–80. https://doi.org/10.1093/jmp/11.1.63.
105. Schaffner KF. Discovery and explanation in biology and medicine, vol. 1. Chicago: University of Chicago Press; 1993.
106. Schaffner KF. Reduction: the Cheshire cat problem and a return to roots. Synthese. 2006;151(3):377–402.
107. Schaffner KF. Ernest Nagel and reduction. J Philos. 2012;109(8/9):534–65.
108. Senn DJ, Sawyer J. Institutional racism: a problem for psychology? Am Psychol. 1971;26(7):671–4. https://doi.org/10.1037//0003-066x.26.7.671.
109. Shorter E. A history of psychiatry: From the era of the asylum to the age of Prozac. New York: John Wiley and Sons; 1997.
110. Siddle R, Haddock G, Tarrier N, Faragher EB. Religious delusions in patients admitted to hospital with schizophrenia. Soc Psychiatry Psychiatr Epidemiol. 2002;37(3):130–8. https://doi.org/10.1007/s001270200005.
111. Smolik P. Validity of nosological classification. Dialogues Clin Neurosci. 1999;1(3):185–90.
112. Spitzer RL. Psychiatric diagnosis: are clinicians still necessary? Compr Psychiatry. 1983;24(5):399–411. https://doi.org/10.1016/0010-440x(83)90032-9.
113. Stanghellini G. The grammar of the psychiatric interview. A plea for the second-person mode of understanding. Psychopathology. 2007;40(2):69–74. https://doi.org/10.1159/000098486.
114. Stanghellini G. Philosophical resources for the psychiatric interviews. In: Fulford KWM, Davies M, G.T. Gipps R et al. (eds) The Oxford handbook of philosophy and psychiatry. Oxford University Press; 2013. https://doi.org/10.1093/oxfordhb/9780199579563.001.0001.
115. Suhail K, Ghauri S. Phenomenology of delusions and hallucinations in schizophrenia by religious convictions. Mental Health, Religion & Culture. 2010;13(3):245–59. https://doi.org/10.1080/13674670903313722.
116. Tilbury F, Rapley M. 'There are orphans in Africa still looking for my hands': African women refugees and the sources of emotional distress. Health Sociol Rev. 2014;13(1):54–64. https://doi.org/10.5172/hesr.13.1.54.
117. Tymieniecka A. The human being in action: the irreducible element in man Part II investigations at the intersection of philosophy and psychiatry (Analecta Husserliana). Springer; 1978.
118. Vaillant GE. Positive mental health: is there a cross-cultural definition? World Psychiatry. 2012;11(2):93–9. https://doi.org/10.1016/j.wpsyc.2012.05.006.
119. Van Praag HM, Asnis GM, Kahn RS, Brown SL, Korn M, Friedman JM, Wetzler S. Nosological tunnel vision in biological psychiatry. A plea for a functional psychopathology. Ann N Y Acad Sci. 1990;600:501–10. https://doi.org/10.1111/j.1749-6632.1990.tb16905.x.
120. Wade JC. Institutional racism: an analysis of the mental health system. Am J Orthopsychiatry. 1993;63(4):536–44. https://doi.org/10.1037/h0079479.
121. Widiger TA, Sankis LM. Adult psychopathology: issues and controversies. Annu Rev Psychol. 2000;51:377–404. https://doi.org/10.1146/annurev.psych.51.1.377.
122. Wittgenstein L. Tractatus Logico-Philosophicus. Cosimo Classics; 2007.
123. Wittgenstein L. Philosophical investigations. 4th ed. London: Wiley-Blackwell; 2009.

Chapter 22
Psychiatry and Society

22.1 Society and Mental Illness

The reciprocal relationship between psychiatry and society is complex and includes several distinct, though not entirely independent, facets. It is unique in medicine since no other medical specialty has such a strong, wide, and deep contact and interaction with society, its elements, and its needs.

Even the early name of psychiatrists, i.e., "alienists," reflects this relationship since it refers to dealing with people alienated from society. The major area of contact between the society and psychiatry is the responsibility of the latter to protect not only the patient but also the society from the consequences of mental disorder, and this often involves complicated situations with conflicting values and interests. In everyday psychiatric practice, the full array of medical ethics apply, but additionally, one should have in mind that mental patients, no matter their diagnosis, constitute a special vulnerable population also at a higher risk for abuse; not only do they have difficulty in claiming their legal rights but they are also often susceptible to weak defenses and poor judgment and are easy victims in need for special protection and care [1, 2].

On the other hand, society was not always humane in its attitude toward mental patients and the way it cared for them. Such a humane practice developed only after the late eighteenth century, and in many countries, it is still the norm even today. The stigma and neglect of patients are still widely occurring even within the most advanced and educated parts of society worldwide.

K. N. Fountoulakis, *Psychiatry*, https://doi.org/10.1007/978-3-030-86541-2_22

22.2 Mental Illness and Social Revolutions

The rise of the civil rights movement, feminism, and labor movement in the 1960s marked a focus on any kind of perceived oppression. It is interesting that since the beginning, these movements borrowed rhetoric from psychiatry and especially from psychoanalysis and psychosocial theories, and soon many activists came to consider psychiatry in the frame of the struggle of liberation vs. oppression. According to this concept, psychiatric patients were considered to be an oppressed minority. This led to the emergence of anti-psychiatry [3, 4]. The relationship of these movements with psychoanalysis is complex. On the one hand, they adopted the psychoanalytical tools in order to interpret the world; on the other hand, they were openly hostile to the clinical practice of psychoanalysis, which was the dominant psychiatric approach of the time, when essentially the term "psychoanalysis" was almost identical to "psychiatry." However, they soon reconciled, since psychoanalysis was and still is by nature extremely flexible and adaptive to societal demands. A bitter comment here is that no scientific discipline could be that flexible! Then, they joined forces against the emergence of the empirical approach in psychiatry, as well as the emergence of psychopharmacotherapy.

Similarly, the emergence of the counterculture and its associated social developments in the 1960s played a major role in the relationships between psychiatry and society. With the "flower children," human potentialists, and drug and sexual revolutions being the prominent players of the period, the emphasis shifted toward hedonism, anti-intellectualism (any cognitive approach is considered as "mind-fucking"), and self-actualization. Especially the anti-intellectual movement created a chasm between humanistic psychology and brain–mind sciences [5].

During that period, a number of radical intellectuals discussed the issue of mental disorder and its social dimension. Michel Foucault (1926–1984; Fig. 11.31) argued that since the eighteenth century and due to the fact that the Enlightenment pushed civilization to embrace reason, society lost its tolerance toward madness, and as a result, there was a social confinement and isolation of mental patients [6]. According to him and to Erving Goffman (1922–1982), psychiatric treatment in hospitals is essentially evil [6, 7]. Foucault was essentially an anarchist and worked on a wide variety of intellectual topics of his time, and his approach was to synthesize history, philosophy, epistemology of the human sciences, and genealogy of the main social, political, and economic institutions of modernity [8], in an admixture that was considered to be chaotic by many authors. His style was to focus on the other (hidden) side and try to explain the whole phenomenon on the basis of this concealed side. Though appealing as an approach, the validity of such a method is questionable. His basic idea was that during the twentieth century, there is the rise of "biopower," which is of the technocratic sciences and organized societies with all aspects of human life being subject to it [9]. He considered all institutions of the society to be oppressive. In this frame, the individual person has become conceivable and manageable only as a "human capital." Concerning psychiatry and mental health, Foucault argued that there was no liberation of the mentally ill during the nineteenth century; only an objectification of the concept of their freedom occurred. This was because according to this theory, the concept of mental illness is considered to be the product of this "biopower environment," with technocratic-economic organization of the whole society, and "patients" are in reality human capital, waiting to be restored and to be optimized as a biological capital with the use of medicine [8]. Thus, psychiatry is part of the disciplinary system, and psychiatrists together with other mental health professionals are simply instruments of public order enforcement. The way Foucault uses the historical events surrounding the madness of King George III of England is interesting. While this case could be used as proof that mental illness is not a social construct but instead reflects true illness associated with disability, for Foucault it is simply the proof of the overwhelming power of psychiatry, which is able to remove even Kings from power.

Herbert Marcuse (1898–1979; Fig. 11.31) suggested that psychoanalysis perpetuates and complicates rather than solves the problems; he supported the use of psychedelic drugs as a way to raise consciousness [10]. In 1967, David Cooper (1931–1986; Fig. 11.32) published *Psychiatry and Antipsychiatry*. In that book he proposed that the major obstacle to the recovery of patients with schizophrenia was mainstream psychiatry and accused contemporary mental health care for alienating physicians and debilitating milieu. Cooper was the first to refuse the existence of mental illness [11]. This was later supported by the theory that schizophrenia is a disorder that recently appeared in the history of mankind [12, 13]. A number of books followed that criticized psychiatry in terms of civil rights and with the anti-psychiatric approach [7, 14, 15]. Eventually, the attachment was so deep that the core nosology of psychiatry was challenged [16].

Jean-Paul Sartre (1905–1980; Figs. 10.49 and 11.33), in a twist of reality, preceded in magnitude only by the irrational teachings of St. Augustine, argued that "mental illness is the revolt that the free organism in its total entity invents in order to live in an unbearable situation" [17].

One of the reasons for these ideas was the fact that there are no biological laboratory findings with diagnostic or predictive value, and the independent verification of the diagnosis is not possible. A core characteristic of mental disorders is that unlike many medical conditions, mental disorders are never asymptomatic. This, in turn, has led to the adoption of phenomenological ideas [18]. However, this embrace with philosophical movements brought psychopathology close to existential theories and the anti-psychiatric approach. This is in sharp contrast to the medical approach. While existential and anti-psychiatric ideas consider mental disorder to be a matter of choice, and maybe the ultimate act of freedom against an oppressing reality, the medical model considers mental patients to be the blameless victims of misfortune, similar to other patients suffering from somatic disorders and disabilities and in need for a medically defined way of treatment.

However, at that time there was a hope that the oppressed could unite and support and draw strength from each other, and mental patients were seen as such an oppressed minority group in need for support by civil rights activists [19]. It is interesting that this sensitivity toward psychiatric theory and practice was asymmetrical and cherry-picking and depended not on theory or empirical findings but on the strength of activist groups. For example, while neither the inclusion of homosexuality as a mental disorder nor the theory of the schizophrenogenic mother had any scientific support at all, nobody cared for the latter. Mothers of patients had no activist support; on the contrary they were often considered to be the cause of illness by intellectuals.

This approach and concept put all treatments available for mental patients under dispute. Psychosurgery and ECT were the most attacked, but medications were also considered as harmful, especially with the recognition of serious and visually impressive adverse effects like tardive dyskinesia. Addictions were considered to be the result of the medicopharmaceutical model and approach; on the other hand, the recreational use of addictive substances was encouraged.

Opposition to ECT was growing around the world. In 1962, Ken Kesey (1935–2001; Fig. 11.34) published the book *One Flew Over the Cuckoo's Nest*, in which he analogized society's oppressors with psychiatric ward staff, while the means of social and political oppression were analogized with the physical treatments of mental disorders, especially ECT and psychosurgery [20]. Discussions on a CIA secret program under the name "MKUltra" on mind control spread [21, 22] gave birth to a genuine conspiracy theory, especially since the related official documents were destroyed. A number of books followed, criticizing psychiatry in terms of civil rights and with the antipsychiatric approach [7, 14, 15]. In the late 1980s, Marius Romme (1934–) in the Netherlands established one of the most enduring and hostile-to-all-orthodox therapies group, under the name "Hearing Voices Self-help Movement" [23, 24].

In May 1968 during the Paris uprising, at the department of psychiatry at Paris University, the office of Jean Delay (1907–1987; Fig. 22.1), the father of psychopharmacology, was ransacked by French students. By virtue of his position, and his role in the evolution of psychopharmacology, Jean Delay was a symbol of the order the students were attacking. The next year, in Japan, Japanese students revolted and occupied the department of psychiatry at Tokyo University and held it for 10 years. These reactions and attacks against psychiatric institutions were extreme; however they carried the message that something was changing in society and thus the societal norms utilized by psychiatry should also change. An example of such a change in psychiatric nosology was the relatively recent abandoning of homosexuality as a psychiatric disorder [25, 26], which undoubtedly reflected changes in society [27]. It is interesting that psychiatry was the first institution to adopt these changes; some US states kept homosexuality in the list of penal crimes until after 2000, and the position of various organized religions is unclear. In spite of this, psychiatry still receives criticism for having homosexuality in the list of mental disorders some 50 years ago, instead of receiving credit for being a driving force for societal change on this topic.

Fig. 22.1 Jean Delay (1907–1987)

22.3 The "Social Revolution" in Psychiatry

Since the mid-1970s, the number of patients treated in asylums declined dramatically, and a decrease of up to 80% in asylum beds has been reported in many countries, in spite of the fact that an adequate network of community psychiatry services to cover the entire population has never been developed anywhere in the world. At the same time, there was a move of the legislation towards the better protection of the rights of the mental patients and especially concerning involuntary hospitalization and refusal of treatments. An additional important change was conceptual: from "not guilty by reason of insanity" to "guilty but mentally ill," which put psychiatric treatment under the supervision of justice and courts. "Guilty but mentally ill" has not replaced the "not guilty by reason of insanity," but it is increasingly used as an alternative, and currently the "not guilty" verdict is very difficult to have in many countries.

In 1975 the film *One Flew Over the Cuckoo's Nest* produced by Michael Douglas and directed by Milos Forman with Jack Nicholson in the leading role received several Oscar awards and embedded a negative image of ECT and psychiatry in the attitude of the general public (Figs. 11.34 and 22.2). Around the world it started becoming impossible to use ECT, especially in Switzerland, and in 1999 it was banned in private hospitals in Italy. It is interesting to see whether there will be any similar effect of the recently released film *Joker* (2019), directed by Todd Phillips, who co-wrote the screenplay with Scott Silver and with Joaquin Phoenix as the

Fig. 22.2 One Flew Over the Cuckoo's Nest (1975)

Joker. The film was considered as a masterpiece by both critics and the general public but triggered much controversy concerning the use of stigmatization concepts linking mental illness with extreme violence in order to give "psychological depth" or to make "more human" a villain comic character. Needless to say that there is no mental disorder with the characteristics depicted in the *Joker*.

The interesting thing with *One Flew Over the Cuckoo's Nest* is that both the book and the film are more a politico-social allegory than a story on mental illness and treatment. From a Marxist point of view, the narration clearly depicts a social-like hierarchy, with the "bad" face of authority (represented by nurse Ratchet) leaving no space for the "good-willed" but weak aspect of it (Dr. John Spivey) or for the lower levels of authority (doctors of the ward), who are also intimidated. Mental patients clearly represent the oppressed in general, but more likely the working class, stripped from essential rights, and enjoying a controlled supply of goods (cigarettes, watching TV, etc.). The central character (McMurphy, played by Jack Nicholson) represents the revolting and revisionist part of people of the middle and higher classes (bourgeois origin). In the book, McMurphy is clearly portrayed as such, in the sense that he was a Korean War hero accused of crime, while in the film, he is a petty thief. The film clearly suggests that the change cannot come from the struggle between parts of authority (nurse Ratchet vs. McMurphy) since the "intelligentsia" as reflected in the character of McMurphy has its own agenda and wishes a "soft" and controlled change and redistribution of wealth (like cigarettes and watching TV in the film) and is aimed at creating or keeping part of the power for itself. According to the author of the book, these efforts are doomed to fail, and only genuine and direct revolution from the oppressed themselves could produce radical change. This is depicted in the final scene of the film, with the Indian, one of the patients, breaking through the window and running free away from the ward. An interesting scene is the rejection of democracy by the author. When McMurphy gains the majority of votes to change the TV schedule, nurse Ratchet rejects the result by arguing that chronic and incapacitated patients had not voted. This was an allegory of the problems of democratic function within a capitalistic system and the power struggle within parts of authority, which at the end of the day is meaningless for the oppressed. Unfortunately, neither the audience nor the critics saw the book or the film as an allegory; instead, they focused exclusively on psychiatry and its practice, taking all the narration literally.

The "social revolution" within psychiatry went even further during the 1960s and the 1970s, and even some large and reputable hospitals started holding therapeutic meetings that often included not only patients and mental health professionals but also everyone whose work got him/her with a direct or indirect contact with the ward, e.g., administration staff, gardeners, and rest of hospital staff as well as the patients. It is of course reasonable to expect that in such meetings, lay opinion and attitude as well as anti-rationalism would prevail over scientific and professional knowledge.

In September 1979, the National Alliance on Mental Illness (NAMI) was founded, and it became one of the most potent consumers lobbying force in mental health. Its mission was to promote research and provide care to mental health patients and their families. It also functions as a mutual-help organization.

22.4 Societal and Politico-Economic Forces Influencing Psychiatry

The events of the last few decades had made it clear that psychiatry and mental health and illness are deeply ingrained within the fabric of the general culture. The debates over DSM-III and DSM-5 have proven that. However, this wide interest does not imply that it has a positive rather than a negative connotation. On the contrary, in many ways this public interest and involvement seem to increase stigma and fuel social prejudice and restrictions for mental patients [28, 29].

One of the most important trends since the 1990s was the increasing influence of economic and market or commercial factors on mental health care but also in psychiatric research. The increasing costs especially of the psychosocial services led to the development of organized efforts to keep the cost under control and often to reduce it. At the same time, there is a double pursue of two conflicting trajectories, the "evidence-based" on the one hand and the support for the concept of the "expert patient" on the other hand in the frame of the so-called post-psychiatry (see also Chap. 24). This concept has created many problems throughout medicine, especially in psychiatry. Three conceptions of people treated for mental disorders have emerged so far [30]:

- The patient (traditional passive role)
- The consumer (with the right to express views)
- The "survivor" (a new social movement disputing psychiatry as a whole)

One specific area of fierce debate and controversy is the issue of involuntary treatment. It is currently resolved, rather because of the needs of the society and not because of scientific knowledge. It is regulated by specific mental health laws that vary among countries with approximately 75% of them having such laws. In general, there is consensus in the international legislation that involuntary psychiatric treatment is permitted after recommendation by more than one psychiatrist, when there is a risk to the patient or others, due to the patient's illness and incapability of giving consent, or the patient refuses or lacks insight into the condition.

The logic behind this legislation is that involuntary treatment can potentially prevent harm to self and others. It can assist many patients in attaining their right to health care when they may be unable to decide in their own interests. However, at the same time, it carries the risk of insult of personal liberty and rights, as well as abuse for political, social, and other reasons. The legislation protects human rights by asking, among other things, for an independent assessment of the mental status of the person, which should have personal access to independent advocacy [31, 32]. In 1991, the United Nations adopted the "Principles for the Protection of Persons with Mental Illness and the Improvement of Mental Health Care," and in 2006 they formally agreed to the "Convention on the Rights of Persons with Disabilities." With both these acts, the UN established minimum standards for human rights protection in the mental health field [33].

Starting in the 1960s, in the USSR, Andrei Vladimirovich Snezhnevsky (1904–1987; Fig. 11.24) and his colleagues introduced the diagnosis of "sluggish schizophrenia" for people who dissented politically. This diagnosis was used exclusively in the Soviet Union and in several Eastern Bloc countries, and it was also applied to persons without any psychotic symptoms on the assumptions they will develop in the future. This is considered the prime example of psychiatric abuse for political reasons in the USSR [34–44]. Additionally, during the Khrushchev–Brezhnev era in the former Soviet Union, the KGB had the control of forensic psychiatric institutions and used them arbitrarily to label political opponents of the regime as suffering from psychosis and incarcerated them for long periods in "special psychiatric hospitals" [45]. Because of this, in 1983 the Soviet Psychiatric Association voluntarily withdrew from the World Psychiatric Association (WPA) when facing almost certain expulsion. It was readmitted back to the WPA only after the Gorbachev reforms, when many of these political prisoners were released. China has been accused of manifesting similar practices, but not to the extent it happened in the Soviet Union. These accusations concerned the way members of the Falun Gong movement (a meditative movement drawing on several Eastern religious doctrines) were treated [46, 47].

22.5 Mental Illness, Morality, Religion, and the Law

Another problematic field is both the lay and the philosophical and religious beliefs connecting mental disorder with vice and morality. This started with Plato, and in spite of the modernism revolution, it still dominates lay opinion. Modern societies have redefined vice in a secular way as criminal behavior or immoral conduct, and both terms originate from culture, serve sociocultural purposes, and are culturally determined. However, vice seems to survive within psychiatric concepts and even within diagnostic categories [48, 49]. These diagnostic categories are extremely few taking into consideration the total number of DSM codes. They include Antisocial Personality Disorder, Conduct Disorder, Intermittent Explosive Disorder, and Pedophilia. They are considered to be "vice-laden" [48–50], which means that these concepts are substantively more moral in essence, rather than non-moral and value-free as medical disorders are [51]. It is impressive that all the cardinal ("seven deadly") sins according to St. Augustine, which include gluttony, lust, greed, wrath, sloth, pride, and envy, are more or less reflected in psychiatric diagnostic categories [52]. Historically, socially deviant behaviors faced varied tolerance through the ages, and during the last few centuries, the society developed various institutions and mechanisms to deal with them. Among them were the asylums, and this was the reason why in them, mental patients, criminals, prostitutes, and the very poor could be found. There was little difference between asylums, jails, and almshouses [53, 54] as well as between alienists and jail guards [54–57].

Contrary to what Foucault preached, with the rise of modernity and the age of reason, Western retributive models of justice were based on an admixture of the Roman legal tradition and the Greco–Judeo–Christian metaphysical philosophy pertaining to free will, individual choice, and personal responsibility [58, 59]. Additionally, more recently, there were some data from works in the psychobiology of criminal behavior, implying that it should be considered to be some kind of a mental disorder [60, 61], while at the same time, the post-modern approach returns to the traditional approach and considers mental disorders as misconduct under the jurisdiction of religious, educational, and justice systems but not of medicine [4, 62].

Research has shown that there is a strong and resilient stereotype of dangerousness and a widespread desire for social distance from mental patients of any kind [63–65]. This is in sharp contrast to the findings in the literature, which suggest that severe mental illness does not independently predict future violent behavior. Also, it is not a leading cause of violence. There are some specific mental disorders in the area of antisocial personality and psychopathy and maybe some specific psychotic symptoms that predispose to violent behavior. Additionally, a number of factors related to mental disorder, including substance and alcohol abuse and various personal, social, and economic factors, are strongly related to such behaviors [66–71]. On the contrary, the vast majority of mental patients is never violent [72] but instead are much more likely to be the victims rather than the perpetrators of violence, and they have difficulty in defending themselves and asking for legal protection [70, 73–75]. There is of course a number of high-profile crime cases that attracted the attention of the media, but violence related to mental disorder is restricted within the family [76] or the health-care settings [77], although sometimes they also involve the neighborhood [78].

22.6 The Image of Mental Illness in the Popular Arts and the Mass Media

The resilience and the spread of the above-described beliefs and attitudes are at least partially due to the way psychiatry and mental patients are depicted in the literature, arts, and media and of course the movies. Even health professionals are heavily influenced by these sources and the embarrassment to medical schools, even medical doctors. The film industry seems to produce all kinds of products, from accusatory anti-psychiatric ones to supportive and understanding (more rarely however). Fortunately, films portraying mental patients as blood-thirsty criminals are becoming less frequent, but the *Joker* case reminds us that not only do they keep coming, but they also tend to receive wide acceptance and enthusiastic comments. They started with the 1960 film *Psycho directed* by Alfred Hitchcock in which the killer is a psychotic cross-dresser played by Anthony Perkins, while one of the most celebrated such films was *The Shining* in 1980 by Stanley Kubrick, with Jack Nicholson again playing the main role. There are a number of films also portraying psychiatrists as having problematic personalities, such as *The Silence of the Lambs in* 1991, which is *famous, starring Anthony Hopkins.* There seems to be three major types concerning psychiatrists, the "crazy," the "wonderful," and "Dr. Evil" [79].

The depictions of mental patients are mostly negative with an emphasis on violence or criminality and no coverage of human rights issues as what happens with other vulnerable groups in the community [80–82]. These negative depictions of psychiatry and mental patients occur even in comics and cartoons, and they facilitate the development of a stigmatizing attitude at a young age in the general population [83, 84]. Iconic is the figure of the Marvel Comics villain Doctor Faustus posing in the same way as Freud in the cover of *Amazing Spiderman* Vol 1 Issue 170 (1977) and that of the DC Comics hero Batman frequently posing with a straitjacket in the Arkham Asylum and especially in the collection "Welcome to Gothtopia" (1993). In DC Comics, the evil psychiatrist is named Professor Hugo Strange. Needless to say, there is no evil cardiologist, endocrinologist, or gastroenterologist in comics. The reader can trace these specific images as well as others since, although protected by copyright (and thus cannot be reproduced here), they are widely circulating on the

Internet. Even in the 2014 film *Captain America: The Winter Soldier*, there is a negative depiction of an electroconvulsive-like process representing a mind-control method.

It is impressive how irresponsible the behavior of the mass media is when dealing with issues concerning mental health. This has been proven beyond doubt after the analysis of their coverage of the debate on the efficacy of antidepressants (Fig. 22.3) [85]. *On the other hand, several public figures including actors are openly advocating for mental patients and their treatment; for example,* the death of Brian Keith (1921–1997) by suicide 2 months after his daughter also committed suicide and the death of Robin Williams (1951–2014) shocked the world. Kitty Dukakis, wife of presidential candidate Michael Dukakis, wrote a book on the treatment she received with ECT for her depression [86]. Brooke Shields openly talked on her depression and her successful treatment with antidepressants, and for this she received a public attack from Tom Cruise in a 2005 interview for *Entertainment*

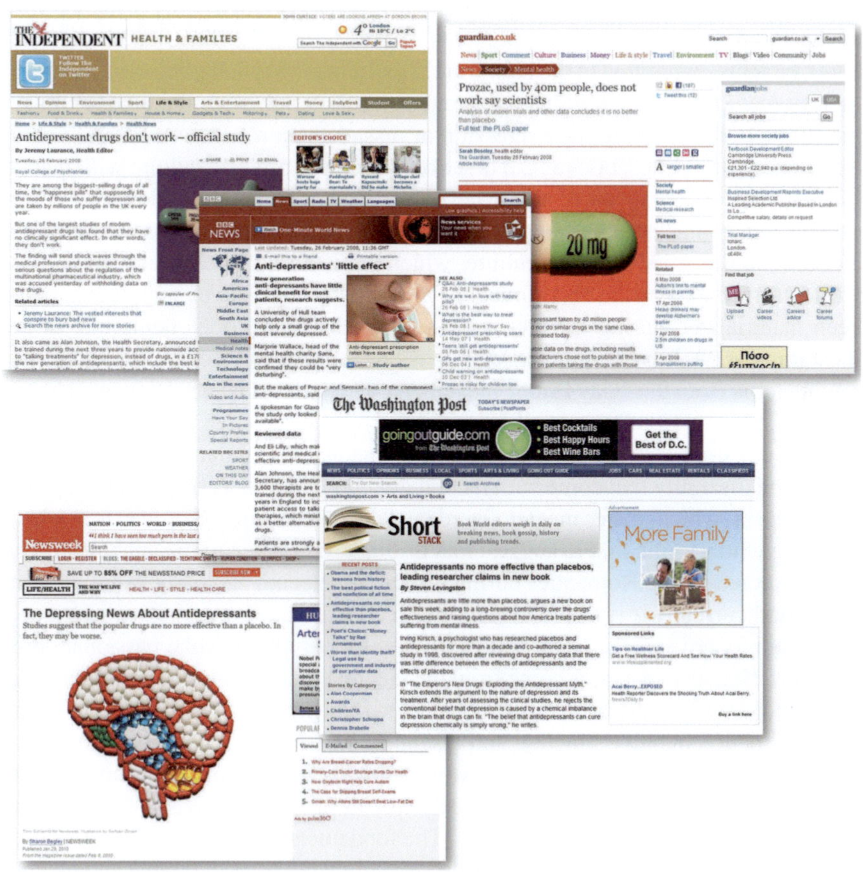

Fig. 22.3 Mass media headlines on the alleged non-efficacy of antidepressants after the publication of Kirsch in 2008

Tonight. In contrast, in spite of the many articles published, no information was ever released concerning the details of the treatment of Catherine Zeta-Jones, wife of Michael Douglas, the producer of *One Flew Over the Cuckoo's Nest*. According to her public interviews, she suffers from a form of bipolar disorder; she was repeatedly hospitalized and went through very difficult times during acute exacerbations.

22.7 The "Mad Identity" and "Mad Culture"

The adoption of existential and anti-psychiatric ideas either openly or covertly by a significant proportion of intellectuals and the society [9] eventually led to the consideration of individuals suffering from mental disorders as a social group with unique characteristics rather than as patients [18].

The first organized collective action and advocacy appeared in the UK in 1845 with the founding of the "Alleged Lunatics" Friends Society by patients with adverse experiences from the treatments of that time [87]. Since 1851 and until 1860, patients committed to the New York City Lunatic Asylum published an annual literary magazine, *The Opal*, which painted mostly a rather benign picture of asylum life but also allowed the expression of dissatisfaction and even radical liberatory rhetoric [88, 89]. This was the first ever attempt to produce culture with mental illness as the essential source and to develop a kind of community identity.

The conceptualization of mental patients as a social group implies that there are an accompanying identity and culture, which both should be characteristic of that particular group. Toward this direction, a number of initiatives and actions have been materialized during the recent decades, with "Mad Pride" being one of the most impressive. It is not only the latest wave of activism but the one, which by borrowing tools from the "Gay Pride" movement, demands social attention and change of social norms [18]. The overall attitude is of a "mad-positive activism" claiming that one can draw strength and identity from "madness" and that amidst all the suffering and disability that come with mental disorder, there is a compensating, glamorous aspect, which makes mental disorder a "dangerous gift." This idea is as old as Aristotle who was the first to suggest that brilliance is related to melancholia, and it was promoted by the Romantics during the late nineteenth and early twentieth centuries, and currently it is a core idea of the "Icarus project."

In the modern society, Mad Pride and Mad Identity both evoke passionate pros and cons. Negative reactions come from the fact that mental illness, even in the form of flamboyant bipolar disorder, is not characterized by creativity or brilliance but by impaired neurocognitive function and disability [90, 91]. Psychotic symptoms like delusions or hallucinations, or even extreme moods as in mania, profoundly affect reality checking, and all domains of behavior destroy reason-based thinking and even the unity and continuity of the self. The fundamental question is whether could also apart from a medical condition to also constitute the basis of culture and identity formation. This is a clear paradox since both of these demand intact reasoning and mental functioning in principle. Therefore, assigning the "social identity" status to mental illness and recognizing that a distinct "mad culture" exists, as a consequence of creativity stemming specifically from "the gift" of mental disorder, could act as a misleading concept both for the general public and for patients, distracting them from proper treatment and care and orienting them toward "alternative" views of the world and unproven treatment options. On the other hand, the question is whether the "culture" issue could fit into the subgroup of patients suffering from milder forms of mental illness, and they are living in the community, often without being under any psychiatric treatment or care. This cynical approach constitutes a compromise, in line with the moral vs. disease dilemma which psychiatry faces since antiquity.

However, even if the issue could fit a subgroup of patients, orienting both patients and the society toward the recognition of a distinct identity and culture could still be problematic because "mad activism" differs radically from activism focusing on the improvement of mental health services and care and the alleviation of the burden that comes with discrimination and stigma. The goal is not so much to reform mental health care or psychiatry but mainly to change the way madness is viewed in the frame of culture, identity, and recognition [92]. It is interesting, and to some extent an oxymoron, that although stemming at least partially from anti-psychiatry and from a social understanding of psychiatry, "mad activism" claims are based on the assumption that mental illness does exist, and there is a categorical distinction between health and illness, that is, mental disorder is a "natural kind." This distinction is a strict prerequisite in order to claim an identity but is in essential conflict with the movements' aim to popularize alternative narratives of psychological, experiential, and emotional diversity and to question the dominant view of madness in society.

In her books and articles, Clare Allan, an ex-patient, wrote about her experiences with mental health services and dismisses the assumption that there is or there could be any pride in suffering from mental illness. According to her, mad activism is a reaction to stigma by the society and an effort by patients to increase self-esteem and alleviate distress by creating an identity that does not exist, as well as by labeling mental illness as a "dangerous gift." On the other hand, she stresses the need for justice, fairness, and participation of patients in decision-making [93].

Mental patients are still victims of what is called "social invisibility." This is a historical concept associated with servitude, slavery, and oppression in society. One such example exists in historical anecdotes that the nobility found it natural to undress in front of their servants, without keeping societal rules they were following before their equals [94]. In this frame, society often deals with mental patients as being "socially invincible," and societal rules, including rights, do not apply for them. This is not a black-and-white phenomenon; more frequently patients are not entirely neglected, but they are experiencing reduced rights and rude or bad behavior by others, and in general, they feel that society has reserved a lower social class level for them [95].

Additionally, it could be true that "mad activism" constitutes a rare and valuable opportunity to approach and try to understand the inner experience and the lives of mental patients and, in this frame, to push and challenge the boundaries of our certainty concerning mental health and illness [96]. Mad activism and pride include a variety of narrations reflecting the experiences, complains, dreams, and desires of persons suffering from various forms of mental disorders and their fight to survive in a world dominated by exclusion, stigma, and dehumanization [97]. They also include radically different demands, from the reform of the mental health-care system to its complete abolition through the depathologization of psychiatric symptoms. In this later line of thinking, the terms "mentalism" and "sanism" were introduced as an analogy with racism and sexism [18, 98–101]. Another such term is "ableism," which refers to the way the society sees persons with some type of disability [18], while the term "survivor" was also coined to replace the term "patient" and, in this way, to reject the medical model as completely harmful [102].

On the other hand, clinicians deal with additional professional and ethical challenges when they face patients who are also activists. In the average mental health professional's mind, mad activism is incompatible with the core of what he/she has been taught during his/her long years of residency and what he/she exercises in everyday clinical practice. Here there is a need for a delicate and difficult balance: while on the one hand the clinician should not abandon his/her scientific background and knowledge and limits should be in place, there is, on the other hand, a need to understand the personal view of the patients and the complex root this type of activism has. Balance is not always achievable, but still it is always necessary to try to. Confronting the patient's ideas as a standard rigid attitude from the side of the therapist is never the right way to treat and care [2].

An interesting example of medical condition related to culture and identity is deafness, with the emergence of the deaf community and the deaf culture [18]. Deafness is an exciting case, but it manifests fundamental differences from mental disorder. In general, it respects mental capacity, cognition, and mental functioning, while at the same time, it is related to a unique way of expression, the sign language. Some deaf persons identify themselves as a "linguistic minority" and although this is not exactly correct, it is a method of communication with a great influence on deaf people's lives [103, 104]. Interestingly, during the recent COVID-19 pandemic, there was concern that the use of face masks could worsen the disability of deaf people because it would make lip reading impossible. This could serve as a wonderful example of how disability is frequently environmentally determined and how changes in the environment could induce disability in unexpected ways.

A broader question is whether culture and identity are absolutely and completely self- defined and declared, objective criteria are of no importance. An additional question is whether any characteristic (especially somatic disorders and conditions) with an important influence on people's everyday life could constitute a sufficient frame for the proclamation of a distinct identity and culture [103]. Such approaches often lead to the creation of "enemies" de novo; in this frame, as mentioned above, the term "sanism" was coined to refer to discrimination and prejudice against patients with mental disorder. The problem that immediately arises is that such radical approaches collide with core characteristics and ways of functioning of the society, e.g., aesthetics, and the way people deal with them in their everyday lives.

Lastly, activism of this kind does not address the question what happens with "culture" when the "condition" that "creates" it is "successfully" and completely "treated" (quotation marks were put deliberately). It is possible that often people think their "way of life" is worth preserving even in cases a "better" alternative is available [105, 106] because they feel the specific unique features of their medical condition identify with their deeper self, give meaning to their lives [107] as well as a sense of belonging and affiliation to a group of others [108].

22.8 The Multiple Facets and Functions of Psychiatry in Modern Societies

In spite of this often negativistic environment, surprisingly, today psychiatry and psychiatrists have accepted an additional role, which is to provide support and guidance to troubled youth without an overt (at present) mental disorder. In the past, this was provided by cultural tradition, which in a given society gives meaning and order to social life by providing persons with shared values and goals. Fixed and firm traditions leave little space for ambivalence and often, but not always, exert some kind of psychological protection. On the contrary, during periods of social transition, when these traditions become fluid and weak, there is some evidence for an increase in mental instability as well. In modern societies transition seems to be an ongoing and never-ending process, with social norms fast changing, leading to ambiguous and discordant customs as well as the co-existence of multiple social groups with different or even opposing values and goals. In this unprecedented environment, the traditional dominant figures and structures including extended families, church, schoolteachers, and the neighborhood can no longer provide guidance to the troubled youth. Subsequently, many teenagers fail to establish internalized values and become both behaviorally normless and existentially purposeless, and mental health professionals are asked to fill the gap between guidance and support [109].

References

1. Fulford KWM, Thornton T, Graham G. From bioethics to values-based practice. In: Fulford KWM, Thornton T, Graham G, editors. The oxford textbook of philosophy and psychiatry. Oxford University Press; 2006.
2. Radden J, Sadler JZ. The virtuous psychiatrist: character ethics in psychiatric practice. New York, NY: Oxford University Press; 2010.
3. Healy D, Savage M, Michael P, Harris M, Hirst D, Carter M, Cattell D, McMonagle T, Sohler N, Susser E. Psychiatric bed utilization: 1896 and 1996 compared. Psychol Med. 2001;31(5):779–90. https://doi.org/10.1017/s0033291701003968.
4. Szasz T. The myth of mental illness: foundations of a theory of personal conduct. New York, NY: Hober-Harper; 1961.
5. Yalom ID. Existential psychotherapy. New York, NY: Basic Books; 1980.

6. Foucault M. Histoire de la folie à l'âge classique. Paris: Plon; 1961.
7. Goffman E. Asylums: essays on the social situation of mental patients and other inmates. New York, NY: Anchor Books; 1961.
8. Leoni F. From madness to mental illness: psychiatry and biopolitics in Michel Faoucault. In: Fulford KWM, Davies M, Gipps R, et al., editors. The Oxford handbook of philosophy and psychiatry. Oxford University Press; 2013. https://doi.org/10.1093/oxfordhb/9780199579563.001.0001.
9. Foucault M. Madness and civilization: a history of insanity in the age of reason. Vintage; 1988.
10. Marcuse H. One-dimensional man. Boston: Beacon Press; 1964.
11. Cooper D. Psychiatry and anti-psychiatry. London: Tavistock; 1967.
12. Hare E. Was insanity on the increase? The fifty-sixth Maudsley lecture. Br J Psychiatry. 1983;142:439–55. https://doi.org/10.1192/bjp.142.5.439.
13. Turner TH. Schizophrenia as a permanent problem. Some aspects of historical evidence in the recency (new disease) hypothesis. Hist Psychiatry. 1992;3(12):413–29. https://doi.org/10.1177/0957154X9200301202.
14. Scheff T. Being mentally ill: a sociological theory. London: Weidenfield & Nicolson; 1966.
15. Scheff T. Labeling madness. New York, NY: Prentice-Hall; 1975.
16. Rosenhan DL. On being sane in insane places. Science. 1973;179(4070):250–8. https://doi.org/10.1126/science.179.4070.250.
17. Laing R. The politics of experience. Harmondsworth: Penguin Books; 1967.
18. Rashed MA. Madness and the demand for recognition. Oxford University Press; 2019. https://doi.org/10.1093/med/9780198786863.001.0001.
19. Fanon F. The wretched of the earth. Harmondsworth: Penguin Books; 1967.
20. Kesey K. One Flew Over the Cuckoo's Nest. London: Methuen; 1962.
21. Collins A. In the sleep room: the story of CIA brainwashing experiments in Canada. Toronto: Key Porter Books; 1988.
22. Marks J. The search for the manchurian candidate. New York, NY: Times Books; 1979.
23. Leucht S, Pitschel-Walz G, Abraham D, Kissling W. Efficacy and extrapyramidal side-effects of the new antipsychotics olanzapine, quetiapine, risperidone, and sertindole compared to conventional antipsychotics and placebo. A meta-analysis of randomized controlled trials. Schizophr Res. 1999;35(1):51–68. https://doi.org/10.1016/s0920-9964(98)00105-4.
24. Romme MA, Escher AD. Hearing voices. Schizophr Bull. 1989;15(2):209–16. https://doi.org/10.1093/schbul/15.2.209.
25. Bayer R. Homosexuality and American psychiatry: the politics of diagnosis. New York, NY: Basic Books; 1981.
26. Lieberman JA. Shrinks: the untold story of psychiatry. Back Bay Books; 2016.
27. Mechanic D. Definitions and perspectives. In: Horowitz A, Scheid T, editors. A handbook for the study of mental health. Cambridge: Cambridge University Press; 1999.
28. Fountoulakis KN. Suicide and the economic situation in Europe: are we experiencing the development of a 'reverse stigma'? Br J Psychiatry. 2016;209(4):273–4. https://doi.org/10.1192/bjp.bp.115.171785.
29. Fountoulakis KN, Souliotis K. The right to choose treatment-without-treatment: respecting civil rights or an unprecedented manifestation of 'reverse stigma'? Acta Neuropsychiatr. 2019;31(1):56–8. https://doi.org/10.1017/neu.2018.24.
30. Pilgrim D, Rogers A. A sociology of mental health and illness. Buckingham: Open University Press; 1993.
31. Sklar R. Starson v. Swayze: the Supreme Court speaks out (not all that clearly) on the question of "capacity". Can J Psychiatr. 2007;52(6):390–6. https://doi.org/10.1177/070674370705200609.
32. World Health Organization. WHO resource book on mental health: human rights and legislation. Geneva: World Health Organization; 2005.
33. United Nations UN Division for social policy and development disability. 2018. https://www.un.org/development/desa/disabilities/. Accessed 29 Apr 2018.

34. Aliev N. "Sluggish schizophrenia" (concerning the article by A.B. Smulevich and P.V. Mikhalev) S.S. Korsakov zhurnal nevropatologii i psikhiatrii, No. 12, 1987, pp. 186-187. Zh Nevropatol Psikhiatr Im S S Korsakova. 1992;92(2):132.
35. Andreyev H. Political dissent and "sluggish" schizophrenia in the Soviet Union. Br Med J (Clin Res Ed). 1986;293(6550):822. https://doi.org/10.1136/bmj.293.6550.822-a.
36. Boiarshinova TN. Psychological diagnosis of sluggish schizophrenia (according to thematic apperception test findings). Zh Nevropatol Psikhiatr Im S S Korsakova. 1975;75(11):1680–2.
37. Gladkova KI. Basic patterns in the formation of sluggish schizophrenia according to epidemiologic findings. Zh Nevropatol Psikhiatr Im S S Korsakova. 1977;77(2):253–6.
38. Makhova TA. Features of the course of sluggish psychopathic-like schizophrenia, its diagnosis and forensic psychiatric significance. Zh Nevropatol Psikhiatr Im S S Korsakova. 1982;82(5):113–9.
39. Merskey H, Shafran B. Political hazards in the diagnosis of 'sluggish schizophrenia'. Br J Psychiatry. 1986;148:247–56. https://doi.org/10.1192/bjp.148.3.247.
40. Sarsembaev KT. Social-work adaptation of patients with sluggish schizophrenia in the late stages of the course of the disease (according to the results of a clinico-epidemiologic study). Zh Nevropatol Psikhiatr Im S S Korsakova. 1985;85(8):1217–21.
41. Smulevich AB. Sluggish schizophrenia in the modern classification of mental illness. Schizophr Bull. 1989;15(4):533–9. https://doi.org/10.1093/schbul/15.4.533.
42. Vilenskii OG, Kolomiichenko LN. Clinical features and work capacity of patients with sluggish schizophrenia and a hypochondriacal syndrome. Zh Nevropatol Psikhiatr Im S S Korsakova. 1983;83(5):728–32.
43. Wilkinson G. Political dissent and "sluggish" schizophrenia in the Soviet Union. Br Med J (Clin Res Ed). 1986;293(6548):641–2. https://doi.org/10.1136/bmj.293.6548.641.
44. Zairov GK, Gurdorzh T, Bavuuzhav O. (Clinico-catamnestic study) Clinical picture, treatment and social-occupational adaptation of patients suffering schizophrenia with a sluggish, uninterrupted course. Zh Nevropatol Psikhiatr Im S S Korsakova. 1977;77(5):722–5.
45. Finlayson J. Political abuse of psychiatry with a special focus on the USSR. Bull R Coll Psychiatr. 2018;11(4):144–5. https://doi.org/10.1192/pb.11.4.144.
46. Lyons D, O'Malley A. The labelling of dissent: politics and psychiatry behind the Great Wall. Psychiatr Bull. 2002;26:434.
47. van Voren R. The WPA World Congress in Yokohama and the issue of political abuse of psychiatry in China. Psychiatr Bull. 2018;26(12):441–2. https://doi.org/10.1192/pb.26.12.441.
48. John ZS. Vice and the diagnostic classification of mental disorders: a philosophical case conference. Philos Psychiat Psychol. 2008;15(1):1–17. https://doi.org/10.1353/ppp.0.0152.
49. Sadler JZ. Vice and mental disorders. In: Fulford KWM, Davies M, Gipps R, et al., editors. The Oxford handbook of philosophy and psychiatry. Oxford University Press; 2013. https://doi.org/10.1093/oxfordhb/9780199579563.001.0001.
50. Sadler JZ. Values and psychiatric diagnosis. Oxford: Oxford University Press; 2005.
51. Frankena WK. Ethics. 2nd ed. Englewood Cliffs, NJ: Prentice-Hall; 1973.
52. Newhauser RE. The seven deadly sins: from communities to individuals (studies in medieval and reformation traditions, volume 123). Leiden: Brill; 2007.
53. Rafter NH. Creating born criminals. Urbana, IL: University of Illinois Press; 1997.
54. Rothman DJ. Conscience and convenience: the asylum and its alternatives in progressive America. Hawthorne, NY: Aldine de Gruyter; 2002.
55. Grob GN. Mental health policy in America: myths and realities. Health Aff (Millwood). 1992;11(3):7–22. https://doi.org/10.1377/hlthaff.11.3.7.
56. Porter R. A social history of madness: the world through the eyes of the insane. New York, NY: Weidenfeld & Nicolson; 1988.
57. Rothman DJ. The discovery of the asylum: social order and disorder in the new republic. Hawthorne, NY: Aldine de Gruyter; 2002.
58. Braithwaite J. Restorative justice: assessing optimistic and pessimistic accounts. Crime Justice. 1999;25:1–127.

59. Wenzel M, Okimoto TG, Feather NT, Platow MJ. Retributive and restorative justice. Law Hum Behav. 2008;32(5):375–89. https://doi.org/10.1007/s10979-007-9116-6.
60. Raine A. The psychopathology of crime: criminal behavior as a clinical disorder. New York, NY: Academic Press; 1993.
61. Raine A. A reply to Dolan's and Cordess' reviews of the psychopathology of crime. Cogn Neuropsychiatry. 2001;6(4):304–7.
62. Szasz TS. The myth of mental illness. New York, NY: Harper and Row; 1974.
63. Kaparounaki CK, Koraka CA, Kotsi ES, Ntziovara AP, Kyriakidis GC, Fountoulakis KN. Greek university student's attitudes and beliefs concerning mental illness and its treatment. Int J Soc Psychiatry. 2019;65(6):515–26. https://doi.org/10.1177/0020764019864122.
64. Link BG, Phelan JC, Bresnahan M, Stueve A, Pescosolido BA. Public conceptions of mental illness: labels, causes, dangerousness, and social distance. Am J Public Health. 1999;89(9):1328–33. https://doi.org/10.2105/ajph.89.9.1328.
65. Pescosolido BA, Monahan J, Link BG, Stueve A, Kikuzawa S. The public's view of the competence, dangerousness, and need for legal coercion of persons with mental health problems. Am J Public Health. 1999;89(9):1339–45. https://doi.org/10.2105/ajph.89.9.1339.
66. Elbogen EB, Johnson SC. The intricate link between violence and mental disorder: results from the National Epidemiologic Survey on Alcohol and Related Conditions. Arch Gen Psychiatry. 2009;66(2):152–61. https://doi.org/10.1001/archgenpsychiatry.2008.537.
67. Fazel S, Gulati G, Linsell L, Geddes JR, Grann M. Schizophrenia and violence: systematic review and meta-analysis. PLoS Med. 2009;6(8):e1000120. https://doi.org/10.1371/journal.pmed.1000120.
68. Metzl JM, MacLeish KT. Mental illness, mass shootings, and the politics of American firearms. Am J Public Health. 2015;105(2):240–9. https://doi.org/10.2105/AJPH.2014.302242.
69. Peterson JK, Skeem J, Kennealy P, Bray B, Zvonkovic A. How often and how consistently do symptoms directly precede criminal behavior among offenders with mental illness? Law Hum Behav. 2014;38(5):439–49. https://doi.org/10.1037/lhb0000075.
70. Stuart H. Violence and mental illness: an overview. World Psychiatry. 2003;2(2):121–4.
71. Taylor PJ, Gunn J. Homicides by people with mental illness: myth and reality. Br J Psychiatry. 1999;174:9–14. https://doi.org/10.1192/bjp.174.1.9.
72. Swanson JW, McGinty EE, Fazel S, Mays VM. Mental illness and reduction of gun violence and suicide: bringing epidemiologic research to policy. Ann Epidemiol. 2015;25(5):366–76. https://doi.org/10.1016/j.annepidem.2014.03.004.
73. Brekke JS, Prindle C, Bae SW, Long JD. Risks for individuals with schizophrenia who are living in the community. Psychiatr Serv. 2001;52(10):1358–66. https://doi.org/10.1176/appi.ps.52.10.1358.
74. Petersilia JR. Crime victims with developmental disabilities. Crim Justice Behav. 2016;28(6):655–94. https://doi.org/10.1177/009385480102800601.
75. Teplin LA, McClelland GM, Abram KM, Weiner DA. Crime victimization in adults with severe mental illness: comparison with the National Crime Victimization Survey. Arch Gen Psychiatry. 2005;62(8):911–21. https://doi.org/10.1001/archpsyc.62.8.911.
76. Solomon PL, Cavanaugh MM, Gelles RJ. Family violence among adults with severe mental illness: a neglected area of research. Trauma Violence Abuse. 2005;6(1):40–54. https://doi.org/10.1177/1524838004272464.
77. Chou KR, Lu RB, Chang M. Assaultive behavior by psychiatric in-patients and its related factors. J Nurs Res. 2001;9(5):139–51. https://doi.org/10.1097/01.jnr.0000347572.60800.00.
78. Logdberg B, Nilsson LL, Levander MT, Levander S. Schizophrenia, neighbourhood, and crime. Acta Psychiatr Scand. 2004;110(2):92–7. https://doi.org/10.1111/j.1600-0047.2004.00322.x.
79. Schneider I. The theory and practice of movie psychiatry. Am J Psychiatry. 1987;144(8):996–1002. https://doi.org/10.1176/ajp.144.8.996.
80. Coverdale J, Nairn R, Claasen D. Depictions of mental illness in print media: a prospective national sample. Aust N Z J Psychiatry. 2002;36(5):697–700. https://doi.org/10.1046/j.1440-1614.2002.00998.x.

81. Diefenbach DL. The portrayal of mental illness on prime-time television. J Community Psychol. 1997;25(3):289–302. https://doi.org/10.1002/(sici)1520-6629(199705)25:3<289::Aid-jcop5>3.0.Co;2-r.
82. Wahl OF. Mass media images of mental illness: a review of the literature. J Community Psychol. 1992;20(4):343–52. https://doi.org/10.1002/1520-6629(199210)20:4<343::Aid-jcop2290200408>3.0.Co;2-2.
83. Fountoulakis K, Kogiopoulos K, Nimatoudis I, Iacovides A, Nikolaou T, Ierodiakonou C. The concept of mental disorder in Greek cinema. Acta Psychiatr Scand. 1998;98(4):336–40. https://doi.org/10.1111/j.1600-0447.1998.tb10093.x.
84. Wahl OF. News media portrayal of mental illness. Am Behav Sci. 2016;46(12):1594–600. https://doi.org/10.1177/0002764203254615.
85. Fountoulakis KN, Hoschl C, Kasper S, Lopez-Ibor J, Moller HJ. The media and intellectuals' response to medical publications: the antidepressants' case. Ann General Psychiatry. 2013;12(1):11. https://doi.org/10.1186/1744-859X-12-11.
86. Dukakis KD. Shock: the healing power of electroconvulsive therapy. New York, NY: Avery Publishing Group; 2006.
87. Hervey N. Advocacy or folly: the alleged Lunatics' friend society, 1845–63. Med Hist. 1986;30(3):245–75. https://doi.org/10.1017/s0025727300045701.
88. Reiss B. Letters from asylumia: the opal and the cultural work of the lunatic asylum, 1851–1860. Am Lit Hist. 2004;16(1):1–28. https://doi.org/10.1093/alh/ajh001.
89. Tenney L. Who fancies to have a revolution here? The opal revisited (1851–1860). Rad Psychol. 2006;5.
90. Fountoulakis K. Creativity and the position of bipolar disorder in popular culture. In: Fountoulakis K, editor. Bipolar disorder: an evidence-based guide to manic depression. Berlin Heidelberg: Springer-Verlag; 2015. p. 685–703. https://doi.org/10.1007/978-3-642-37216-2.
91. Fountoulakis K. Neurocognitive functioning in bipolar disorder: a comprehensive review of recent data. In: Fountoulakis K, editor. Bipolar disorder: an evidence-based guide to manic depression. Berlin Heidelberg: Springer-Verlag; 2015. p. 109–62. https://doi.org/10.1007/978-3-642-37216-2.
92. Schrader S, Jones N, Shattell M. Mad pride: reflections on sociopolitical identity and mental diversity in the context of culturally competent psychiatric care. Issues Ment Health Nurs. 2013;34(1):62–4. https://doi.org/10.3109/01612840.2012.740769.
93. Allan C. Misplaced pride. The Guardian; 2006.
94. Honneth A. Invisibility: on the epistemology of "recognition". Proce Aristotel Soc Suppl. 2001;75:111–26.
95. Haslam N, Loughnan S. Dehumanization and infrahumanization. Annu Rev Psychol. 2014;65:399–423. https://doi.org/10.1146/annurev-psych-010213-115045.
96. LeFrançois B, Menzies R, Reaume G. Mad matters: a critical reader in Canadian mad studies. Canadian Scholars' Press; 2013.
97. Poole J, Ward J. "Breaking open the bone": storying, sanism, and mad grief. In: LeFrançois B, Menzies R, Reaume G, editors. Mad matters: a critical reader in Canadian mad studies. Toronto: Canadian Scholars' Press; 2013. p. 94–104.
98. Chamberlin J. On our own: patient-controlled alternatives to the mental health system. New York, NY: McGraw-Hill; 1978.
99. Chamberlin J. The ex-patients' movement: where we've been and where we're going. J Mind Behav. 1990;11(3/4):323–36.
100. Fabris E. Mad success: what could go wrong when psychiatry employs us as "peers"? In: LeFrançois B, Menzies R, Reaume G, editors. Mad matters: a critical reader in Canadian mad studies. Toronto: Canadian Scholars' Press; 2013. p. 130–40.
101. Menzies R, LeFrançois B, Reaume G. Introducing mad studies. In: LeFrançois B, Menzies R, Reaume G, editors. Mad matters: a critical reader in Canadian mad studies. Toronto: Canadian Scholars' Press; 2013. p. 1–22.
102. Crossley N. Not being mentally ill. Anthropol Med. 2004;11(2):161–80. https://doi.org/10.1080/13648470410001678668.

103. Kymlicka W. Multicultural citizenship: a liberal theory of minority rights. Oxford: Oxford University Press; 1995.
104. Padden C, Humphries T. Inside deaf culture. Harvard University Press; 2005.
105. Margalit A, Halbertal M. Liberalism and the right to culture. Soc Res. 1994;61(3):491–510.
106. Taylor C. Can liberalism be communitarian? Crit Rev. 1994;8(2):257–62.
107. Tamir Y. A strange alliance: isaiah berlin and the liberalism of the fringes. Ethical Theory Moral Pract. 1998;1(2):279–89.
108. Gans C. Nationalism and immigration. Ethical Theory Moral Pract. 1998;1(2):159–80.
109. Millon T. Masters of the mind: exploring the story of mental illness from ancient times to the new millennium. Hoboken, NJ: Wiley; 2004.

Chapter 23
Anti-psychiatry

Psychiatry is the only medical specialty that faces such an intense criticism, which goes as deep as its existence itself. There are no anti-surgery or anti-dermatology movements, and nobody argues that doctors should stop operating or stop dealing with the skin. There is of course a large "alternative medicine" movement, but it rarely goes that deep and is not radically hostile when confronting medicine. A more recent development is the anti-vaccination movement and it is true that during the COVID-19 pandemic a number of peculiar anti-medical movements and attitudes emerged. While these movements, which are critical to mainstream medicine, accuse doctors for their conflicts of interest and for not applying the best "available" treatment, movements against psychiatry claim that psychiatry does not help mental patients and, even more importantly, that mental disorders do not exist and psychiatry creates a myth around their presence as well as for the need for treatment, thus acting as a henchman for dark forces within society who want to suppress people. Maybe only the anti-vaccination movement as well as some movements related to the COVID-19 pandemic that recently emerged are comparable in terms of hostility and radical rejection. But still, these movements do not dispute the existence of infectious diseases or the immune mechanisms vaccination is based on; they accuse some dark political forces for taking advantage of medicine to profit and induce harm. The attacks psychiatry faces are, to some extent, against the medical model for mental disorders [1], although they started probably as a reaction to the expanding sociopolitical influence of psychoanalysis (see also Chap. 11).

As already mentioned, an important development during the 1960s was the rise of the civil rights movement, of feminism, and of the labor movement. All of them, already from the beginning, borrowed rhetoric from psychiatry and especially from psychoanalysis and psychosocial theories. Additionally, in their approach, psychiatry served as a model and as an example for the whole society, and in turn, mental patients were considered to be a minority, oppressed by society through psychiatry.

The most important and astonishing development within this frame was the emergence of anti-psychiatry. It was the product of the interaction between social

K. N. Fountoulakis, *Psychiatry*, https://doi.org/10.1007/978-3-030-86541-2_23

psychiatry and social activism and begun as social activism within the boundaries of mainstream psychiatry. In 1961, Thomas Szasz (1920–1912; Fig. 11.31) published *The Myth of Mental Illness* and argued that the very concept of mental illness was a logical absurdity [2]. According to him, insanity is defined exclusively on the basis of value (ethics, morality, legal issues, and social norm), while in contrast somatic illness is a deviation from biological norms [3, 4]. His target was mainly the psychoanalytic industry in the USA and its problematic model and practice. But since psychoanalysis was identified with mainstream psychiatry of the time in the USA, this approach was perceived as an attack on psychiatry itself [5]. At the time of the publication of his first book in 1961, psychopharmacology had just appeared, and it was by no means dominant. When the efforts to create a reliable classification system emerged in the 1970s, and after the Rosenhan "Thud" study in 1969–1970 (see Chap. 11) [6], Szasz went even further and placed psychiatry in the company of alchemy and astrology and defined it as pseudoscience [7]. His views constituted a form of behavioral relativism, which considered any behavior including psychopathological symptoms as meaningful and valid if viewed from the proper perspective. Apart from criticizing the views of contemporary psychiatry, Szasz also criticized the way other foundations of the society were dealing with mental patients. He particularly criticized and challenged the legal approach that considered that insanity identifies with irrationality as far as the justice system is concerned. According to his view, the so-called mental patients are perfectly accountable for their deeds. With this, he challenged the proposal that mental illness is a radical failure of rationality and autonomy [3, 8] and was aligned with the Sartre and the existentialist ideas.

Michel Foucault (1926–1984; Fig. 11.31) argued that since the eighteenth century, because the Enlightenment pushed civilization to lose tolerance toward madness, there was a social confinement and isolation of mental patients [9]. According to him and to Erving Goffman (1922–1982), psychiatric treatment in hospitals is essentially evil [9, 10]. Foucault worked on a wide variety of intellectual topics of his time, and his approach was to synthesize history, philosophy, epistemology of the human sciences, and genealogy of the main social, political, and economic institutions of modernity [11] in an admixture, which was considered to be chaotic by many authors. His style was to focus on the other (hidden) side and try to explain the whole phenomenon on the basis of this concealed side. Though appealing as an approach, the validity of such a method is questionable. His basic idea was that during the twentieth century, there is the rise of "biopower," which is of the technocratic sciences and organized societies with all aspects of human life being subject to it [12]. In this frame, the individual person has become conceivable and manageable only as a "human capital." Concerning psychiatry and mental health, Foucault argued that there was no liberation of the mentally ill during the nineteenth century; only an objectification of the concept of their freedom occurred. This was because according to this theory, the concept of mental illness is considered to be the product of this "biopower environment" with technocratic–economic organization of the whole society, and "patients" are in reality human capital waiting to be restored and to be optimized as a biological capital with the use of medicine [11]. Thus, psychiatry is part of the disciplinary system, and psychiatrists together with other mental health professionals are simply instruments of public order enforcement. The way Foucault uses the historical events surrounding the madness of King George III of England is interesting. While this case could be used as proof that mental illness is not a social construct but instead reflects true illness associated with disability, for Foucault it is simply the proof of the overwhelming power of psychiatry, which is able to remove even kings from power.

Herbert Marcuse (1898–1979; Fig. 11.31) suggested that psychoanalysis perpetuates and complicates rather than solves the problems and supported the use of psychedelic drugs as a way to raise consciousness [13]. In 1967, David Cooper (1931–1986; Fig. 11.32) published *Psychiatry and Anti-psychiatry*, introducing the term "anti-psychiatry." In this book he proposed that the major obstacle to the recovery of patients with schizophrenia was mainstream psychiatry and accused contemporary mental health care for alienating physicians and debilitating milieu. Cooper was the first to refuse the existence of mental illness [14]. This was later supported by the theory that schizophrenia is a disorder that recently appeared in the history of mankind, which, however, was never adequately supported by empirical research [15, 16]. A number of books followed, which criticized psychiatry in terms of civil rights and with the anti-psychiatric approach [10, 17, 18]. Eventually the very core nosology of psychiatry was challenged with the famous Rosenhan study [6]. In 1962 David Cooper established an experimental therapeutic community in Shenley Hospital, Hertfordshire, UK, which later became known as Villa 21 or "the anti-hospital" [19]. It was an "ultra-permissive" experiment, which sought to restore power and autonomy to patients by eliminating staff–patient distinctions and giving patients democratic control of the ward [14]. The experiment was a bitter failure, although Cooper never accepted that, and came to an end in 1966.

Overall, radically different views have been expressed [20], and all of them are in conflict with the medical model and with each other. Mental illness has been proposed to be as follows:

- Some kind of maladaptive learned behaviors [21].
- Merely a social construct and arbitrary labeling [22, 23].
- Such labeling processes actively inhibit recovery [6, 24].
- Some kind of suppressive mechanism of sociopolitical control [9, 12, 25].
- A kind of spiritual experience [26].
- Expression of family problems [27, 28] or life problems [2, 7].

The adoption of existential and anti-psychiatric ideas either openly or covertly by a significant proportion of intellectuals and the society eventually led to the consideration of individuals suffering from mental disorders as a social group with unique characteristics rather than as patients [29]. This is in accordance with the interpretation that in philosophical terms, all the above represent some type of "social behaviorism," that is, social forces are the only causes that could induce mental illness, and this happens de novo; mental illness would not have been there without the effect of the social environment. It is astonishing that although many of the authors seem to have a background in materialism and more specifically in dialectic materialism, their approach to mental illness seems closer to a Cartesian-like dualistic (biology plays no role) or an idealistic (mental phenomena are primary) concept of mental life. Today both are considered unscientific theories. The much celebrated Kendell vs. Szasz debate [1, 7, 30] did not clarify things for those who chose to believe in anti-scientific dogmata.

The rise of "existentialist philosophy" in Europe probably also played a role. The term comes from the Latin word *exisisto* (meaning I exist), which in turn comes from two other Latin words, *ex* and *sister* (meaning "to stand out," "to become," "to emerge"). The Greek analog *εξίσταμαι* (existame) means I separate from the others, I protest. According to this school of thought, existence is not a static objective state but a flowing experience of being, and life is defined only in terms of existence, that is, of subjective inner experience. Existential philosophy and its offspring "humanistic psychology" are the direct product of disillusionment from the catastrophic experience of life in the first half of the twentieth century and more specifically with the experience of WWII. Because of this, many Europeans turned their back on science and modernism and focused on self-understanding and psychological well-being.

According to the existential approach, mental illness is the consequence of the individual's inability to relate to self. Without self, there is no identity, and therefore there can be no "being-in-the-world" (dasein). Thus, there can be no fulfillment of instinctual drives, no meaningful relationship with others, and no introspect. This creates an "ontological anxiety," a frightening alienation from self which eventually could lead to "nothingness" and mental disorder.

Central to the existential theory is *phenomenology*, a term that comes from the Greek words *phenomenon* (meaning observable event or object) and *logos.* According to Immanuel Kant (1724–1804; Fig. 7.5), the mind can know only what the senses allow ("phenomenon") but not the meaning and the qualities of the thing ("noumenon"). Edmund Husserl (1859–1938; Fig. 10.3) developed the "epoche" method (from the Greek word *αποχή*, meaning "to abstain" from everyday beliefs and common biases), which dictates that the observer should be genuinely unbiased. On the other hand, Søren Aabye Kierkegaard (1813–1855; Fig. 8.19) argued for the absolute free will of humans and divided their choices into "inauthentic" (those pleasing the will of the masses – "tyranny of the plebs") and "authentic" (genuine free will of the person). Existential anxiety is the result of the struggle of the person to exercise his/her free will. Failure of this leads to the manifestation of mental disorder.

Although the existentialist dilemma is appealing as a theory, and enjoyed much popularity especially among lay people and in the arts, it cannot be considered to constitute a valid and robust theory of mental life, and more particularly with implications in the concept and definition of mental illness. It resembles more a skillful literary depiction, but as a theory, it lacks coherence and scientific structure.

Another problem with humanitarian theories is that they consider the individual as completely free of the malevolent distortions of society. This is not only invalid but potentially dangerous and is not supported by any kind of empirical data. Existential and humanistic theories seem to reflect banal lay moral concepts in a more elaborate and shining way.

Jean-Paul Sartre (1905–1980; Figs. 10.49 and 11.33), the leading figure of existential philosophy, in his foreword to the best-seller book *The Divided Self* by Ronald David Laing (1927–1989; Fig. 11.32) noted that "mental illness is the revolt that the free organism in its total entity invents in order to live in an unbearable situation" [27]. According to Laing, patients with schizophrenia are potentially better off than "healthy normal" people, because the latter have constructed a false self to present to the world, while patients are those who have refused to do so. In accordance with a version of existential theories, Laing suggested that patients with schizophrenia are less alienated from themselves and that the psychotic experience is part of a healing spiritual journey leading to a higher form of sanity [27, 31, 32].

Laing went on and like Cooper, he founded a therapeutic community at Kingsley Hall in London where the same democratic approach with no differentiation between staff and patients was applied. There, the best-known patient to be treated was Mary Barnes (1923–2001), who later wrote the story of her life there [33]. He was also the celebrated character of the 2017 film *Mad to Be Normal* (Fig. 23.1). At that time there was a hope that the oppressed throughout the world could unite and support

Fig. 23.1 *Mad to Be Normal* (2017), a film on the life and work of Ronald David Laing (1927–1989)

and draw strength from each other. Mental patients were seen as such an oppressed minority group in need for support by civil rights activists [34].

This approach and concept put all treatments available for mental patients under dispute. Psychosurgery and electroconvulsive therapy (ECT) were the most attacked, but also medications were considered as harmful especially with the recognition of serious and visually impressive adverse effects like tardive dyskinesia. Addictions were considered to be a result of the medico-pharmaceutical model and approach, although the use of addictive drugs was promoted as leading to a higher level of insight and to a journey inside the self. Opposition to ECT was growing around the world. In 1962 Ken Kesey (1935–2001; Fig. 11.34) published the book *One Flew Over the Cuckoo's Nest* in which he analogized society's oppressors with psychiatric ward staff and the means of social and political oppression with physical treatments of mental disorders, especially ECT and psychosurgery [35]. Unfortunately, some of these allegations were at least partially true in everyday clinical practice where the humane treatment of patients was missing. Psychiatry is loaded with problematic treatments, like all medicine, but for psychiatry, the criticism is radical.

In May 1968 during the Paris Uprising, the office of Jean Delay in the department of psychiatry at the University of Paris was ransacked by French students. By virtue of his position, Jean Delay was a symbol of the order the students were attacking. Additionally, his role in the evolution of psychopharmacology made him a special target as the proud father of the dreaded camisole chimique (chemical straitjacket). The next year, in Japan, following visits by Szasz and others, Japanese students revolted and occupied the department of psychiatry at Tokyo University, at that time chaired by Hiroshi Utena, and held it for 10 years, thus bringing research to a halt.

Claims that oppressed minorities exist were often extreme and maybe dictated by personal interests or acquaintances. Nobody ever gave a fight against the ridiculous concept of "schizophrenogenic mother," developed by Frieda Fromm-Reichmann (1889–1957) and Gregory Bateson (1904–1980) [36, 37], which put the blame on families and especially on mothers in a tragic and unfair way and caused severe problems and burden without having any scientific support at all [38–40]. These mothers of patients with schizophrenia were literally accused by psychoanalysts of being the direct cause of the mental illness of their children and were indeed oppressed. But instead, several prominent civil rights activists engaged in movements for the legalization of pedophilia and the decriminalization of child abuse, especially in France. Pedophilia is included among the DSM-5 and ICD-10 diagnostic categories, with the arbitrary age threshold for minors set to 13 years (while in most countries, the legal age is 14–16), but in 1977 a petition calling for the decriminalization of all consenting relations between adults and minors was signed by Jean-Paul Sartre, Simone de Beauvoir, Michel Foucault, Jacques Derrida, Roland Barthes, and the child psychoanalyst Françoise Dolto (1908–1988), among others [41]. As Michael Foucault argued, the central idea was that prohibition of such relationships was a direct violation of children's civil rights and an abuse by itself [42, 43]. In this frame, the notion of child protection was considered to be a kind of patriarchal oppression and denounced as a danger to the child's personal integrity

[44]. Foucault also argued that the whole idea of "consent" was a trap because nobody really gives his/her consensus until the act is performed, and thus the act itself is a measure of consent [42, 43]. These arguments were hosted by prestigious newspapers and journals [45, 46] and generally were setting the threshold age at 12 [45], but in a 1979 letter to the newspaper *Libération* in support of Gérard R., an accused child sex criminal awaiting his trial, they defended sexual encounters with minors as young as 6 years old [46]. Opponents of this movement were often accused as "reactionary," and pedophiles were considered as victims.

The first "psychiatric survivor" groups had emerged by this time, often united by their opposition to ECT. Since then, many such groups emerged [47, 48]. The big blow to ECT was given in 1975 by the film *One Flew Over the Cuckoo's Nest*, which was based on the novel by Ken Kesey. The film was produced by Michael Douglas and was directed by Milos Forman, with Jack Nicholson playing the leading role. It won all five main Oscars at that year and embedded a negative image of ECT and psychiatry in the attitude of the general public (Fig. 11.34). In the USA, legislatures drafted motions to ban the use of ECT. Around the world it started becoming effectively impossible to use ECT. Many books critical of ECT were published, and Thomas Szasz commented on Leonard Roy Frank's (1932–2015) "History of Shock Treatment" [49] "what the rock and the stake were to the inquisition, what the concentration camp and the gas chamber were to national socialism, the mental hospital and electroshock are to institutional Psychiatry." Interestingly, while this opposition made it difficult for mental patients to receive ECT in public settings, it led to the emergence of several private and very expensive settings exclusively dedicated to the application of ECT. A review of the field [50] reported that out of 22 Hollywood films featuring ECT, 18 depicted negative outcomes, including zombification and death [51], while all of the 13 films in the Hindu cinema depicted it being given by force in order to punish, to change a person's identity, or to induce insanity [52].

The "social revolution" within psychiatry went even further, and even some large and reputable hospitals started holding therapeutic meetings that often included not only patients and mental health professionals but also everyone whose work got him/her a direct or indirect contact with the ward, e.g., administration staff, gardeners, and the rest of hospital staff as well as the patients. Many participants in these meetings were against all forms of coercion. It is of course reasonable to expect that in such meetings, lay opinion and attitude and contra-rationalism would prevail over scientific and professional knowledge. The prevailing attitude often was that any kind of restriction of patient is unacceptable, and basing on experimental data was correlated with experimenting in concentration camps.

In the late 1980s, Marius Romme (1934–) in the Netherlands established one of the most enduring and hostile-to-all-orthodox therapies group under the name "Hearing Voices Self-Help Movement" [53, 54].

Another pole hostile to psychiatry is the Church of Scientology (Fig. 23.2), which in alliance with Thomas Szasz in 1969 founded a specific organization, the Citizens Commission on Human Rights (CCHR), entirely devoted to the anti-psychiatry movement as well as a "museum" in Los Angeles to show the "evils" of

Fig. 23.2 Left: The Scientology symbol, composed of the letter S, which stands for Scientology, and the ARC and KRC triangles, two important concepts in Scientology (detail from the front of the Flag Building, Clearwater, Florida). Center: A Scientologist introduces the E-meter to a potential student. Right: A Scientology demonstration

Fig. 23.3 Ron Hubbard (1911–1986) the founder of the Church of Scientology

psychiatry and psychology [55–58]. It was founded by the science fiction author L. Ron Hubbard (1911–1986; Fig. 23.3) and is strongly and openly supported by a number of public figures, including several famous Hollywood actors. As a religion, Scientology preaches that humans are immortal beings who have forgotten their true nature and past lives. Thus, it recommends that they should undergo a process through which they will re-experience painful or traumatic events from their past, so as to free themselves. In this frame it rejects psychiatric diagnosis as well as treatment, especially with medication.

Anti-psychiatrists and other critics of psychiatry systematically accuse psychiatry for entering an evil collaboration with the pharma industry against the interests of the patients [59–62]. Overall, the anti-psychiatry message is that psychiatric treatments are damaging and dangerous rather than helpful, and there is no such thing as mental disorder and mental patients [48, 63, 64]. It frequently holds aggressive demonstrations outside major psychiatric meetings and congresses, although the intensity has faded away the last few years (Fig. 23.4). On the other hand, anti-psychiatry undoubtedly led to much greater sensitivity to the needs and rights of the patient and led to the refinement of the values of psychiatry.

While it seems that anti-psychiatry is currently dead and its arguments are obsolete [65, 66], the essence of its concepts and concerns tend to survive through various new movements like "critical psychiatry" and "post-psychiatry" [67]. But mainly they survive through the inclusion of many anti-psychiatric features in contemporary mainstream everyday clinical practice as well as mental health policy [20]. Essentially today there is no need for an anti-psychiatric movement; there are many mainstream psychiatrists both academics as well as clinicians who believe that there is no mental illness the way classification systems or standard clinical psychiatry suggest. These opinions are incorporated into the wider mainstream psychiatry and are often taught as part of official curricula. Many psychoanalysts are in fact following a social behaviorism approach with an existential flavor. The key common denominator is the opposition to the reduction in the value load for mental life with the advancement in psychiatry and the neuroscience [68–73], which are feared by some authors to lead to abusively dehumanizing consequences [1].

The greatest problem with anti-psychiatry is that it spread a wide bitterness and conflict among mental health professionals and the general public while also strongly accusing the relatives of the patients as directly responsible for the patients' mental condition, among other things, with the introduction of the concept of schizophrenogenic mother [38, 40]. It had a profound effect on public mental health because it encouraged the utilization of "alternative" treatments of all kinds, of minimal or zero efficacy and often with high cost.

Fig. 23.4 Citizens Commission on Human Rights (part of Scientology) demonstration outside the World Psychiatric Association congress in Lisbon on August 21–24, 2019. An interesting element is the use of the cross not only on the coffin but also as a central element of the scene by a non-Christian Church (it is recognized as a separate religion by the Portuguese government since 2007)

References

1. Fulford KWM, Van Staden CW. Values-based practice: Topsy-Turvy take-home messages from ordinary language philosophy (and a few next steps). In: Fulford KWM, Davies M, Gipps R, et al., editors. The Oxford handbook of philosophy and psychiatry. Oxford University Press; 2013. https://doi.org/10.1093/oxfordhb/9780199579563.001.0001.
2. Szasz T. The myth of mental illness: foundations of a theory of personal conduct. New York, NY: Hober-Harper; 1961.
3. Bortolotti L. Rationality and sanity: the role of rationality judgements in understanding psychiatric disorders. In: Fulford KWM, Davies M, Gipps R, et al., editors. The Oxford handbook of philosophy and psychiatry. Oxford University Press; 2013. https://doi.org/10.1093/oxfordhb/9780199579563.001.0001.
4. Fulford KWM, Thornton T, Graham G. Natural classification, realism and psychiatric science. In: Fulford KWM, Thornton T, Graham G, editors. The Oxford textbook of philosophy and psychiatry. Oxford University Press; 2006.
5. Healy D, Savage M, Michael P, Harris M, Hirst D, Carter M, Cattell D, McMonagle T, Sohler N, Susser E. Psychiatric bed utilization: 1896 and 1996 compared. Psychol Med. 2001;31(5):779–90. https://doi.org/10.1017/s0033291701003968.
6. Rosenhan DL. On being sane in insane places. Science. 1973;179(4070):250–8. https://doi.org/10.1126/science.179.4070.250.
7. Szasz TS. The myth of mental illness. New York, NY: Harper and Row; 1974.
8. Edwards RB. Mental health as rational autonomy. J Med Philos. 1981;6(3):309–22. https://doi.org/10.1093/jmp/6.3.309.
9. Foucault M. Histoire de la folie à l'âge classique. Paris: Plon; 1961.
10. Goffman E. Asylums: essays on the social situation of mental patients and other inmates. New York, NY: Anchor Books; 1961.
11. Leoni F. From madness to mental illness: psychiatry and biopolitics in Michel Faoucault. In: Fulford KWM, Davies M, Gipps R, et al., editors. The Oxford handbook of philosophy and psychiatry. Oxford University Press; 2013. https://doi.org/10.1093/oxfordhb/9780199579563.001.0001.
12. Foucault M. Madness and civilization: a history of insanity in the age of reason. Vintage; 1988.
13. Marcuse H. One-dimensional man. Boston: Beacon Press; 1964.
14. Cooper D. Psychiatry and anti-psychiatry. London: Tavistock; 1967.
15. Hare E. Was insanity on the increase? The fifty-sixth Maudsley lecture. Br J Psychiatry. 1983;142:439–55. https://doi.org/10.1192/bjp.142.5.439.
16. Turner TH. Schizophrenia as a permanent problem. Some aspects of historical evidence in the recency (new disease) hypothesis. Hist Psychiatry. 1992;3(12):413–29. https://doi.org/10.1177/0957154X9200301202.
17. Scheff T. Being mentally ill: a sociological theory. London: Weidenfield & Nicolson; 1966.
18. Scheff T. Labeling madness. New York, NY: Prentice-Hall; 1975.
19. Wall O. The birth and death of villa 21. Hist Psychiatry. 2013;24(3):326–40. https://doi.org/10.1177/0957154X13483049.
20. Fulford KWM, Thornton T, Graham G. Philosophical problems in mental health care and research. In: Fulford KWM, Thornton T, Graham G, editors. The Oxford textbook of philosophy and psychiatry. Oxford University Press; 2006.
21. Eysenck HJ. Classifi cation and the problems of diagnosis. In: Eysenck HJ, editor. Handbook of abnormal psychology. London: Pitman Medical Publishing Company Ltd.; 1960.
22. Scheff TJ. The labelling theory of mental illness. Am Sociol Rev. 1974;39(3):444–52.
23. Sedgwick P. Illness--mental and otherwise. Stud Hastings Cent. 1973;1(3):19–40.
24. Allott P, Loganathan L, Fulford KWM. Discovering hope for recovery. Can J Commun Ment Health. 2002;21(2):13–33. https://doi.org/10.7870/cjcmh-2002-0014.
25. Foucault M, Howard R, Cooper D. Madness and civilization. Taylor & Francis. 1961; https://doi.org/10.4324/9780203278796.

26. Campbell EJ, Scadding JG, Roberts RS. The concept of disease. Br Med J. 1979;2(6193):757–62. https://doi.org/10.1136/bmj.2.6193.757.
27. Laing R. The politics of experience. Harmondsworth: Penguin Books; 1967.
28. Laing RD. The divided self. London: Tavistock; 1960.
29. Rashed MA. Madness and the demand for recognition. Oxford University Press; 2019. https://doi.org/10.1093/med/9780198786863.001.0001.
30. Kendell RE. The concept of disease and its implications for psychiatry. Br J Psychiatry. 1975;127:305–15. https://doi.org/10.1192/bjp.127.4.305.
31. Laing RD. Th e divided self. Harmondsworth: Penguin; 1965.
32. Laing RD, Esterson A. Sanity, madness and the family. Harmondsworth: Penguin; 1970.
33. Barnes M, Berke J. Mary Barnes: two accounts of a journey through madness. Other Press; 2002.
34. Fanon F. The wretched of the earth. Harmondsworth: Penguin Books; 1967.
35. Kesey K. One flew over the Cuckoo's nest. London: Methuen; 1962.
36. Bateson G, Jackson DD, Haley J, Weakland J. Toward a theory of schizophrenia. Behav Sci. 2007;1(4):251–64. https://doi.org/10.1002/bs.3830010402.
37. Fromm-Reichmann F. Notes on the development of treatment of schizophrenics by psychoanalytic psychotherapy. Psychiatry. 1948;11(3):263–73. https://doi.org/10.1080/00332747.1948.11022688.
38. Harrington A. The fall of the schizophrenogenic mother. Lancet. 2012;379(9823):1292–3. https://doi.org/10.1016/s0140-6736(12)60546-7.
39. Morris JA. The fall of the schizophrenogenic mother. Lancet. 2012;380(9837):110. https://doi.org/10.1016/S0140-6736(12)61180-5.
40. Neill J. Whatever became of the schizophrenogenic mother? Am J Psychother. 1990;44(4):499–505. https://doi.org/10.1176/appi.psychotherapy.1990.44.4.499.
41. Mottier V. Sexuality: a very short introduction. Oxford University Press; 2008.
42. Foucault M. Sexuality morality and the law. New York, NY: Routledge; 1988.
43. Foucault M, Hocquenghem G, Danet J. Semiotext(e) special intervention series 2: loving boys/loving children semiotext(e) magazine. New York, NY; 1980.
44. Déchaux J-H. Le sacre de l'enfant. Regards sur une passion contemporaine (Consecrating the Child). Rev Fr Sociol. 2014;55(3):537–61. https://doi.org/10.3917/rfs.553.0537.
45. Aragon L, Ponge F, Barthes R, de Beauvoir S, Belladona J, Bon M, Sartre J, et al. Letter. Le Monde; 26 Jan 1977.
46. Elles P, Bruckner P, Moustaki G, et al. Letter. Liberasion; 1979.
47. The Antipsychiatry Coalition. http://www.antipsychiatry.org/. 2018. Accessed 29 Apr 2018
48. Burns T. Psychiatry: a very short introduction. Oxford University Press; 2006.
49. Frank L. The history of shock treatment. San Francisco; 1978.
50. Maughan D, Molodynski A. An international perspective on the acceptability and sustainability of electroconvulsive therapy. BJ Psych Int. 2016;13(1):10–2. https://doi.org/10.1192/s2056474000000891.
51. McDonald A, Walter G. Hollywood and ECT. Int Rev Psychiatry. 2009;21(3):200–6. https://doi.org/10.1080/09540260902747888.
52. Andrade C, Shah N, Venkatesh BK. The depiction of electroconvulsive therapy in Hindi cinema. J ECT. 2010;26(1):16–22. https://doi.org/10.1097/YCT.0b013e3181d017ba.
53. Leucht S, Pitschel-Walz G, Abraham D, Kissling W. Efficacy and extrapyramidal side-effects of the new antipsychotics olanzapine, quetiapine, risperidone, and sertindole compared to conventional antipsychotics and placebo. A meta-analysis of randomized controlled trials. Schizophr Res. 1999;35(1):51–68. https://doi.org/10.1016/s0920-9964(98)00105-4.
54. Romme MA, Escher AD. Hearing voices. Schizophr Bull. 1989;15(2):209–16. https://doi.org/10.1093/schbul/15.2.209.
55. Coco G Scientologists really, really hate psychiatrists. Vice. 2013. https://www.vice.com/en_us/article/nn4mx7/scientologists-really-really-hate-psychiatrists. Accessed 29 Apr 2018.
56. Kent SA, Manca TA. A war over mental health professionalism: scientology versus psychiatry. Ment Health Relig Cult. 2014;17(1):1–23. https://doi.org/10.1080/13674676.2012.737552.

57. Miscavige D. Citizens commission on human rights expands its activities to expose and handle psychiatric abuse in clearwater. 2015. Tampa Bay via New Center. https://www.scientology.org/scientology-today/clearwater-citizens-commission-on-human-rights-center-opening.html. Accessed 29 Apr 2018.
58. Westbrook DA. The enemy of my enemy is my friend. Nova Religio J Alternat Emerg Relig. 2017;20(4):37–61. https://doi.org/10.1525/nr.2017.20.4.37.
59. Ban TA. Academic psychiatry and the pharmaceutical industry. Prog Neuro-Psychopharmacol Biol Psychiatry. 2006;30(3):429–41. https://doi.org/10.1016/j.pnpbp.2005.11.014.
60. Swenson K. Selling sickness: how the world's biggest pharmaceutical companies are turning us all into patients. Can J Commun. 2009;34(2). https://doi.org/10.22230/cjc.2009v34n2a2231.
61. Whitaker R. Anatomy of an epidemic: magic bullets, psychiatric drugs, and the astonishing rise of mental illness in America. New York, NY: Random House; 2010.
62. Wright D. The medicalization of society: on the transformation of human conditions into treatable disorders. By Peter Conrad. (Baltimore: Johns Hopkins University Press, 2007. xiv + 204 p., notes, bibl., index. ISBN 0-8018-8585-X $22). Scientia Canadensis: Can J Hist Sci Technol Med. 2010;33(1):119. https://doi.org/10.7202/1000847ar.
63. O'Brien AP, Woods M, Palmer C. The emancipation of nursing practice: applying anti-psychiatry to the therapeutic community. Aust N Z J Ment Health Nurs. 2001;10(1):3–9. https://doi.org/10.1046/j.1440-0979.2001.00183.x.
64. Weitz D. Call me antipsychiatry activist--not "consumer". Ethical Hum Sci Serv. 2003;5(1):71–2.
65. Hoff P, Fulford BK, Sadler JZ. History and philosophy. Curr Opin Psychiatry. 2009;22(6):552–3. https://doi.org/10.1097/YCO.0b013e3283326891.
66. Hoff P, Fulford BK, Sadler JZ. History, philosophy and practical psychiatry: neither gap nor contradiction, but mutual enrichment. Curr Opin Psychiatry. 2012;25(6):485. https://doi.org/10.1097/YCO.0b013e3283590bfd.
67. Bracken P, Thomas P. Challenges to the modernist identity of psychiatry! User empowerment and recovery. In: Fulford KWM, editor. The Oxford handbook of philosophy and psychiatry. Oxford University Press; 2013. p. 123.
68. Boorse C. On the distinction between disease and illness. Philos Public Aff. 1975;5(1):49–68.
69. Boorse C. What a theory of mental health should be. J Theory Soc Behav. 1976;6(1):61–84. https://doi.org/10.1111/j.1468-5914.1976.tb00359.x.
70. Boorse C. A second rebuttal on health. J Med Philos. 2014;39(6):683–724. https://doi.org/10.1093/jmp/jhu035.
71. Hare RM. Descriptivism. Proc Br Acad. 1963;49:115–34.
72. Wakefield JC. The concept of mental disorder. On the boundary between biological facts and social values. Am Psychol. 1992;47(3):373–88. https://doi.org/10.1037//0003-066x.47.3.373.
73. Wakefield JC. The DSM's theory-neutral nosology is scientifically progressive: response to Follette and Houts (1996). J Consult Clin Psychol. 1998;66(5):846–52. https://doi.org/10.1037//0022-006x.66.5.846.

Chapter 24
Post-psychiatry

While modern psychiatry follows the rules of science and the pathways of medicine, the environment in the field is shaped also by patient advocacy organizations that have expanded with the opening of societies to accept differentiality.

Historically, mental patients started to be organized and collectively pursue their goals since the seventeenth century [1], but strong and influential organizations appeared only in the recent decades [2]. Carers have also created groups, and a carers' rights movement was also developed. These groups work to challenge the stereotypical thinking, fight stigma, and eliminate exclusion associated with mental conditions and work with national governments, the World Health Organization, and other national and international bodies. A movement for global mental health has emerged [3]. In principle, academic psychiatry and psychiatric organizations and associations are eager to consult with such patient and carers' organizations, provided that they are willing to accept the medical framing of their experiences, although this is not mandatory and is not always the case.

This movement is split to some groups aligning themselves with mainstream psychiatry and collaborating on research projects and campaigns to improve services and to end the stigma, whereas others reject the medical/scientific model altogether for the sake of a more subjective and personalized understanding of mental health.

As already discussed, many authors have criticized not only the specific criteria and method for psychiatric diagnosis but the very utility of diagnosis itself. Alternatives to diagnosis have been proposed on the basis of a broader description of the patients' complains and history [4]. In the frame of a post-modern approach, the scientific way to reach a specific diagnosis is considered to be mistaken and misleading [5], and this is in accordance with the prevailing attitude in everyday clinical practice, which is resistant to the evidence-based approach and appreciates the opinion of colleagues rather than the scientific method to a larger extend than this happens in other specialties [6]. A more recent evolution in the anti-psychiatry movement concerns a revisiting of the separation between biological psychiatry and

K. N. Fountoulakis, *Psychiatry*, https://doi.org/10.1007/978-3-030-86541-2_24

the psychological component of psychiatry. Peter Roger Breggin (1936–), in his book *Toxic Psychiatry* (1993), argues against neurobiology and is in favor of psychosocial interventions in a humane context [7].

This is not new, however; instead, it has a long tradition. Since antiquity there has been a long discussion on whether mental events follow the rules of nature. The belief that mental life is not subject neither to natural laws nor to the scientific method was systematically developed and shaped during the nineteenth and twentieth centuries and reached its glory with Freudian and existentialist theories. Although there are fundamental differences between these two approaches, and while psychoanalysis started as an effort to bring science into mental health, they both suggest that "understanding" (in German *verstehen*) of the content through insightful and empathic involvement (in German *einfuhlung*, meaning "into feeling") is essential, while describing the mechanism is insignificant.

After Immanuel Kant (1724–1804; Fig. 7.5), it was Johann Gustav Droysen (1808–1884; Fig. 24.1) who suggested for the first time that *verstehen* is the prime method for the humanities. Wilhelm Dilthey (1833–1911; Fig. 24.2) expanded this approach to include volition, morality, etc., and suggested that the a priori Kantian

Fig. 24.1 Johann Gustav Droysen (1808–1884)

Fig. 24.2 Wilhelm Dilthey (1833–1911)

principles through which humans shape the world change with time and are reshaped in the course of the history of human society. Dilthey was a strong advocator of the different "quality" of human sciences and distinguished between the need for "understanding" in human sciences and "explaining" in the sciences of nature. Heinrich Rickert (1863–1936;s Fig. 24.3) suggested that the meta-experience (*nacherleben*) of the others is mediated by the understanding of their cultural values. According to him, this understanding does not concern real objects or events but pure mental constructs. In scientific terms, however, there is very low reliability and subsequently low validity in this kind of approaches.

Karl Jaspers (1883–1969; Fig. 10.17) and his pupil Kurt Schneider (1887–1967; Fig. 10.31), although existentialists themselves, kept distance from these approaches, developed a pragmatic method for approaching psychopathology and arriving at a diagnosis, and made a distinction concerning the diagnosis of mental disorders (*dass-sein* or *dasein*) and their content (*so-sein*) [8, 9]. However this was not the case with most of French post-modernist scholars who eventually put equal emphasis and attributed similar value to science and to trivial banal experienced by lay people every day [10], and through the popularity of their ideas, they frequently led to disastrous results for public health [11–14].

Fig. 24.3 Heinrich Rickert (1863–1936)

One of the reasons for these ideas was the fact that there are no biological laboratory findings with diagnostic or predictive value, and as a result, the independent verification of the diagnosis is not possible. A core characteristic of mental disorders is that unlike many medical conditions, mental disorders are never asymptomatic. This, in turn, has led to the adoption of phenomenological ideas [15]. However, this embrace with philosophical movements brought psychopathology close to existential theories and the anti-psychiatric approach. This is in sharp contrast to the medical approach. While existential and anti-psychiatric ideas consider mental disorder to be a matter of choice and maybe the ultimate act of freedom against an oppressing reality, the medical model considers mental patients to be the blameless victims of misfortune, similar to other patients suffering from somatic disorders and disabilities and in need for a medically defined way of treatment.

The core characteristic of scientific innovations and the resulting applied technology is that they are both instrumental, that is, they have nothing to do with opinions, values, relationships, or priorities [16]. Accordingly, within the medical model, mental illness is not a spiritual or moral issue [17]. This is something many scholars dislike because of its moral neutrality. Thus, idealism is chosen as their philosophical approach instead in accordance with ideas originally phrased by Plato more than 2000 years ago.

While the medico-scientific model does not ignore non-technical issues and is interested in registering narrations pertaining to relationships, meanings, or values [18], it downgrades them as being of secondary importance only. In this frame, the patient organizations are expected to be of help in the management of practical everyday issues from fund-raising to awareness promotion, but they have no essential role in the determination of the core concepts and elements of science behind psychiatry and related treatments themselves. This limitation in the role of these organizations is targeted by many scholars who considered it to be authoritarian [19, 20]. These scholars challenge the modernist approach, which was brought by the European Enlightenment and caused a shift from religious revelation to human reason as the path to the solution of every human problem [16]. The scientific modernist revolution of the nineteenth century claimed that all human problems, including personal or societal, could be framed, analyzed, and solved in a scientific way [17, 21]. However progress itself as a concept has been disputed, and the ability of humans to adjust to the new technocratic environment has been questioned [22] with influences from existentialism but also with strong elements of naturalistic fallacy.

According to these scholars, the problem lies in the stance the Enlightenment took toward mental illness because it labeled irrational behavior and considered it deeply problematic [23, 24]. It is interesting how strongly appealing these interpretations are, and how widely spread and accepted they are in spite of the fact that the historical data are against such interpretations. Asylums and stigma existed for centuries before the Enlightenment; concerns and measures about problematic behavior can be found as early as the ancient Roman laws. Foucault tried to synthesize history, philosophy, epistemology of the human sciences, and genealogy of the main social, political, and economic institutions of modernity [25], but the end result was rather chaotic and based on an idiosyncratic interpretation of the evidence. He considered mental illness and psychiatry to be the product of conflicts among different societal forces, with medicine being eventually triumphant. Society utilized this in order to organize suppression, making psychiatry part of the disciplinary system [25, 26].

Additionally, today there are many people who have experienced the modern ways of psychiatry and are unsatisfied. It has been shown that there is a discrepancy in the views and the priorities between patients and therapists as well as in the hierarchy of issues and goals [27, 28]. It seems that while the priorities of the therapists conform to a medical and disease-oriented model with control of the symptoms and restoration of functioning as the top priorities, patients consider acceptance, shared experience and shared identity, emotional support, finding meaning and purpose in life, having control, choices, security and safety, and similar personal issues as more important [29]. The unsatisfied patients came to be self-determined as "survivors" and consider psychiatry and mental health services to be toxic and evil [19]. Groups of "survivors" were established like "the Icarus Project" or the Hearing Voices Network (HVN), and they consider recovery to be a personal journey [30, 31]. Events like the "Mad Pride," which was modeled on the Gay Pride, are now a regular feature in many countries across the world, especially outside psychiatric congresses (Fig. 24.4) [32, 33].

All the above together set an environment within which psychiatry is called to abandon its scientific and medical orientation and adopt an approach based on a holist individualized assessment of the person and the interpretation and understanding of his condition [34]. No matter how appealing this is, it is in sharp contrast with the accumulated evidence during the recent decades of scientific research. The main issue for many scholars is the power balance and how much scientific knowledge empowers the psychiatrist in comparison with and against the patient [24, 35]. This is true, knowledge of all kinds is power, and those in need are at a disadvantageous position in every way one can conceive it. Power and authority abuse should be punishable, and conduct should always be monitored; however, this does not mean that the core of science is subject to idiosyncratic or voluntaristic interpretations.

With anti-psychiatry getting out of fashion and its main arguments collapsing, today most movements criticizing psychiatry talk about a "post-psychiatric" approach [19, 20, 35–37], and essentially they suggest being open-minded to a variety of perspectives, usually by rephrasing old anti-psychiatric ideas in a new more modern frame, for example, "philosophy of psychotherapy," which rejects scientific validity for the sake of "usefulness" and perceived "utility" [38, 39]. But in reality, "open-minded" means rejecting the scientific method that poses rules and methods and tries to clarify which perspective is valid and which is not.

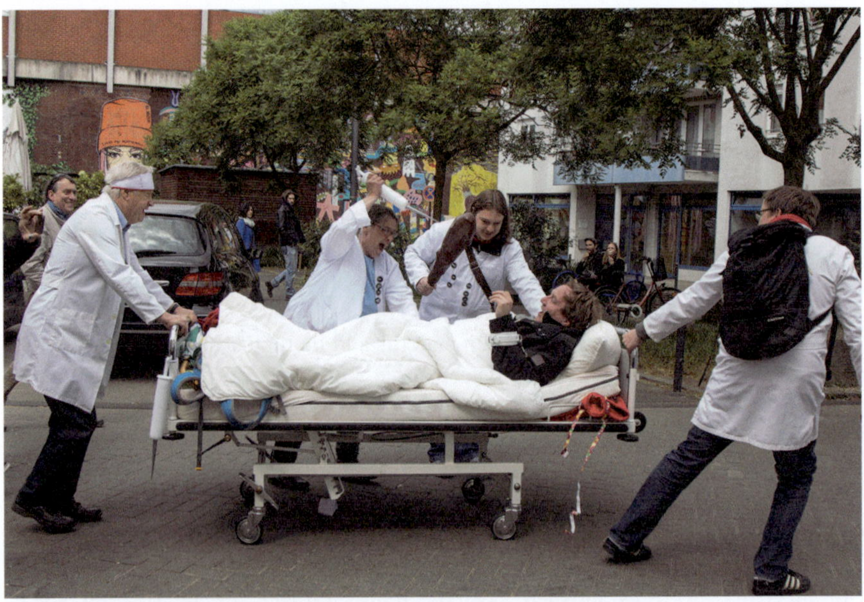

Fig. 24.4 Bed Push at Mad Pride Parade in Cologne, May 16th, 2016. © Superbass/CC-BY-SA-4.0

Additionally these movements push for a central position of service users in the decision process and the recognition of ideological implications of psychiatric practice [40].

This is a wider phenomenon in medicine today, that particularly in chronic illness, the patient himself/herself is the "expert" in his/her own symptoms and illness experience [41]. At the extreme of this approach, medicine as a whole is regarded as evil, and some scholars argue that modern medicine, through its technology, dominates our lives and actually undermines our own capacity to self-heal, and in this way, it eventually destroys our health [42, 43]. In this frame, the placebo effect is misinterpreted and is attributed to metaphysical powers as if it gives patients the ability to unconsciously muster up his/her own ability to self-heal [19, 44].

Of course this is in direct conflict with the "evidence-based approach," which is also pushed forward today [45]. While many authors claim that this concept of "post-psychiatry" is not anti-medical or anti-psychiatry, it merely challenges the dominant modernist, technological version of psychiatry, as well as an imbalanced distribution of authority [23]. Moreover, they claim that in this way, the service users are no longer silenced [19], and the resulting theories and practice are perfectly in accordance with the old anti-psychiatric arguments and concepts.

On the other hand, it seems that in everyday clinical practice, the clinician should probably follow a balanced approach incorporating values into his/her technocratic skills and practice [46]; this is necessary because these movements at least partially reflect the fact that a lot of patients have an unsatisfactory experience from mental health care and often from their contact with mental health professionals both in terms of time allocated but often also in terms of attitude and behavior.

References

1. Campbell P. The service user/survivor movement. In: Reynolds J, Muston R, Heller T, Leach J, McCormick M, Wallcraft J, editors. Mental health still matters. Milton Keynes: Open University/Palgrave Macmillan; 2009. p. 46–52.
2. Wallcraft J, Read J, Sweeney A. On our own terms: users and survivors of mental health services working together for support and change. London: Sainsbury Centre for Mental Health; 2003.
3. Patel V, Prince M. Global mental health: a new global health field comes of age. JAMA. 2010;303(19):1976–7. https://doi.org/10.1001/jama.2010.616.
4. Boyle M. Diagnosis. In: Newnes C, Dunn C, Holmes G, editors. This is madness: a critical look at psychiatry and the future of mental health services. Herefordshire: PCCS Books; 1999. p. 75–90.
5. Roberts GA. Narrative and severe mental illness: what place do stories have in an evidence-based world? Adv Psychiatr Treat. 2018;6(6):432–41. https://doi.org/10.1192/apt.6.6.432.
6. Rees H, Sipos A, Spence M, Harrison G. Attitudes of psychiatrists to evidence-based guidelines: a questionnaire survey. Psychiatr Bull. 2002;26:421–4.
7. Breggin P. Toxic psychiatry: drugs and electroconvulsive therapy: the truth and the better alternatives. London: HarperCollins; 1993.
8. Jaspers K. General psychopathology (trans: Hoenig J, Hamilton M). Chicago: The Johns Hopkins University Press; 1997.

9. Schneider K. Klinische Psychopathologie. [Cinical psychopathology.]. Oxford: Georg Thieme; 1950.
10. Kaparounaki CK, Koraka CA, Kotsi ES, Ntziovara AP, Kyriakidis GC, Fountoulakis KN. Greek university student's attitudes and beliefs concerning mental illness and its treatment. Int J Soc Psychiatry. 2019;65(6):515–26. https://doi.org/10.1177/0020764019864122.
11. Dawkins R. Postmodernism disrobed. Nature. 1998;394(6689):141–3. https://doi.org/10.1038/28089.
12. Kuntz M. The postmodern assault on science. If all truths are equal, who cares what science has to say? EMBO Rep. 2012;13(10):885–9. https://doi.org/10.1038/embor.2012.130.
13. Sokal A. Intellectual impostures: postmodern philosophers' abuse of science. London: Profile; 1999.
14. Wolin R. The seduction of unreason: the intellectual romance with fascism: from Nietzsche to postmodernism. Princeton: Princeton University Press; 2019.
15. Rashed M. Madness and the demand for recognition. A philosophical inquiry into identity and mental health activism. Oxford University Press; 2019. https://doi.org/10.1093/med/9780198786863.001.0001.
16. Ellul J. The technological society. New York, NY: Vintage Books; 1964.
17. Phillips J. Introduction. In: Phillips J, editor. Philosophical perspectives on technology and psychiatry. Oxford: Oxford University Press; 2009. p. 1–19.
18. Fulford KWM, Thornton T, Graham G. From bioethics to values-based practice. In: Fulford KWM, Thornton T, Graham G, editors. The Oxford textbook of philosophy and psychiatry. Oxford University Press; 2006.
19. Bracken P, Thomas P. Challenges to the modernist identity of psychiatry: user empowerment and recovery. In: Fulford KWM, Davies M, Gipps R, et al., editors. The Oxford handbook of philosophy and psychiatry. Oxford University Press; 2013. https://doi.org/10.1093/oxfordhb/9780199579563.001.0001.
20. Cohen CI, Timimi S, editors. Liberatory psychiatry: philosophy, politics and mental health. Cambridge: Cambridge University Press; 2008. https://doi.org/10.1017/CBO9780511543678.
21. Lewis B. Moving beyond prozac, DSM, and the new psychiatry. Ann Arbor, MI: University of Michigan Press; 2006. https://doi.org/10.3998/mpub.93209.
22. Giddens A. Modernity and self-identity. Self and society in the late modern age. Cambridge: Polity Press; 1991.
23. Buck-Zerchin DS. Seventy years of coercion in psychiatric institutions, experienced and witnessed. In: Kallert T, Mezzich J, Monahan J, editors. Coercive treatment in psychiatry. Wiley; 2011. p. 235–43. https://doi.org/10.1002/9780470978573.ch14.
24. Foucault M, Howard R, Cooper D. Madness and civilization. Taylor & Francis; 1961. https://doi.org/10.4324/9780203278796.
25. Leoni F. From madness to mental illness: psychiatry and biopolitics in Michel Faoucault. In: Fulford KWM, Davies M, Gipps R, et al., editors. The Oxford handbook of philosophy and psychiatry. Oxford University Press; 2013. https://doi.org/10.1093/oxfordhb/9780199579563.001.0001.
26. Foucault M. Madness and civilization: a history of insanity in the age of reason. Vintage; 1988.
27. Faulkner A, Layzell S. Strategies for living: a report of user-led research into people's strategies for living with mental distress. London: Mental Health Foundation; 2000. https://doi.org/10.7748/mhp.1.6.7.s13.
28. Rogers A, Pilgrim D, Lacey R. Experiencing psychiatry: users' views of services. Basingstoke: Macmillan Press; 1993.
29. Hornstein G. Narratives of madness, as told from within. Chronicle of Higher Education; 2002. p. B7–10.
30. Romme M, Escher S. Accepting voices. London: MIND Publications; 1993.
31. Romme M, Escher S. INTERVOICE: accepting and making sense of voices. In: Stastny P, Lehmann P, editors. Alternatives beyond psychiatry. Berlin: Peter Lehmann Publishing; 2007. p. 131–6.

32. Mitchell-Brody M. The Icarus project: dangerous gift s, iridescent visions and mad community. In: Stastny P, Lehmann P, editors. Alternatives beyond psychiatry. Berlin: Peter Lehmann Publishing; 2007. p. 137–45.
33. Stastny P, Lehmann P. Alternatives beyond psychiatry. Berlin: Peter Lehmann Publishing; 2007.
34. Bracken P, Thomas P. Postpsychiatry: a new direction for mental health. BMJ. 2001;322(7288):724–7. https://doi.org/10.1136/bmj.322.7288.724.
35. Bracken P, Thomas P. Postpsychiatry. Oxford University Press; 2005. https://doi.org/10.1093/med/9780198526094.001.0001.
36. Campbell P. The history of the user movement in the United Kingdom. In: Heller J, Reynolds R, editors. Mental health matters: a reader. Basingstoke: Macmillan; 1996.
37. Hoff P, Fulford BK, Sadler JZ. History, philosophy and practical psychiatry: neither gap nor contradiction, but mutual enrichment. Curr Opin Psychiatry. 2012;25(6):485. https://doi.org/10.1097/YCO.0b013e3283590bfd.
38. Erwin E. Philosophy and psychotherapy (perspectives on psychotherapy series). Sage; 1997.
39. Gergen K. An invitation to social construction. 3rd ed. Sage; 2015.
40. Double DB. Integrating critical psychiatry into psychiatric training. In: Newnes C, Holmes G, Dunn C, editors. This is madness too. Ross-on-Wye: PCCS Books; 2001.
41. Fountoulakis KN, Souliotis K. The right to choose treatment-without-treatment: respecting civil rights or an unprecedented manifestation of 'reverse stigma'? Acta Neuropsychiatr. 2019;31(1):56–8. https://doi.org/10.1017/neu.2018.24.
42. Illich I. The industrialization of medicine. Ciba Found Symp. 1975;32:157–69. https://doi.org/10.1002/9780470720141.ch11.
43. Illich I. Limits to medicine. Medical nemesis: the expropriation of health. Harmondsworth: Penguin; 1976.
44. Moerman DE. Meaning, medicine, and the "placebo effect". Cambridge: Cambridge University Press; 2002. https://doi.org/10.1017/cbo9780511810855.
45. Shaw J, Baker M. "Expert patient"—dream or nightmare? BMJ. 2004;328(7442):723.
46. Fulford KWM, Thornton T, Graham G. From bioethics to value-based practice in psychiatric diagnosis. In: Fulford KWM, Thornton T, Graham G, editors. The Oxford textbook of philosophy and psychiatry. Oxford University Press; 2006.

Chapter 25
Conspiracy Theories

I decided to write this chapter the last moment. It was September–October of 2020, and the COVID-19 pandemic was at its highest point, or that was what we were thinking at that time, so were conspiracy theories concerning the virus and the outbreak in general. Later this would focus specifically on vaccines. It was more than clear that in the years to come, one of the new topics, if not the real battlegrounds that will occupy the attention of society, will be the conspiracy theories. Psychiatry will likely be involved in this, since the main characteristics of the problem are irrationality, intense emotions including anger, and problematic behaviors that pose dangers to public health and safety, including vaccination denial, following harmful alternative medicine ways, violent acts [1, 2], and more recently failure to follow the COVID-19 lockdown and safety measures. This new "battlefront" is so new, so uncharted and difficult, and so peculiar that the side of conspiracists includes at least three Nobel laureates, editors of prominent medical journals, and a significant number of reputed scientists, while an additional significant number walks a tightrope between radical unconventional thinking and conspiracism.

25.1 Definition

The rhetoric of conspiracy can be found already in Ancient Greece and Rome [3], and some anthropologists suggest that it is a universal phenomenon, present in almost every human culture and society, regardless of the level of civilization advance or the economic–societal structure [4–7].

Although conspiracy theories seem to exist since the antiquity [8], the term started appearing in journals only since 1870 [9, 10], but the earliest documented example was an article in 1909 in the American Historical Review [10]. The term entered the lay culture and everyday language in the late 1960s as a result of the wide suspicion concerning the report of the Warren Commission on the murder of

K. N. Fountoulakis, *Psychiatry*, https://doi.org/10.1007/978-3-030-86541-2_25

President John F. Kennedy and after the New York Times published five stories using it [11].

One frequent confusion in the literature is between fake news and conspiracy theories. While the two are strongly correlated, they constitute different concepts and entities. Fake news are the deliberate distribution of news that do not correspond to real events, while conspiracy theories are a system of beliefs and interpretations of events, some of which, but not all, could be fake. The word "conspiracy" is derived from the Latin words *con-* (meaning with, together) and *spirare* (meaning "to breathe"). A core characteristic of conspiracy theories is the belief that conspiracies are the prime force that shapes history [12]. Conspiracy theories point to the existence of sinister and powerful groups, often with political motivation [13], and they involve some type of prejudice and circular reasoning since the evidence to support the theory is insufficient or even absent, and often the theory itself is in sharp contrast to logical reasoning [14]. It is impressive that both negative evidence and the absence of evidence are considered as evidence that the theory is true [14, 15], and a characteristic of these theories is that they strongly resist falsification. In a way, believing in conspiracy theories seems similar to religious or supernatural beliefs and is more a matter of faith rather than a matter of evidence [13–18].

On the other hand, it is undoubtedly true that often the distrust in the fundamental pillars of modern society, including the government and science, is proven to be justified. There are a number of "true conspiracies" that have been revealed, and others emerge every now and then. Examples include the Watergate and the Iran–Contra scandal, or the faulty intelligence on which the war in Iraq was based. Another example is the Tuskegee Syphilis Study of the 1930s–1970s. It is also true that in 1901, vaccinations in St Louis caused the death of 13 children. Throughout history there are several proven cases of unethical human experimentation and reckless medical procedures, including the thalidomide case in the late 1950s [8, 19, 20]. More recently, in July 15, 2020, Dr. Anthony Fauci, the director of the National Institute of Allergy and Infectious Diseases, and a key member of the White House coronavirus task force, in an interview with *CBS Evening News* anchor Norah O'Donnell, which was published in InStyle, confirmed that the earlier advice to the public not to wear masks was due to the shortage of masks and explained that masks do actually work [21]. These conflicting positions fueled distrust in governmental actions and advice. Whether these "true conspiracies" play any role in the adoption of conspiracy theories is a matter of debate, but still they provide some ethical justification for the distrust.

On the other hand, the example of smoking is striking; governments all over the world took decisive steps to reduce tobacco consumption as soon as research revealed the health risks associated with smoking, and the big financial interests and efforts by the powerful tobacco industry had no effect on changing the resolutions.

David Robert Grimes calculated the time a conspiracy could hold, on the basis of the number of people involved, and how this determined the exposure of the PRISM surveillance program, the Tuskegee Syphilis Study, and the FBI forensic scandal. According to his model, a moon landing hoax would require the involvement of

411,000 people and would be exposed within 3.68 years. A vaccination conspiracy would require at least 22,000 people (drug companies not included) and would be exposed in approximately 3 years [22].

25.2 Epidemiology

There is a concern that with the widespread use of the Internet and the social media, followers of conspiracy theories proliferated and gained the status of a cultural phenomenon [23–28]. There is no reliable estimation of how much conspiracy theories are accepted by the general population, but there are many reports suggesting that depending on the specific theory, more than one-third of people are followers [29].

Studies from Europe reported that depending on the particular theory, conspiracies enjoy wide acceptance. A 2017 study in France found that 48% of participants believed without evidence that political and media elites are conspiring to replace white people with immigrants [30]. In the UK, conspiracies related to COVID-19 were absolutely believed by one-tenth of the population, with an additional 15% having strong acceptance and one-quarter having some degree of acceptance. Only half of the population seemed to generally reject COVID-19-related conspiracies [31]. Data from university students from Greece suggested that almost one-third accept and one-fifth are open to conspiracy theories concerning COVID-19, while in total two-thirds do not clearly reject them. Students of theoretical studies manifested higher rates [32, 33].

Studies from the USA suggest that while blacks and Hispanics were more than three times more likely than whites to believe in AIDS origin conspiracy theories, this did not reduce the likelihood of participating in research or receiving medical treatment. One in seven African Americans was reported to believe that HIV was created by the government to control the black population. One in three said they believed that HIV was produced in a government laboratory, and more than half said there was a cure for HIV/AIDS that was being withheld from the poor [34]. Additionally it seems that the percentages of believers tend to increase, and while in 1999, 27.8% of blacks, 23.6% of Hispanics, and 8% of whites were accepting this conspiracy theory, the percentages rose to 34.1%, 21.9%, and 8.4%, respectively, in 2003 [35]. Similar acceptance was reported concerning the "chemtrail conspiracy theory" [7] with an estimated 17% believing that the theory is true or partly true [36]. Impressively, it seems that 56% believe that doctors want to vaccinate children, even though they know vaccines are harmful; 37% agree that the FDA refuses to release the cure of cancer; and 54% believe that fluoridation is a secret plot to poison people. This clearly means that in the USA, less than half of citizens definitely reject medical conspiracy theories, no matter how extreme they are [37].

Among health-care workers in Ecuador, 24.2% believed that COVID-19 was developed intentionally in a lab [38]. In Soweto, South Africa, one-third believed in a conspiracy theory about the origins of HIV [39]. A variety of conspiracy theories are reported to be highly accepted in Arab countries [40]. One-third of GPs in India believe in conspiracy theories concerning AIDS [41].

In the realm of the Internet and social media, conspiracy often dominates. This was more than obvious in the first few months of the COVID-19 outbreak when the majority (82%) of circulating news concerning the outbreak were false [42]. In Instagram, two-thirds of posts concerning water fluoridation are anti-fluoride activists [43]. In YouTube, majority of uploads have anti-HPV vaccine content, and it has more views and more likes. In these videos, a strong rhetoric concerning civil liberties and conspiracy accusations, together with half-truth and selective presentation of information, in often professional-level video production is the rule rather than the exception [44, 45].

The above paint a dark picture concerning the attitudes of the public and the dynamic of conspiracy theories. They also sound the alarm on the long-term consequences on all areas of human activity, including health, politics, and democracy itself. However, in spite of the fact that research findings from different and distant parts of the world seem to agree, one should always have in mind that there is always the possibility of bias in those research studies, and prevalence could be greatly exaggerated. More followers of conspiracy theories or vulnerable groups of the population might be more prone to participating and registering their thoughts, and also they might be more actively engaging in discussions and posting. Thus, they could represent a very active and maybe dynamic and aggressive but very small minority, which, however, could still be very influential and dangerous, especially in times of difficulty, like the COVID-19 outbreak [46].

In the modern era, the phenomenon of "superspreader" moments has been observed. Such an example is the covering of the New York Post of the interview by Roger Stone, a former advisor to President Trump. In that interview Stone said that he would not trust a vaccine funded by Gates. Following this, the specific conspiracy theory enjoyed a tremendous popularity in the social media and the Internet, since the covering by the New York Post worked as if a mainstream media had verified the story [47].

25.3 List of Conspiracy Theories

Conspiracy theories cover a wide range of topics, and in the Internet era, they tend to proliferate [48]. They could concern isolated events (e.g., the Kennedy assassination) or long-term secret mechanisms and processes, which usually involve a very demanding goal (e.g., controlling the entire world) and a relatively simple evil group of people (e.g., Freemasons). "Superconspiracy theories" include multiple conspiracies in a single unifying theory in an hierarchical way [17, 18].

The most important of them are listed below in alphabetical order and also in Table 25.1:

Table 25.1 List of most important conspiracy theories (in alphabetical order)

• 5G and electromagnetic fields
• Adrenochrome
• Aliens
• Alternative therapy suppression and the big Pharma
• Anti-mask
• Chemtrails
• Covid-19
• Deaths of prominent people
• Energy weapons
• Flat and hollow earth
• Fluoridation of water
• Global warming
• Human immunodeficiency virus/acquired immune deficiency syndrome (HIV/AIDS)
• Jews and Zionism (anti-semitism), including holocaust denial
• Moon landing denial
• Political (deep state)
• Racial
• Secret societies
• Targeted persons
• Vaccinations (anti-vaxxers)

5G and Electromagnetic Fields

Conspiracy theories on electromagnetic fields begun in the 1970s and especially when in 1979 there was a study published suggesting that children living near power lines develop cancer more often [49]. There were some later studies agreeing with these results [50, 51], including a meta-analysis which found only nine studies worthy of consideration [52]. Interestingly, that meta-analysis reported that only 3 out of 30 risk ratios calculated within studies were significant and also 2 out of 6 meta-analytical findings. These claims were rejected by subsequent studies [53–55]. At that time there was a generalized panic concerning the electromagnetic emissions of various household devices, including TV sets, microwave ovens, even hair dryers, and electric blankets.

These fears disappeared only to be replaced by the fear for the new technology of cellphones and wireless communication networks and borrowed rhetoric from the attempts of the tobacco industry to conceal the harmful effects of smoking. The public's concern about advances in technology and radiation emissions was documented since the mid-1970s [56]. Through the recent decades, skepticism came to include a significant number of prestigious people. In 2015, 190 scientists sent a letter to the UN and the WHO and, in 2017, to the European Union, warning about the risks to public health posed by 5G. Several books were published promoting this theory (Fig. 25.1), and throughout the world, including Brussels and California, the

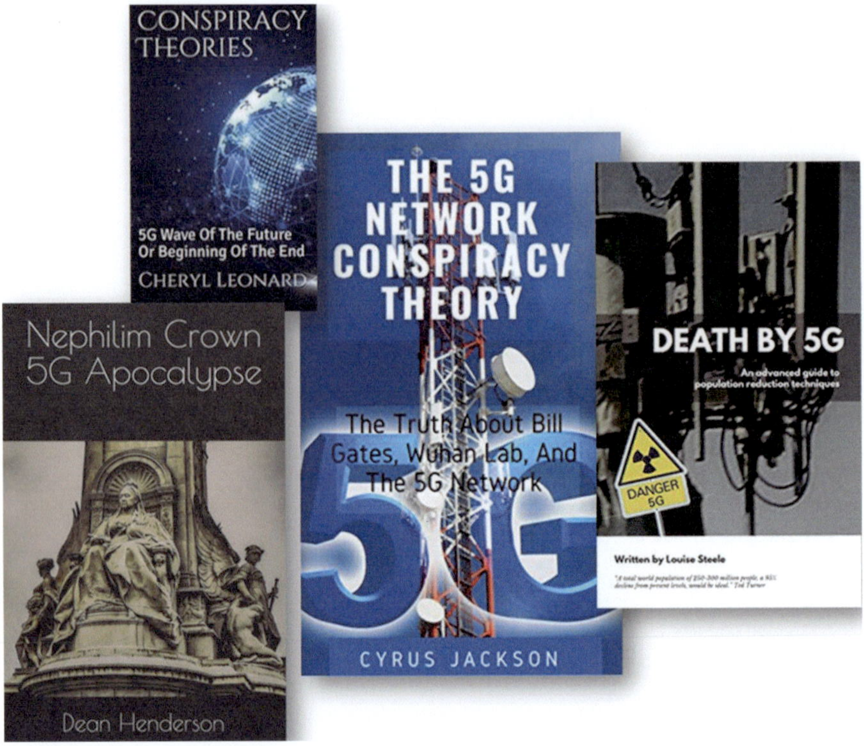

Fig. 25.1 Book covers promoting 5G conspiracy theories

administration started being hesitant in pushing forward with the 5G technology. Some activist groups have even proposed a new disease "electromagnetic hypersensitivity" or "microwave sickness" with exclusively subjective and self-reported symptoms like fatigue, dizziness, and nausea [57, 58]. The business opportunity to sell protective devices was grabbed, and a number of products were developed, including the "belly armor" for pregnant women (Fig. 25.2) or the carbon mineraloid called shungite, which is supposed to "harmonize" electromagnetic waves. The "alternative" and "harmonizing" product market has increased exponentially in the few recent years.

The point is that what we know today on the mechanisms of interaction between the human body and electromagnetic radiation does not leave much place for such assumptions, and "there are good reasons to believe that weak fields can have no significant biological effect at the cell level -and no strong reason to believe otherwise" [59]. Much of the problem lies on the psychological impact of the word "radiation," and this is probably why conspiracy theories and fear tend to move and search for their next subject. There are now more than two decades of widespread, round-the-clock use of cellphones and exposure to this kind of fields, and no cancer epidemic emerged. Probably it is now time to find a new object of fear.

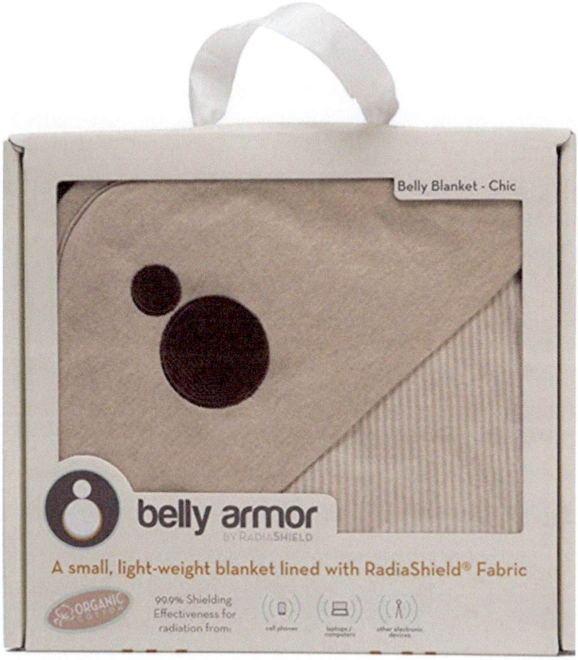

Fig. 25.2 Commercial product of "belly armor" that women can use to protect their ovaries and the fetus when working with PCs, laptops, tablets, mobile phones, or any equipment that supposedly emits radiation

More recently, a connection between the spread of COVID-19 and 5G emerged. Although physics of such a relationship are illogical, the idea gained much ground, and people started even damaging 5G towers around the world to stop the pandemic. In spite of its incompatibility with science, this theory enjoyed wide acceptance by the public [23].

Adrenochrome

Adrenochrome is a chemical compound with the molecular formula $C_9H_9NO_3$. It is the product of the oxidation of adrenaline. Research in the 1950s and 1960s suggested that it is psychotomimetic and neurotoxic. Even a theory that it was responsible for the development of schizophrenia has been developed, but soon it was proven false [60, 61].

The conspiracy theory on adrenochrome suggests that the "liberal elites" have a secret child trafficking ring and that they extract adrenochrome from the blood of these children, or from an adrenal gland in the brain (no such glands exist) after torturing them in "farms." It is related to the "PizzaGate theory" in 2016, according to which Hillary Clinton and Hollywood celebrities were running a child trafficking ring.

Aliens

A very popular conspiracy theory often depicted in films and TV shows like the *Men in Black* or the *X-Files* is that governments, especially the US government, has been contacted by an extraterrestrial life species and either conceal negotiations and ongoing relationships or cover up UFO incidents and secretly study alien artifacts in areas like Roswell and Area 51. Many authors, the first of whom was Erich von Däniken, have proposed theories of alien influence throughout human history and even that humans were the result of alien genetic engineering of earth life. The claims by Zecharia Sitchin that the Anunnaki from Sumerian mythology (Fig. 25.3) were actually a race of extraterrestrials also enjoyed much popularity.

The "reptilian" conspiracy was popularized by David Icke, who claimed that shape-shifting reptilian aliens control Earth. He even suggested that the Bush family, Margaret Thatcher and the British Royal family, are such creatures [17, 62]. It is possible that this theory came from earlier pseudohistorical legends developed during the exploration of Africa.

Alternative Therapy Suppression and the Big Pharma

Commercials for alternative cures take advantage and also promote such theories since it is the only way to give credit to their products. They heavily invest on manipulating fears and conspiratorial suspicions, especially concerning the

Fig. 25.3 Statues of Anunnaki, which were Mesopotamian deities, supposedly to be an alien serpentoid race

relationship between the government and the Big Pharma, and promise that by choosing their alternative cures, clients would win back both their liberty and health [63].

The "Big Pharma" conspiracy theory is a group of conspiracy theories suggesting that the medical community, the government, and the large pharmaceutical companies cooperate for sinister purposes to earn profit without caring that this has definitely an adverse effect on public health. According to these theories, treatment for major diseases or vaccines, including HIV or cancer, already exist but are concealed from the public or research on them is delayed deliberately. In a series of YouTube videos, Peter C. Gøtzsche, co-founder of the Cochrane Collaboration and former leader of the Nordic Cochrane Center, suggested that medication treatment is a leading cause of death and characterized the pharmaceutical industry as a criminal organization [64, 65]. He also suggested that mental disorders do not exist, and their conceptualization is not based on scientific data [66–68], and even more recently, he strongly supported COVID-19 conspiracy theories in an elegant way but still promoting denial through semi-truths and incomplete interpretations [69]. His membership of Cochrane was terminated by its Governing Board of Trustees on September 25, 2018.

According to conspiracy author Kevin Trudeau, there exist all-natural cures for serious illnesses, including cancer, arthritis, AIDS, acid reflux disease, mental disorders, obesity, diabetes, neurological and autoimmune disorders, and others, and all of them are being deliberately suppressed from the public by a sinister alliance of the Food and Drug Administration and the pharmaceutical companies [70, 71].

While it is true that there are proven cases of cheating from the side of the pharmaceutical industry, these cases do not fall under a general big plan orchestrated in agreement among major players in order to impose harm on people for the sake of profit. This is the difference between a scandal and a conspiracy theory.

Anti-mask

Opposition against the use of cloth facemasks emerged also during the Spanish flu pandemic in 1918, but this event was largely forgotten (Fig. 25.4). At that time the leading figure of the opposition was the lawyer, suffragette, and civil rights activist, Mrs. E.C. Harrington. It came to surface again with the COVID-19 pandemic when the Internet was flooded with false claims that wearing a mask will lead to lower oxygen or higher carbon dioxide levels in the blood and then to cardiorespiratory disorder or even cancer and eventually to death. Even ridiculous claims that masks contain a hidden microchip aiming to spy on people emerged. Arguments pertaining to the violation of civil rights were also dominant (Fig. 25.5). The fact that wearing a mask makes identification and security controls more difficult did not discourage conspiracists from claiming that this was yet another measure aiming to silence and control the population and train people to obey.

Fig. 25.4 A call to protest by the Anti-mask League, published in the *San Francisco Chronicle* on Jan. 25, 1919

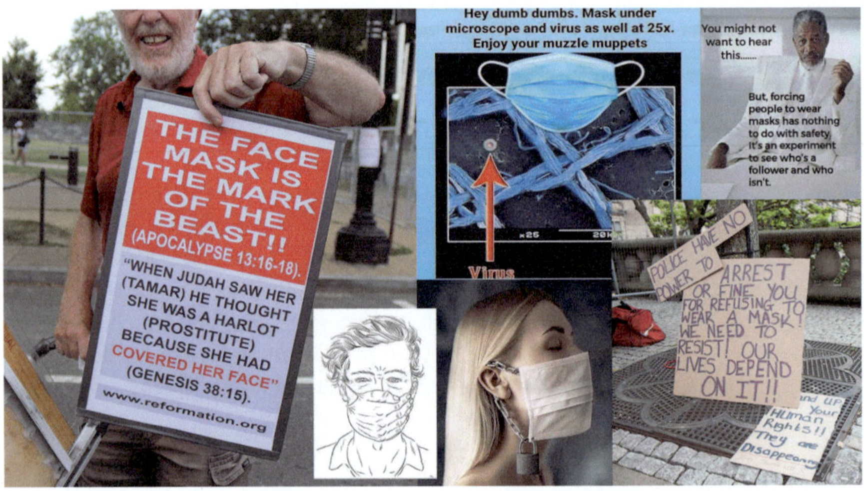

Fig. 25.5 Left: During the COVID-19 pandemic, Independence Day Walk at Constitution Avenue and 17th Street, NW, Washington DC on July 4, 2020 (by Elvert Barnes Photography). Center, up: An image circulating on the Internet arguing that the COVID-19 virus is too small for the mask fabric to stop. Center, down: An image circulating on the Internet arguing that the measure of mandatory wearing of masks during the COVID-19 pandemic is a threat to civil rights. Right, up: An image circulating on the Internet showing the actor Morgan Freeman from the 2003 film *Bruce Almighty* posing as God, arguing that the measure of mandatory wearing of masks during the COVID-19 pandemic is a threat to civil rights. Right, down: Demonstration against COVID-19 vaccination (photo by Karen Eliot)

Chemtrails

When a jet plane flies, the engines leave a water condensation trail (contrail) behind. The chemtrail conspiracy theory argues that this trail includes a number of chemical or biological agents, or a toxic mix of aluminum, strontium, and barium, in accordance with some secret government policies at the worldwide level [36]. While the reason for this spraying of chemicals in the atmosphere is a matter of debate among believers, theories vary from protection of the earth from solar radiation, weather modification, sinister psychological manipulation, and human population control.

These conspiracy theories emerged after the US Air Force released a report about weather modification in 1996. Every action by the US Federal authorities to reject the claims served only their intensification. Frequently, photographs of barrels installed in the passenger space of aircrafts to simulate passengers' weight for flight test purposes are used as proof of the theory. These barrels are filled with water, and this method is used simply because by pumping water from barrel to barrel, one can easily change the center of the gravity of the plane and test its behavior (Fig. 25.6).

Fig. 25.6 Left, up: Barrels inside an airplane during a test flight. Photos like this one are used to support the theory that planes loaded with barrels containing chemicals are spraying the population. The tubes are connected to the barrels so that changes in the location of the weight during the test flight is possible. Left, down: Trails on the sky caused by flights. Center, down: Poster of a pro-"Chemtrails" video on the Internet. Right up and down: Protests against "chemtrails" and "geo-engineering" in Germany

COVID-19

Not surprisingly, if one reviews the public response to the 1918 Spanish flu pandemic, the recent COVID-19 pandemic seems to be a déjà vu in many ways. It resulted not only in misinformation but also in the appearance of novel conspiracy theories concerning the nature, the origin, and the existence of the pandemic itself. The new element was that probably even covert operations by a number of countries utilized the Internet and the social media for spreading fake news and creating conspiracy theories. The spreading of real as well as fake news was so intense that the World Health Organization (WHO) has called the situation an infodemic. In a way, the pandemic challenged the competency of the digital generation to survive in the very environment it considered to be its comfortable home [72].

Overall, there seems to be a circular relationship with those people who rely significantly on unreliable sources of news as well as on information in the Internet, being those who are most likely to be negatively affected by fake news concerning COVID-19. This has a significant impact on their health behavior and adherence to preventive and protective measures against the pandemic [73]. Interestingly, there was a theory that included also elements from the anti-vaccination conspiracy and high-technology conspiracies and suggested an evil plan by Bill Gates to control humans and to reduce human population (Fig. 25.7) [47]. This was related to another characteristic, that theories related to COVID-19 were strongly related to believing in other theories by the same person and to a negative attitude toward and lack of trust to the government and institutions in general (Fig. 25.8) [74].

The number of well-known and reputed scientists who risked and lost their credibility by supporting alternative theories or theories based on soft evidence or even conspiracies is notably high. A closer look might suggest that although these scientists were considered to be brilliant before the pandemic, they were all too often radical, were attention-seeking, and were pushing forward with their ideas in an aggressive and arrogant way and often with anger. This was the case with their involvement with COVID-19 since they used words like "fiasco" or "fraud" to intimidate their intellectual opponents and insisted or never apologized, even when the number of deaths had skyrocketed. The major theories were that the virus was engineered [75] and it escaped or was stolen from a research laboratory [76]. The blame and accusations for different things was put mainly on Bill Gates and the Rockefellers.

Maybe the most prominent was Luc Montagnier (born 1932; Fig. 25.9), who won the Nobel Prize in 2008 for his work on HIV. He was a very controversial figure in the scientific community even before the pandemic for being anti-vaccination and pro-homeopathy and arguing that "water has memory." He argued that the virus is man-made and accused biologists for having inserted HIV genetic sequence from HIV in the genetic material of COVID-19 [77]. In the past Luc

Fig. 25.7 Left: The image of the entrance of the Bill and Melinda Gates foundation, modified and used to spread the theory that the foundation's aim is global population reduction. Right: The real entrance of the foundation

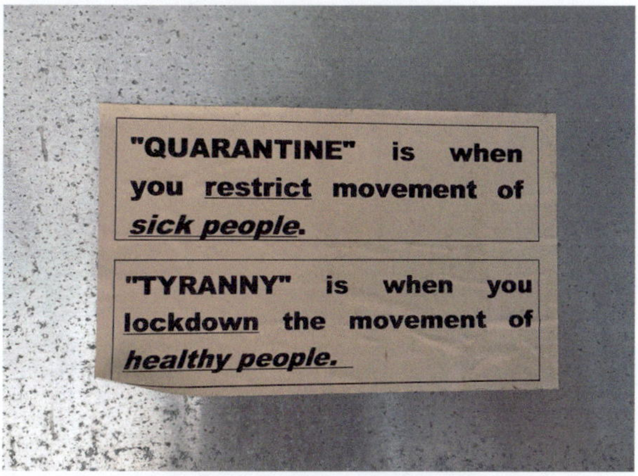

Fig. 25.8 COVID-19 conspiracy protest stickers on poles, Beaumaris, Wales UK (photo by Philip Mallis)

Montagnier had claimed that a good immune system was all what was needed to be protected against HIV. In 2017 he accused the French government of implementing mandatory vaccinations to 'poison the next generation little by little," which led 106 academic scientists to write an open letter "calling him to order." Another absurd idea of his was the emission of electromagnetic waves by DNA and bacteria [78] which among others could be used to diagnose HIV [79]. Also, he suggested that DNA can be generated from its teleported "quantum imprint" involving also the theory of "water memory." This was combined with support to homeopathy [80].

Interestingly, conspiracy theories seemed to emerge also at least partially in the frame of covert actions of one country against another, and in this frame, there were theories that the virus was engineered by Israel, Iran, China, the CIA, countries of sub-Saharan Africa, or religious or ethnic groups. Such allegations were expressed

Fig. 25.9 Luc Montagnier, 2008 Nobel Prize Laureate for Physiology or Medicine

even by official statesmen and politicians around the globe. On the opposite side, a number of religious groups and institutions, including Ultra-Orthodox Jews, the Islamic missionary movement Tablighi Jamaat, the head of Fatima Masumeh Shrine in Iran, the River of Grace Community Church in Gyeonggi Province in South Korea, the Church of Greece, and others around the world, claimed that their faith would protect them from the pandemic and refused to cancel ceremonies or take precautions.

Deaths of Prominent People

The deaths of prominent people triggered a variety of conspiracy theories. This holds true both for people who died by assassination like the US Presidents John F. Kennedy and Abraham Lincoln, as well as Martin Luther King, Jr., and for persons who died by accident, including Diana, Princess of Wales. From more than 1000 books written for the assassination of JFK, more than 90% deal with

LATE EDITION

The Onion

TU STULTUS ES

Take a Moment to Remember Where You Are

Friday, November 22, 1963 FINEST NEWS SOURCE IN AMERICA ★★★ Price Seven Cents

KENNEDY SLAIN BY CIA, MAFIA, CASTRO, LBJ, TEAMSTERS, FREEMASONS

President Shot 129 Times from 43 Different Angles

DALLAS, Tex.—President Kennedy was assassinated Friday by operatives of the CIA, the Giancana crime syndicate, Fidel Castro, Vice President Johnson, Parkland Hospital, where doctors with ties to Johnson's inner circle performed a staged autopsy. They pronounced him dead at 2:18 p.m. CST.

Fig. 25.10 The Nov 23, 1963, front page of *The Onion* exposing the contradictory and heterogeneous nature of conspiracy theories concerning JFK assassination

conspiracy theories (the mother of all conspiracies, Fig. 25.10), which involved players unrelated or even hostile to each other, including the CIA, the KGB, Fidel Castro, the Mafia, or the sitting Vice President Lyndon B. Johnson [81, 82].

Another famous conspiracy theory concerns Elvis Presley (1935–1977). According to it, Elvis did not die in 1977 but instead had staged his death. Frequently popular newspapers and magazines report on Elvis sightings around the USA, while a popular theory suggests he was abducted by aliens.

Energy Weapons

Conspiracy theorists claim that government agents are utilizing directed energy weapons and electronic surveillance to harass members of the population and cite research into psychotronic weapons, the Cuban Health Attacks, and the microwave auditory effect as proof of their theory.

This conspiracy theory appeared after the California forest fires of 2018, and one of the earliest promoters of that claim was a flat-earth YouTuber going by the name ODD Reality. The theory suggests that those fires were not caused by natural phenomena or even simple arson but by evil elements of the government, specifically through the use of directed energy weapons (DEWs). Their aim was to put fire, destroy homes, and take lives. One of the basic arguments is that unexplainably certain building and structures are spared by fire, and this is a sign of man-made intentional damage. The DEW theory is inextricably tied to skepticism about the very idea of climate change and considers it an attempt by the government to justify wildfires and conceal the use of DEWs. It is also tied to the theory that there is a weather control system called HAARP (in fact HAARP is a US Air Force research program of the outer atmosphere).

Flat and Hollow Earth

Maybe the most ridiculous of all, the Flat Earth theory [83], first emerged in the nineteenth-century England, despite the Earth's spherical nature having been known since at least the Hellenistic period more than 2000 years ago. Interestingly, this should be considered within a certain frame, that of generalized opposition to science. The so-called Zetetic (meaning "seeker" or "skeptic") societies were flourishing in England during the nineteenth century, and they were reserved exclusively toward science and were strongly affiliated with religious communities and a literal interpretation of the Bible (Figs. 25.11 and 25.12). The Universal Zetetic Society (UZS) was founded in 1890 and begun publishing the journal

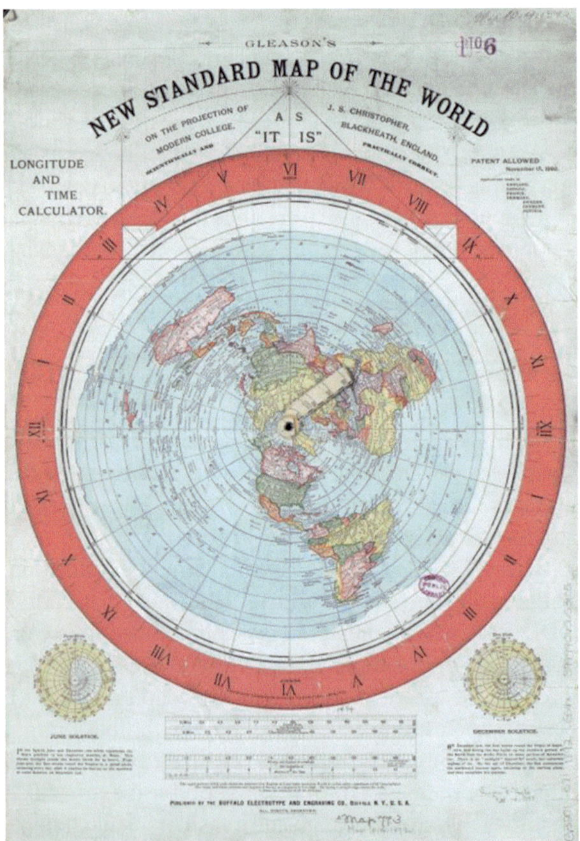

Fig. 25.11 A "flat-Earth" map drawn by Alexander Gleason in 1892

Fig. 25.12 A "flat-Earth" map drawn by Orlando Ferguson in 1893

titled *The Earth Not a Globe Review*. In 1849, Samuel Birley Rowbotham (1816–1884), a nineteenth-century religious fundamentalist, was the first to write a book on the topic, the *Earth Not a Globe*. He utilized arguments including light refraction in order to prove that the evidence that the earth is a globe was based on "optical illusions." He even calculated that the sun had a 32-mile diameter and was 700 miles away from the earth.

In the recent years, the conspiracy often implicates the NASA and also suggests that the GPS is rigged to trick aircraft pilots to believe that they are flying around a globe.

The conspiracy theory of Hollow Earth suggests that the earth is spherical but hollow and that there is access to the interior at a number of different locations, including the North Pole and South Pole. It is not a theory with many followers, but groups advocating it tend to appear from time to time during the last couple of centuries.

Fluoridation of Water

Water fluoridation refers to the addition of fluoride to public water supply in order to reduce tooth decay. The procedure was officially utilized for the first time in January 25, 1945, at the Grand Rapids, Michigan, and was widely adopted by the

USA in 1951. It is based on the works of Frederick McKay (1874–1959), Greene Vardiman Black (1836–1915), and H. V. Churchill, who studied the cause of the "Colorado brown stain" and found that this stain was because of fluoride, which however also protected teeth from developing cavities [84, 85]. The cost of water fluoridation is very low and is about 1 USD per person per year. The combination of low cost and high benefit is of utmost important since tooth decay is still a major public health concern in most countries, with a prevalence of 60–90% [86]. The efficacy of fluoridation was calculated to be a 25–35% reduction in teeth cavities by a Cochrane review [87], but many authors suggested that the use of fluoride in toothpastes and other mouthwashes has made water fluoridation obsolete [88]. Still the WHO supports it [86], while the European Commission rather favors topical use of fluorides [89]. In the USA, the CDC includes water fluoridation in the list of the 10 most important public health achievements of the twentieth century [90].

Interestingly, opposition to water fluoridation started before its application. In Grand Rapids, there were complains of swollen gums and sore throat weeks before the beginning of fluoridation when the authorities mistakenly had informed the population that it had already started. Across the USA, groups of people opposed fluoridation with the Fluoride Action Network and Citizens for Safe Drinking Water lining it to cancer, mental retardation, birth defects, declining birth rates, and heart disease. Conspiracy theorists claimed that this is a way to dispose industrial waste or to hide the failure to provide dental care to the poor [91, 92]. More bizarre theories included that of the John Birch Society in the 1960s, which suggested that fluoridation was a communist plot to weaken the American population, a view shared by the character General Jack Ripper in the 1964 film *Dr. Strangelove* [92]. A myth that the Nazis were experimenting with fluoride, which supposedly gradually reduces the persons' individual power to resist domination and oppression and make them submissive, involved the name of Charles Elliott Perkins, a supposed chemist appointed to study the Nazi laboratories after WWII, and a member of the Australian Parliament, Harley Rivers Dickinson (1938–2008), who supposedly brought the issue to the Parliament (Fig. 25.13).

Global Warming

Global warming denialism rejects the evidence on climate change as a result of human activity and either suggests it is solely a naturally occurring phenomenon or it does not exist at all and that the data have been fabricated. It includes opposition to the installation of alternative sources of energy like solar panels or windmills.

Fig. 25.13 Left: Poster circulating on the Internet, trying to link water fluoridation with population control, civil rights suppression, and Hitler. Right: Illustration in a 1955 flier by the Keep America Committee, alleging that fluoridation is a dangerous Communist plot

Human Immunodeficiency Virus/Acquired Immune Deficiency Syndrome (HIV/AIDS)

There is a number of conspiracy theories concerning the HIV/AIDS health problem, and each one of them has unique characteristics concerning race, religion, lay beliefs, and, as usual, civil rights and freedom [93]. One such theory, very frequent among African-Americans, is that the US government is using HIV in order to reduce the black population. There are anecdotal reports that a rumor was circulated by the KGB in the 1980s, suggesting that the CIA had engineered the virus. These are combined with a belief that the demographics of the black community are targeted by the government, which leads to less frequent use of condoms and, in turn, to higher rates of HIV infection, thus creating a vicious cycle [94, 95]. Denialism of the existence of HIV or the relationship between HIV infection and AIDS is another very frequent theory, especially in sub-Saharan African countries. A number of scientists, journalists, and lay authors, including Stefan Lanka, Joan Shenton, and Celia Farber, gave interviews or wrote books supporting HIV/AIDS denialism. In 1988, Eleni Papadopulos-Eleopulos published an article in the non-peer-reviewed journal *Medical Hypotheses*, arguing in favor of an oxidizing hypothesis and against HIV infection as the cause of AIDS [96]. The same authors in 1993 disputed the validity of the Western blot test for HIV [97]. This article was published in the prestigious journal *Nature Biotechnology*, which at that time was edited by Harvey

Bialy, a known denialist himself. Harvey Bialy, from his position as editor of such a prestigious journal, played a significant role in the dissemination of AIDS denialism theories and especially of Duesberg theories [98].

Undoubtedly, Peter Duesberg (born 1936) was the star of HIV/AIDS denialism and the one who formulated the hypothesis that AIDS was not due to HIV but instead was the result of recreational drug use, malnutrition, poor sanitation, and hemophilia or, most importantly, the effect of AIDS treatment itself [99–122] (Fig. 25.14). Gay community magazines significantly supported Duesberg theories and helped in their dissemination to the public. The scientific community held a debate in the journal *Science* in 1988 [100, 123], but in spite of the intellectual victory against denialism, the debate contributed significantly to the rising popularity of Duesberg theories, since the message at the end of the day was that a reputable journal hosted them, and they were significantly enough to attract fire from opponent scientists. Other scientists associated with HIV/AIDS denialism include biochemists David Rasnick (born 1948) and Kary Mullis (1944–2019). Rasnick went

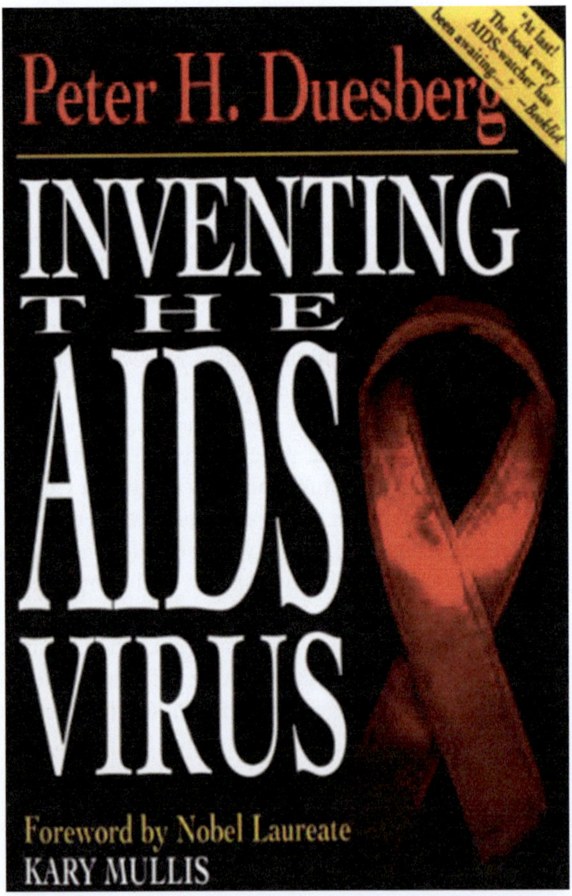

Fig. 25.14 Cover of the 1996 book *Inventing the AIDS Virus* by Peter Duesberg, promoting the AIDS denial conspiracy theory

to South Africa, where he discouraged AIDS patients from taking anti-retroviral drugs and instead promoted treatment with vitamin cocktails. He also proposed a complete ban on HIV testing in South Africa and denied the existence of an HIV epidemic in South Africa. In 2008, the South African court ruled that the vitamin trials conducted by Rasnick were illegal and banned the conduct of clinical trials or the advertisement of vitamin supplements for the treatment of AIDS [124]. Mullis was awarded a Nobel Prize for his role in the development of the polymerase chain reaction (PCR) in 1993, a discovery that revolutionized laboratory testing. Interestingly he rejected the scientific evidence supporting climate change, ozone depletion, and the link between HIV and AIDS and asserted his belief in astrology [125, 126].

As a consequence, an HIV/AIDS denialism theory was crystalized, suggesting that AIDS was due to sexual behavior, recreational drugs, malnutrition, poor sanitation, and hemophilia or, most importantly, the effect of AIDS treatment itself [127–129]. No matter how irrational this may seem, it is widely accepted in these countries and is politically supported by many, including the former South African President Thabo Mbeki, whose HIV policy is considered to be related to a large number of HIV infections and deaths because of AIDS [130–135]. The consequences are not only local or concern the individual alone; interrupted treatment increases the likelihood of the emergence of novel HIV strains that are resistant to treatment [136]. This policy was stopped in 2009 when Aaron Motsoaledi became the Health Minister of South Africa [137] and clearly and openly condemned this irrational policy in 2016.

The story of Maria Papagiannidou-St Pierre, who was a senior Greek journalist, is striking. She was born in 1965 and was diagnosed with HIV in 1985; she has been suffering from AIDS since 1995 but was doing extremely well under medication treatment. In 2006 she started following the HIV denialism and in 2007 stopped treatment. She went on to write several books denying the existence of AIDS [138] (Fig. 25.15) and the need for treatment. She died a few years later from AIDS complications. For her anti-AIDS activism, she was proclaimed "woman of the year" by a lifestyle magazine, thus fueling the glamorous lifestyle aspect of conspiracism.

Jews and Zionism (Anti-Semitism and Holocaust Denial)

Conspiracy theories often include Jews in the role of the aggressor, responsible for the disasters or threats and for plotting against the nation or the religion. In this frame denial of the holocaust exists as a conspiracy theory, with all the relevant characteristics from complete denial to having being orchestrated by the Jews themselves in order to be able to create the state of Israel. They are often included as official or semi-official point of view by certain countries and political parties. These theories are an example of how belief in naive theories can serve as a strong and universal explanation for all the problems of society, especially during times of crisis [139].

Fig. 25.15 Cover of the 2009 book *Goodbye AIDS-Did It Ever Exist?'* by Maria Papagiannidou-Sen Pier, promoting the AIDS denial conspiracy theory

Moon Landing

Conspiracists deny the landing on moon of the Apollo spaceships and suggest that the NASA has staged the landings in a film studio. Thy involve Stanley Kubrick, the director of the film *2001: A Space Odyssey* as the mastermind behind the film shootings, in an outstandingly naïve correlation. In 2001 there was a documentary on *Fox* Television Network supporting the moon landing denial. Even the French parody mockumentary *The Dark Side of the Moon*, with staged interviews and satirically criticized conspiracy theories, is taken by their followers at face value. The placing of retroflectors on the moon's surface did little to persuade deniers.

Political (Deep State)

There is a conspiracy theory suggesting that elected government officials have limited power, and subsequently the will of the people is often pushed to the side. This is because there is a hidden bureaucratic organization at the national level that serves the interests of a small but powerful elite. The concept originated probably in the Middle East but today enjoys worldwide acceptance and serves perfectly as an excuse for those politicians who fail to satisfy their promises made during the election campaigns. The deep state was a key issue in the political campaign of Donald Trump.

Racial

Almost all human races have their own conspiracy theory that is often identical to that of other races, like a mirror image. For whites, blacks are the threat and vice versa. The white genocide conspiracy theory suggests that immigration and integration of colored immigrants and refugees in countries with a majority of white population, in combination with low fertility rates and abortions of the white population, are a coordinated plan to turn whites into a minority in their own country [140–142]. Then, mirror theories exist concerning a black genocide or the replacement of black with white populations in cities where they are currently the majority [143–146]. They also include “the plan” of white powerful persons to take back the Washington DC local government by replacing low-income black residents who flew the city with wealthy whites [147].

Secret Societies

The New World Order theory suggests that there are a number of international groups of elites, including organizations such as the Federal Reserve System, the Council on Foreign Relations, the European Union, the United Nations, the World Bank, and the International Monetary Fund, existent associations, and groups like the Bilderberg Group, Freemasons, or university societies or even fictitious groups like the Illuminati. It is supposed that these groups control governments, the mass media, and the economic activity around the globe. They stage events and plan wars, economic crises, and even pandemics. Secret society conspiracy theories are often related to anti-semitism [16]. One interesting fictional group is “Antifa,” which, as the conspiracy theorists claim, is a single far-left militant group.

Targeted Persons

A number of persons and families have been targeted with the involvement of their names in conspiracy theories. The Rockefeller family is such an example, and its name is included in a variety of theories with aspects of anti-semitism. Hungarian-American investor George Soros is also the subject of conspiracy theories often linked to anti-semitism, because he supported financially many political, social, educational, and scientific causes. In the first few months of 2020, conspiracy theories emerged about Bill Gates and the COVID-19 outbreak, suggesting that the virus

Fig. 25.16 Poster circulating on the Internet and promoting the theory that Bill Gates funds the development of vaccines with malevolent purposes to control the world population and reduce its size

was his creation [47]. The reason was that he funds research on treatments, vaccines and health technology. Conspiracy theorists argued that Gates would use vaccines to insert an electronic chip into the human body to control people (Fig. 25.16). The basis of all these theories is that they suggest that the support these persons provide to various social causes is done for nefarious purposes.

Vaccinations (Anti-vaxxers)

Lay persons lack specialized knowledge in order to be able to understand how exactly vaccines and vaccination policy work, and additionally they lack understanding of basic statistical and epidemiological concepts (Fig. 25.17), including herd immunity. Thus, the development of the first vaccine by Edward Jenner (1749–1823) triggered a wave of conspiracy theories. Interestingly the terms vaccine and vaccination are derived from *Variolae vaccinae* (smallpox of the cow), and Jenner used these words to denote cowpox and its protective effect against smallpox. The term was generalized by Louis Pasteur in 1881 to cover all the new protective inoculations then being developed. However at the time of Jenner, people were discussing the possibility of developing horns or other adverse effects of the vaccine because of its relationship with cows [148, 149] (Fig. 25.18).

Already since its appearance, vaccination was the focus of reactions (Fig. 25.19). The need for universal vaccination, the entering of an external fluid into one's body without being sick, and the mandatory nature of the measure caused many people to react negatively. In the early nineteenth century, in Europe and North America,

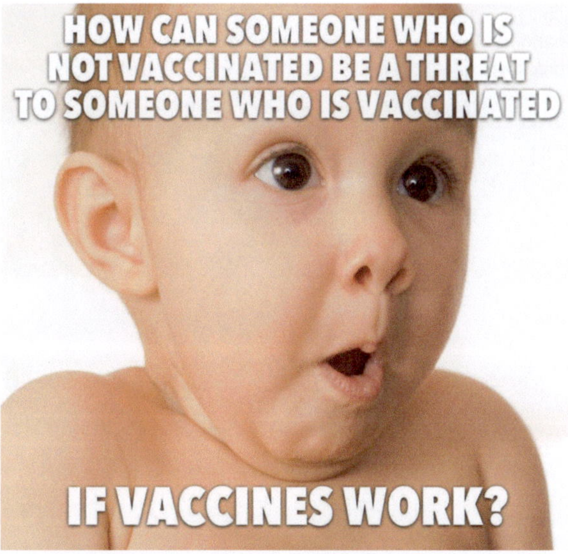

Fig. 25.17 Typical manichaeistic naïve way of thinking exposing also ignorance concerning the mechanism through which vaccines protect the population from epidemics

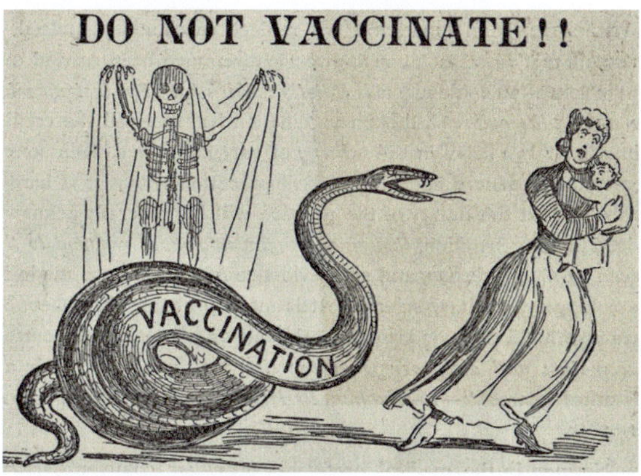

Fig. 25.18 Cartoon from 1892 with the label: "Do Not Vaccinate!!" The Historical Medical Library of the College of Physicians of Philadelphia

there were reactions to a number of oppressive laws and political decisions, and for many, mandatory vaccination was considered in that frame, which is of another oppressive measure. The Anti-vaccination League was created shortly after the passage of the Vaccination Act of 1853 in Great Britain. Others followed the "naturalistic fallacy" approach and considered vaccination a "foreign assault on traditional order," an unnatural and impermissible intervention. A prominent figure of the nineteenth century anti-vax movement was George Bernard Shaw (1856–1950),

Fig. 25.19 Cartoon against anti-vaccination, anti-scientific attitudes, and conspiracism. It was probably created in the 1930s by an unknown artist

who received the Nobel Prize in Literature in 1925 and also believed in homeopathy. But at that time, only two vaccines (against smallpox and rabies) were available (Fig. 25.20).

Since the 1970s, the anti-vax movement got stronger. Probably this was the effect of the proliferation of available vaccines after WWII. In 1974, John Wilson and M. Kulenkampff reported neurological complications in children 24 h after being vaccinated with the DPT vaccine. The large media coverage and the resulting drop in vaccination rates led to an explosive increase in deaths from Pertussis in the UK and other countries [150, 151]. The documentary *DTP: Vaccine Roulette* was aired in the USA by the WRY-TV channel in 1982, and for many it shaped the modern anti-vax movement. To this direction was also the contribution of Robert Mendelsohn [152], one of the first anti-vaccine pediatricians who was also against water fluoridation, coronary bypass surgery, and screening examinations to detect breast cancer. Lea Thompson promoted anti-vax issues in her TV show, prompted parents to form the group Dissatisfied Parents Together (DPT), and wrote the book *A Shot in the Dark*. The celebrities' list of anti-vaxxers include Lisa Bonet of *The Cosby Show*, who stated on *The Phil Donahue Show* that vaccines could "introduce alien microorganisms into our children's blood …," and Heather Whitestone, the first deaf Miss

Fig. 25.20 Protesters at an anti-vaccination rally at Toronto City Hall on November 13, 1919 (City of Toronto Archives)

America in 1994 who erroneously stated that she had lost her hearing because of vaccines. But all these were rather marginal until the game-changer 1998 article in the Lancet by Andrew Wakefield who claimed that the MMR vaccine causes autism [153]. Eventually the claim was dismissed, the paper was retracted from the journal, and most of Wakefield's co-authors denounced it, but the damage had already been done [154]. An even worse case occurred in 2014 when Brian Hooker published a reanalysis of the data from a 2004 CDC study and argued that the CDC had missed that African-American boys given the MMR vaccine before the age of 36 months had increased chances of developing autism spectrum disorder [155, 156]. This paper was also later retracted, but it inspired the 2016 film *Vaxxed*. Currently there is a large number of anti-vax literature authored by pediatricians, journalists, and celebrities.

A different story is that of Pakistan and Nigeria, where there is a widely circulating conspiracy theory that the polio vaccine is a ploy designed the by the CIA to make Muslim men sterile. The negative attitude toward vaccination was further increased in Pakistan after it was revealed that a fake vaccination campaign was used as a cover during the operation to kill Osama Bin Laden. This led to fatal mass shootings and bombings at vaccine clinics in both countries [157].

25.4 Mental Illness Denialism

The issue of the existence or not and the definition of mental illness was discussed in Chap. 13, and the various philosophical opinions were discussed also in Chaps. 11, 15, 21, 22, 23, and 24. The question which will be discussed here, is whether under the light of what has been analyzed in these chapters, the denial of the existence of mental illness as a medical disorder in need for specific treatment according to the biomedical model, manifests characteristics of conspiracy theories, and whether this denial should be included in the list of conspiracy theories next to anti-vaccination movement, to AIDS denial and to COVID-19 conspiracy theories. If mental illness denial presents similar characteristics, then failure to include it in this list only reflects stigma and the marginalization of mental patients.

Historically, although the concept of mental illness and the establishment of psychiatry as part of medical disciplines were not completed until the eighteenth century, discussions on the existence or not of mental disorder begun after WWII (see also Chap. 11) and more specifically in the 1960s. With the rise of the civil rights movement, of feminism, and of the labor movement, psychiatry served as a model for them and also as an example for the whole society, and in turn mental patients were considered to be a minority, oppressed by society through psychiatry [158]. In 1961, Thomas Szasz (1920–1912; Fg. 11.31) published *The Myth of Mental Illness* and argued that the very concept of mental illness was a logical absurdity defined exclusively on the basis of value (ethics, morality, legal issues, and social norm) [159–161] and challenged the proposal that mental illness is a radical failure of rationality and autonomy [159, 162]. He went even further and placed Psychiatry in the company of alchemy and astrology and defined it as pseudoscience [163]. In agreement with Szasz, Michel Foucault (1926–1984; Fig. 11.31) considered psychiatric treatment to be essentially evil [164, 165] and argued that an objectification of the concept of the patients' freedom occurred since the Enlightenment. This was because according to this theory, the concept of mental illness is considered to be the product of this "biopower environment" with technocratic–economic organization of the whole society, and "patients" are in reality human capital waiting to be restored and to be optimized as a biological capital with the use of medicine [166]. According to both Szasz and Foucault, psychiatry is part of the disciplinary system, and psychiatrists together with other mental health professionals are simply instruments of public order enforcement. David Cooper (1931–1986; Fig. 11.32) and Ronald David Laing (1927–1989; Fig. 11.32) followed on a similar path [167]. Jean-Paul Sartre (1905–1980; Figs. 10.49 and 11.33), the leading figure of existential philosophy, in his foreword to the best-seller book *The Divided Self* by Ronald David Laing (1927–1989; Fig. 11.32) noted that "mental illness is the revolt that the free organism in its total entity invents in order to live in an unbearable situation" [168].

Details can be found in Chap. 13, but what is important is that today, the theory of constructivism suggests that social forces rather than scientific data are those that shape our conceptualization of mental disorders [169, 170]. According to it, mental disorders do not really exist, but instead they were invented in order to serve sociopolitical aims and demands as well as the commercialization of medicine and more particularly of psychiatry, especially by the pharmaceutical industry [171–174]. Additionally, psychiatric treatments are damaging and dangerous rather than helpful [175–177]. This view has been associated with anti-psychiatry activists, and the existentialist approach should be considered as part of this third approach. These arguments are absolute, and they suggest that there is absolutely nothing more in mental disorder apart from social control plus the deleterious effect of modern civilization and commercial interests. These anti-medical approaches reject the scientific method and instead adopt an empirical idiosyncratic view and interpretation close to lay persons' beliefs. Such approaches are very difficult to tackle because they a priori reject the scientific way; they do not engage in discussions that follow the rules of science, or even at least those of analytical philosophy. In other words, rational arguments do not play any role.

The rejection of psychiatry and of mental illness as medical disorders as well as of its treatments enjoys wide acceptance in the world of arts and literature. For example, a review of the field [178] reported that out of 22 Hollywood films featuring ECT, 18 depicted negative outcomes, including zombification and death [179], while all of the 13 films in the Hindu cinema depicted it being given by force in order to punish, to change a person's identity, or to induce insanity [180]. In 1962 Ken Kesey (1935–2001; Fig. 11.34) published the book *One Flew Over the Cuckoo's Nest*, in which he analogized society's oppressors with psychiatric ward staff and the means of social and political oppression with physical treatments of mental disorders, especially ECT and psychosurgery [181]. Later a film was based on this novel and received several Oscar awards.

Issues of substance abuse have also been the subject of various conspiracy theories. It is frequently alleged that marijuana is a safe drug and was only outlawed under pressure from the paper industry, as the hemp plant was a competitor. By contrast, conspiracy theorists typically accept that cocaine is a dangerous drug, but many believe that the crack cocaine epidemic across the USA in the 1980s was actually due to the US government's plan to specifically target African-Americans and keep them addicted while at the same time profiting from the illegal trade to finance paramilitary groups in Nicaragua [182].

The greatest problem with anti-psychiatry is that it spread a wide bitterness and conflict among mental health professionals and the general public, and also it strongly accused the relatives of patients as directly responsible for the patients' mental condition, among other things, with the introduction of the concept of schizophrenogenic mother [183, 184]. It had a profound effect on public mental health because it encouraged the utilization of "alternative" treatments of all kinds, with minimal or zero efficacy and often with high cost.

With the arrival of evidence-based psychiatry, meta-analyses started to be conducted. The general outcome was that most treatment options might be overvalued and that there was a publication bias and some inconsistences or even manipulation in the reporting of results, but these were not dominant findings, and the gross picture of the treatment efficacy was generally unchallenged. But this was more or less the same picture in all medicine.

But then, in 2008, a meta-analysis was published and suggested that anti-depressants fall well below the criteria for clinical relevance, efficacy reaches clinical relevance only in trials involving the most severely depressed patients and alternative therapies (e.g., exercise and psychotherapy) may be a better treatment choice for depression [185, 186]. The impact on the mass media was huge (Fig. 11.54). The main author of this article, Irwin Kirsch, went on and wrote a book entitled *The Emperor's New Drugs: Exploding the Antidepressant Myth* [187] and accused the FDA of conspiracy [185]. The Kirsch dataset [186] has been re-analyzed by two other groups [188–190], which independently reported results different from those reported by Kirsch. While the Kirsch report enjoyed wide coverage by the major international media, there was not a similar coverage of the rejection of its results. Similarly the Kirsch paper enjoyed several times higher a citation number in comparison with the rejection papers.

The view that anti-depressants do not work was also promoted by Marcia Angell who was former Editor-in-Chief of the NEJM, and in 1997 she was named one of the 25 most influential Americans for that year by *Time*. She authored a book accusing the pharma industry of deceiving the public [191]. She wrote two articles in the *New York Review of Books* as comments on Irwin Kirsch's book *The Emperor's New Drugs: Exploding the Antidepressant Myth*, Robert Whitaker's *Anatomy of an Epidemic: Magic Bullets, Psychiatric Drugs, and the Astonishing Rise of Mental Illness in America*, and Daniel Carlat's *Unhinged: The Trouble With Psychiatry—A Doctor's Revelations About a Profession in Crisis*' [192, 193]. A detailed analysis of her texts can be found elsewhere [194], but what is important as a conclusion is the cherry-picking of data and arguments, while at points, the text is completely misleading and adopts an anti-psychiatric rhetoric, e.g., when she writes that "instead of developing a drug to treat an abnormality, an abnormality was postulated to fit a drug," while the text is not in accordance with the evidence when she writes that "Altogether, there were forty-two trials of the six drugs. Most of them were negative," since Kirsch et al. had received from the FDA data from 47 trials, data were sufficient for analysis only for 35 of them (not 42), and of course most of the studies were not negative; in fact in 18 (out of 35), the effect size was not only significant but also above 0.30 [186, 188, 190]. The overall effect size was 0.32, which points to a non-perfect but clear superiority of the medication vs placebo [186, 188, 190, 195, 196]. Then she forgot her strong support of evidence-based medicine and argued that "… and exercise have been shown to be as effective as drugs for depression, and their effects are longer-lasting…".

The above is impressive when written from a former Editor-in-Chief of a great journal, as the NEJM is. But she went further to suggest that "….medication changed the natural history of mental illness to the worse" and that "There are no citations of scientific studies to support (DSMs) decisions."

Another voice against the use of anti-depressants came from Ann Blake Tracy, Executive Director of the International Coalition For Drug Awareness under the title *Has Anti-depressant Myth Bubble Burst?* [197]. In this article the British newspaper

The Guardian is cited [198]. The article is a patchwork of issues and ideas directly or indirectly related to the antidepressant debate.

Recently, similar reservations on the usefulness of antipsychotics emerged, and many authors insist on the possible harmful effects of long-term antipsychotic treatment [199–201] and even of the short-term initial treatment [202]. The criticism on positive data is harsh [67, 200]. As soon as the first reports on ventricular enlargement in patients with schizophrenia were published [203], one of the first comments was that this was a consequence of anti-psychotic medication treatment [204]. Since then, several authors have attributed these findings to treatment with antipsychotics, cannabis use, diabetes, and hypertension [205–208]. Detailed discussion and dismiss of these claims one by one can be found elsewhere [209].

The attack on antidepressants was relaunched by the Danish physician and former board member of the Cochrane, Peter Gotzsche, who suggested that there is more or less a conspiracy by the "BigPharma," which he labeled "organized crime" [64, 65] and by medical journals to hide the truth concerning anti-depressants [210]. However, his views were all too frequently published by the Lancet or the BMJ to justify a conspiracy by journals. He even involved the Cochrane in these accusations [211] and was eventually expelled from it in 2018. He argued that psychiatric medications are a leading cause of death, and the truth is hidden from the public in an organized way [66]. His opinion is a clear denial of the efficacy of psychopharmacological treatment options, although he does not deny the existence of mental illness per se.

People outside psychiatry see that "psychiatrists disagree." The fact that this "disagreement" is an essential argument by itself, even in the article by Marcia Angel, reflects this limited exposure of psychiatry as a scientific discipline and part of medicine in the highly reputable scientific and medical journals. Instead it is discussed as if it is a matter of personal opinion or even taste. In her article Marcia Angell shows an unexpectedly poor knowledge and understanding of the issue, along with signs of a deliberately negativistic way of arguing. The backbone of her articles is not supported by the data, not even by Kirsch's work. They are against psychiatry as a scientific medical field, in line with anti-psychiatric texts.

Finally, as with all conspiracy and denialist theories, the metaphysical–religious–supernatural component could not be absent, and it is represented mainly by the church of scientology, for which an extended discussion is included in Chap. 23.

The characteristics of the denial of mental illness and its medication treatment are summarized in Table 25.2, also in comparison with how these characteristics appear in other medical denial theories and with the holocaust denial theory. It is more than evident that mental illness denial accumulates many more conspiracy theory characteristics than vaccines or COVID-19 do, and probably only because of their deep and pervasive roots in the lay culture that mental illness and treatment denial are not yet included in the list of conspiracy theories.

Table 25.2 Comparison of the characteristics of medical denial conspiracy theories and mental illness denial

	HIV/ AIDS denial	Anti-Vax	Climate change denial	COVID-19	Mental illness denial	Holocaust denial
There is no real disease/ problem			x	x	x	x
Different cause of the disease/ problem	x		x		x	x
Treatment is useless	x	x			x	
Treatment is harmful	x	x			x	
Treatment causes the disease	x				x	
It is an oppression/civil rights issue	x	x	x	x	x	
Cherry-picking of data	x	x	x	x	x	x
Disease/problem is man-made	x	x		x	x	
Specific agencies involved	x	x		x	x	x
Reflects an organized criminal activity	x	x			x	
Emotionally loaded debates	x	x	x	x	x	x
Strong opinion by lay persons		x	x	x	x	x
Correlates with other theories	x	x	x	x	x	x
Religious component				x	x	x
Racial component	x	x		x		x

Holocaust denial is included for comparison reasons with conspiracism in general. Mental illness denial definitely has the qualities to be considered a conspiracy theory

25.5 Interpretation

The history of conspiracy theories reveals a heterogeneous amalgam of elements, which is very difficult to understand and to put into a single unifying comprehensive frame. Still, understanding the essence of these theories is of utmost importance in order to be able to tackle them and avoid their adverse impact on population behavior in various fields, including health and politics.

The core question seems impossible: How is it possible and brilliant (not just clever) persons, including scientists of the level of Nobel laureate or editors of prominent journals, follow conspiracy theories and go against the data and the methodology they were following for most of their scientific life? How is it that rational argument is so ineffective in tackling these theories? Why is logical thinking short-circuited so easily in order to serve irrationality? It is impossible to understand which mental mechanism permits the Nobel laureate who developed the method of PCR and revolutionized laboratory testing to deny the relationship between HIV and AIDS and also believe in astrology. So, while discussing the psychological and sociopolitical components and in our effort to interpret conspiracism, we need to bear in mind that neither cognitive abilities nor education or special training are adequate assumptions for this interpretation.

First of all, there is the hypothesis that for evolutionary reasons, humans tend to see patterns in random series of events, which helps in the identification of hidden threats [212–217]. In accordance with this, it has been reported that believers of conspiracy theories are even more likely to detect nonexistent patterns in random data [218] (Fig. 25.21).

Second, the general environment and dynamics that serve the spreading of conspiracy theories are important. While in the past conspiracy theories were limited to fringe audiences, during the last few decades, they have become commonplace in the mass media and the Internet, and they have emerged as a cultural phenomenon [17, 27, 28]. It seems that false news travel faster than reliable news on Twitter

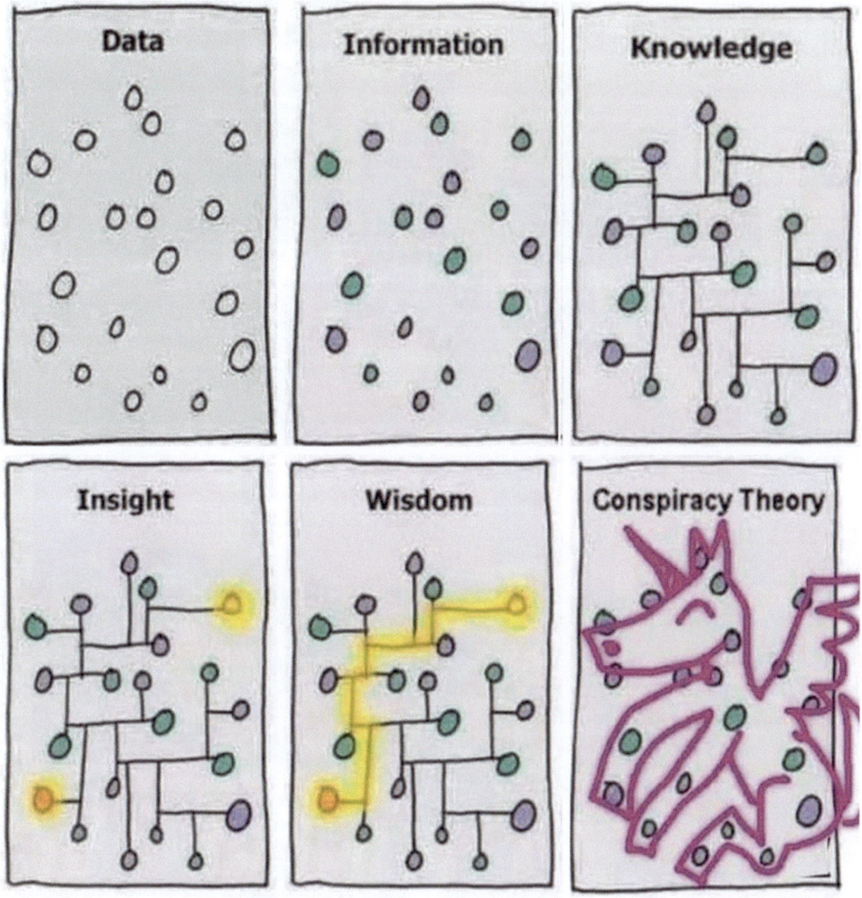

Fig. 25.21 Image circulating on the Internet, which visualizes the tendency of followers of conspiracy theories to find peculiar connections and to interpret the data in an unconventional way far away from reality

Fig. 25.22 James Randi during the 2013 "The Amazing Meeting" (TAM), an annual conference that focused on exposing fraud and charlatanism

Fig. 25.23 The WHO cartoon for infodemic

[219]. There seems also to be an inclination of the mass media and more recently of the Internet to promoting and supporting impossible but flamboyant stories, no matter how strongly they scream out that they are fake or hoaxes. This is clearly shown in the story of the magician and pro-science skeptic James Randi (1928–2020; Fig. 25.22) who, in 1988, sent his late husband, Jose Alvarez, a performance artist, to Australia to pose as a 2000-year-old spirit named "Carlos." The mass media easily adopted the "Great Carlos" story, and Alvarez performed in front of big audiences, before Randi revealed the hoax and embarrassed the whole media system of Australia.

Recently, fake news about COVID-19 became a second pandemic (announced as "infodemic" by the WHO; Fig. 25.23). It is now more evident than ever, that for the average person, no matter its role in society or its education, it is rather impossible to understand elementary concepts of the scientific way of thinking and critically processing the information before accepting conclusions. However, average persons seem to fall under certain categories. The categories are the "seeds" who are adamantly convinced of the truth of the conspiracy theory, the "agnostics" who are negative to any such theory, and "others," who is open to either approach. It seems that the dynamics of these groups are complex, but under certain unusual conditions, even a small number of seeds could influence a large number of "others" [220]. This is combined with the "uncertainty vacuum" in our current knowledge concerning many issues, and the statistical way science deals with data and the complex way interpretation and conclusions are derived, including their restrictions. This allows superficially reputable sources to jump in without real expertise, including academics with some credentials and good at crunching numbers but lacking a deep understanding of the underlying science and the essence of things [47]. This environment is fertile for the development of conspiracy theories as an "intellectual construct," which brings order to a perceived chaos and gives the naïve sense of privileged access to special knowledge or way of thinking that separates believers from the masses [221], by giving the "illusion of explanatory depth" [222]. These academics become the guru of the faith, and they are valuable for the theories because they lend them their fame and they also fill in the missing details, but most often with bad science [223].

Third, a core psychological characteristic is that humans find it difficult to believe that big events like the assassination of prominent people or pandemics might be the result of casual events or randomness. This is the so-called proportionality bias, and it leads to the inevitable conclusion that these events are the result of grand conspiracies [224, 225]. This way of thinking seems to be prominent in small children [226] but also in adults under severe stress [224, 227, 228] and when facing uncertainty [227]. Some authors argued that the substrate involves strong individualistic/hierarchical worldviews [229] and maybe psychotic-like mental functioning with schizotypal features and cold affect, a psychological condition that many authors call 'Machiavellianism" [25, 230–237]. It might also involve illusory pattern perception [218], with lower crystallized intelligence [238], collective narcissism with moral absolutism [239–242], and paranoia [233, 239, 243, 244]. Often the case is disbelieving a mainstream explanation rather than believing in a specific alternative [245].

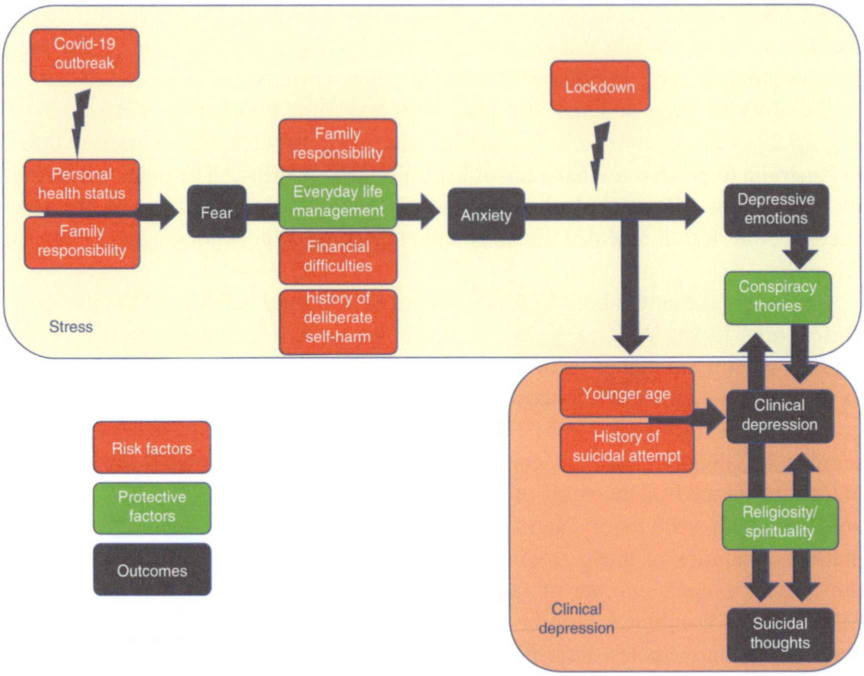

Fig. 25.24 The developed multiple vulnerabilities model representing the mechanism through which the COVID-19 outbreak in combination with the lockdown could lead to depression through stress. A number of variables act as risk factors (red) or as protective factors (green). From Fountoulakis et al., J Affective Dis 2021, Jan 15;279:624–629

The fourth and most important characteristic of conspiracy beliefs is that they resist any presentation of data against and also persuasion by reasoning. This suggests that they are not simply the product of an evolutionary tendency to give meaning to random clues, but instead, some kind of psychological dysfunction is also necessary [231, 246]. The presence of cognitive rigidity, accompanied by affective dysfunction, is supported by the inability of believers to distinguish between genuine conspiracy and satirical–sarcastic or parodistic stories [247], but this is frequent in fanatic persons together with the presence of a monological belief system [248–250]. Just like them, believers in conspiracy theories interpret the evidence against them as proof of oppression [22], and they constantly engage in a confirmation bias in their way of thinking [251]. This is in accordance with the finding of a classical study which reported that whenever individuals strongly adhere to beliefs that turn out to be untrue, this causes discomfort [252].

It is important to clarify that believing in conspiracy theories differs radically from delusional ideas, although research findings are conflicting and inconclusive [253, 254], and heterogeneity is great [255, 256]. This difference might not be clear concerning some very bizarre conspiracy theories, but it seems true for the overwhelming majority of them. While both delusional ideas and conspiratorial thinking share a strong resistance to rational arguments and data checking, delusions are the product of the individual patient, while conspiracy theories come from an outside source, often in the form of a "guru," and the person adopts the position of the faithful follower, almost like in a religious way [17, 257, 258]. Delusional ideas have a direct and specific influence on the person itself, and the person is their center. On the contrary, conspiracy theories are more general; they concern a threat that is neither personalized nor immediate, and it is rarely taken into consideration when it comes to the everyday life. While the content of the delusional idea is present here and now and the person acts in response to this, followers of conspiracy theories are rather passive observers as if the threat is not real and constitutes a vague threat for a big group of people [259]. This is probably the reason why behavioral problems due to conspiracism are far fewer than expected, taking into consideration the percentage of believers in the population.

While conspiracy theories are based on intuitive and sentimental rather than analytical thinking [260], it seems that the balanced presence of much intuitive and minimal counterintuitiveness makes the most successful theories [261]. This is probably the explanation why quasi-religious or supernatural beliefs are combined with pseudoscientific or marginal scientific theories and often lead not only to contradictory claims and beliefs [258, 262–264] but also to complex theories that combine a significant number of unrelated and heterogeneous elements [265]. It is rare that a person believes in a single conspiracy theory [13, 238].

Taking into consideration all the above, it seems that the cognitive style alone is not a satisfactory explanation [266–268], and additionally, conspiracy theories appear more frequently when major catastrophic events occur, like fires, epidemics, and wars [269], probably because of the feelings of helplessness they cause [266]. It seems that stress and anxiety constitute a common substrate for conspiracy theories [270] as people have the tendency to over-recognize danger and threat [271–274] as well as to classify other persons into allies and foes [275, 276]. Maybe this is the mechanism behind the high rates of conspiracy beliefs among members of stigmatized minority groups [277, 278]. Interestingly, believing in conspiracy theories may satisfy key psychological motives [25] and often provide reassurance and a sense of safety by abolishing the stressful and chaotic reality, while sometimes it provides some kind of identity [8, 13, 279, 280]. A recent study proposed that beliefs in conspiracy theories acted as a maladjusting coping mechanism that might had protected believers from feelings of depression, helplessness, and despair during the COVID-19 pandemic (Fig. 25.24) [281]. For many followers it might even constitute the ideological underpinning of an emerging social movement [282] (Table 25.3).

At the end of the day, one could argue that finally Jean-Paul Sartre was right in a way. Sartre argued that insanity is the ultimate act of freedom; the human mind chooses madness over an unbearable and stressful reality (see also Chaps. 11, 21, and 22). Sartre was mistaken concerning psychosis, but research might suggest that his idea is directly relevant to the mechanism of the development of conspiracy theories.

Table 25.3 Comparison of characteristics between conspiracy theories and delusional ideas

Characteristic	Conspiracy theories	Delusional ideas
Originates from the person	No	Yes
Resistance to rational arguments	Yes	Yes
Affects everyday life	No	Yes
Induces fear or anxiety	Rarely	Often
Induces anger	Yes	Sometimes
Induces sense of safety	Often	No
Provides identity	Often	Rarely
Bizarre	Yes	Yes

Apart from psychological explanations, there seems to be a strong sociopolitical element [13, 283]. In social and political terms, the core problem is not the problematic interpretation of facts but what sources people trust [47]. While knowledge of history and involvement in politics were thought to influence such beliefs, a recent study revealed that 72% of African-Americans and 94% of Latinos have never heard of the Tuskegee Syphilis Study [284], or they do not understand the details and the historical meaning [285], suggesting that such knowledge cannot explain the high levels of mistrust in the official sources of information and the government.

While some authors argue that in democratic societies conspiracy theories could serve an ethical purpose by enforcing monitoring of sensitive activities [27], research suggests that conspiracists' ideation could be psychologically harmful or pathological [231, 246], but also their use by governments, political parties, or politicians could be sinister. It is frequent that political leaders utilize conspiracy theories to increase consolidation among their followers, leading to what was called "fusion paranoia" by Michael Kelly, a Washington Post journalist, or to the "paranoid style" in American politics as said by Richard Hofstadter [244]. When they lose elections, conspiracy theories serve the aim of explaining the results and keep their support in place. This in turn is in accordance with the ideas of Karl Popper, who argued that totalitarianism was always founded on "conspiracy theories" since they provided imaginary plots leading to tribalism, chauvinism, or racism. According to Popper, although genuine conspiracies do exist, their impact on history is small, and it is highly unlikely that they are successful or remain a secret for a long time [286]. Recent research clearly shows that the conspiratorial narratives directly undermine attitudes and behaviors that lie at the heart of democracy [287]. For many, Donald Trump's presidency has marked a new era in the utilization of these theories in politics [288] (Fig. 25.25).

In sociological terms, there is a strong element of individualism, and the image of the "courageous independent scientist resisting orthodoxy" (hero scientist), which fascinates the masses with reference to historical figures like Galileo Galilei and Giordano Bruno [289]. However, the essence of this image is not to provide alternative scientific explanations. Being critical is easy; actually having a better theory is very difficult and rarely true. At the core there is a

Fig. 25.25 The Eye of Providence, or the all-seeing eye of God, seen here on the US$1 bill, has been taken by some as evidence of a conspiracy involving the founders of the United States and the Illuminati. In essence, as a Freemasonry symbol, it appeared later in 1797 with the publication of Thomas Smith Webb's Freemasons Monitor

pervasive denial of science as a whole and in general [290], but education has no independent role in this. It seems there is an interplay of multiple psychological factors with education, but education has little effect [291]. On the other hand, many scientists simply manipulate conclusions on the basis that being critical is easy in order to obtain fame and citations. However, the effect of true social problems should not be overlooked or underappreciated. Big social and economic inequalities lead to suspiciousness and collective paranoia [278], and often, trivial socioeconomic changes are wrapped in a conspiracy explanation [292–294].

In terms of journalism, it seems that all kinds of data and news manipulation are in place, from various forms of reconfiguration to complete fabrication of news, with misinformation coming from politicians and celebrities enjoying the majority of mass and social media engagement. The media are filled with quackery and charlatans (cultropreneurs) offering fake cures in close alliance with "living icons," persons who provide the proof of concept with their lives as examples, and "praise singers," who are usually journalists, artists, and film-makers who promote the first two groups. Modern Internet made it easy to combine different theories in a perplex network (the hate multiverse) through what were called "wormhole links," further embedding their power and their ability to penetrate public opinion [47].

Celebrities are an interesting social group. Some of them, like Donald Trump, are also politicians. Many engage in the preaching of conspiracy theories, and it is not clear whether they honestly believe them or they use them to gain publicity. Concerning medical conspiracy theories, chances are they do believe them, as many of them are very vocal in promoting them, including Jenny McCarthy, Jim Carrey, Robert De Niro, and Bill Maher [295–297]. It is not unrelated that gossiping is a strong element that accompanies celebrity and also is an important adaptation in human evolution [298, 299]. What is even more alarming is that research has shown a positive association between anti-vaccination attitudes and celebrity admiration and interest, which was unrelated to the mistrust in vaccine benefits themselves [300]. This suggests that conspiracism is strongly related to a tendency to trust a certain type of sources of information, probably on the basis of emotional/instinctual choice.

Overall, research so far suggests that beliefs in conspiracism are driven by psychological mechanisms triggered by feelings of fear, anxiety, or depression. The most probable explanation is that these beliefs constitute a rather immature, but still normal, coping mechanism, which is dominant during childhood, and through which the person attempts to relief dysphoria and tension because of an unbearable sense of reality. Evolutionarily developed mental processes and features, including various biases, seeking of strong authoritarian protective figures, and the tendency to recognize threats, are recruited in order to formulate an alternative and less existentially threatening version of reality. These mechanisms resemble more those that are behind the formulation of religious beliefs or fanatic attachments to political ideologies rather than those behind clinical mental disorders and serve very deep psychological needs also by rejecting reason.

25.6 Politics

Conspiracism does not seem to be not linked to a specific political ideology or movement, either left or right, although specific theories might relate to a specific political ideology (e.g., climate change with right-wing and opposition to genetically modified foods with left-wing) [301]. However there seems to be a certain relationship with a tendency "against the system" as this system is reflected in all forms of establishment, from the government to the elites [12, 26]. In this frame, there seems to be a link with hypernationalism and religious fundamentalism [7], populism [302], political extremism of any kind [303], and radicalization of fringe groups [304]. It seems strongly related with the way people understand their political environment and respond to it. For many, believing in a conspiracy theory and being "against the system" makes them the "brave few" who resist and fight (Fig. 25.26). More specifically they seem to correlate with viewing the political system as less responsive to the needs and the demands of the citizens, and in many ways, these beliefs undermine those attitudes and behaviors at the core of the democratic process [287].

Jewish conspiracy theories are embedded in a very flexible way in the narratives of all kinds of political movements, and they are modified to serve the particular ideology. Recently similar theories have emerged concerning the Chinese [305, 306].

Fig. 25.26 Poster circulating on the Internet promoting the idea of the brave fighter against the corrupt system vs. the submissive mass

There is also the opposite view that conspiracy theories play a very important role in democracy, that is to keep a healthy democratic check on powerful elites, but this approach has little support [27, 307]. On the contrary, there is strong evidence that political forces and countries have weaponized conspiracy theories. The most notorious example is the 1905 fabrication of the "Protocols of the Elders of Zion," which is of unknown origin (some believe it was the Okhrana, the Russian secret service; Fig. 25.27). Another such example is the fabrication by the intelligence services in East Germany (German Democratic Republic (GDR)) of the theory that HIV was made in a military lab in the USA as part of the propaganda during the Cold War [308].

One of the most interesting cases is that of the QAnon, which constitutes a phenomenon with new unique characteristics, and while it is basically a conspiracy theory network, it seeks political power and manifests several characteristics mostly seen in religious cults. It appeared in October 2017 with a post on the anonymous imageboard 4chan by "Q," whose identity is still unknown. The "Q" refers either to the Q clearance used by the US Department of Energy or it is an influence from the novel *Q* by Luther Blissett (nom de plume for Roberto Bui, Giovanni Cattabriga, Federico Guglielmi, and Luca de Meo) [309]. The "Anon" refers to any anonymous or pseudonymous Internet poster. The basic narrative of QAnon is that many liberal Hollywood actors, Democratic politicians, and other "members of the elite" participate in the "cabal," a group that includes Barack Obama, Hillary Clinton, and George Soros, and is planning a coup d'état and is also running an international child sex trafficking ring related to adrenochrome harvesting. This narrative includes various other claims, including that the German Chancellor Angela Merkel is Adolf Hitler's granddaughter, many Hollywood celebrities are pedophiles, and that the

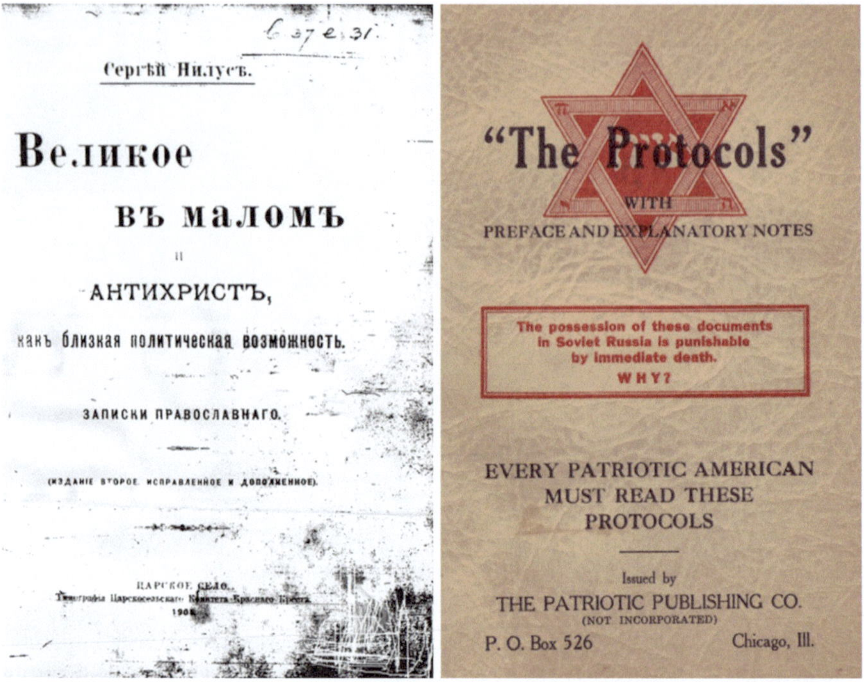

Fig. 25.27 Left: Cover of the first publication by Serge Nilus of *The Protocols of Zion* in 1905 in Russia. Right: A 1934 imprint by the so-called Patriotic Publishing Co. Some consider it to be cover action by Nazi agents operating in the USA at that time since the author and publisher do not exist

Rothschild family leads a satanic cult. The existence of a worldwide cabal of Satan-worshiping pedophiles who rule the world is central to QAnon theories, and it is combined with the belief that a great event, known as "the Storm" is imminent. During that event thousands of members of the cabal will be arrested, and the US military will take over the country by force. The narration has elements of religious millenarianism and apocalypticism, which has led several authors to suggest that it is essentially a religious movement in development. The *Time* Magazine included the "anonymous Q" among the 25 Most Influential People on the Internet for the year 2018.

One unique and innovative feature of the QAnon is that it provides followers with "clues" rather than presenting a theory directly, thus leaving space for an interaction, which makes the whole process addictive like a video game. In this process, posts have become more cryptic and vague, thus allowing everyone to project their own beliefs and tailor and develop the theory as they like and simultaneously have the feeling that they are participating in a historical process that is of an enormous importance.

QAnon interfered with the Donald Trump re-election campaign, and Trump himself was reported to have had amplified QAnon messaging by retweeting or

Fig. 25.28 The notorious photo of US Vice-President Mike Pence with members of the SWAT team of Broward County, Florida, on November 30, 2018. The man at the left of the image is displaying a red and black "Q" patch used by believers of the QAnon conspiracy theory on his uniform. The photo was tweeted, removed, and then substituted in Pence's feed

mentioning them several times a day. When challenged on August 19, 2020, during a press conference, he kept a neutral-positive attitude, but this was not the case with other members of the administration. More specifically Vice-President Mike Pence characterized it clearly as a conspiracy theory, while the FBI classified it as a potential domestic terror threat, which was something happening for the first time concerning a conspiracy theory.

An interesting event was the photo of the US Vice-President Mike Pence taken on November 30, 2018, with members of the SWAT team of Broward County, Florida (Fig. 25.28). In that photo, one of the SWAT officers was carrying the red and black "Q" patch used by QAnon followers on his uniform. The photograph triggered much discussions and reactions; the photo has been removed from the Tweet profile of the Vice-President, and the officer was disciplined.

25.7 The Role of the Internet

Many believe that a decisive factor that gave a new boost in the dissemination of conspiracy theories was the rise of electronic technology and, most especially, the social networks in the Internet [310]. It is reasonable to suggest that in human history, it was often technology that made the difference, and so this is the case here. Modern communication and networking technology have freed the spreading of rumors from their physical limitations and the need for proximity and launched them into the digital era.

The modern Internet-based network creates a huge number of echo chambers from which people can chose whom to listen to. This acts like an amplifier for the already embedded worldviews and silences the opposite arguments. It also brings

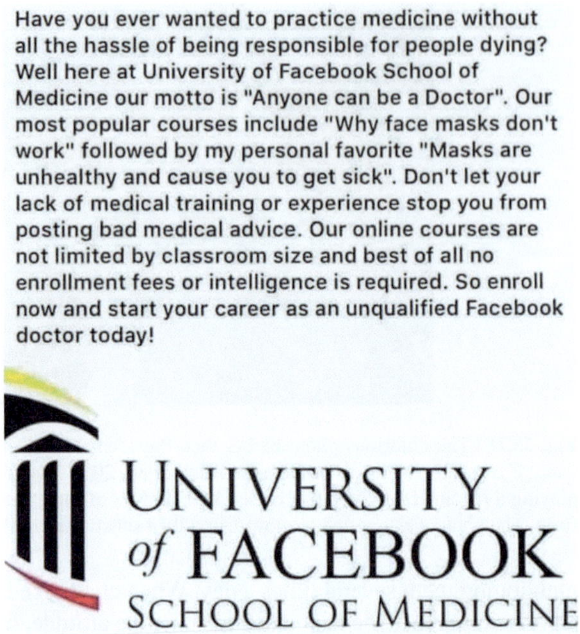

Fig. 25.29 A humoristic image circulating on the Internet, making fun of people that rely on search of unreliable sources on the Internet in order to develop strong medical theories and then argue fiercefully in favor of them on Facebook

together people and worldviews previously marginal and isolated and networks them at a global scale. This removes filters and makes continuous and fast evolution and adaptation of conspiracy theories possible as well as their interconnection and the creation of mega-theories.

As the social media platforms are made and are oriented toward the maximization of user engagement, the spreading of rumors and gossip as well as the spreading of fake news and conspiracy theories is much stronger than truth and rational arguments [47] (Fig. 25.29).

25.8 The Role of Bibliometrics and Publicity

The involvement of prominent scientists, including Nobel laureates and editors of high-impact factor journals, in the formulation and promotion of conspiracy theories, especially medical ones, is impressive but also worrying.

There seems to be a gray zone between genuine scientific dispute and conspiracism. Interestingly, many scientists, although in many topics they seem to follow the conspiracy path, also have significant contribution to science per se, and this seems awkward and an oxymoron. This blending is problematic, because in this way, they lend their reliability and personal scientific weight to conspiracy theories, and they justify them. Often, they are the "guru" or the "flag" of conspiracists, and they enjoy much publicity, sometimes to the degree of true worship.

The accusation of deliberate and organized fraud constitutes the fine but hard line that separates hard scientific discussion from conspiracy theories. The discussion that the X treatment option does not really work is a matter of scientific debate. Claims that the editor of a specific journal has taken side, or that a specific group of people or a pharma company have interests, and they push toward a specific conclusion might be true and are within the expected problems human nature and human society face all the time. We live in an imperfect world, and much of our struggle is to improve it. But to say that all the journals or all the pharma industry together with government agencies at a worldwide scale are in a sinister hidden agreement with qualities resembling organized crime, this is definitely conspiracism.

The question is, what makes these prominent men and women of science "cross the Rubicon" and violate the principles they have served for many years with significant success?

First of all, with the passing years, many prominent scientists acquire a narcissistic self-perception because of their success in the field. Gradually they stop truly inquiring; they tend to develop more general theories and seek deeper truth, but this comes at the expense of the methodology they utilize. There seems also to be some fatigue from continuous intellectual confrontations, and a sense of too much work for too little outcome appears. Cherry-picking and arbitrary interpretations replace rigorous work, and authority and prestige obtained with hard work over years kill any attitude to discuss and negotiate. This is not unusual as human nature pushes away from the tiresome and non-rewarding meticulous work toward a more relaxed, self-satisfying approach that provides clearer answers and simultaneously avoids the intellectual battlegrounds. However, this phenomenon, alone, is not sufficient to explain why some slip into the conspiracy realm.

A second element is that a small minority of scientists, usually those with the most radical approaches since the beginning of their career, eventually adopts the conspiracy rhetoric and essentially become conspiracists themselves. In retrospect, for the vast majority of them, one can see that their career was always characterized by a narcissistic flavor, anger, and unusual rigidity and hostility when supporting their findings and theories. They seem to lack the ability to self-control when things do not turn out as they think they should. Their views, even those accepted for consideration by the scientific community, were usually highly controversial; moreover, their scientific interests seem not to be balanced, but still they contributed to the field, and this is why they are considered successful, and their support of conspiracism is important. In many of them, in spite of the big success of their scientific contributions, one could often see a superficial knowledge and analysis and arbitrary conclusions even concerning some of their important contributions. Intellectuals often line up with denialists, and it seems frequent that similar views are shared by members of the scientific community at key-positions in the publication system [194].

The third element is some kind of sense of omnipotence, in combination with craving for citations and publicity. The big problem is that the way scientific publications and citations work today is much different from how they used to be and functioned 40 or 50 years ago. Bibliometrics are dominant in the way scientific work is assessed, and at the same time, the number of journals has proliferated, leading to problems in the quality of papers and the peer-review system. Reliable peer-reviewers are hard to find, and all citations are of equal value, regardless of what journal they appear in or whether they reflect a positive or negative comment. Most publications come from low-level authors, usually young PhD candidates or enthusiastic junior researchers in desperate need for citations. An extreme view would say that currently bibliometrics do not reflect scientific value, but they are rather a referendum among these young authors who, on the one hand, are enthusiastic with radical viewpoints and, at the same time, lack the depth of knowledge to judge the essence of papers.

A fourth and very dark element is attention and publicity craving, as well as deliberate chasing of citations by utilizing provocation. Of course, these are very serious ethical accusations and cannot be attributed to specific persons, since they cannot be proven. However, this could often be the case, especially with the pressure researchers face today to improve their metrics or to be able to receive grants. One cannot avoid noticing that many very successful scientific careers have been built on the basis of provocative and often arbitrary dispute of mainstream medicine, based on the very bad science these same individuals herald they fight against.

The metrics of the antidepressant debate are revealing these dynamics. The initial paper of Irwin Kirsch, who triggered a wave of rejection for antipsychotics and made the headlines in many newspapers and news media around the world (Fig. 11.54), enjoyed more than 2700 citations, and the Emperor's new drugs, in which he accused the industry and the FDA for conspiring [311], received more than 800 citations. For the reason of comparison, the two meta-analyses of the same dataset by the author, which were the first to shoot down Kirsch's claims, received only 68 [189] and 141 [188] citations, respectively, and the media were absolutely indifferent to their results. An opinion article in a low-impact journal by John Ioannidis supporting Kirsch's conclusions [312] enjoyed 267 citations, while in comparison, an article by the author of the current book in the top psychiatric journal in terms of impact factor received only 6 [313]. The full article clarifying the issue of antidepressants by David Nutt [314] received 31 citations, and the negative comment to this article by Peter C. Gøtzsche [68] received 56 (both published in the Lancet). Gotzche's book on *Deadly Psychiatry* [66] received 105 citations, and it is to be noted that his attack on mammography enjoyed thousands of them. One should be fair and mention that in spite of his extreme views, Gotzsche has a significant contribution to the refinement of methodology for evidence-based medicine, with a total of more than 100,000 citations, and when he was expelled from the Cochrane in 2018 because of his conduct, several colleagues supported him openly and accused the Cochrane of "slander, administrative incompetence and character assassination" [315].

What the COVID-19 crisis has shown is that many known conspiracy-inclined scientists also supported the COVID denialism theories, thus proving that there is a general tendency toward conspiracism and radical opposition as their aim rather than scientific discussion and interaction. The COVID crisis was also a judgment day for many. It revealed a profile of negative psychological as well as scientific characteristics, dominated by attention and publicity craving, in previously considered bright and strictly scientifically oriented scholars.

Last but not least, there seems to be a similar deep and disturbing problem with the journals and the publication field, but also with universities and the academia. It seems that the problematic bibliometric environment has contaminated the policy of journals that are more than ever oriented toward citation harvesting, while universities seem to lack the mechanisms to go beyond superficial bibliometrics when judging a curriculum vitae. The lack of adequate peer review is also evident more than ever. There is no other explanation when one sees reputed journals with impact factor above 6.5, published by top publishing houses to include papers authored by scientists affiliated with a major mainstream US university and which claim that:

> Serpentinization-induced lithospheric long-wavelength magnetic anomalies in Proterozoic bedrocks in a weakened geomagnetic field mediate the aberrant transformation of biogenic molecules in COVID-19 via magnetic catalysis' and that 'We propose here that carbon dioxide-rich water-peridotite (a ferromagnesian silicate) interactions generate abnormal lithospheric long-wavelength magnetic anomalies (LWMAs) via serpentinization, during conditions with increased terrestrial water storage and atmospheric carbon dioxide, and a weakened geomagnetic field. Furthermore, we provide evidence supporting a hypothesis, which posits, COVID-19 is a pathologic manifestation of resonant LWMAs-induced magnetic catalysis of iron oxides-silicate-like minerals from biogenic molecules and the coronavirus from endogenous viral elements, with the virus particles capable of replication and transmission to other hosts. We propose that those LWMAs are associated with the production of iron oxides-silicate rock minerals in tectonic plates with Proterozoic cratons. Thus, severe COVID-19 outbreaks are/will predominately occur in Eurasia and the Americas and are governed by the spatiotemporal dynamics of terrestrial water storage and the semiannual oscillation of the weakening geomagnetic magnetic field. We propose that the ferromagnetic-like iron stores in humans are the unifying determinant for COVID-19-induced morbidity and mortality. Furthermore, we propose that Nephrite-Jade amulets (a calcium-ferromagnesian silicate) developed by Neolithic Chinese Medicine to prevent thoracic organ disease, may prevent COVID-19 [316].

25.9 Impact of Conspiracy Theories on Behavior

The true impact of conspiracy theories on the behavior of the population and more specifically on political and health-related behavior has been the focus of discussions, but the question remains unanswered. It is more than certain that medical conspiracy theories have at least some negative effect on public health, but the true extent of this effect specifically has not been determined with certainty yet. It seems that the impressive percentages of believers in the general population are not translated directly into the same percentage of problematic behaviors. However some kind of relationship almost certainly exists.

It has been shown that medical mistrust could be associated with testing practices [317] and having a detectable viral load but not disclosing patients' HIV status to all their sexual partners [318]. Conflicting results also exist, making the interpretation difficult [319]. Concerning vaccination, data from the UK suggest that in parents, there is a strong negative correlation between believing in conspiracy theories and intention to vaccinate their children. Even simple exposure to anti-vaccination- and conspiracy-promoting material had an adverse effect on the attitudes toward vaccination [320]. Pakistan is one of the sole three remaining countries where polio has not been eradicated, and this is to a large extent the result of the widespread conspiracy theory that the polio vaccine is a ploy designed by the CIA to make Muslim men sterile [157]. Even in the middle of the COVID-19 pandemic, utilization of health-protective behaviors was negatively influenced by the belief in conspiracy theories [73], and with the obvious urgent need for an effective vaccine, 14%–25% of the US and French adults reported that they would not use one if it became available [321, 322]. This is in accordance with the explanation that conspiracism was behind the failure of the 2008 HPV vaccination campaign in Romania [323].

Similar negative effects have been recorded concerning the effect of conspiracy theories on social and political behaviors [324], including racist and discriminative behaviors [325] as well as delinquency [326] and even violence [327], which are justified by conspiracism as legitimate reactions.

25.10 Change

Since the belief in conspiracy theories is almost by definition resilient to rational arguments and the presentation of data, almost to a delusionary extent, and since such attempts are usually followed by further strengthening of these beliefs, the most important question is how modern society can deal with this potential threat to multiple societal functions, including public health and democracy.

Psychiatrists and psychologists will find themselves involved in this new and emerging battlefield, and the area is still uncharted. Overall, there are some data suggesting that rational and ridiculing arguments could have a beneficial effect, especially on people who are not hard believers; contrarily empathizing with the believers had no effect [2, 328].

Concerning medical conspiracies, one important action should be the change in the way benefits and possible adverse events are presented in official sites and are disseminated by official and governmental agencies. Currently there seems to be an over-presentation and unnecessary emphasis on the negative consequences, partially because of the need to avoid legal action if such an event occurs [329]. Some health information campaigns were proven to be successful and effective in changing the behaviors related to conspiracy theories [330].

However the fact that conspirational thinking is the prime cause behind these beliefs causes the modification of the message to be of limited effectiveness as an intervention [229]. On the contrary, there are data suggesting that these beliefs are further strengthened as a reaction to attempts to refute them on the basis of rational arguments, as these efforts are interpreted as orchestrated malevolent attacks on truth. This is known as the "backfire effect" [331–333]. Some authors argue on the presence of a stronger "elusive backfire effect," that is, people are persuaded by the presentation of data and the analytical discussion of theories [334], but this probably is not a universal effect and at any case, its potential is rather weak. In this frame, education seems to play a protective role [291], but surprisingly, in several cases it might also play a negative role, because individuals can reaffirm their suspicion by learning about real historical conspiracies [335, 336]. Research has shown that while many African-Americans have heard of the Tuskegee Syphilis experiments, most of them cannot fully understand its consequences [285].

Some authors propose a more proactive and aggressive action and that governments should impose a kind of censorship, with control of potential sources of disinformation over the Internet, since there lies the main core of conspiracism [73, 337, 338]. In this frame they also concluded that the authorities ought to engage in counterspeech and dialogue, which they termed "cognitive infiltration" [339]. This could include, among other things, the mandatory screening of films in certain occasions the citizen deals with the administration [340]. However it is evident that mandatory activities and suppression of conspiracist speech and censorship are highly problematic, controversial, and potentially dangerous actions, although the deliberate spread of false information cannot be protected the same way as free speech [341]. Other voices argue on the need to satisfy the principle of autonomy in medical ethics, as well as public engagement and communication [342], but again apart from theoretical value, the real-world efficacy of these proposals remains to be shown.

A very good practice could be the emphasis on rhetorical and emotional tools rather than rational and analytical arguments, with the engagement of celebrities eager to spread the message. Scare tactics returned mixed results as anti-smoking campaigns have clearly shown [343, 344], but advertisement-like and lifestyle campaigns seem more promising. One explanation could be that fear and anxiety predicts the general tendency to believe in conspiracy theories, and therefore scare tactics often backfire [270]. A balanced information in an appealing frame and with visual cues seems more efficacious [345, 346], and the creation of hearing groups with the sharing of personal experience as well as of networking could be of particular value because of the powerful effect of information transmission among groups and communities [260, 347–349]. Therefore, the strengthening of community links and support could be of great importance.

It is interesting that even under conditions of accelerating mortality, communities might assimilate health information rapidly, but unfortunately the change in behaviors is slow and incomplete, and the outcome is questionable [350].

Overall, research so far suggests that the efforts to fight conspiracy theories and their adverse effects on human behavior, especially health-related behaviors, should utilize rational but sentimentally wrapped messages delivered by celebrities rather than strict rational information by reliable sources and authorities.

Is this a negative view of the human nature? It might well be, but there is nothing new to it. We already know the way humans function psychologically and what the limits of human mental life are.

References

1. March E, Springer J. Belief in conspiracy theories: the predictive role of schizotypy, Machiavellianism, and primary psychopathy. PLoS One. 2019;14(12):e0225964. https://doi.org/10.1371/journal.pone.0225964.
2. van Prooijen JW, Douglas KM. Belief in conspiracy theories: basic principles of an emerging research domain. Eur J Soc Psychol. 2018;48(7):897–908. https://doi.org/10.1002/ejsp.2530.
3. Roisman J. The rhetoric of conspiracy in ancient Athens. Los Angeles: University of California Press; 2006.
4. Chagnon N. Yanomamo: the fierce people. New York, NY: Holt, Rinehart and Winston; 1983.
5. Evans-Pritchard EE. Witchcraft, oracles and magic among the azande. London: Clarendon Press; 1963.
6. von Rueden C, van Vugt M. Leadership in small-scale societies: some implications for theory, research, and practice. Leadersh Q. 2015;26(6):978–90. https://doi.org/10.1016/j.leaqua.2015.10.004.
7. West H, Sanders T. Transparency and conspiracy: ethnographies of suspicion in the new world order. Durham, NC: Duke University Press; 2003.
8. Andrade G. Medical conspiracy theories: cognitive science and implications for ethics. Med Health Care Philos. 2020;23(3):505–18. https://doi.org/10.1007/s11019-020-09951-6.
9. Anonymous. The medico-psychological association. J Ment Sci. 2018;16(73):134–42. https://doi.org/10.1192/bjp.16.73.134.
10. Johnson A, Ray PO. The repeal of the Missouri compromise: its origin and authorship. Am Hist Rev. 1909;14(4):835–6. https://doi.org/10.2307/1837085.
11. deHaven-Smith L. Conspiracy theory in America. University of Texas Press; 2014.
12. Mintz FP. The liberty lobby and the American right: race, conspiracy, and culture. Westport, CT: Greenwood; 1985.
13. Goertzel T. Belief in conspiracy theories. Polit Psychol. 1994;15(4):731–42. https://doi.org/10.2307/3791630.
14. Byford J. Conspiracy theories: a critical introduction. Houndmills, Basingstoke, Hampshire, New York: Palgrave Macmillan; 2011.
15. Keeley BL. Of conspiracy theories. J Philos. 1999;96(3):109–26. https://doi.org/10.2307/2564659.
16. Aaronovitch D. Voodoo histories: the role of the conspiracy theory in shaping modern history. Jonathan Cape; 2009.
17. Barkun M. A culture of conspiracy: apocalyptic visions in contemporary America. Berkeley: University of California Press; 2003.
18. Barkun M. Chasing phantoms: reality, imagination, and homeland security since 9/11. Chapel Hill: University of North Carolina Press; 2011.
19. McNeill P. The ethics and politics of human experimentation. Cambridge: University of Cambridge Press; 1993.
20. Moreno JD. Undue risk: secret state experiments on humans. New York, NY: Routledge; 2013.
21. Fauci A. Dr. Fauci says, “with all due modesty, i think i’m pretty effective”. InStyle; 2020.
22. Grimes DR. On the viability of conspiratorial beliefs. PLoS One. 2016;11(1):e0147905. https://doi.org/10.1371/journal.pone.0147905.
23. Ahmed W, Vidal-Alaball J, Downing J, Lopez Segui F. COVID-19 and the 5G conspiracy theory: social network analysis of twitter data. J Med Internet Res. 2020;22(5):e19458. https://doi.org/10.2196/19458.
24. Camp G, editor. Selling fear conspiracy theories and end-times paranoia. Baker Books; 1997.
25. Douglas KM, Sutton RM, Cichocka A. The psychology of conspiracy theories. Curr Dir Psychol Sci. 2017;26(6):538–42. https://doi.org/10.1177/0963721417718261.
26. Enders A. Conspiratorial thinking and political constraint. Public Opin Q. 2019;83(3):510–33.
27. Fenster M. Conspiracy theories: secrecy and power in American culture. 2nd ed. University of Minnesota Press; 2008.

28. Goldberg RA. Enemies within: the culture of conspiracy in modern America. Yale University Press; 2001.
29. Tingley D, Wagner G. Solar geoengineering and the chemtrails conspiracy on social media. Palgrave Commun. 2017;3(1):12. https://doi.org/10.1057/s41599-017-0014-3.
30. Eager PW. From freedom fighters to terrorists: women and political violence. Ashgate Publishing, Ltd; 2013.
31. Freeman D, Waite F, Rosebrock L, Petit A, Causier C, East A, Jenner L, Teale AL, Carr L, Mulhall S, Bold E, Lambe S. Coronavirus conspiracy beliefs, mistrust, and compliance with government guidelines in England. Psychol Med. 2020:1–13. https://doi.org/10.1017/S0033291720001890.
32. Kaparounaki CK, Patsali ME, Mousa DV, Papadopoulou EVK, Papadopoulou KKK, Fountoulakis KN. University students' mental health amidst the COVID-19 quarantine in Greece. Psychiatry Res. 2020;290:113111. https://doi.org/10.1016/j.psychres.2020.113111.
33. Patsali ME, Mousa DV, Papadopoulou EVK, Papadopoulou KKK, Kaparounaki CK, Diakogiannis I, Fountoulakis KN. University students' changes in mental health status and determinants of behavior during the COVID-19 lockdown in Greece. Psychiatry Res. 2020;292:113298. https://doi.org/10.1016/j.psychres.2020.113298.
34. Conspiracy theories of HIV/AIDS. Lancet. 2005;365(9458):448. https://doi.org/10.1016/S0140-6736(05)17875-1
35. Russell SL, Katz RV, Wang MQ, Lee R, Green BL, Kressin NR, Claudio C. Belief in AIDS origin conspiracy theory and willingness to participate in biomedical research studies: findings in whites, blacks, and Hispanics in seven cities across two surveys. HIV Clin Trials. 2011;12(1):37–47. https://doi.org/10.1310/hct1201-37.
36. Shearer C, West M, Caldeira K, Davis SJ. Quantifying expert consensus against the existence of a secret, large-scale atmospheric spraying program. Environ Res Lett. 2016;11:8.
37. Oliver JE, Wood T. Medical conspiracy theories and health behaviors in the United States. JAMA Intern Med. 2014;174(5):817–8. https://doi.org/10.1001/jamainternmed.2014.190.
38. Chen X, Zhang SX, Jahanshahi AA, Alvarez-Risco A, Dai H, Li J, Ibarra VG. Belief in a COVID-19 conspiracy theory as a predictor of mental health and Well-being of health care workers in Ecuador: cross-sectional survey study. JMIR Public Health Surveill. 2020;6(3):e20737. https://doi.org/10.2196/20737.
39. Hogg R, Nkala B, Dietrich J, Collins A, Closson K, Cui Z, Kanters S, Chia J, Barhafuma B, Palmer A, Kaida A, Gray G, Miller C. Conspiracy beliefs and knowledge about HIV origins among adolescents in Soweto, South Africa. PLoS One. 2017;12(2):e0165087. https://doi.org/10.1371/journal.pone.0165087.
40. Gray M. Conspiracy theories in the Arab world: sources and politics. Routledge; 2010.
41. Fatal ignorance. AIDS Asia. 1996;3(2):5.
42. Islam MS, Sarkar T, Khan SH, Mostofa Kamal AH, Hasan SMM, Kabir A, Yeasmin D, Islam MA, Amin Chowdhury KI, Anwar KS, Chughtai AA, Seale H. COVID-19-related Infodemic and its impact on public health: a global social media analysis. Am J Trop Med Hyg. 2020; https://doi.org/10.4269/ajtmh.20-0812.
43. Basch CH, Milano N, Hillyer GC. An assessment of fluoride related posts on Instagram. Health Promot Perspect. 2019;9(1):85–8. https://doi.org/10.15171/hpp.2019.11.
44. Briones R, Nan X, Madden K, Waks L. When vaccines go viral: an analysis of HPV vaccine coverage on YouTube. Health Commun. 2012;27(5):478–85. https://doi.org/10.1080/10410236.2011.610258.
45. Ekram S, Debiec KE, Pumper MA, Moreno MA. Content and commentary: HPV vaccine and youtube. J Pediatr Adolesc Gynecol. 2019;32(2):153–7. https://doi.org/10.1016/j.jpag.2018.11.001.
46. Sutton RM, Douglas KM. Agreeing to disagree: reports of the popularity of Covid-19 conspiracy theories are greatly exaggerated. Psychol Med. 2020:1–3. https://doi.org/10.1017/S0033291720002780.
47. Ball P, Maxmen A. The epic battle against coronavirus misinformation and conspiracy theories. Nature. 2020;581(7809):371–4. https://doi.org/10.1038/d41586-020-01452-z.

48. MacDonald R. Web of conspiracy: a guide to conspiracy theory sites on the internet. Emerald Group Publishing Limited; 2009. https://doi.org/10.1108/02640470910934894.
49. Wertheimer N, Leeper E. Electrical wiring configurations and childhood cancer. Am J Epidemiol. 1979;109(3):273–84. https://doi.org/10.1093/oxfordjournals.aje.a112681.
50. Draper G, Vincent T, Kroll ME, Swanson J. Childhood cancer in relation to distance from high voltage power lines in England and Wales: a case-control study. BMJ. 2005;330(7503):1290. https://doi.org/10.1136/bmj.330.7503.1290.
51. Linet MS, Hatch EE, Kleinerman RA, Robison LL, Kaune WT, Friedman DR, Severson RK, Haines CM, Hartsock CT, Niwa S, Wacholder S, Tarone RE. Residential exposure to magnetic fields and acute lymphoblastic leukemia in children. N Engl J Med. 1997;337(1):1–7. https://doi.org/10.1056/NEJM199707033370101.
52. Zhao L, Liu X, Wang C, Yan K, Lin X, Li S, Bao H, Liu X. Magnetic fields exposure and childhood leukemia risk: a meta-analysis based on 11,699 cases and 13,194 controls. Leuk Res. 2014;38(3):269–74. https://doi.org/10.1016/j.leukres.2013.12.008.
53. Amoon AT, Crespi CM, Ahlbom A, Bhatnagar M, Bray I, Bunch KJ, Clavel J, Feychting M, Hemon D, Johansen C, Kreis C, Malagoli C, Marquant F, Pedersen C, Raaschou-Nielsen O, Roosli M, Spycher BD, Sudan M, Swanson J, Tittarelli A, Tuck DM, Tynes T, Vergara X, Vinceti M, Wunsch-Filho V, Kheifets L. Proximity to overhead power lines and childhood leukaemia: an international pooled analysis. Br J Cancer. 2018;119(3):364–73. https://doi.org/10.1038/s41416-018-0097-7.
54. Health and Food Scientific Committee. Final opinion on Potential health effects of exposure to electromagnetic fields. 2015.
55. Pedersen C, Johansen C, Schuz J, Olsen JH, Raaschou-Nielsen O. Residential exposure to extremely low-frequency magnetic fields and risk of childhood leukaemia, CNS tumour and lymphoma in Denmark. Br J Cancer. 2015;113(9):1370–4. https://doi.org/10.1038/bjc.2015.365.
56. Tyler PE. Overview of electromagnetic radiation research: past, present, and future. Ann N Y Acad Sci. 1975;247(1):6–14. https://doi.org/10.1111/j.1749-6632.1975.tb35979.x.
57. Carpenter DO. The microwave syndrome or electro-hypersensitivity: historical background. Rev Environ Health. 2015;30(4):217–22. https://doi.org/10.1515/reveh-2015-0016.
58. Hocking B. Microwave sickness: a reappraisal. Occupat Med (Oxford, Engl). 2001;51(1):66–9. https://doi.org/10.1093/occmed/51.1.66.
59. Adair RK. Constraints on biological effects of weak extremely-low-frequency electromagnetic fields. Phys Rev A. 1991;43(2):1039–48. https://doi.org/10.1103/physreva.43.1039.
60. Hoffer A, Osmond H, Smythies J. Schizophrenia; a new approach. II. Result of a year's research. J Ment Sci. 1954;100(418):29–45. https://doi.org/10.1192/bjp.100.418.29.
61. Smythies J. The adrenochrome hypothesis of schizophrenia revisited. Neurotox Res. 2002;4(2):147–50. https://doi.org/10.1080/10298420290015827.
62. Fritze RH. Invented knowledge: false history, fake science and pseudo-religions. London: Reaktion Books; 2009.
63. Yair G. The conspiracist strategy: lessons from American alternative health promotions. Eur J Am Stud. 2019;14(2).
64. Gotzsche P. Deadly medicines and organised crime: how big pharma has corrupted healthcare. CRC Press; 2013.
65. Gotzsche PC. Big pharma often commits corporate crime, and this must be stopped. BMJ. 2012;345:e8462. https://doi.org/10.1136/bmj.e8462.
66. Gotzsche P. Deadly psychiatry and organised denial. ArtPeople; 2015.
67. Gøtzsche P. Psychiatry ignores an elephant in the room. Mad in America. 2017. https://www.madinamerica.com/2017/09/psychiatry-ignores-elephant-room/.
68. Gotzsche PC. Why I think antidepressants cause more harm than good. Lancet Psychiatry. 2014;1(2):104–6. https://doi.org/10.1016/S2215-0366(14)70280-9.
69. Gøtzsche PC. The coronavirus pandemic: can we handle such epidemics better? J R Soc Med. 2020;113(5):171–5. https://doi.org/10.1177/0141076820924587.

70. Shermer M. Cures and Cons. Natural scams "he" doesn't want you to know about. Scientific American online. 2006. https://www.scientificamerican.com/article/cures-and-cons/.
71. Trudeau K. Natural cures "they" don't want you to know about. Alliance Publishing; 2005.
72. Desta TT, Mulugeta T. Living with COVID-19-triggered pseudoscience and conspiracies. Int J Public Health. 2020;65(6):713–4. https://doi.org/10.1007/s00038-020-01412-4.
73. Allington D, Duffy B, Wessely S, Dhavan N, Rubin J. Health-protective behaviour, social media usage and conspiracy belief during the COVID-19 public health emergency. Psychol Med. 2020:1–7. https://doi.org/10.1017/S003329172000224X.
74. Georgiou N, Delfabbro P, Balzan R. COVID-19-related conspiracy beliefs and their relationship with perceived stress and pre-existing conspiracy beliefs. Pers Individ Dif. 2020;166:110201. https://doi.org/10.1016/j.paid.2020.110201.
75. Andersen KG, Rambaut A, Lipkin WI, Holmes EC, Garry RF. The proximal origin of SARS-CoV-2. Nat Med. 2020;26(4):450–2. https://doi.org/10.1038/s41591-020-0820-9.
76. Calisher C, Carroll D, Colwell R, Corley RB, Daszak P, Drosten C, Enjuanes L, Farrar J, Field H, Golding J, Gorbalenya A, Haagmans B, Hughes JM, Karesh WB, Keusch GT, Lam SK, Lubroth J, Mackenzie JS, Madoff L, Mazet J, Palese P, Perlman S, Poon L, Roizman B, Saif L, Subbarao K, Turner M. Statement in support of the scientists, public health professionals, and medical professionals of China combatting COVID-19. Lancet. 2020;395(10226):e42–3. https://doi.org/10.1016/S0140-6736(20)30418-9.
77. Perez JC, Montagnier L. Covid-19, sars and bats coronaviruses genomes peculiar homologous rna sequences. Int J Res GRANTHAALAYAH. 2020;8(7):217–63. https://doi.org/10.29121/granthaalayah.v8.i7.2020.678.
78. Montagnier L, Aissa J, Ferris S, Montagnier JL, Lavallee C. Electromagnetic signals are produced by aqueous nanostructures derived from bacterial DNA sequences. Interdiscip Sci. 2009;1(2):81–90. https://doi.org/10.1007/s12539-009-0036-7.
79. Montagnier L, Aissa J, Lavallee C, Mbamy M, Varon J, Chenal H. Electromagnetic detection of HIV DNA in the blood of AIDS patients treated by antiretroviral therapy. Interdiscip Sci. 2009;1(4):245–53. https://doi.org/10.1007/s12539-009-0059-0.
80. Enserink M. French Nobelist escapes "intellectual terror" to pursue radical ideas in China. Science. 2010;330(6012):1732. https://doi.org/10.1126/science.330.6012.1732.
81. Broderick JF, Miller DW. The assasination of JFK. In: MacDonald R, editor. Web of conspiracy: a guide to conspiracy theory sites on the internet, vol. 27. Emerald Group Publishing Limited; 2009. p. 190–1. https://doi.org/10.1108/02640470910934894.
82. Bugliosi V. Reclaiming history: the assassination of president John F. Kennedy. W. W. Norton & Company; 2007.
83. Garwood C. Flat earth, the history of an infamous. Macmillian: Idea; 2007.
84. Peterson J. Solving the mystery of the Colorado Brown Stain. J Hist Dent. 1997;45(2):57–61.
85. Unde MP, Patil RU, Dastoor PP. The untold story of fluoridation: revisiting the changing perspectives. Indian J Occup Environ Med. 2018;22(3):121–7. https://doi.org/10.4103/ijoem.IJOEM_124_18.
86. Petersen PE, Lennon MA. Effective use of fluorides for the prevention of dental caries in the 21st century: the WHO approach. Community Dent Oral Epidemiol. 2004;32(5):319–21. https://doi.org/10.1111/j.1600-0528.2004.00175.x.
87. Iheozor-Ejiofor Z, Worthington HV, Walsh T, O'Malley L, Clarkson JE, Macey R, Alam R, Tugwell P, Welch V, Glenny AM. Water fluoridation for the prevention of dental caries. Cochrane Database Syst Rev. 2015;2015(6):CD010856. https://doi.org/10.1002/14651858.CD010856.pub2.
88. Pizzo G, Piscopo MR, Pizzo I, Giuliana G. Community water fluoridation and caries prevention: a critical review. Clin Oral Investig. 2007;11(3):189–93. https://doi.org/10.1007/s00784-007-0111-6.
89. SCHER. Fluoridation. 2011. https://ec.europa.eu/health/scientific_committees/opinions_layman/fluoridation/en/about.htm#7.
90. Centers for Disease Control and Prevention. Ten great public health achievements--United States, 1900-1999. MMWR Morb Mortal Wkly Rep. 1999;48(12):241–3.

91. Armfield JM. When public action undermines public health: a critical examination of antifluoridationist literature. Aust New Zealand Health Policy. 2007;4:25. https://doi.org/10.1186/1743-8462-4-25.
92. Freeze R, Lehr J. The fluoride wars: how a modest public health measure became America's longest-running political melodrama. Wiley; 2009.
93. Dickinson D. Myths or theories? Alternative beliefs about HIV and AIDS in South African working class communities. Afr J AIDS Res. 2013;12(3):121–30. https://doi.org/10.2989/16085906.2013.863212.
94. Bogart LM, Bird ST. Exploring the relationship of conspiracy beliefs about HIV/AIDS to sexual behaviors and attitudes among African-American adults. J Natl Med Assoc. 2003;95(11):1057–65.
95. Thorburn S, Bogart LM. Conspiracy beliefs about birth control: barriers to pregnancy prevention among African Americans of reproductive age. Health Educ Behav. 2005;32(4):474–87. https://doi.org/10.1177/1090198105276220.
96. Papadopulos-Eleopulos E. Reappraisal of AIDS--is the oxidation induced by the risk factors the primary cause? Med Hypotheses. 1988;25(3):151–62. https://doi.org/10.1016/0306-9877(88)90053-9.
97. Papadopulos-Eleopulos E, Turner VF, Papadimitriou JM. Is a positive western blot proof of HIV infection? Bio/technol (Nature Publishing Company). 1993;11(6):696–707. https://doi.org/10.1038/nbt0693-696.
98. Epstein S. Impure science. University of California Press; 1996.
99. Baumann E, Bethell T, Bialy H, Duesberg PH, Farber C, Geshekter CL, Johnson PE, Maver RW, Schoch R, Stewart GT, et al. AIDS proposal. Group for the scientific reappraisal of the HIV/AIDS hypothesis. Science. 1995;267(5200):945–6. https://doi.org/10.1126/science.7863334.
100. Duesberg P. HIV is not the cause of AIDS. Science. 1988;241(4865):514–7. https://doi.org/10.1126/science.3399880.
101. Duesberg P. Defective viruses and AIDS. Nature. 1989;340(6234):515. https://doi.org/10.1038/340515a0.
102. Duesberg P. Virus hunting and the scientific method. Science. 1991;251(4995):724. https://doi.org/10.1126/science.2025322.
103. Duesberg P. AIDS: the alternative view. Lancet. 1992;339(8808):1547.
104. Duesberg P. HIV-free AIDS reports. Science. 1992;257(5078):1848. https://doi.org/10.1126/science.1411497.
105. Duesberg P. Aetiology of AIDS. Lancet. 1993;341(8859):1544. https://doi.org/10.1016/0140-6736(93)90691-9.
106. Duesberg P. HIV and AIDS. Science. 1993;260(5115):1705–6. https://doi.org/10.1126/science.8511572.
107. Duesberg P. HIV and the aetiology of AIDS. Lancet. 1993;341(8850):957–8. https://doi.org/10.1016/0140-6736(93)91247-j.
108. Duesberg P. AIDS data. Science. 1995;268(5209):350–2.
109. Duesberg P. Commentary: non-HIV hypotheses must be studied more carefully. BMJ. 1996;312(7025):210–1. https://doi.org/10.1136/bmj.312.7025.210.
110. Duesberg P, Bialy H. HIV results in the frame. HIV an illusion. Nature. 1995;375(6528):197.; author reply 198. https://doi.org/10.1038/375197a0.
111. Duesberg P, Koehnlein C, Rasnick D. The chemical bases of the various AIDS epidemics: recreational drugs, anti-viral chemotherapy and malnutrition. J Biosci. 2003;28(4):383–412. https://doi.org/10.1007/BF02705115.
112. Duesberg P, Rasnick D. The AIDS dilemma: drug diseases blamed on a passenger virus. Genetica. 1998;104(2):85–132. https://doi.org/10.1023/a:1003405220186.
113. Duesberg PH. Human immunodeficiency virus and acquired immunodeficiency syndrome: correlation but not causation. Proc Natl Acad Sci U S A. 1989;86(3):755–64. https://doi.org/10.1073/pnas.86.3.755.

114. Duesberg PH. Duesberg replies. Nature. 1990;346(6287):788. https://doi.org/10.1038/346788b0.
115. Duesberg PH. AIDS epidemiology: inconsistencies with human immunodeficiency virus and with infectious disease. Proc Natl Acad Sci U S A. 1991;88(4):1575–9. https://doi.org/10.1073/pnas.88.4.1575.
116. Duesberg PH. Duesberg on AIDS and HIV. Nature. 1991;350(6313):10. https://doi.org/10.1038/350010a0.
117. Duesberg PH. HIV as target for zidovudine. Lancet. 1992;339(8792):551. https://doi.org/10.1016/0140-6736(92)90368-d.
118. Duesberg PH. The HIV gap in national AIDS statistics. Bio/Technol (Nature Publishing Company). 1993;11(8):955–6. https://doi.org/10.1038/nbt0893-955.
119. Duesberg PH. HIV latency. Bio/Technol (Nature Publishing Company). 1993;11(3):247. https://doi.org/10.1038/nbt0393-247a.
120. Duesberg PH. Is HIV the cause of AIDS? Lancet. 1995;346(8986):1371–2. https://doi.org/10.1016/s0140-6736(95)92389-6.
121. Duesberg PH, Nicholson JM, Rasnick D, Fiala C, Bauer HH. WITHDRAWN: HIV-AIDS hypothesis out of touch with South African AIDS—a new perspective. Med Hypotheses. 2009; https://doi.org/10.1016/j.mehy.2009.06.024.
122. Duesberg PH, Zhou RP, Goodrich D. Cancer genes by illegitimate recombination. Ann N Y Acad Sci. 1989;567:259–73. https://doi.org/10.1111/j.1749-6632.1989.tb16477.x.
123. Blattner W, Gallo RC, Temin HM. HIV causes AIDS. Science. 1988;241(4865):515–6. https://doi.org/10.1126/science.3399881.
124. Anti-AIDS vitamin advertising banned. Nature. 2008;453(7198):969. https://doi.org/10.1038/453969c.
125. Mullis K. Dancing naked in the mind field. Vintage Books; 1998.
126. Mullis KB. A hypothetical disease of the immune system that may bear some relation to the acquired immune deficiency syndrome. Genetica. 1995;95(1–3):195–7. https://doi.org/10.1007/BF01435010.
127. Fourie P, Meyer M. The politics of AIDS denialism. New York, NY: Routledge; 2010.
128. Institute of Medicine Committee for the Oversight of AIDS Activities. Confronting AIDS: update 1988. Washington DC: National Academies Press; 1988.
129. Kalichman S. The psychology of AIDS denialism pseudoscience, conspiracy thinking, and medical mistrust. Eur Psychol. 2014;19(1):13–22.
130. Chigwedere P, Seage GR 3rd, Gruskin S, Lee TH, Essex M. Estimating the lost benefits of antiretroviral drug use in South Africa. J Acquir Immune Defic Syndr. 2008;49(4):410–5. https://doi.org/10.1097/qai.0b013e31818a6cd5.
131. Cohen J. The Duesberg phenomenon. Science. 1994;266(5191):1642–4. https://doi.org/10.1126/science.7992043.
132. The Durban declaration. Nature. 2000;406 (6791):15–16. https://doi.org/10.1038/35017662.
133. Nattrass N. AIDS and the scientific governance of medicine in post-apartheid South Africa. Afr Aff. 2008;107(427):157–76. https://doi.org/10.1093/afraf/adm087.
134. Smith TC, Novella SP. HIV denial in the internet era. PLoS Med. 2007;4(8):e256. https://doi.org/10.1371/journal.pmed.0040256.
135. Watson J. Scientists, activists sue South Africa's AIDS 'denialists'. Nat Med. 2006;12(1):6. https://doi.org/10.1038/nm0106-6a.
136. Tang MW, Shafer RW. HIV-1 antiretroviral resistance: scientific principles and clinical applications. Drugs. 2012;72(9):e1–25. https://doi.org/10.2165/11633630-000000000-00000.
137. Motsoaledi A. Aaron Motsoaledi: South African minister of health. Interview by Clare Kapp. Lancet. 2009;374(9692):776. https://doi.org/10.1016/S0140-6736(09)61580-4.
138. Papagiannidou M. Goodbye AIDS: did it ever exist? Impact Investigative Media Productions; 2009.
139. Kofta M, Soral W, Bilewicz M. What breeds conspiracy antisemitism? The role of political uncontrollability and uncertainty in the belief in Jewish conspiracy. J Pers Soc Psychol. 2020;118(5):900–18. https://doi.org/10.1037/pspa0000183.

140. Fekete L. The Muslim conspiracy theory and the Oslo massacre. Race Class. 2012;53:30–47.
141. Kaplan J. Encyclopedia of white power: a sourcebook on the radical racist right. AltaMira Press; 2000.
142. Perry B. White genocide': white supremacists and the politics of reproduction. In: Ferber A, editor. Home-grown hate: gender and organized racism. Routledge; 2003. p. 75–85.
143. Agins DB. Maya Angelou: diversity makes for a rich tapestry. Berkeley Heights, NJ: Enslow Publishers; 2006.
144. Smith M. Birth control and the negro woman. Ebony. 1968;23(5):29.
145. Weisbord RG. Birth control and the black American: a matter of genocide? Demography. 1973;10(4):571–90. https://doi.org/10.2307/2060884.
146. Wright N. Black power vs. black genocide. The Black Scholar. 1969;1(2):47–52.
147. Henig JR, Rich WC. Mayors in the middle: politics, race, and mayoral control of urban schools. Princeton University Press; 2004.
148. Baxby D. Edward Jenner's inquiry; a bicentenary analysis. Vaccine. 1999;17(4):301–7. https://doi.org/10.1016/s0264-410x(98)00207-2.
149. Dube E, Vivion M, MacDonald NE. Vaccine hesitancy, vaccine refusal and the anti-vaccine movement: influence, impact and implications. Expert Rev Vaccines. 2015;14(1):99–117. https://doi.org/10.1586/14760584.2015.964212.
150. Dyer C. Whooping cough vaccine on trial again. Br Med J (Clin Res Ed). 1987;295(6605):1053–4. https://doi.org/10.1136/bmj.295.6605.1053.
151. Kulenkampff M, Schwartzman JS, Wilson J. Neurological complications of pertussis inoculation. Arch Dis Child. 1974;49(1):46–9. https://doi.org/10.1136/adc.49.1.46.
152. Mendelsohn R. The medical time bomb of immunization against disease. East West J. 1984.
153. Wakefield AJ, Murch SH, Anthony A, Linnell J, Casson DM, Malik M, Berelowitz M, Dhillon AP, Thomson MA, Harvey P, Valentine A, Davies SE, Walker-Smith JA. Ileal-lymphoid-nodular hyperplasia, non-specific colitis, and pervasive developmental disorder in children. Lancet. 1998;351(9103):637–41. https://doi.org/10.1016/s0140-6736(97)11096-0.
154. Goldacre B. Bad science. London: Harper Collins; 2008.
155. DeStefano F, Bhasin TK, Thompson WW, Yeargin-Allsopp M, Boyle C. Age at first measles-mumps-rubella vaccination in children with autism and school-matched control subjects: a population-based study in metropolitan Atlanta. Pediatrics. 2004;113(2):259–66. https://doi.org/10.1542/peds.113.2.259.
156. Hooker BS. Measles-mumps-rubella vaccination timing and autism among young African American boys: a reanalysis of CDC data. Transl Neurodeg. 2014;3:16. https://doi.org/10.1186/2047-9158-3-16.
157. Andrade GE, Hussain A. Polio in Pakistan: political, sociological, and epidemiological factors. Cureus. 2018;10(10):e3502. https://doi.org/10.7759/cureus.3502.
158. Fanon F. The wretched of the earth. Harmondsworth: Penguin Books; 1967.
159. Bortolotti L. Rationality and sanity: the role of rationality judgements in understanding psychiatric disorders. In: Fulford KWM, Davies M, Gipps R, et al., editors. The Oxford handbook of philosophy and psychiatry. Oxford University Press; 2013. https://doi.org/10.1093/oxfordhb/9780199579563.001.0001.
160. Fulford KWM, Thornton T, Graham G. Natural classification, realism and psychiatric science. In: Fulford KWM, Thornton T, Graham G, editors. The Oxford textbook of philosophy and psychiatry. Oxford University Press; 2006.
161. Szasz T. The myth of mental illness: foundations of a theory of personal conduct. New York, NY: Hober-Harper; 1961.
162. Edwards RB. Mental health as rational autonomy. J Med Philos. 1981;6(3):309–22. https://doi.org/10.1093/jmp/6.3.309.
163. Szasz TS. The myth of mental illness. New York, NY: Harper and Row; 1974.
164. Foucault M. Histoire de la folie à l'âge classique. Paris: Plon; 1961.
165. Goffman E. Asylums: essays on the social situation of mental patients and other inmates. New York, NY: Anchor Books; 1961.

166. Leoni F. From madness to mental illness: psychiatry and biopolitics in Michel Faoucault. In: Fulford KWM, Davies M, Gipps R, et al., editors. The Oxford handbook of philosophy and psychiatry. Oxford University Press; 2013. https://doi.org/10.1093/oxfordhb/9780199579563.001.0001.
167. Cooper D. Psychiatry and anti-psychiatry. London: Tavistock; 1967.
168. Laing R. The politics of experience. Harmondsworth: Penguin Books; 1967.
169. Scheff TJ. The labelling theory of mental illness. Am Sociol Rev. 1974;39(3):444–52.
170. Sedgwick P. Illness--mental and otherwise. Stud Hastings Cent. 1973;1(3):19–40.
171. Ban TA. Academic psychiatry and the pharmaceutical industry. Prog Neuro-Psychopharmacol Biol Psychiatry. 2006;30(3):429–41. https://doi.org/10.1016/j.pnpbp.2005.11.014.
172. Swenson K. Selling sickness: how the world's biggest pharmaceutical companies are turning us all into patients. Can J Commun. 2009;34(2). https://doi.org/10.22230/cjc.2009v34n2a2231.
173. Whitaker R. Anatomy of an epidemic: magic bullets, psychiatric drugs, and the astonishing rise of mental illness in America. New York, NY: Random House; 2010.
174. Wright D. The medicalization of society: on the transformation of human conditions into treatable disorders. By Peter Conrad. (Baltimore: Johns Hopkins University Press, 2007. xiv + 204 p., notes, bibl., index. ISBN 0-8018-8585-X $22). Scientia Canadensis: Canad J Hist Sci Technol Med. 2010;33(1):119. https://doi.org/10.7202/1000847ar.
175. Burns T. Psychiatry: a very short introduction. Oxford University Press; 2006.
176. O'Brien AP, Woods M, Palmer C. The emancipation of nursing practice: applying anti-psychiatry to the therapeutic community. Aust N Z J Ment Health Nurs. 2001;10(1):3–9. https://doi.org/10.1046/j.1440-0979.2001.00183.x.
177. Weitz D. Call me antipsychiatry activist--not "consumer". Ethical Hum Sci Serv. 2003;5(1):71–2.
178. Maughan D, Molodynski A. An international perspective on the acceptability and sustainability of electroconvulsive therapy. BJPsych Int. 2016;13(1):10–2. https://doi.org/10.1192/s2056474000000891.
179. McDonald A, Walter G. Hollywood and ECT. Int Rev Psychiatry. 2009;21(3):200–6. https://doi.org/10.1080/09540260902747888.
180. Andrade C, Shah N, Venkatesh BK. The depiction of electroconvulsive therapy in Hindi cinema. J ECT. 2010;26(1):16–22. https://doi.org/10.1097/YCT.0b013e3181d017ba.
181. Kesey K. One flew over the Cuckoo's nest. London: Methuen; 1962.
182. Webb G. Dark alliance: the CIA, the contras, and the cocaine explosion. Seven Stories Press; 2019.
183. Harrington A. The fall of the schizophrenogenic mother. Lancet. 2012;379(9823):1292–3. https://doi.org/10.1016/s0140-6736(12)60546-7.
184. Neill J. Whatever became of the schizophrenogenic mother? Am J Psychother. 1990;44(4):499–505. https://doi.org/10.1176/appi.psychotherapy.1990.44.4.499.
185. Kirsch I. Antidepressants and the placebo response. Epidemiol Psichiatr Soc. 2009;18(4):318–22. https://doi.org/10.1017/s1121189x00000282.
186. Kirsch I, Deacon BJ, Huedo-Medina TB, Scoboria A, Moore TJ, Johnson BT. Initial severity and antidepressant benefits: a meta-analysis of data submitted to the Food and Drug Administration. PLoS Med. 2008;5(2):e45. https://doi.org/10.1371/journal.pmed.0050045.
187. Kirsch I. The emperor's new drugs: exploding the antidepressant myth. London: The Bodley Head; 2009.
188. Fountoulakis KN, Moller HJ. Efficacy of antidepressants: a re-analysis and re-interpretation of the Kirsch data. Int J Neuropsychopharmacol. 2011;14(3):405–12. https://doi.org/10.1017/S1461145710000957.
189. Fountoulakis KN, Veroniki AA, Siamouli M, Moller HJ. No role for initial severity on the efficacy of antidepressants: results of a multi-meta-analysis. Ann General Psychiatry. 2013;12(1):26. https://doi.org/10.1186/1744-859X-12-26.
190. Horder J, Matthews P, Waldmann R. Placebo, prozac and PLoS: significant lessons for psychopharmacology. J Psychopharmacol. 2011;25(10):1277–88. https://doi.org/10.1177/0269881110372544.

191. Angell M. The truth about the drug companies: how they deceive us and what to do about it. Random House; 2004.
192. Angell M. The epidemic of mental illness: why? 2011. http://www.nybooks.com/articles/archives/2011/jun/23/epidemic-mental-illness-why/?pagination=false. Accessed 21 Aug 2011.
193. Angell M. The illusions of psychiatry. 2011. http://www.nybooks.com/articles/archives/2011/jul/14/illusions-of-psychiatry/?pagination=false.
194. Fountoulakis KN, Hoschl C, Kasper S, Lopez-Ibor J, Moller HJ. The media and intellectuals' response to medical publications: the antidepressants' case. Ann General Psychiatry. 2013;12(1):11. https://doi.org/10.1186/1744-859X-12-11.
195. Fountoulakis KN, Moeller H. Antidepressant drugs and the response in the placebo group: the real problem lies in our understanding of the issue. J Psychopharmacol. 2011;26(5):744–50.
196. Fountoulakis KN, Möller H-J. Antidepressants vs. placebo: not merely a quantitative difference in response. Int J Neuropsychopharmacol. 2011;14(10):1435–7. https://doi.org/10.1017/s1461145711000964.
197. Tracy A. Has anti-depressant myth bubble burst? 2008. http://www.rense.com/general81/1v.htm. Accessed 25 Aug 2011.
198. The Guardian. The creation of the Prozac myth. 2008. http://www.guardian.co.uk/society/2008/feb/27/mentalhealth.health1. Accessed 25 Aug 2011.
199. Moncrieff J. Antipsychotic maintenance treatment: time to rethink? PLoS Med. 2015;12(8):e1001861. https://doi.org/10.1371/journal.pmed.1001861.
200. Moncrieff J. Inconvenient truths about antipsychotics: a response to Goff et al. Mad in America. 2017. https://www.madinamerica.com/2017/05/inconvenient-truths-about-antipsychotics-a-response-to-goff-et-al/.
201. Murray RM, Quattrone D, Natesan S, van Os J, Nordentoft M, Howes O, Di Forti M, Taylor D. Should psychiatrists be more cautious about the long-term prophylactic use of antipsychotics? Br J Psychiatry. 2016;209(5):361–5. https://doi.org/10.1192/bjp.bp.116.182683.
202. Moncrieff J. Does antipsychotic withdrawal provoke psychosis? Review of the literature on rapid onset psychosis (supersensitivity psychosis) and withdrawal-related relapse. Acta Psychiatr Scand. 2006;114(1):3–13. https://doi.org/10.1111/j.1600-0447.2006.00787.x.
203. Johnstone EC, Crow TJ, Frith CD, Husband J, Kreel L. Cerebral ventricular size and cognitive impairment in chronic schizophrenia. Lancet. 1976;2(7992):924–6. https://doi.org/10.1016/s0140-6736(76)90890-4.
204. Marsden CD. Cerebral atrophy and cognitive impair ment in chronic schizophrenia. Lancet. 1976;308(7994):1079. https://doi.org/10.1016/s0140-6736(76)90984-3.
205. Ho BC, Andreasen NC, Nopoulos P, Arndt S, Magnotta V, Flaum M. Progressive structural brain abnormalities and their relationship to clinical outcome: a longitudinal magnetic resonance imaging study early in schizophrenia. Arch Gen Psychiatry. 2003;60(6):585–94. https://doi.org/10.1001/archpsyc.60.6.585.
206. Murray RM. Mistakes I have made in my research career. Schizophr Bull. 2017;43(2):253–6. https://doi.org/10.1093/schbul/sbw165.
207. Vita A, De Peri L, Deste G, Barlati S, Sacchetti E. The effect of antipsychotic treatment on cortical gray matter changes in schizophrenia: does the class matter? A meta-analysis and meta-regression of longitudinal magnetic resonance imaging studies. Biol Psychiatry. 2015;78(6):403–12. https://doi.org/10.1016/j.biopsych.2015.02.008.
208. Zipursky RB, Reilly TJ, Murray RM. The myth of schizophrenia as a progressive brain disease. Schizophr Bull. 2013;39(6):1363–72. https://doi.org/10.1093/schbul/sbs135.
209. Fountoulakis KN, Moeller HJ, Kasper S, Tamminga C, Yamawaki S, Kahn R, Tandon R, Correll CU, Javed A. The report of the joint WPA/CINP workgroup on the use and usefulness of antipsychotic medication in the treatment of schizophrenia. CNS Spectr. 2020:1–75. https://doi.org/10.1017/S1092852920001546.
210. Gøtzsche P. It's unlikely that 'antidepressants' have a true effect on depression. Mad in America. 8 Jan 2018. https://www.madinamerica.com/2018/2001/antidepressants-effect-depression/.

211. Gøtzsche P. The review on antidepressant withdrawal that cochrane won't publish. Mad in America. 11 Feb 2020. https://www.madinamerica.com/2020/2002/review-cochrane-wont-publish/.
212. Guthrie S. Faces in the clouds: a new theory of religion. Oxford University Press; 1995.
213. Haselton MG, Buss DM. Error management theory: a new perspective on biases in cross-sex mind reading. J Pers Soc Psychol. 2000;78(1):81–91. https://doi.org/10.1037//0022-3514.78.1.81.
214. Heider F, Simmel M. An experimental study of apparent behavior. Am J Psychol. 1944;57(2):243–59. https://doi.org/10.2307/1416950.
215. Imhoff R, Bruder M. Speaking (Un-)truth to power: conspiracy mentality as a generalised political attitude. Eur J Personal. 2014;28(1):25–43. https://doi.org/10.1002/per.1930.
216. Kelemen D. The scope of teleological thinking in preschool children. Cognition. 1999;70(3):241–72. https://doi.org/10.1016/s0010-0277(99)00010-4.
217. van Prooijen JW, van Vugt M. Conspiracy theories: evolved functions and psychological mechanisms. Perspect Psychol Sci. 2018;13(6):770–88. https://doi.org/10.1177/1745691618774270.
218. van Prooijen JW, Douglas KM, De Inocencio C. Connecting the dots: illusory pattern perception predicts belief in conspiracies and the supernatural. Eur J Soc Psychol. 2018;48(3):320–35. https://doi.org/10.1002/ejsp.2331.
219. Vosoughi S, Roy D, Aral S. The spread of true and false news online. Science. 2018;359(6380):1146–51. https://doi.org/10.1126/science.aap9559.
220. Zehmakan NA, Galam S. Rumor spreading: a trigger for proliferation or fading away. Chaos. 2020;30(7):073122. https://doi.org/10.1063/5.0006984.
221. Barkun M. Conspiracy theories as stigmatized knowledge. Diogenes. 2016; https://doi.org/10.1177/0392192116669288.
222. Rozenblit L, Keil F. The misunderstood limits of folk science: an illusion of explanatory depth. Cogn Sci. 2002;26(5):521–62. https://doi.org/10.1207/s15516709cog2605_1.
223. Sperber D. Introduction. In: Metarepresentations: a multidisciplinary perspective. Oxford: Oxford University Press; 2000.
224. Brotherton R. Suspicious minds: why we believe conspiracy theories. New York, NY: Bloomsbury; 2013.
225. Ebel-Lam AP, Fabrigar LR, MacDonald TK, Jones S. Balancing causes and consequences: the magnitude-matching principle in explanations for complex social events. Basic Appl Soc Psychol. 2010;32(4):348–59. https://doi.org/10.1080/01973533.2010.519245.
226. Piaget J, Inhelder B. The psychology of the child. 1969.
227. Malinowski B. Magic, science and religion. New York, NY: Waveland; 1992.
228. Rosset E. It's no accident: our bias for intentional explanations. Cognition. 2008;108(3):771–80.
229. Hornsey MJ, Harris EA, Fielding KS. The psychological roots of anti-vaccination attitudes: a 24-nation investigation. Health Psychol. 2018;37(4):307–15. https://doi.org/10.1037/hea0000586.
230. Barron D, Furnham A, Weis L, Morgan KD, Towell T, Swami V. The relationship between schizotypal facets and conspiracist beliefs via cognitive processes. Psychiatry Res. 2018;259:15–20. https://doi.org/10.1016/j.psychres.2017.10.001.
231. Barron D, Morgan K, Towell T, Altemeyer B, Swami V. Associations between schizotypy and belief in conspiracist ideation. Personal Individ Differ. 2014;70:156–9. https://doi.org/10.1016/j.paid.2014.06.040.
232. Dagnall N, Drinkwater K, Parker A, Denovan A, Parton M. Conspiracy theory and cognitive style: a worldview. Front Psychol. 2015;6:206. https://doi.org/10.3389/fpsyg.2015.00206.
233. Darwin H, Neave N, Holmes J. Belief in conspiracy theories. The role of paranormal belief, paranoid ideation and schizotypy. Personal Individ Differ. 2011;50(8):1289–93. https://doi.org/10.1016/j.paid.2011.02.027.
234. Douglas KM, Sutton RM. Does it take one to know one? Endorsement of conspiracy theories is influenced by personal willingness to conspire. Br J Soc Psychol. 2011;50(3):544–52. https://doi.org/10.1111/j.2044-8309.2010.02018.x.

235. Raab MH, Ortlieb SA, Auer N, Guthmann K, Carbon CC. Thirty shades of truth: conspiracy theories as stories of individuation, not of pathological delusion. Front Psychol. 2013;4:406. https://doi.org/10.3389/fpsyg.2013.00406.
236. Swami V, Weis L, Lay A, Barron D, Furnham A. Associations between belief in conspiracy theories and the maladaptive personality traits of the personality inventory for DSM-5. Psychiatry Res. 2016;236:86–90. https://doi.org/10.1016/j.psychres.2015.12.027.
237. van Elk M. Perceptual biases in relation to paranormal and conspiracy beliefs. PLoS One. 2015;10(6):e0130422. https://doi.org/10.1371/journal.pone.0130422.
238. Swami V, Coles R, Stieger S, Pietschnig J, Furnham A, Rehim S, Voracek M. Conspiracist ideation in Britain and Austria: evidence of a monological belief system and associations between individual psychological differences and real-world and fictitious conspiracy theories. Br J Psychol. 2011;102(3):443–63. https://doi.org/10.1111/j.2044-8295.2010.02004.x.
239. Cichocka A, Marchlewska M, de Zavala AG. Does self-love or self-hate predict conspiracy beliefs? Narcissism, self-esteem, and the endorsement of conspiracy theories. Soc Psychol Personal Sci. 2015;7(2):157–66. https://doi.org/10.1177/1948550615616170.
240. Cichocka A, Marchlewska M, Golec de Zavala A, Olechowski M. 'They will not control us': Ingroup positivity and belief in intergroup conspiracies. Br J Psychol. 2016;107(3):556–76. https://doi.org/10.1111/bjop.12158.
241. Leone L, Giacomantonio M, Lauriola M. Moral foundations, worldviews, moral absolutism and belief in conspiracy theories. Int J Psychol. 2019;54(2):197–204. https://doi.org/10.1002/ijop.12459.
242. Lukic P, Zezelj I, Stankovic B. How (Ir)rational is it to believe in contradictory conspiracy theories? Eur J Psychol. 2019;15(1):94–107. https://doi.org/10.5964/ejop.v15i1.1690.
243. Cichocka A, Marchlewska M, Golec de Zavala A, Olechowski M. 'They will not control us': Ingroup positivity and belief in intergroup conspiracies. Br J Psychol (Lond, Engl: 1953). 2016;107(3):556–76. https://doi.org/10.1111/bjop.12158.
244. Hofstadter R. The paranoid style in American politics. New York, NY: Vintage; 2012.
245. Wood MJ, Douglas KM. Online communication as a window to conspiracist worldviews. Front Psychol. 2015;6:836. https://doi.org/10.3389/fpsyg.2015.00836.
246. Freeman D, Bentall RP. The concomitants of conspiracy concerns. Soc Psychiatry Psychiatr Epidemiol. 2017;52(5):595–604. https://doi.org/10.1007/s00127-017-1354-4.
247. Bessi A, Coletto M, Davidescu GA, Scala A, Caldarelli G, Quattrociocchi W. Science vs conspiracy: collective narratives in the age of misinformation. PLoS One. 2015;10(2):e0118093. https://doi.org/10.1371/journal.pone.0118093.
248. Galliford N, Furnham A. Individual difference factors and beliefs in medical and political conspiracy theories. Scand J Psychol. 2017;58(5):422–8. https://doi.org/10.1111/sjop.12382.
249. Hagen K. Conspiracy theorists and monological belief systems. Argumentation. 2018;3:2.
250. Wood MJ. Propagating and debunking conspiracy theories on twitter during the 2015-2016 zika virus outbreak. Cyberpsychol Behav Soc Netw. 2018;21(8):485–90. https://doi.org/10.1089/cyber.2017.0669.
251. Klayman J, Ha Y-w. Confirmation, disconfirmation, and information in hypothesis testing. Psychol Rev. 1987;94(2):211–28. https://doi.org/10.1037/0033-295x.94.2.211.
252. Festinger L. A theory of cognitive dissonance. A theory of cognitive dissonance. Stanford University Press; 1957.
253. Goreis A, Voracek M. A systematic review and meta-analysis of psychological research on conspiracy beliefs: field characteristics, measurement instruments, and associations with personality traits. Front Psychol. 2019;10:205. https://doi.org/10.3389/fpsyg.2019.00205.
254. Jasinskaja-Lahti I, Jetten J. Unpacking the relationship between religiosity and conspiracy beliefs in Australia. Br J Soc Psychol. 2019;58(4):938–54. https://doi.org/10.1111/bjso.12314.
255. Klein C, Clutton P, Dunn AG. Pathways to conspiracy: the social and linguistic precursors of involvement in Reddit's conspiracy theory forum. PLoS One. 2019;14(11):e0225098. https://doi.org/10.1371/journal.pone.0225098.

256. Klein C, Clutton P, Polito V. Topic modeling reveals distinct interests within an online conspiracy forum. Front Psychol. 2018;9:189. https://doi.org/10.3389/fpsyg.2018.00189.
257. Birchall C. Just because you're paranoid, doesn't mean they're not out to get you. Cult Mach. 2004;6.
258. Franks B, Bangerter A, Bauer MW. Conspiracy theories as quasi-religious mentality: an integrated account from cognitive science, social representations theory, and frame theory. Front Psychol. 2013;4:424. https://doi.org/10.3389/fpsyg.2013.00424.
259. Van Prooijen JW, Van Lange PAM. The social dimension of belief in conspiracy theories. In: Van Prooijen JW, Van Lange PAM, editors. Power, politics, and paranoia: why people are suspicious of their leaders. Cambridge: Cambridge University Press; 2014.
260. Swami V, Voracek M, Stieger S, Tran US, Furnham A. Analytic thinking reduces belief in conspiracy theories. Cognition. 2014;133(3):572–85. https://doi.org/10.1016/j.cognition.2014.08.006.
261. Boyer PT. The naturalness of religious ideas. Los Angeles: University of California Press; 1994.
262. Norenzayan A, Atran S, Faulkner J, Schaller M. Memory and mystery: the cultural selection of minimally counterintuitive narratives. Cogn Sci. 2006;30(3):531–53. https://doi.org/10.1207/s15516709cog0000_68.
263. Slone J. Theological incorrectness: why religious people believe what they shouldn't. Oxford: Oxford University Press; 2004.
264. Wood MJ, Douglas KM, Sutton RM. Dead and alive. Soc Psychol Personal Sci. 2012;3(6):767–73. https://doi.org/10.1177/1948550611434786.
265. Banks N. AIDS, opium, diamonds, and empire: the deadly virus of international greed. New York, NY: I Universe; 2010.
266. Abalakina-Paap M, Stephan WG, Craig T, Gregory WL. Beliefs in conspiracies. Polit Psychol. 2002;20(3):637–47. https://doi.org/10.1111/0162-895x.00160.
267. Dieguez S, Wagner-Egger P, Gauvrit N. Nothing happens by accident, or does it? A low prior for randomness does not explain belief in conspiracy theories. Psychol Sci. 2015;26(11):1762–70. https://doi.org/10.1177/0956797615598740.
268. Jimenez AV, Mesoudi A, Tehrani JJ. No evidence that omission and confirmation biases affect the perception and recall of vaccine-related information. PLoS One. 2020;15(3):e0228898. https://doi.org/10.1371/journal.pone.0228898.
269. McCauley C, Jacques S. The popularity of conspiracy theories of presidential assassination: a Bayesian analysis. J Pers Soc Psychol. 1979;37(5):637–44. https://doi.org/10.1037/0022-3514.37.5.637.
270. Green R, Douglas KM. Anxious attachment and belief in conspiracy theories. Personal Individ Differ. 2018;125:30–7. https://doi.org/10.1016/j.paid.2017.12.023.
271. Brotherton R, French CC. Intention seekers: conspiracist ideation and biased attributions of intentionality. PLoS One. 2015;10(5):e0124125. https://doi.org/10.1371/journal.pone.0124125.
272. Neuberg SL, Kenrick DT, Schaller M. Human threat management systems: self-protection and disease avoidance. Neurosci Biobehav Rev. 2011;35(4):1042–51. https://doi.org/10.1016/j.neubiorev.2010.08.011.
273. Ohman A, Flykt A, Esteves F. Emotion drives attention: detecting the snake in the grass. J Exp Psychol Gen. 2001;130(3):466–78. https://doi.org/10.1037//0096-3445.130.3.466.
274. Rozin P, Fallon AE. A perspective on disgust. Psychol Rev. 1987;94(1):23–41. https://doi.org/10.1037/0033-295X.94.1.23.
275. Cikara M, Botvinick MM, Fiske ST. Us versus them: social identity shapes neural responses to intergroup competition and harm. Psychol Sci. 2011;22(3):306–13. https://doi.org/10.1177/0956797610397667.
276. Tooby J, Cosmides L. Conceptual foundations of evolutionary psychology. In: Buss D, editor. Handbook of evolutionary psychology. London: Wiley; 2015.

277. Davis J, Wetherell G, Henry PJ. Social devaluation of African Americans and race-related conspiracy theories. Eur J Soc Psychol. 2018;48(7):999–1010. https://doi.org/10.1002/ejsp.2531.
278. Swami V, Coles R. The truth is out there. The Psychol. 2010;23(7):560–3.
279. Baigent M, Leigh R, Lincoln H. The messianic legacy. Henry Holt & Co.; 1987.
280. Sternisko A, Cichocka A, Van Bavel JJ. The dark side of social movements: social identity, non-conformity, and the lure of conspiracy theories. Curr Opin Psychol. 2020;35:1–6. https://doi.org/10.1016/j.copsyc.2020.02.007.
281. Fountoulakis K, Apostolidou M, Atsiova M, Filippidou A, Florou A, Gousiou D, Katsara A, Mantzari S, Padouva-Markoulaki M, Papatriantafyllou E, Sacharidi P, Tonia A, Tsagalidou E, Zymara V, Prezerakos P, Koupidis S, Fountoulakis N, Chrousos G. Self-reported changes in anxiety, depression and suicidality during the COVID-19 lockdown in Greece. J Affect Dis. 2021;279:624–9.
282. Franks B, Bangerter A, Bauer MW, Hall M, Noort MC. Beyond "monologicality"? Exploring conspiracist worldviews. Front Psychol. 2017;8:861. https://doi.org/10.3389/fpsyg.2017.00861.
283. Douglas KM, Sutton RM. The hidden impact of conspiracy theories: perceived and actual influence of theories surrounding the death of Princess Diana. J Soc Psychol. 2008;148(2):210–21. https://doi.org/10.3200/SOCP.148.2.210-222.
284. Mays VM, Coles CN, Cochran SD. Is there a legacy of the U.S. Public Health Syphilis Study at Tuskegee in HIV/AIDS-related beliefs among heterosexual African-Americans and Latinos? Ethics Behav. 2012;22(6):461–71. https://doi.org/10.1080/10508422.2012.730805.
285. Freimuth VS, Quinn SC, Thomas SB, Cole G, Zook E, Duncan T. African Americans' views on research and the Tuskegee Syphilis Study. Soc Sci Med. 2001;52(5):797–808. https://doi.org/10.1016/s0277-9536(00)00178-7.
286. Popper K. Open society and its enemies, book II. London: Routledge and Kegan Paul; 1945.
287. Ardevol-Abreu A, Gil de Zuniga H, Gamez E. The influence of conspiracy beliefs on conventional and unconventional forms of political participation: the mediating role of political efficacy. Br J Soc Psychol. 2020;59(2):549–69. https://doi.org/10.1111/bjso.12366.
288. Hellinger D. Conspiracy and conspiracy theories in the age of trump. New York, NY: Palgrave; 2019.
289. Goertzel T. Conspiracy theories in science. EMBO Rep. 2010;11(7):493–9. https://doi.org/10.1038/embor.2010.84.
290. Denying science. Nat Med. 2006;12(4):369. https://doi.org/10.1038/nm0406-369.
291. van Prooijen JW. Why education predicts decreased belief in conspiracy theories. Appl Cogn Psychol. 2017;31(1):50–8. https://doi.org/10.1002/acp.3301.
292. Foster GM. Limited good or limited goods: observations on Acheson. Am Anthropol. 1974;76(1):53–7.
293. Pasek J, Sood G, Krosnick JA. Misinformed about the affordable care act? Leveraging certainty to assess the prevalence of misperceptions. J Commun. 2015;65(4):660–73. https://doi.org/10.1111/jcom.12165.
294. Quadagno J. Right-wing conspiracy? Socialist plot? The origins of the patient protection and affordable care act. J Health Polit Policy Law. 2014;39(1):35–56. https://doi.org/10.1215/03616878-2395172.
295. Bearman P. Just-so stories: vaccines, autism, and the single-bullet disorder. Soc Psychol Q. 2009;73:112–5. https://doi.org/10.1177/0190272510371672.
296. Gottlieb SDJB. Vaccine resistances reconsidered: vaccine skeptics and the Jenny McCarthy effect. BioSocieties. 2016;11:152–74.
297. Sharfstein JM. Vaccines and the trump administration. JAMA. 2017;317(13):1305–6. https://doi.org/10.1001/jama.2017.2311.
298. Dunbar R. Grooming, gossip and the evolution of language. Salem: Harvard University Press; 1996.

299. McAndrew FT, Milenkovic MA. Of tabloids and family secrets: the evolutionary psychology of gossip1. J Appl Soc Psychol. 2002;32(5):1064–82. https://doi.org/10.1111/j.1559-1816.2002.tb00256.x.
300. Martinez-Berman L, McCutcheon L, Huynh HP. Is the worship of celebrities associated with resistance to vaccinations? Relationships between celebrity admiration, anti-vaccination attitudes, and beliefs in conspiracy. Psychol Health Med. 2020:1–10. https://doi.org/10.1080/13548506.2020.1778754.
301. Lewandowsky S, Gignac GE, Oberauer K. The role of conspiracist ideation and worldviews in predicting rejection of science. PLoS One. 2013;8(10):e75637. https://doi.org/10.1371/journal.pone.0075637.
302. Castanho Silva B, Vegetti F, Littvay L. The elite is up to something: exploring the relation between populism and belief in conspiracy theories. Swiss Polit Sci Rev. 2017;23(4):423–43. https://doi.org/10.1111/spsr.12270.
303. van Prooijen J-W, Krouwel APM, Pollet TV. Political extremism predicts belief in conspiracy theories. Soc Psychol Personal Sci. 2015;6(5):570–8. https://doi.org/10.1177/1948550614567356.
304. Bartlett J, Miller C. The power of unreason: conspiracy theories, extremism and counter-terrorism. London: Demos; 2010.
305. Nefes TS. Scrutinizing impacts of conspiracy theories on readers' political views: a rational choice perspective on anti-semitic rhetoric in Turkey. Br J Sociol. 2015;66(3):557–75. https://doi.org/10.1111/1468-4446.12137.
306. Swami V. Social psychological origins of conspiracy theories: the case of the Jewish conspiracy theory in Malaysia. Front Psychol. 2012;3:280. https://doi.org/10.3389/fpsyg.2012.00280.
307. Dentith M. The philosophy of conspiracy theories. New York, NY: Palgrave; 2014.
308. Jeppsson A. How East Germany fabricated the myth of HIV being man-made. J Int Assoc Provid AIDS Care. 2017;16(6):519–22. https://doi.org/10.1177/2325957417724203.
309. Blissett L. Q. Einaudi; 1999.
310. Clarke S. Conspiracy theories and the internet: controlled demolition and arrested development. Episteme. 2012;4(2):167–80. https://doi.org/10.3366/epi.2007.4.2.167.
311. Kirsch I. The emperor's new drugs: exploding the antidepressant myth. New York, NY: Basic Books; 2010.
312. Ioannidis JPA. Effectiveness of antidepressants: an evidence myth constructed from a thousand randomized trials? Philos Ethics Humanit Med. 2008;3(1):14. https://doi.org/10.1186/1747-5341-3-14.
313. Fountoulakis KN, Moller HJ. Are antidepressants clinically useful? Conclusion of a decade of debate. World Psychiatry. 2014;13(2):201–2. https://doi.org/10.1002/wps.20112.
314. Nutt DJ, Goodwin GM, Bhugra D, Fazel S, Lawrie S. Attacks on antidepressants: signs of deep-seated stigma? Lancet Psychiatry. 2014;1(2):102–4. https://doi.org/10.1016/S2215-0366(14)70232-9.
315. Ioannidis JPA. Cochrane crisis: secrecy, intolerance and evidence-based values. Eur J Clin Investig. 2019;49(3):e13058. https://doi.org/10.1111/eci.13058.
316. Bility MT, Agarwal Y, Ho S, Castronova I, Beatty C, Biradar S, Narala V, Periyapatna N, Chen Y, Nachega J. Can Traditional Chinese Medicine provide insights into controlling the COVID-19 pandemic: Serpentinization-induced lithospheric long-wavelength magnetic anomalies in Proterozoic bedrocks in a weakened geomagnetic field mediate the aberrant transformation of biogenic molecules in COVID-19 via magnetic catalysis. Sci Total Environ. 2020; https://doi.org/10.1016/j.scitotenv.2020.142830.
317. Patev AJ, Hood KB, Speed KJ, Cartwright PM, Kinman BA. HIV conspiracy theory beliefs mediates the connection between HIV testing attitudes and HIV prevention self-efficacy. J Am Coll Heal. 2019;67(7):661–73. https://doi.org/10.1080/07448481.2018.1500472.
318. Quinn KG, Kelly JA, DiFranceisco WJ, Tarima SS, Petroll AE, Sanders C, Lawrence JSS, Amirkhanian YA. The health and sociocultural correlates of AIDS genocidal beliefs and

medical mistrust among African American MSM. AIDS Behav. 2018;22(6):1814–25. https://doi.org/10.1007/s10461-016-1657-6.
319. Ojikutu BO, Amutah-Onukagha N, Mahoney TF, Tibbitt C, Dale SD, Mayer KH, Bogart LM. HIV-related mistrust (or HIV conspiracy theories) and willingness to use PrEP among black women in the United States. AIDS Behav. 2020;24(10):2927–34. https://doi.org/10.1007/s10461-020-02843-z.
320. Jolley D, Douglas KM. The effects of anti-vaccine conspiracy theories on vaccination intentions. PLoS One. 2014;9(2):e89177. https://doi.org/10.1371/journal.pone.0089177.
321. Coconel Group. A future vaccination campaign against COVID-19 at risk of vaccine hesitancy and politicisation. Lancet Infect Dis. 2020;20(7):769–70. https://doi.org/10.1016/S1473-3099(20)30426-6.
322. Reiter PL, Pennell ML, Katz ML. Acceptability of a COVID-19 vaccine among adults in the United States: how many people would get vaccinated? Vaccine. 2020;38(42):6500–7. https://doi.org/10.1016/j.vaccine.2020.08.043.
323. Craciun C, Baban A. "Who will take the blame?": understanding the reasons why Romanian mothers decline HPV vaccination for their daughters. Vaccine. 2012;30(48):6789–93. https://doi.org/10.1016/j.vaccine.2012.09.016.
324. Jolley D, Douglas KM. The social consequences of conspiracism: exposure to conspiracy theories decreases intentions to engage in politics and to reduce one's carbon footprint. Br J Psychol. 2014;105(1):35–56. https://doi.org/10.1111/bjop.12018.
325. Jolley D, Meleady R, Douglas KM. Exposure to intergroup conspiracy theories promotes prejudice which spreads across groups. Br J Psychol. 2020;111(1):17–35. https://doi.org/10.1111/bjop.12385.
326. Jolley D, Douglas KM, Leite AC, Schrader T. Belief in conspiracy theories and intentions to engage in everyday crime. Br J Soc Psychol. 2019;58(3):534–49. https://doi.org/10.1111/bjso.12311.
327. Jolley D, Paterson JL. Pylons ablaze: examining the role of 5G COVID-19 conspiracy beliefs and support for violence. Br J Soc Psychol. 2020;59(3):628–40. https://doi.org/10.1111/bjso.12394.
328. Orosz G, Kreko P, Paskuj B, Toth-Kiraly I, Bothe B, Roland-Levy C. Changing conspiracy beliefs through rationality and ridiculing. Front Psychol. 2016;7:1525. https://doi.org/10.3389/fpsyg.2016.01525.
329. Okuhara T, Ishikawa H, Okada M, Kato M, Kiuchi T. Contents of Japanese pro- and anti-HPV vaccination websites: a text mining analysis. Patient Educ Couns. 2018;101(3):406–13. https://doi.org/10.1016/j.pec.2017.09.014.
330. Bode L, Vraga EK. See something, say something: correction of global health misinformation on social media. Health Commun. 2018;33(9):1131–40. https://doi.org/10.1080/10410236.2017.1331312.
331. Carey JM, Chi V, Flynn DJ, Nyhan B, Zeitzoff T. The effects of corrective information about disease epidemics and outbreaks: evidence from Zika and yellow fever in Brazil. Sci Adv. 2020;6(5):eaaw7449. https://doi.org/10.1126/sciadv.aaw7449.
332. Nyhan B, Reifler J. When corrections fail: the persistence of political misperceptions. Polit Behav. 2010;32(2):303–30. https://doi.org/10.1007/s11109-010-9112-2.
333. Nyhan B, Reifler J, Richey S, Freed GL. Effective messages in vaccine promotion: a randomized trial. Pediatrics. 2014;133(4):e835–42. https://doi.org/10.1542/peds.2013-2365.
334. Wood T, Porter E. The elusive backfire effect: mass attitudes' steadfast factual adherence. Polit Behav. 2019;41:135–63.
335. Kata A. A postmodern Pandora's box: anti-vaccination misinformation on the internet. Vaccine. 2010;28(7):1709–16. https://doi.org/10.1016/j.vaccine.2009.12.022.
336. Nelson JC, Adams G, Branscombe NR, Schmitt MT. The role of historical knowledge in perception of race-based conspiracies. Race Soc Probl. 2010;2(2):69–80. https://doi.org/10.1007/s12552-010-9031-1.

337. O'Neill O. Some limits of informed consent. J Med Ethics. 2003;29(1):4–7. https://doi.org/10.1136/jme.29.1.4.
338. Weiser PJ. The future of internet regulation. UC Davis L Rev. 2009;43:529.
339. Sunstein CR, Vermeule A. Conspiracy theories. Causes and Cures. 2009;17(2):202–27.
340. Evers S, Farewell VT, Halloran PF. Public awareness of organ donation. CMAJ. 1988;138(3):237–9.
341. Lidsky LB. Where's the harm: free speech and the regulation of lies. Wash & Lee L Rev. 2008;65:1091.
342. Heldman AB, Schindelar J, Weaver JB. Social media engagement and public health communication: implications for public health organizations being truly "social". Public Health Rev. 2013;35(1):13. https://doi.org/10.1007/bf03391698.
343. Backer TE, Rogers E, Sopory P. Designing health communication campaigns: what works? Newbury Park, CA: Sage; 1992.
344. Hill D, Chapman S, Donovan R. The return of scare tactics. Tob Control. 1998;7(1):5–8. https://doi.org/10.1136/tc.7.1.5.
345. Garcia-Retamero R, Cokely ET. Communicating health risks with visual aids. Curr Dir Psychol Sci. 2013;22(5):392–9. https://doi.org/10.1177/0963721413491570.
346. Leshner G, Cheng IH. The effects of frame, appeal, and outcome extremity of antismoking messages on cognitive processing. Health Commun. 2009;24(3):219–27. https://doi.org/10.1080/10410230902804117.
347. Acocella I. The focus groups in social research: advantages and disadvantages. Qual Quant. 2011;46(4):1125–36. https://doi.org/10.1007/s11135-011-9600-4.
348. Jimenez AV, Stubbersfield JM, Tehrani JJ. An experimental investigation into the transmission of antivax attitudes using a fictional health controversy. Soc Sci Med. 2018;215:23–7. https://doi.org/10.1016/j.socscimed.2018.08.032.
349. Kitzinger J. Qualitative research. Introducing focus groups. BMJ. 1995;311(7000):299–302. https://doi.org/10.1136/bmj.311.7000.299.
350. Abramowitz S, McKune SL, Fallah M, Monger J, Tehoungue K, Omidian PA. The opposite of denial: social learning at the onset of the Ebola emergency in Liberia. J Health Commun. 2017;22(Suppl 1):59–65. https://doi.org/10.1080/10810730.2016.1209599.

Chapter 26
The Future of Psychiatry

> In spite of becoming more technocratic, psychiatry will probably become even more interpersonally demanding in the future, simply because the need to communicate the benefits of technocratic advances will increase, and eventually its success will make the real difference.

Psychiatry has an interesting past, and much debate currently exists on what its future will be. This discussion concerns mostly not a prediction of what is going to happen and how psychiatry will evolve but rather what stakeholders in mental health should do and what path of actions they should follow to shape a particular future on the basis of the unmet needs that we chart today. Everybody asks for more funds in his specific area of expertise, which he/she considers to be at the core of the future developments in psychiatry and mental health. But is this realistic?

In order to formulate a fair prediction of the future, we need to have at least an elementary understanding of the mechanisms, opposing forces, and general environment, both academic but also professional in the field of psychiatry and mental health in general.

The main impression one has after reviewing the history of psychiatry is that there has always been a reaction to any action, a counter-theory to any theory, and strengthening of the opposite view when a specific current obtained acceptance or scientific support. It is impressive that the anti-scientific current within mental health increases in strength in parallel with the scientific current, and this is unlikely to change in the foreseen future.

26.1 The Academic Future

For all those following the scientific way, it is clear that psychiatry is grounded in clinical neuroscience, which includes not only medicine, biology, biochemistry, and physics but also psychology. According to this viewpoint, a better outcome for the

K. N. Fountoulakis, *Psychiatry*, https://doi.org/10.1007/978-3-030-86541-2_26

patients is most likely to originate from clinical and translational neuroscience [1]. Psychology and social sciences have a place in this field but only as long as and to the extent they follow the hard scientific way, but not some way that resembles but is not science [2].

At the same time, academic research should deal with all kinds of conflicts of interest, including academic and economic interests.

Scientific progress is taken for granted today, and there is a widespread belief that no matter the obstacles, the lack of resources, or opposition of any kind, the scientific method will continue to be productive and fruitful and will lead to deeper knowledge, better understanding, and eventually improved care. According to this view, the traditional boundaries between psychiatry and neurology are slowly disappearing [3].

Although such a bright view is not always in accordance with the dim reality, it is the short span of human life rather than the nature of the scientific progress that often makes things appear pessimistic. Thus, it is more than certain that the advances promised by neurosciences and biological psychiatry and psychopharmacology will eventually emerge. However, it seems unlikely that there will be a "game-changer" discovery within the next decade, although breakthrough discoveries usually occur without prior notification. It is important that the US National Institute of Mental Health (NIMH) has developed the Research Domain Criteria (RDoC) project to promote "research to validate dimensions defined by neurobiology and behavioural measures that cut across current disorder categories and can inform future revisions of our diagnostic system," but until now, no clinically useful progress has been made [4, 5]. Thus, there is an oxymoron in place: while academic knowledge is exploding, true clinical advances are still to come.

This is partially because relative to most other specialties, the number of research-intensive departments of psychiatry is relatively small [1], and this has a number of consequences, including lack of research background for most psychiatrists, extremely few mentors, resistance of clinical practice to evidence-based change, and failure to recruit a fair share of the best and brightest medical students [6]. As a result there is a pervasive belief among clinicians that neuroscience is not relevant to patient care [7] in spite of attempts to include neuroscience in many residency programs [8, 9]. At the same time, training in unproven approaches like psychoanalysis or exotic types of psychotherapy is still widespread, and the hope that somehow after a century of failure they will obtain a reliable and scientific character is still strong [10–12]. A significant problem is that too little training in biological psychiatry and psychopharmacology is included in current residency curricula, and too much of it is left to the will and the interests of the pharma industry.

In the near future, the diagnostic approach is highly unlikely to be replaced by a purely biological or investigative approach. It is questionable even, the extent to which such an approach will come to be relevant. Instead, the diagnosis will continue to rely on descriptive psychopathology [13], with a huge contribution of arbitrary and hypothetical biopsychosocial and cultural formulations. This also reflects the necessity to satisfy the human need of patients to be "heard and understood" in order to be able to establish a therapeutic alliance; therefore adding value to technocratic practice in psychiatry seems unavoidable [14]. Personalized and precision psychiatry is the ultimate goal of research, with the search for the identification of relevant biomarkers being a priority. These biomarkers are expected to improve response rates and reduce time to response by selecting a treatment option that is more likely to work. However, the most likely change in the diagnostic assessment practices in the next few years will probably be the increased use of measurement-based care in routine clinical practice. This involves the use of psychometric and neuropsychological scales to assist treatment planning and treatment outcome [15]. Currently, their use is neither standard nor frequent [16] in spite of persistent recommendations [15, 17]. In reality, there is a real danger, which is the clinical practice to adopt this kind of measurements in order to cover the lack of true hard laboratory indices to guide the decision-making. These measurements are neither as reliable nor as valid as laboratory examinations are for the other specialties of medicine. In fact, and with a few exceptions, they are far less reliable and valid in comparison with the careful clinical assessment when made by a well-trained and experienced clinician.

Digital technology will rapidly change a number of things, especially in everyday clinical practice and without the need for any advancement of our knowledge. Examples are the digital delivery of cognitive behavioral therapy (CBT) [18] or virtual reality exposure therapy [19] and the development of chatbots, which are computer algorithms based on artificial intelligence, to deliver some kind of psychotherapy [20].

On the other hand, so far, advances in biological psychiatry and psychopharmacology and even the adoption of the evidence-based approach rather fueled the postmodern and anti-scientific approaches instead of discouraging them [21]. Many ideological supporters of these theories utilize a problematic version of the evidence-based approach in order to reject psychopharmacology. While theory is known to have an effect on the quality and the essence of data gathered [22], it is also true that through the years, these opposing theories have transformed several times and adapted to the new developments, while the irrational core remained intact and is flourishing. They originated from the existential theories and political liberalization of the 1960s and 1970s to a populist rhetoric at present, but with everlasting strong influence on the mass media and popular arts, especially the cinema. Therefore, this pattern is also expected to last at least for the near future and until the "game-changer" neurobiological discovery occurs, although it is highly possible that even such a discovery might fuel further irrationality. This is the distinct nature of psychiatry within medicine and society.

Of course, the other side of the coin is also true; the Big Pharma is in a constant effort to manipulate research. However, it should be made clear that neither of these efforts can be hidden from the informed and experienced eye of the unbiased researcher since in these cases, a clear violation of a specific rule will always be present.

A deep culture change is of prime importance in order to ensure that our field is really ready to embrace new findings as they emerge. This change primarily concerns the enhancement of communication and collaboration between researchers and clinicians, which will expedite the yield of research finding in clinical practice and will provide the research endeavor with feedback from the real world.

26.2 The Future of Mental Health Care

The number of available mental health-care professionals and especially psychiatrists is a crucial factor for the future of mental health. Currently in most of the world, the numbers are disappointing (Table 20.1; Fig. 20.1).

There is a hidden problem which is widespread in medicine, but it is more profound in psychiatry. It is called the "science to practice gap" and concerns the discrepancy between what we scientifically know as a field and how the average clinician works in his everyday clinical practice. Today we clearly know that psychiatry has many treatments that work, and the data suggest that they are similar in efficacy to those of the rest of medicine for chronic medical conditions [23]. While the academic future is probably genomic, neuroscientific, and digital, with the introduction of highly advanced methods in the frame of precision and personalized psychiatry and on the basis of the patients' neurobiological characteristics, a number of questions arise. The first question is of course when this will happen and what until then? The second question is what kind of ethical dilemmas will arise, and the third is whether this will address the needs and the wishes of patients, service users, families, friends, communities, and the society [24].

Better care delivery resulting in better mental health for all needs a broad alliance of all disciplines and professionals active in the field. It requires the bridging of clinical psychiatry with neuroscience, epidemiology, and psychological, behavioral, and social sciences as well as health economics, among others, but the word "science" should accompany this alliance as the crucial determinant and not simply as a decoration.

An important challenge that recently emerged is that patients are informed but often are over- and misinformed by the Internet about mental illnesses and their treatments. Although in most cases this source of information is highly problematic, it is true that it has increased patient autonomy and produced a controversial democratizing effect on the hierarchical doctor–patient relationship. As a consequence, excellence in psychiatric care in the coming decades will rely on the psychiatrists' skill in incorporating the patients and their families in the clinical decision making. The change in the psychiatrist's clinical role over the past two decades as part of deinstitutionalization worldwide [25, 26] requires that the psychiatrist of the immediate future will have to build alliances, learn to work as part of a team together with other health professionals, and alter the traditional isolation of psychiatry [27, 28].

Concerns on the future of Psychiatry made the World Psychiatric Association (WPA) appoint a joint WPA-Lancet Psychiatry Commission on the Future of Psychiatry [29]. The commission reported mainly on the changes at the professional level and at everyday clinical practice rather than at the academic level. It stressed the unmet needs and especially the need for integration of mental with physical health care and suggested that cultural factors play an increasingly significant role, since the psychiatrist of the future will deal with older, more urban, and migrant populations and possibly with more patients suffering from a variety of addiction disorders. This means that communication skills, vigilance for the protection of the rights of patients, as well as training in mental health legislation and the use of digital technology will be important qualities. It also stressed that unfortunately, a dramatic reduction in stigmatization is unlikely to occur in the near future.

Other authors pointed to the possible effects of climate change and the effects it will have on the public health in combination with increased aging of the population and mental somatic disorder comorbidity [30], with a special emphasis on suicidality [31, 32]. Additionally, a number of unconventional treatment approaches are expected to strengthen, including the networking with peers in support groups [33], with some of them directly stemming out of service-user experiential and first-hand knowledge and developing outside the psychiatric paradigm, in a social–existential perspective [34, 35]. The truth is that people living with mental health problems experience psychiatry as just one part of a complex system of care and support. This part might become even smaller in the future given the global scarcity of resources that leads to less contact between patients and clinicians and less time per visit.

There is a wide consensus that in the near future, if not already, the average psychiatrist will find himself/herself in the position of engaging with critics and detractors both from within the profession and outside it [36]; dealing with transcultural, psychiatry, and value-based practices [37]; and testing the foundations, way of operating, and legitimacy of the discipline. Since the knowledge, political status, and arguments of critics have already been developing in decades [38, 39], it is highly unlikely that the average clinician with the usual training during residency, as it happens today, will be adequately educated to cope up with this obligation. As said, "It is absolutely clear that Psychiatry cannot afford to spend the future retreating into a digital and biomedical bunker, but instead it needs to reflect on politics, rights, and the renegotiation of the contract between Psychiatry and society" [24, 40]. In spite of becoming more technocratic, psychiatry will probably become even more interpersonally demanding in the future [41], simply because the need to communicate the benefits of technocratic advances will increase, and eventually its success will make the real difference.

Two more issues will be increasingly important.

The first is the presence of stigma for mental illness and mental patients. It will continue to cause discrimination leading to negative media portrayals of patients, their families, and psychiatrists but also of mental health settings of any kind. Negative attitudes toward psychiatry and psychiatrists make it difficult to work effectively [42]. More importantly it will continue to lead to governmental unwillingness to invest resources in mental health, thus creating a vicious cycle [43].

The second issue concerns the delicate balance between human rights on the one hand and the prevention of violence on the other. Different groups in society pull to opposite directions, and this will intensify. The United Nations have issued the UN Convention on the Rights of Persons with Disabilities (CRPD), which according to some interpretations might be incompatible with many aspects of involuntary treatment [44], while on the other hand, there is increasing pressure to report patients who might pose a threat to public safety.

26.3 The Future of Psychiatry as a Profession

Psychiatry as a profession has many different faces depending mainly on the setting and the specific role. A psychiatrist could act as consultant for other mental health professionals, could often give technical guidance without actually seeing patients at a regular basis, or on the contrary, he/she could work in private practice where there is indirect negotiation with the individual patient.

The threats to psychiatry as a profession and to its practitioners are many [45], but maybe the most important is a deep deprofessionalization that is not specific or isolated to psychiatry. It is an ongoing process within the medical profession, fueled by social, political, and economic factors, changing expectations, and marked by incidents of medical malpractice.

While medical professionalism emerged as a result of a class structure within the society [46], this has changed during the last few decades, and subsequently medical professionalism must change with it [47–49]. In this frame psychiatry faces additional problems because of the stigma attributed not only to patients but also to psychiatrists themselves [42].

As with other specialties of medicine, the professional psychiatrist of the present as well as of the future should be able to deal with a number of conflicts of interest. For the average clinician, these are mainly economic conflicts of interest, but they are multiple, and within this big discussion, both medication and psychotherapy treatments are sensitive issues. The influence of the economic benefits from applying endless psychotherapy to patients that do not need this treatment is not inferior to that of the pharma industry concerning overprescribing medications.

References

1. Reynolds CF 3rd, Lewis DA, Detre T, Schatzberg AF, Kupfer DJ. The future of psychiatry as clinical neuroscience. Acad Med. 2009;84(4):446–50. https://doi.org/10.1097/ACM.0b013e31819a8052.
2. Fountoulakis KN. Being strictly scientific is the only way forward for psychiatry. Acta Psychiatr Scand. 2021;143(3):187–8. https://doi.org/10.1111/acps.13284.
3. Detre T, McDonald MC. Managed care and the future of psychiatry. Arch Gen Psychiatry. 1997;54(3):201–4. https://doi.org/10.1001/archpsyc.1997.01830150019002.
4. Cuthbert BN. The RDoC framework: facilitating transition from ICD/DSM to dimensional approaches that integrate neuroscience and psychopathology. World Psychiatry. 2014;13(1):28–35. https://doi.org/10.1002/wps.20087.
5. Cuthbert BN. Translating intermediate phenotypes to psychopathology: the NIMH Research Domain Criteria. Psychophysiology. 2014;51(12):1205–6. https://doi.org/10.1111/psyp.12342.

6. Kupfer DJ, Hyman SE, Schatzberg AF, Pincus HA, Reynolds CF 3rd. Recruiting and retaining future generations of physician scientists in mental health. Arch Gen Psychiatry. 2002;59(7):657–60. https://doi.org/10.1001/archpsyc.59.7.657.
7. Ross DA, Travis MJ, Arbuckle MR. The future of psychiatry as clinical neuroscience: why not now? JAMA Psychiat. 2015;72(5):413–4. https://doi.org/10.1001/jamapsychiatry.2014.3199.
8. Akil M, Etkin A. Transforming neuroscience education in psychiatry. Acad Psychiatry. 2014;38(2):116–20. https://doi.org/10.1007/s40596-014-0065-3.
9. Ross DA, Rohrbaugh R. Integrating neuroscience in the training of psychiatrists: a patient-centered didactic curriculum based on adult learning principles. Acad Psychiatry. 2014;38(2):154–62. https://doi.org/10.1007/s40596-014-0055-5.
10. Askay R, Farquhar-Lacewing J. Being unconscious: Heidegger and Freud. In: Fulford KWM, Davies M, Gipps R, et al., editors. The oxford handbook of philosophy and psychiatry. Oxford University Press; 2013. https://doi.org/10.1093/oxfordhb/9780199579563.001.0001.
11. Hopkins J. Understanding and healing: psychiatry and psychoanalysis in the era of neuroscience. In: Fulford KWM, Davies M, Gipps R, et al., editors. The Oxford handbook of philosophy and psychiatry. Oxford University Press; 2013. https://doi.org/10.1093/oxfordhb/9780199579563.001.0001.
12. Lacewing M. Could psychoanalysis be a science? In: Fulford KWM, Davies M, Gipps R, et al., editors. The Oxford handbook of philosophy and psychiatry. Oxford University Press; 2013. https://doi.org/10.1093/oxfordhb/9780199579563.001.0001.
13. Escott-Price V, Sims R, Bannister C, Harold D, Vronskaya M, Majounie E, Badarinarayan N, Gerad/Perades, IGAP Consortia, Morgan K, Passmore P, Holmes C, Powell J, Brayne C, Gill M, Mead S, Goate A, Cruchaga C, Lambert JC, van Duijn C, Maier W, Ramirez A, Holmans P, Jones L, Hardy J, Seshadri S, Schellenberg GD, Amouyel P, Williams J. Common polygenic variation enhances risk prediction for Alzheimer's disease. Brain. 2015;138(Pt 12):3673–84. https://doi.org/10.1093/brain/awv268.
14. Fulford KWM, Thornton T, Graham G. From bioethics to value-based parctice in psychiatric diagnosis. In: Fulford KWM, Thornton T, Graham G, editors. The Oxford textbook of philosophy and psychiatry. Oxford University Press; 2006.
15. Harding KJ, Rush AJ, Arbuckle M, Trivedi MH, Pincus HA. Measurement-based care in psychiatric practice: a policy framework for implementation. J Clin Psychiatry. 2011;72(8):1136–43. https://doi.org/10.4088/JCP.10r06282whi.
16. Zimmerman M, McGlinchey JB. Why don't psychiatrists use scales to measure outcome when treating depressed patients? J Clin Psychiatry. 2008;69(12):1916–9. https://doi.org/10.4088/jcp.v69n1209.
17. Valenstein M, Adler DA, Berlant J, Dixon LB, Dulit RA, Goldman B, Hackman A, Oslin DW, Siris SG, Sonis WA. Implementing standardized assessments in clinical care: now's the time. Psychiatr Serv. 2009;60(10):1372–5. https://doi.org/10.1176/ps.2009.60.10.1372.
18. Andersson G, Cuijpers P, Carlbring P, Riper H, Hedman E. Guided internet-based vs. face-to-face cognitive behavior therapy for psychiatric and somatic disorders: a systematic review and meta-analysis. World Psychiatry. 2014;13(3):288–95. https://doi.org/10.1002/wps.20151.
19. Morina N, Ijntema H, Meyerbroker K, Emmelkamp PM. Can virtual reality exposure therapy gains be generalized to real-life? A meta-analysis of studies applying behavioral assessments. Behav Res Ther. 2015;74:18–24. https://doi.org/10.1016/j.brat.2015.08.010.
20. Fitzpatrick KK, Darcy A, Vierhile M. Delivering cognitive behavior therapy to young adults with symptoms of depression and anxiety using a fully automated conversational agent (woebot): a randomized controlled trial. JMIR Ment Health. 2017;4(2):e19. https://doi.org/10.2196/mental.7785.
21. Fulford KWM, Thornton T, Graham G. Knowledge, research and evidence-based medicine. In: Fulford KWM, Thornton T, Graham G, editors. The Oxford textbook of philosophy and psychiatry. Oxford University Press; 2006.
22. Fulford KWM, Thornton T, Graham G. Psychopathology and the theory dependence of data. In: Fulford KWM, Thornton T, Graham G, editors. The Oxford textbook of philosophy and psychiatry. Oxford University Press; 2006.

23. Leucht S, Hierl S, Kissling W, Dold M, Davis JM. Putting the efficacy of psychiatric and general medicine medication into perspective: review of meta-analyses. Br J Psychiatry J Ment Sci. 2012;200(2):97–106. https://doi.org/10.1192/bjp.bp.111.096594.
24. Carr S. Renegotiating the contract. Lancet Psychiatry. 2017;4(10):740–1. https://doi.org/10.1016/S2215-0366(17)30365-6.
25. Chow WS, Priebe S. Understanding psychiatric institutionalization: a conceptual review. BMC Psychiatry. 2013;13:169. https://doi.org/10.1186/1471-244X-13-169.
26. Goffman E. The characteristics of total institutions. In: Etzioni A, editor. Complex organisations: a sociological reader. New York, NY: Holt, Rinehart and Winston; 1961. p. 312–40.
27. Fakhoury W, Priebe S. The process of deinstitutionalization: an international overview. Curr Opin Psychiatry. 2002;15(2):187–92. https://doi.org/10.1097/00001504-200203000-00011.
28. Lancet Global Mental Health Group, Chisholm D, Flisher AJ, Lund C, Patel V, Saxena S, Thornicroft G, Tomlinson M. Scale up services for mental disorders: a call for action. Lancet. 2007;370(9594):1241–52. https://doi.org/10.1016/S0140-6736(07)61242-2.
29. Bhugra D, Tasman A, Pathare S, Priebe S, Smith S, Torous J, Arbuckle MR, Langford A, Alarcon RD, Chiu HFK, First MB, Kay J, Sunkel C, Thapar A, Udomratn P, Baingana FK, Kestel D, Ng RMK, Patel A, Picker L, McKenzie KJ, Moussaoui D, Muijen M, Bartlett P, Davison S, Exworthy T, Loza N, Rose D, Torales J, Brown M, Christensen H, Firth J, Keshavan M, Li A, Onnela JP, Wykes T, Elkholy H, Kalra G, Lovett KF, Travis MJ, Ventriglio A. The WPA-lancet psychiatry commission on the future of psychiatry. Lancet Psychiatry. 2017;4(10):775–818. https://doi.org/10.1016/S2215-0366(17)30333-4.
30. Sartorius N. More than one scenario exists for the future of psychiatry. Lancet Psychiatry. 2017;4(10):738–9. https://doi.org/10.1016/S2215-0366(17)30361-9.
31. Fountoulakis KN, Chatzikosta I, Pastiadis K, Zanis P, Kawohl W, Kerkhof AJ, Navickas A, Hoschl C, Lecic-Tosevski D, Sorel E, Rancans E, Palova E, Juckel G, Isacsson G, Jagodic HK, Botezat-Antonescu I, Rybakowski J, Azorin JM, Cookson J, Waddington J, Pregelj P, Demyttenaere K, Hranov LG, Stevovic LI, Pezawas L, Adida M, Figuera ML, Jakovljevic M, Vichi M, Perugi G, Andreassen OA, Vukovic O, Mavrogiorgou P, Varnik P, Dome P, Winkler P, Salokangas RK, From T, Danileviciute V, Gonda X, Rihmer Z, Forsman J, Grady A, Hyphantis T, Dieset I, Soendergaard S, Pompili M, Bech P. Relationship of suicide rates with climate and economic variables in Europe during 2000-2012. Ann General Psychiatry. 2016;15:19. https://doi.org/10.1186/s12991-016-0106-2.
32. Fountoulakis KN, Savopoulos C, Zannis P, Apostolopoulou M, Fountoukidis I, Kakaletsis N, Kanellos I, Dimellis D, Hyphantis T, Tsikerdekis A, Pompili M, Hatzitolios AI. Climate change but not unemployment explains the changing suicidality in Thessaloniki Greece (2000-2012). J Affect Disord. 2016;193:331–8. https://doi.org/10.1016/j.jad.2016.01.008.
33. Lazareva I. Cameroon's 'auntie army' of rape survivors battles sex abuse of girls. 2017. http://www.reuters.com/article/us-cameroon-women-rape-idUSKCN18C0BU.
34. Beresford P, Nettle M, Perring R. Towards a social model of madness and distress? Exploring what service users say. York: Joseph Rowntree Foundation; 2010.
35. Slade M, Amering M, Farkas M, Hamilton B, O'Hagan M, Panther G, Perkins R, Shepherd G, Tse S, Whitley R. Uses and abuses of recovery: implementing recovery-oriented practices in mental health systems. World Psychiatry. 2014;13(1):12–20. https://doi.org/10.1002/wps.20084.
36. Burstow B. Psychiatry and the business of madness: an ethical and epistemological accounting. Basingstoke: Palgrave Macmillan; 2015.
37. Fulford KW. Values-based practice: a new partner to evidence-based practice and a first for psychiatry? Mens Sana Monogr. 2008;6(1):10–21. https://doi.org/10.4103/0973-1229.40565.
38. Rose D. Service user/survivor-led research in mental health: epistemological possibilities. Disabil Soc. 2017;32(6):773–89. https://doi.org/10.1080/09687599.2017.1320270.
39. Rose D, Perry E, Rae S, Good N. Service user perspectives on coercion and restraint in mental health. BJPsych Int. 2017;14(3):59–61. https://doi.org/10.1192/s2056474000001914.
40. Spandler H, McKeown M. Exploring the case for truth and reconciliation in mental health services. Ment Health Rev J. 2017;22(2):83–94. https://doi.org/10.1108/Mhrj-01-2017-0011.

41. Giacco D, Amering M, Bird V, Craig T, Ducci G, Gallinat J, Gillard SG, Greacen T, Hadridge P, Johnson S, Jovanovic N, Laugharne R, Morgan C, Muijen M, Schomerus G, Zinkler M, Wessely S, Priebe S. Scenarios for the future of mental health care: a social perspective. Lancet Psychiatry. 2017;4(3):257–60. https://doi.org/10.1016/S2215-0366(16)30219-X.
42. Farooq K, Lydall GJ, Bhugra D. Research into recruitment: critical gaps in the literature. Int Rev Psychiatry. 2013;25(4):366–70. https://doi.org/10.3109/09540261.2013.822853.
43. Bhugra D. Alienated alienists: a new hope? Lancet Psychiatry. 2014;1(4):257–9. https://doi.org/10.1016/S2215-0366(14)70283-4.
44. United Nations. Convention on the rights of persons with disabilities (CRPD). 2007. https://www.un.org/development/desa/disabilities/convention-on-the-rights-of-persons-with-disabilities.html.
45. Bhugra D, Gupta S, Smyth G, Webber M. Through the prism darkly: how do others see psychiatrists? Aust Psychiat Bull R Aust N Z Coll Psychiatr. 2010;18(1):7–11. https://doi.org/10.3109/10398560903274399.
46. Shock M. Medicine at the Centre of the nation's affairs. BMJ (Clin Rese Ed). 1994;309(6970):1730–3. https://doi.org/10.1136/bmj.309.6970.1730.
47. Committee RP. Medical professionalism in the new millennium: a physicians' charter. Radiology. 2006;238(2):383–6. https://doi.org/10.1148/radiol.2382052538.
48. Tallis RC. Doctors in society: medical professionalism in a changing world. Clin Med (Lond). 2006;6(1):7–12. https://doi.org/10.7861/clinmedicine.6-1-7.
49. Working Party of the Royal College of Physicians. Doctors in society. Medical professionalism in a changing world. Clin Med (Lond). 2005;5(6 Suppl 1):S5–40.

MIX
Papier aus verantwortungsvollen Quellen
Paper from responsible sources
FSC® C105338

If you have any concerns about our products,
you can contact us on
ProductSafety@springernature.com

In case Publisher is established outside the EU,
the EU authorized representative is:
Springer Nature Customer Service Center GmbH
Europaplatz 3, 69115 Heidelberg, Germany

Printed by Libri Plureos GmbH
in Hamburg, Germany